American Geophysical Union | ANTARCTIC RESEARCH SERIES

Antarctic Research Series Volumes

1. Biology of the Antarctic Seas I *Milton O. Lee (Ed.)*
2. Antarctic Snow and Ice Studies *M. Mellor (Ed.)*
3. Polychaeta Errantia of Antarctica *O. Hartman (Ed.)*
4. Geomagnetism and Aeronomy *A. H. Waynick (Ed.)*
5. Biology of the Antarctic Seas II *G. A. Llano (Ed.)*
6. Geology and Paleontology of the Antarctic *J. B. Hadley (Ed.)*
7. Polychaeta Myzostomidae and Sedentaria of Antarctica *O. Hartman (Ed.)*
8. Antarctic Soils and Soil Forming Processes *J. C. F. Tedrow (Ed.)*
9. Studies in Antarctic Meteorology *M. J. Rubin (Ed.)*
10. Entomology of Antarctica *J. L. Gressit (Ed.)*
11. Biology of the Antarctic Seas III *G. A. Llano, W. L. Schmitt (Eds.)*
12. Antarctic Bird Studies *O. L. Austin, Jr. (Ed.)*
13. Antarctic Ascidiacea *P. Kott (Ed.)*
14. Antarctic Cirripedia *W. A. Newman, A. Ross (Eds.)*
15. Antarctic Oceanology I *L. Reid (Ed.)*
16. Antarctic Snow and Ice Studies II *A. P. Crary (Ed.)*
17. Biology of the Antarctic Seas IV *G. A. Llano, I. E. Wallen (Eds.)*
18. Antarctic Pinnipedia *W. H. Burt (Ed.)*
19. Antarctic Oceanology II: The Australian-New Zealand Sector *D. E. Hayes (Ed.)*
20. Antarctic Terrestrial Biology *G. A. Llano (Ed.)*
21. Recent Antarctic and Subantarctic Brachiopods *M. W. Foster (Ed.)*
22. Human Adaptability to Antarctic Conditions *E. K. Eric Gunderson (Ed.)*
23. Biology of the Antarctic Seas V *D. L. Pawson (Ed.)*
24. Birds of the Antarctic and Sub-Antarctic *G. E. Watson (Ed.)*
25. Meteorological Studies at Plateau Station, Antarctica *J. Businger (Ed.)*
26. Biology of the Antarctic Seas VI *D. L. Pawson (Ed.)*
27. Biology of the Antarctic Seas VII *D. L. Pawson (Ed.)*
28. Biology of the Antarctic Seas VIII *D. L. Pawson, L. S. Korniker (Eds.)*
29. Upper Atmosphere Research in Antarctica *L. J. Lanzerotti, C. G. Park (Eds.)*
30. Terrestrial Biology II *B. Parker (Ed.)*
31. Biology of the Antarctic Seas IX *L. S. Kornicker (Ed.)*
32. Biology of the Antarctic Seas X *L. S. Kornicker (Ed.)*
33. Dry Valley Drilling Project *L. D. McGinnis (Ed.)*
34. Biology of the Antarctic Seas XI *L. S. Korniker (Ed.)*
35. Biology of the Antarctic Seas XII *D. Pawson (Ed.)*
36. Geology of the Central Transantarctic Mountains *M. D. Turner, J. F. Splettstoesser (Eds.)*
37. Terrestrial Biology III *B. Parker (Ed.)*
38. Biology of the Antarctic Seas XIII [crinoids, hydrozoa, copepods, amphipoda] *L. S. Korniker (Ed.)*
39. Biology of the Antarctic Seas XIV *L. S. Kornicker (Ed.)*
40. Biology of the Antarctic Seas XV *L. S. Korniker (Ed.)*
41. Biology of the Antarctic Seas XVI *L. S. Korniker (Ed.)*
42. The Ross Ice Shelf: Glaciology and Geophysics *C. R. Bentley, D. E. Hayes (Eds.)*
43. Oceanology of the Antarctic Continental Shelf *S. Jacobs (Ed.)*
44. Biology of the Antarctic Seas XVII [benthic satiation, brittle star feeding, pelagic shrimps, marine birds] *L. S. Korniker (Ed.)*
45. Biology of the Antarctic Seas XVIII, Crustacea Tanaidacea of the Antarctic and the Subantarctic 1. On Material Collected at Tierra del Fuego, Isla de los Estados, and the West Coast of the Antarctic Peninsula *L. S. Korniker (Ed.)*
46. Geological Investigations in Northern Victoria Land *E. Stump (Ed.)*

47	Biology of the Antarctic Seas XIX [copepods, teleosts] *L. S. Korniker (Ed.)*
48	Volcanoes of the Antarctic Plate and Southern Oceans *W. E. LeMasurier, J. W. Thomson (Eds.)*
49	Biology of the Antarctic Seas XX, Antarctic Siphonophores From Plankton Samples of the United States Antarctic Research Program *L. S. Kornicker (Ed.)*
50	Contributions to Antarctic Research I *D. H. Elliot (Ed.)*
51	Mineral Resources Potential of Antarctica *J. F. Splettstoesser, G. A. M. Dreschhoff (Eds.)*
52	Biology of the Antarctic Seas XXI [annelids, mites, leeches] *L. S. Korniker (Ed.)*
53	Contributions to Antarctic Research II *D. H. Elliot (Ed.)*
54	Marine Geological and Geophysical Atlas of the Circum-Antarctic to 30ES *D. E. Hayes (Ed.)*
55	Molluscan Systematics and Biostratigraphy Lower Tertiary La Meseta Formation, Seymour Island, Antarctic Peninsula *J. D. Stilwell, W. J. Zinsmeister*
56	The Antarctic Paleoenvironment: A Perspective on Global Change, Part One *J. P. Kennett, D. A. Warnke (Eds.)*
57	Contributions to Antarctic Research III *D. H. Elliot (Ed.)*
58	Biology of the Antarctic Seas XXII *S. D. Cairns (Ed.)*
59	Physical and Biogeochemical Processes in Antarctic Lakes *W. J. Green, E. L Friedmann (Eds.)*
60	The Antarctic Paleoenvironment: A Perspective on Global Change, Part Two *J. P. Kennett, D. A. Warnke (Eds.)*
61	Antarctic Meteorology and Climatology: Studies Based on Automatic Weather Stations *D. H. Bromwich, C. R. Steams (Eds.)*
62	Ultraviolet Radiation in Antarctica: Measurements and Biological Effects *C. S. Weiler, P. A. Penhale (Eds.)*
63	Biology of the Antarctic Seas XXIV, Antarctic and Subantarctic Pycnogonida: Ammotheidae and Austrodecidae *S. D. Cairns (Ed.)*
64	Atmospheric Halos *W. Tape*
65	Fossil Scleractinian Corals From James Ross Basin, Antarctica *H. F. Filkorn*
66	Volcanological and Environmental Studies of Mt. Erebus *P. R. Kyle (Ed.)*
67	Contributions to Antarctic Research IV *D. H. Elliot, G. L. Blaisdell (Eds.)*
68	Geology and Seismic Stratigraphy of the Antarctic Margin *A. K. Cooper, P. F. Barker, G. Brancolini (Eds.)*
69	Biology of the Antarctic Seas XXIV, Antarctic and Subantarctic Pycnogonida: Nymphonidae, Colossendeidae, Rhynchothoraxida, Pycnogonidae, Phoxichilidiidae, Endeididae, and Callipallenidae *S. D. Cairns (Ed.)*
70	Foundations for Ecological Research West of the Antarctic Peninsula *R. M. Ross, E. E. Hofmann, L. B. Quetin (Eds.)*
71	Geology and Seismic Stratigraphy of the Antarctic Margin, Part 2 *P. F. Barker, A. K. Cooper (Eds.)*
72	Ecosystem Dynamics in a Polar Desert: The McMurdo Dry Valleys, Antarctica *John C. Priscu (Ed)*
73	Antarctic Sea Ice: Biological Processes, Interactions and Variability *Michael P. Lizotte, Kevin R. Arrigo (Eds.)*
74	Antarctic Sea Ice: Physical Processes, Interactions and Variability *Martin O. Jeffries (Ed.)*
75	Ocean, Ice and Atmosphere: Interactions at the Continental Margin *Stanley S. Jacobs, Ray F. Weiss (Eds.)*
76	Paleobiology and Paleoenvironments of Eocene Rocks, McMurdo Sound, East Antarctica *Jeffrey D. Stilwell, Rodney M. Feldmann (Eds.)*

THE ANTARCTIC RESEARCH SERIES

The Antarctic Research Series, published since 1963 by the American Geophysical Union, now comprises more than 70 volumes of authoritative original results of scientific work in the high latitudes of the southern hemisphere. Series volumes are typically thematic, concentrating on a particular topic or region, and may contain maps and lengthy papers with large volumes of data in tabular or digital format. Antarctic studies are often interdisciplinary or international, and build upon earlier observations to address issues of natural variability and global change. The standards of scientific excellence expected for the Series are maintained by editors following review criteria established for the AGU publications program. Priorities for publication are set by the Board of Associate Editors. Inquiries about published volumes, work in progress or new proposals may be sent to Antarctic Research Series, AGU, 2000 Florida Avenue NW, Washington, DC 20009 (http://www.agu.org), or to a member of the Board.

BOARD OF ASSOCIATE EDITORS

Rodney M. Feldmann, Chairman, *Paleontology*
Robert A. Bindschadler, *Glaciology*
David H. Bromwich, *Meteorology and Upper Atmosphere Physics*
Nelia W. Dunbar, *Geology*
Stanley S. Jacobs, *Oceanography*
Jerry D. Kudenov, *Marine/Polychaete Biology*
John C. Priscu, *Terrestrial Biology*

Volume 77 | ANTARCTIC RESEARCH SERIES

The West Antarctic Ice Sheet
Behavior and Environment

Richard B. Alley and Robert A. Bindschadler
Editors

American Geophysical Union
Washington, D.C.
2001

THE WEST ANTARCTIC ICE SHEET: BEHAVIOR AND ENVIRONMENT
Richard B. Alley and Robert A. Bindschadler, Editors

Published under the aegis of the Board of Associate Editors, Antarctic Research Series

Library of Congress Cataloging-in-Publication Data
The west Antarctic ice sheet : behavior and environment / Richard B. Alley and Robert A. Bindschadler, editors.
    p. cm. -- (Antarctic research series ; v. 77)
    Includes bibliographical references.
    ISBN 0-87590-957-4
    1. Ice sheets -- Antarctica. I. Alley, Richard B. II. Bindschadler, R. A. (Robert A.) III. Series.
GB2597 .W36 2000
551.31'2'09989--dc21

                                                00-061074

ISBN 0-87590-957-4
ISSN 0066-4634

Copyright 2001 by the American Geophysical Union
2000 Florida Avenue, N.W.
Washington, DC 20009

    Figures, tables, and short excerpts may be reprinted in scientific books and journals if the source is properly cited.

    Authorization to photocopy items for internal or personal use, or the internal or personal use of specific clients, is granted by the American Geophysical Union for libraries and other users registered with the Copyright Clearance Center (CCC) Transactional Reporting Service, provided that the base fee of $01.50 per copy plus $0.50 per page is paid directly to CCC, 222 Rosewood Dr., Danvers, MA 01923. 0066-4634/01/$01.50+0.50.
    This consent does not extend to other kinds of copying, such as copying for creating new collective works or for resale. The reproduction of multiple copies and the use of full articles or the use of extracts, including figures and tables, for commercial purposes requires permission from the American Geophysical Union.

Published by
American Geophysical Union
2000 Florida Avenue, N.W.
Washington, D.C. 20009

Printed in the United States of America

CONTENTS

Preface
R. B. Alley and R. A. Bindschadler   xi

The West Antarctic Ice Sheet and Sea-level Change
R. B. Alley and R. A. Bindschadler   1

Setting

Morphology and Surface Characteristics of the West Antarctic Ice Sheet
Mark Fahnestock and Jonathan Bamber   13

The Lithospheric Setting of the West Antarctic Ice Sheet
I. W. D. Dalziel and L. A. Lawver   29

History

Evolution of the West Antarctic Ice Sheet
John B. Anderson and Stephanie S. Shipp   45

The Glacial Geologic Terrestrial Record from West Antarctica With Emphasis on the Last Glacial Cycle
Harold W. Borns, Jr.   59

West Antarctic Ice Sheet Elevation Changes
Eric J. Steig, James L. Fastook, Christopher Zweck, Ian D. Goodwin, Kathy J. Licht, James W. C. White, and Robert P. Ackert, Jr.   75

Interactions

The El Niño-Southern Oscillation Modulation of West Antarctic Precipitation
David H. Bromwich and Aric N. Rogers   91

Geologic Controls on the Initiation of Rapid Basal Motion for West Antarctic Ice Streams: A Geophysical Perspective Including New Airborne Radar Sounding and Laser Altimetry Results
D. D. Blankenship, D. L. Morse, C. A. Finn, R. E. Bell, M. E. Peters, S. D. Kempf, S. M. Hodge, M. Studinger, J. C. Behrendt, and J. M. Brozena   105

Flow

Onset of Streaming Flow in the Siple Coast Region, West Antarctica
Robert Bindschadler, Jonathan Bamber, and Sridhar Anandakrishnan   123

Ice Stream Shear Margins
C. F. Raymond, K. A. Echelmeyer, I. M. Whillans, and C. S. M. Doake   137

Basal Zone of the West Antarctic Ice Streams and its Role in Lubrication of Their Rapid Motion
Barclay Kamb   157

The Contribution of Numerical Modelling to our Understanding of the West Antarctic Ice Sheet
C. L. Hulbe and A. J. Payne — 201

Case Studies

Rutford Ice Stream, Antarctica
C. S. M. Doake, H. F. J. Corr, A. Jenkins, K. Makinson, K. W. Nicholls, C. Nath, A. M. Smith, and D. G. Vaughan — 221

A Review of Pine Island Glacier, West Antarctica:
Hypotheses of Instability vs. Observations of Change
David G. Vaughan, Andrew M. Smith, Hugh F. J. Corr, Adrian Jenkins, Charles R. Bentley, Mark D. Stenoien, Stanley S. Jacobs, Thomas B. Kellogg, Eric Rignot, and Baerbel K. Lucchitta — 237

Ice Streams B and C
I. M. Whillans, C. R. Bentley, and C. J. van der Veen — 257

The Flow Regime of Ice Stream C and Hypotheses Concerning its Recent Stagnation
S. Anandakrishnan, R. B. Alley, R. W. Jacobel, and H. Conway — 283

PREFACE

Sea-level rise from greenhouse warming is of considerable economic importance. Projections, however, typically include great uncertainty because the stability of the West Antarctic ice sheet cannot be guaranteed even over the brief time scale of human economies. Thus, the low-probability/high-impact collapse of West Antarctic ice has stimulated vigorous research over the last 30 years, which we summarize in this volume.

Major results to date include the following: the West Antarctic ice sheet has largely or completely disappeared after it formed, but at an unknown rate; the West Antarctic ice sheet shares important similarities with, but some differences from, past ice sheets that changed greatly and rapidly on widely separated occasions; portions of the West Antarctic ice sheet are changing rapidly now while averages over the whole ice sheet show little change; and, some models of the ice sheet project stability while others suggest that rapid changes remain possible. It should come as no surprise that we lack a consensus prediction of ice-sheet stability.

The road toward a consensus prediction is long, owing to the vast difficulties of characterizing a subcontinental block of ice more than 3 km thick in places, of reconstructing its history, learning how it behaves, and understanding its linkages with the surrounding oceans and atmosphere and the geology beneath. Despite the difficulties, recent progress has been gratifyingly rapid, and our view of the ice sheet and its environment is becoming increasingly clear.

Workers from opposite sides of the Atlantic Ocean have been journeying to opposite sides of West Antarctica to fashion views of the ice sheet. Annual workshops of the European Filchner-Ronne Ice Shelf Project (FRISP) and U.S. West Antarctic Ice Sheet (WAIS) Initiative teams have charted the progress by each group. Increasing interaction between the groups led to a joint FRISP/WAIS meeting in September, 1998 at the University of Maine in Orono. An outgrowth of this historic meeting was the decision to commission world experts to review and advance the state of knowledge of the WAIS within the present book. We editors eagerly awaited the insights in each new paper, and we trust that these papers will prove illuminating to those interested in sea level, climate, ice sheets, glacial geology, ocean sediments, and related subjects.

We thank the European and U.S. National Science Foundations for funding, H. Borns and co-workers at the University of Maine for hosting us, and numerous dedicated reviewers for raising the quality of the papers within this volume.

R. B. Alley
The Pennsylvania State University

R. A. Bindschadler
NASA/Goddard Space Flight Center

Editors

THE WEST ANTARCTIC ICE SHEET AND SEA-LEVEL CHANGE

R.B. Alley

Environment Institute and Department of Geosciences, The Pennsylvania State University, University Park, Pennsylvania

R.A. Bindschadler

Oceans and Ice Branch, NASA/Goddard Space Flight Center, Greenbelt, Maryland

A collapse of the West Antarctic ice sheet is considered possible if not highly likely over the next few centuries, with potential to raise global sea level approximately 5 m, and larger sea-level rise possible if changes propagate into the East Antarctic ice sheet. Continuation of recent West Antarctic ice-sheet retreat may contribute to sea-level rise for some time even if complete ice-sheet collapse does not occur. While there is no proof that collapse of the West Antarctic ice sheet is or is not imminent, a growing body of evidence shows that some ice sheets have the potential for rapid collapse, and that regional dynamic changes are occurring within the West Antarctic ice sheet. Fundamental advances in our understanding are being made by improved observations of the ice sheet and its environment, reconstruction of the ice sheet's history, and modeling of the ice sheet's future evolution. However, further advances are required to learn the future of the West Antarctic ice sheet with the confidence needed to assist policy-makers.

INTRODUCTION

In 1968, John Mercer argued from geologic evidence that the West Antarctic ice sheet (Figure 1) may have disappeared in the geologically recent past. Glaciological insights suggested that the ice sheet could shrink and disappear again. If too much future warming were to occur for any reason including "industrial pollution of the atmosphere", then "the unstable West Antarctic Ice Sheet will become a threat to coastal areas of the world within 6 m of sea level" [*Mercer*, 1968, p. 223], perhaps causing "...major dislocations in coastal cities, and submergence of low-lying areas such as much of Florida and the Netherlands" [*Mercer*, 1978, p. 325].

Weertman [1974] followed Mercer's warnings by identifying a glaciological mechanism that might allow ice-sheet collapse. Based on a simple model, he argued that a marine ice sheet (one grounded well below sea level) such as the West Antarctic ice sheet has only two stable configurations: grounded to the edge of the continental shelf, or completely collapsed. In theory, the slightest perturbation could drive the ice sheet to a slow advance or to an accelerating retreat. *Thomas and Bentley* [1978] used computer modeling to solve a more-complete set of equations than addressed by *Weertman* [1974]. Discussing changes over ice-age cycles, they suggested that "...for a suitably precarious marine ice sheet a small change in climate could trigger growth or decay which, once started, may be irreversible" [p. 164].

Hughes [1972, 1973] produced the first science plan for studying this radical idea of marine-ice-sheet instability. Many individual and coordinated research projects have taken place over the three decades since Mercer's first paper on this subject. The list of projects includes the Ross Ice Shelf Project (RISP, which penetrated the ice shelf during the 1977-78 and 1978-79 seasons) and the Ross Ice Shelf Glaciological and Geophysical Survey (RIGGS, with field seasons 1973-1977), the Siple Coast Project (SCP, 1983-1991), and the ongoing Filchner-Ronne Ice Shelf Project (FRISP, for which a subcommittee of the Scientific Committee on Antarctic Research Working Group on Glaciology was established in 1983) and the West Antarctic Ice Sheet initiative (WAIS, started in 1991). The hard science produced by the earlier of

Copyright 2001 by the American Geophysical Union

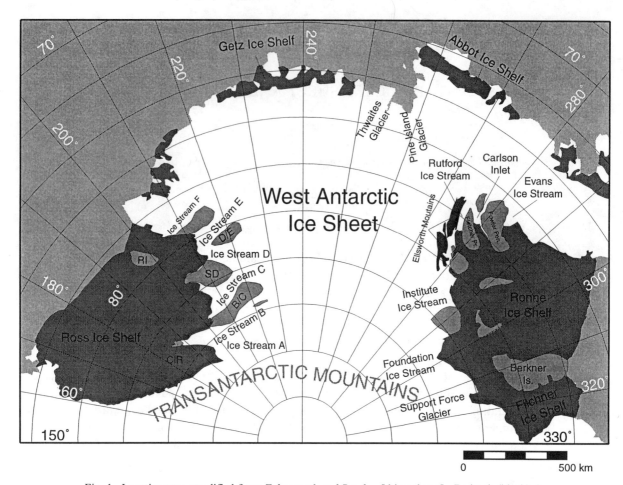

Fig. 1. Location map, modified from *Fahnestock and Bamber* [this volume]. Bedrock (black), ice shelves (dark), and ice rises and inter-ice-stream ridges (lighter) are shown, as are ocean (still lighter) and other grounded ice (white). The Siple Coast encompasses ice streams A-F and inter-stream ridges. In the Ross Ice Shelf, CIR and RI are the Crary Ice Rise and Roosevelt Island, respectively. On the Siple Coast draining into the Ross Ice Shelf, B/C and D/E are the interstream ridges between ice streams B and C, and ice streams D and E, respectively, and SD is Siple Dome, the ridge between ice streams C and D.

these research programs showed that questions being asked about the West Antarctic ice sheet were very difficult, and that the tools and the basic data needed to answer these questions were not yet available.

Indeed, West Antarctica has proven to be one of the most complex and interesting regions on the globe [e.g., *Oppenheimer*, 1998]. The West Antarctic ice sheet is a subcontinental-sized mass of ice, locally more than 4 km thick, that would raise global sea level approximately 5 m if melted [e.g., *Drewry*, 1983]; global sea-level change could be even larger if changes propagated from West Antarctica into the East Antarctic ice sheet. Retreat of the West Antarctic ice sheet has been occurring [e.g., *Conway et al.*, 1999; *Anderson and Shipp*, this volume; *Borns*, this volume], and continuation of this trend could affect global sea levels significantly for some time even if the trend were to stop before total ice-sheet collapse [*Bindschadler*, 1998].

The West Antarctic ice sheet exhibits a more-complex pattern of flow than its larger, colder and slower neighbor, the East Antarctic ice sheet. Extreme contrasts and feedbacks exist in the West Antarctic ice sheet--adjacent regions can have hundredfold differences in ice-flow rates despite similar gravitational driving stresses. Significant parts of the ice sheet have changed greatly in less than a century for reasons we have yet to understand, so we cannot guarantee predictability of the behavior of the West Antarctic ice sheet even on the time scale of human economic processes. Global sea-level rise this century has averaged ~2 mm/yr or slightly less, and has had significant economic effects on some coastal regions [*Houghton et al.*, 1996]. Loss of the West Antarctic ice sheet over ~2500 years would at least double this rate, depending on propagation of changes into the East Antarctic ice sheet. Faster collapse, perhaps in as little as a single century, remains possible [e.g., *Oppenheimer*, 1998; *Bindschadler*, 1998] if not likely [*Bentley*, 1998].

Study of the West Antarctic environment is important for many other reasons. The marginal seas of West Antarctica are both highly productive and scenic, making the region economically significant. Ice and sediment cores reveal the history of atmospheric and oceanic processes well beyond West Antarctica. Complex interactions of the ocean with atmosphere, sea ice, ice shelves and the grounded ice sheet centered on West Antarctic continental shelves provide one of the two major sources of cold bottom water in the world's oceans. This deep-water link helps drive the global circulation of the oceans, affecting climate worldwide [e.g., *Broecker*, 1994; *Wang et al.*, 1999]. Most West Antarctic research to date has addressed the sea-level link, so we will focus our discussion on that topic, but we expect that West Antarctica will be the site of major future advances in other topics.

SEEING MORE CLEARLY

Initially, the overriding difficulties of studying the West Antarctic environment were observational. When Mercer focused attention on the West Antarctic ice sheet, we knew only its general outline, plus a few surface elevations and ice thicknesses along traverse lines. A series of fundamental observational advances has entirely changed this situation, and we are becoming data-rich in some regions. The amount of data needed to enable predictions increases with the complexity of the system concerned, so these advances bring us much closer to being able to assess future changes in West Antarctica.

Here, we highlight some of the advances. The individual chapters of this volume provide greater insight to many of these advances and their implications for our understanding of the West Antarctic ice sheet. The chapters also contain more-complete reference lists.

Satellite-based observations [reviewed *by Fahnestock and Bamber*, this volume] are the fastest way to survey a continent. Over Antarctica, radar altimetry, and soon laser altimetry, produce surface-elevation maps and allow assessment of ice-flow directions and changes in surface elevation [*Wingham et al.*, 1998]. Imagery allows mapping of features including grounding lines and ice margins that record ongoing and past ice-sheet processes. Repeat imagery reveals ice velocities in regions of recognizable surface features moving with the ice that survive long enough to move significantly [*Bindschadler et al.*, 1996]. The shading captured in images allows photoclinometry [*Bindschadler and Vornberger*, 1994; *Scambos and Fahnestock*, 1998] and thus interpolation of elevations without the complexity of geophysical migration. Measurement of energy emissions at microwave and other wavelengths can be used to quantify surface temperatures [*Comiso*, 1994] and to monitor near-surface snow conditions [*Shuman et al.*, 1993]. Actively broadcasting imaging radar is useful for feature-mapping even through cloud cover [*Bindschadler et al.*, 1987], and interferometric use of repeat imagery is revolutionizing the mapping of ice-flow velocities by providing data with 25-m resolution even in regions lacking visible features [*Joughin et al.*, 1998; 1999]. In the near future, joint interpretation of satellite-based gravity measurements and laser altimetry may help place important constraints on the history and mass balance of the ice sheet [*Bentley and Wahr*, 1998]. The long-standing difficulty with satellite-based observations of Antarctica is that the orbits of most sensor platforms are not truly polar and often do not allow observations of all or part of Antarctica. Pointable sensors, or those with wider fields of view, partially alleviate this limitation.

Aerogeophysical results are illustrated by *Blankenship et al.* [this volume]. Airborne radar remains the only practical method to map ice thickness over broad areas [*Drewry*, 1983; *Shabtaie* and *Bentley*, 1987; *Blankenship et al.*, this volume]. Combining airborne radar with laser altimetry, aerogravity and aeromagnetics provides key insights to the geological [*Dalziel and Lawver*, this volume] and glaciological setting. Phase-coherent radar offers the possibility of mapping elements of the basal water system, which is central to basal lubrication of fast flow [*Bentley et al.*, 1998; *Blankenship et al.*, this volume].

Internal layers observed in surface-based and some airborne radar records [*Raymond et al.*, this volume] are (primarily) isochrones [*Whillans*, 1976]. Their deformation integrates the history of ice flow and demonstrates interesting non-linear effects of flow [*Nereson et al.*, 1998; *Vaughan et al.*, this volume].

Other surface glaciological and geophysical techniques are proving increasingly valuable. Motion surveys using GPS techniques supplement remote-sensing results and serve as the primary data in regions lacking satellite coverage. Active and passive seismic techniques [e.g., *Blankenship et al.*, 1986; *Smith*, 1997; *Anandakrishnan et al.*, this volume] allow mapping and characterization of subglacial tills, sedimentary units, and other features that control fast flow in many places.

Emplacement of long-term markers in shallow holes bored in the ice-sheet surface [*Hamilton et al.*, 1998] allows GPS tracking of ice motion and calibration of mass-balance determinations by repeat altimetry. Deeper borings allow ice thermometry for rheological and paleoclimatological studies, and provide access to the bed for studies of water, till and substrate that are central to stability questions [*Kamb,* this volume].

Offshore, improved ability to core and image sediments is providing new views of the history of the ice sheet [*Anderson and Shipp,* this volume]. Onshore, improved ice-core analyses and exposure-age dating are doing the same [*Borns,* this volume; *Steig et al.,* this volume].

In the marine realm, global projects such as the World Ocean Circulation Experiment (WOCE) and its regional component focused on the southern oceans are contributing to our understanding of processes there [e.g., *Gordon et al.,* 1999]. The ice shelves are confined to continental-shelf areas, so conditions in the deep ocean must first be transferred to the shelf waters before being able to affect the undersides of ice shelves. Oceanographic observations and models, especially as part of FRISP, are showing how ocean and ice sheet interact beneath the ice shelves and adjacent sea ice [*Jenkins and Bombosch,* 1995; *Jenkins et al.,* 1997; *Gerdes et al.,* 1999].

Our understanding of the atmospheric environment is also becoming clearer [*Bromwich and Rogers,* this volume]. The network of automatic weather stations remains sparse, but coverage is much better than even a decade ago. Satellite data and reanalysis products are improving interpretation of mass input to the ice sheet and our understanding of circulation patterns [*Bromwich and Rogers,* this volume].

And, this is not a complete list of recent observational advances. Much of the work has required breakthroughs of methodology. Superb and dedicated researchers have answered these challenges with innovative approaches and techniques. We now stand on the verge of achieving a deep understanding of the West Antarctic environment. We next discuss some of the evidence that contributes to our improving understanding.

SOME ARGUMENTS ON THE POSSIBILITY OF WEST ANTARCTIC ICE-SHEET COLLAPSE

Public and political concerns about global warming have forced the Antarctic community to address the spectre of West Antarctic ice-sheet collapse just as our knowledge base is expanding, and before we are capable of accurate prediction. Because scientific discussions often impact arguments for or against reduction in emissions of greenhouse gases, there are large economic implications, and opinions often become polarized. It is easy to support one extreme or the other through selective reading of the literature. Review of the relevant literature shows how complex the system really is, and how difficult it is to draw conclusions about the likelihood of West Antarctic ice sheet collapse, whether from human or natural causes.

History of the West Antarctic Ice Sheet

Geologically young diatom shells and short-half-life cosmogenic ^{10}Be are present in sediments beneath the West Antarctic ice sheet at the Upstream B (UpB) camp on ice stream B hundreds of kilometers from the nearest modern open water [*Scherer et al.,* 1998]. Aeolian processes deliver materials to the ice sheet [*Burckle and Potter,* 1996], but several characteristics of the UpB samples imply that this site was ice-free [*Scherer et al.,* 1998]. This in turn indicates that the ice sheet was significantly smaller or completely lost sometime after it formed [*Anderson and Shipp,* this volume]. However, these data do not tell exactly when or how rapidly the ice sheet shrank.

Several lines of circumstantial evidence point to marine isotope stage 11 about 400,000 years ago [*Droxler et al.,* 1999] as the most-likely time of shrinkage, and suggest that the ice sheet was lost entirely, possibly causing loss of some ice from East Antarctica and a sea-level stand much higher than from the West Antarctic ice sheet alone. A stage-11 West Antarctic ice-sheet collapse does not preclude a stage-5 collapse about 120,000 years ago as originally postulated by Mercer, but the possibility that a likely stage-5 sea-level high-stand was largely derived from the Greenland ice sheet [*Cuffey and Marshall,* 2000] may indicate West Antarctic ice-sheet stability during stage-5 warmth. Neither stage 11 nor stage 5 is an exact analog for projected greenhouse warming.

History of other ice sheets.

Ice sheets can exist under a great range of conditions on Earth. It is rather easy to construct a model ice sheet that exhibits no significant changes over centuries, but it also is easy to construct a model ice sheet that changes greatly on such timescales [e.g., *MacAyeal,* 1993]. These models suggesting stability and instability would obey the physical laws and conditions governing ice sheets, as we understand them, but emphasize different aspects of the environment. Until fairly recently, there were many speculations on rapid changes in ice sheets but few hard data.

This situation changed with the recognition and characterization of the Heinrich events in north Atlantic sediments. Heinrich events were first observed as layers of concentrated ice-rafted debris deposited when the surface waters of the north Atlantic were exceptionally cold and fresh [*Heinrich,* 1988; *Bond et al.,* 1992]. Numerous studies have demonstrated that at least most of the Heinrich layers were deposited in centuries or less, and across much of the north Atlantic are dominated by material with Hudson Bay affinities [reviewed by *Broecker*, 1994; *Alley and Clark*, 1999]. The leading model is that a surge of the ice sheet in Hudson Bay disgorged large quantities of debris-laden icebergs into the north Atlantic, cooling and freshening the surface water, raising sea level, and depositing the Heinrich layers [*MacAyeal*, 1993; *Alley and MacAyeal*, 1994].

It remains, however, that the sea-level impact of Heinrich events is not known accurately [*Marshall and Clarke*, 1997]. Perhaps more importantly, the reconstructed Heinrich surges emanated from an ice sheet that no longer exists and that differed from the modern West Antarctic ice sheet in significant ways. Similarities between the former Laurentide ice sheet in Hudson Bay and the modern West Antarctic ice sheet include a bed below sea level resting on sedimentary rather than crystalline rocks in many places. However, the detailed geometry and characteristics of the substrates differ, and Hudson Bay is at lower latitude than is West Antarctica. Furthermore, there is a suggestion that Heinrich events were triggered by cooling rather than warming [reviewed by *Alley and Clark*, 1999]. We thus cannot tell at present whether the ice-sheet collapses that led to Heinrich events provide useful information about the future of the West Antarctic ice sheet.

Recent Changes of the West Antarctic ice sheet.

There is no longer any doubt that regions within the West Antarctic ice sheet can change rapidly. A wealth of information accumulated over recent decades shows that regions of the West Antarctic ice sheet are grossly out of balance, thinning or thickening at rates on the order of 1 m/yr, much faster than the rate at which ice accumulates on their surface [*Alley and Whillans*, 1991; *Bindschadler et al.*, 1996].

For example, in the Pine Island Bay region [see *Vaughan et al.*, this volume] ice-shelf loss and grounding-line retreat appear to be coupled to inland thinning of the fast-moving portions of the ice sheet [*Rignot et al.*, 1998; *Shepherd et al.*, 1999]. Evidence of rapid changes is not as common in the drainages feeding the Filchner-Ronne Ice Shelf, but some significant changes are indicated [*Doake et al.*, this volume].

In the Siple Coast region, ice flow in the lower reaches of ice stream C all but stopped just over a century ago [*Retzlaff and Bentley*, 1993], but inland tributaries feeding the ice stream remain active [*Anandakrishnan and Alley*, 1997a], forming an enlarging "ice bulge" [*Joughin et al.*, 1999]. In contrast, the head region of ice stream B is thinning as the ice stream extends into inland ice [*Shabtaie and Bentley*, 1987; *Shabtaie et al.*, 1988]. Crary Ice Rise is a young feature [*Bindschadler et al.*, 1990]. Ice-stream margins migrate and jump frequently [*Clarke and Bentley*, 1995; *Bindschadler and Vornberger*, 1998; *Echelmeyer and Harrison*, 1999; *Jacobel et al.*, in press]. Looped flow lines on the surface of the Ross Ice Shelf record varying output of ice streams [*Casassa et al.*, 1991; *Fahnestock et al.*, in press]. Were the regions of fastest thinning to become widespread, it is easy to envision the runaway collapse postulated by *Weertman* [1974] and *Thomas and Bentley* [1978].

Over larger areas, however, many of these changes average out. The thinning of ice stream B is nearly balanced by the thickening of ice stream C [*Shabtaie et al.*, 1988]. At the scale that matters to sea level, the ice sheet may be thinning slowly but is not far from balance today [*Wingham et al.*, 1998]. Nevertheless, the large loss of West Antarctic ice since the last glacial maximum accounts for an average contribution to sea-level rise of nearly 1 mm/yr over the most recent 11,000 years [*Bindschadler*, 1998].

Model Predictions of Behavior of the West Antarctic Ice Sheet

Much of the credibility of the early suggestions of West Antarctic ice-sheet collapse came from the ability of simple models [*Weertman*, 1974; *Thomas and Bentley*, 1978] to simulate ice-sheet collapses. Those models omitted some essential physical processes and so could not be considered predictive, but they did serve to motivate considerable research.

Some of the modeling exercises that followed have reached the opposing conclusion, finding that the ice sheet is relatively stable except under very large changes in boundary conditions [reviewed in *Houghton et al.*, 1996]. Such modeling involves sophisticated solutions of heat and mass balance, often incorporates ice-shelf physics, and parameterizes surface mass balance [see *Hulbe and Payne*, this volume; *Steig et al.*, this volume].

However, the "fast" physical processes that produce the extreme lubrication of some ice streams are not well-represented in any of these models. Some fail to resolve ice streams because of the large grid spacing required to allow efficient computation. Features resembling ice streams may develop for thermal reasons, with thawing in deep troughs and freezing to higher bedrock [e.g., *Huybrechts*, 1990; *Payne*, 1995]. However, there is a tendency for some of these models to have used continental-average geothermal fluxes rather than the higher geothermal fluxes indicated for the Siple Coast by the limited available data [e.g., *Alley and Bentley*, 1988; see *Blankenship et al.*, this volume]. The strong dependence of basal conditions on geothermal flux raises questions about the actual controlling physical processes beneath the real ice sheet vs. those processes simulated in the models [*Doake et al.*, this volume; *Whillans et al.*, this volume].

An attempt to include the "fast" physics was made by *MacAyeal* [1992]. He found that glacial-interglacial cycles caused the model West Antarctic ice sheet to oscillate with changes of a significant fraction of the ice-sheet volume in as little as a few centuries, and with complete collapses during two out of ten simulated 100,000-year cycles but not during the other eight cycles. However, because the combined behavior of subglacial sediment and water remains quite complex and poorly understood (despite major advances, as outlined by *Kamb* [this volume] and *Anandakrishnan et al.* [this volume], among others), caution is required in interpreting the results.

Summary of Evidence on Possible Collapse of the West Antarctic Ice Sheet

We conclude that the West Antarctic ice sheet has grown and shrunk since it formed, and may have disappeared entirely, but we do not know how rapidly the large changes occurred. Another marine ice sheet, the portion of the Laurentide ice sheet in Hudson Bay, changed greatly and rapidly in the past, but differed in some ways from the modern West Antarctic ice sheet. The portion of the Laurentide ice sheet in Hudson Bay also exhibited many millennia of stability between its rapid changes. The triggers of those past changes may have been quite different from expected future forcings of the West Antarctic ice sheet. Portions of the West Antarctic ice sheet have changed significantly in decades to centuries, but ongoing changes averaged over the whole ice sheet are relatively small. Some models of the West Antarctic ice sheet indicate great stability on an ice-sheet scale, whereas others allow instability. Clearly, we cannot yet assess West Antarctic ice-sheet stability with confidence.

COLLAPSE MECHANISMS

Numerous mechanisms can be envisioned by which the West Antarctic ice sheet could significantly increase its rate of mass transfer to the ocean and affect sea level, including the surface-climate changes of ice-age cycles, internal instabilities, and ice-shelf mechanisms. These mechanisms serve as foci for research on stability of the West Antarctic ice sheet.

Surface Climate Over Ice-Age Cycles

The time scale for surface changes to propagate through an ice sheet to the bed is typically millennia or longer [*Whillans*, 1978; *Alley and Whillans*, 1984]. Thus, the large ice sheets have not completed responding to the ending of the last ice age between about 20,000 and 10,000 years ago, although the relatively small or zero measured modern imbalance [*Wingham et al.*, 1998] may indicate that much of the response has been completed. The end of an ice age involves warming and increased accumulation at the surface. The increased accumulation initially tends to cool the bed in broad regions, because ice-sheet flow adjusts so that the downward motion of ice at the surface balances the snow accumulation, and this downward motion transports cold surface ice nearer the bed, thus cooling the bed. The increased surface temperature tends to warm the bed, by making the ice transported toward the bed warmer, and by reducing the temperature difference between the surface and deep in the Earth and so reducing the rate of energy transfer upward. Broadly, these effects tend to offset [*Alley and Whillans*, 1984; *Huybrechts*, 1990], but in detail they will not.

Basal melting might increase millennia or longer after the end of an ice age, allowing faster sliding, which would create more heat and promote more melting, thus triggering a collapse. For relatively thin ice with low surface slope as in much of the West Antarctic ice sheet, enhanced basal lubrication may be the most likely mechanism for enhanced ice discharge. However, increased basal melting is not sufficient for increased ice discharge; geological constraints (a soft-sediment or a very smooth bed) must also be satisfied [e.g., *Blankenship et al.*, this volume]. Furthermore, it is not clear that the bed of the ice sheet is warming significantly over wide regions today.

Internal Instabilities

Negative feedbacks stabilize most ice-sheet regions against small perturbations. However, numerous positive feedbacks exist in ice sheets, and under appropriate conditions these might amplify even quite small perturbations. Some data from the West Antarctic ice sheet can be interpreted to indicate that positive feedbacks are amplifying small changes. For example, the stagnation of ice stream C and the thinning of ice stream B may be related to the details of the interactions of these two ice streams [*Alley et al.*, 1994; *Payne*, 1995; *Anandakrishnan et al.*, this volume]. Additionally, the widespread evidence for jumping ice-stream margins and other instabilities discussed above indicates that there are great complexities in the system.

Positive feedbacks involving ice-stream margins may be active. Available data indicate that the fast-moving ice of the Siple Coast ice streams is restrained significantly or dominantly by the slow-moving ridges between, with much of the driving stress for ice flow opposed by restraint generated at the beds beneath the ridges near the ice streams and transmitted across the ice-stream margins [*Raymond et al.*, this volume; *Whillans et al.*, this volume]. Ice-stream marginal drag probably also is significant or dominant along at least portions of the Pine Island and Thwaites Glaciers [*Vaughan et al.*, this volume] and Rutford ice stream [*Doake et al.*, this volume], among others.

Ice-stream widening would increase the stress on the margins and on the beds just outboard of the margins. The creep rate of ice in margins increases with the third power of the stress, so increasing the stress can greatly increase the ice velocity [e.g., *Raymond et al.*, this volume]. Such a velocity increase in turn might trigger one or several additional stabilizing or destabilizing feedbacks.

Looking first at stabilizing responses to ice-stream widening and speed increase, if the bed has numerous sticky spots, or is a till whose strength permits only a slow increase of strain rate with increasing stress (a low stress exponent in a power-law-creep relationship for till deformation), extra stress from ice-stream widening may be transmitted to the ice-stream bed. If the ice begins to thin, colder ice will move nearer to the bed, which may begin to freeze, generating more sticky spots [*Payne*, 1995; *B. Kamb and H.E. Engelhardt*, Chapman Conference, Orono, ME, Sept. 1998]. Additionally, extra ice passing down the ice streams may enhance grounding around ice rises in the ice shelf down-glacier [*Bindschadler et al.*, 1990], increasing backpressure on the grounded ice and slowing its flow. Finally, the ice may thin but be unable to propagate significant thinning to inland regions because of geological controls [*Bindschadler et al.*, this volume; *Blankenship et al.*, this volume]. In any of these cases, a perturbation to ice-stream velocity will be damped, tending to maintain stability.

On the other hand, enhanced velocity may trigger a string of positive feedbacks in which increased motion creates increased friction generating more basal water, enhancing lubrication and increasing motion still further. This acceleration mechanism could cause the fast motion to spread inland and into interstream ridges, triggering ice-sheet collapse. Ice streams, as large features with wet beds, may be examples of the action of this mechanism, which highlights the question of what is limiting ice-stream size and speed [e.g., *Kamb*, this volume].

Ice-Shelf Mechanisms

While the mechanisms discussed above do not explicitly involve the marine character of the West Antarctic ice sheet, this factor should not be overlooked. Marine character may have been important during previous retreats of the West Antarctic ice sheet [e.g., *Ackert et al.*, 1999; *Conway et al.*, 1999]. Also, the surges of the Laurentide ice sheet that led to Heinrich events emanated from a marine ice sheet.

In the classic *Weertman* [1974] model, ice flowing from the grounded portion of the West Antarctic ice sheet floats in surrounding seas as ice shelves. Weertman assumed that ice thickness decreases with increasing distance from the ice-sheet center in part because the bed becomes shallower away from the ice-sheet center. Floating ice shelves spread and thin under gravity's load at a rate roughly proportional to the third power of the ice thickness. If the grounding line moves inland so that thicker ice floats, this thicker ice will spread more rapidly than the downstream ice, thinning the newly floating ice. Weertman postulated that this thinning will affect grounded ice just up-glacier of the grounding line, by steepening the surface slope, increasing the driving stress and so increasing the ice-flow velocity. That grounded ice then would thin, driving the grounding line farther inland. This positive-feedback loop could cause complete ice-sheet collapse.

Stability can be achieved if the ice shelf is not free to spread, because it runs aground on high spots in its bed or because it occupies an embayment and must shear past its sides. In this view, the great ice shelves have allowed the West Antarctic ice sheet to persist to today. However, recent events in the Antarctic Peninsula show that ice shelves are vulnerable to climatic change [*Rott et al.*, 1996; *Vaughan and Doake*, 1996]. Those ice shelves along the Antarctic Peninsula that have disintegrated recently did not abut marine ice, and so do not provide tests of the Weertman hypothesis, but they show that warmth can remove ice shelves rapidly. Such a mechanism figured in Mercer's original warning of possible West Antarctic ice-sheet collapse.

Work by *Anandakrishnan and Alley* [1997b] on tidal cycles is most directly interpreted as showing that increased backpressure from the ocean side does slow the forward motion of grounded ice, at least on (anomalous) ice stream C and over short times. Numerous data sets [e.g., *Jezek et al.*, 1985; *MacAyeal et al.*, 1987; 1989] show that the stress state at the grounding zone would be more extensional if the Ross Ice Shelf were not present.

Beneath portions of the West Antarctic ice sheet, the bed morphology is much more complex than envisioned in the *Weertman* [1974] model. A bedrock sill may stabilize Pine Island Glacier (as one likely stabilized the Hudson Bay ice of the Laurentide ice sheet prior to Heinrich events), with possibility of retreat from the sill into deeper regions as modeled by Weertman [*Vaughan et al.*, this volume]. However, some other West Antarctic grounding zones now occur in regions that locally do not deepen up-glacier [e.g., *Shabtaie et al.*, 1987].

Where data are available, the local stress state up-glacier of the grounding line is not dominated by longitudinal stresses transmitted from the grounding line [*Whillans et al.*, this volume], so loss of an ice shelf would not immediately change ice motion inland. Rather, inland adjustments would have to result from propagation of changes up-glacier from the grounding line. Data and models indicate that changes at the margin of the ice sheet create waves of adjustment that move inland, affecting broad regions of an ice sheet [e.g. *Alley and Whillans*, 1984; *Steig et al.*, this volume].

The analogy of the stress state in a reservoir upstream of a dam on a river illustrates how this situation may be more precipitous than present measurements indicate. The stresses within the reservoir are not dominated by stresses transmitted directly from the dam. However, were the dam suddenly removed, the entire reservoir would respond rather quickly. Clearly, an ice stream is not a water reservoir, and it is possible that a reduction of ice-shelf restraint would not have great effects inland [*Hindmarsh*, 1996], in the same way that removal of a dam would not quickly have great effect on the river upstream of a reservoir. Perhaps ice-shelf loss would cause local ungrounding along ice streams but continued grounding of interstream ridges, essentially re-forming the ice shelf; however, this outcome is not certain.

Forcings

In the near term, we must be concerned not only with internal instabilities, but also with responses with long time constants to past forcings, and responses with short time constants to recent or future forcings. The main forcings are sea-level change, temperature change, and mass-balance change (including changes in snow accumulation on the surface and melting beneath ice shelves), although geological forcing is also possible [*Blankenship et al.*, 1993].

The forcing history is becoming increasingly clear. Sea-level studies outside of Antarctica and along the Antarctic coast [*Berkman et al.*, 1998; *Conway et al.*, 1999] are constraining that parameter. Great progress has been made in reconstructing the post-glacial-maximum history of ice-sheet retreat and thinning [*Ackert et al.*, 1999; *Conway et al.*, 1999; *Anderson and Shipp*, this volume; *Borns*, this volume; *Steig et al.*, this volume]. It is likely that sea-level change was important in triggering post-glacial retreat of the West Antarctic ice sheet, but that dynamical processes have continued that retreat even after slow-down of global sea-level rise coupled with local isostatic response caused fall in local sea level [*Conway et al.*, 1999].

Reconstruction of histories of temperature and snowfall in West Antarctica from ice-core and other data has been complicated by the dynamic response of the ice sheet. The multi-disciplinary approach to modern research in West Antarctica is now beginning to resolve this issue [*Steig et al.*, this volume]. Ongoing ice-coring studies [e.g., *Taylor*, 1999] will certainly contribute further.

Future changes in West Antarctic surface conditions often are assumed to be related simply to global projections [e.g., *Houghton et al.*, 1996]. West-Antarctic surface-temperature change may be taken to equal global-average changes, or to exceed them by a constant multiplicative factor, and snow accumulation often is taken to increase some percentage per degree warming.

However, such assumptions are problematic. El Niño has strongly modulated snow accumulation in one sector of West Antarctica, but the sign of the correlation has switched during the instrumental period [*Bromwich and Rogers*, this volume]. This example makes the true complexity of the whole system apparent. The antiphase response of temperature inferred from ice isotopes at Taylor Dome vs. Byrd Station to the millennial climate oscillations of the most recent deglaciation [*Steig et al.*, 1998] also argues that Antarctic climate change is unlikely to have a simple, direct dependence on global mean forcing. Fortunately, the tools that revealed the El Niño signal [*Bromwich and Rogers*, this volume] should help greatly in understanding the Antarctic climatic system and predicting its future.

The complex Antarctic response to millennial climate variability has probably been caused in part by changes in oceanic circulation and deep-water formation on West Antarctic continental shelves adjacent to and beneath ice shelves [*Broecker*, 1998]. Because millennial climate changes are recognized globally [reviewed by *Alley and Clark*, 1999], this suggests that Antarctic changes may have global consequences [*Wang et al.*, 1999]. If ice shelves are important in West Antarctic stability, these questions become critical because of the great sensitivity of ice shelves to oceanic processes. Modern sub-ice-shelf melting rates range from negative to more than 10 m per year [*Jenkins et al.*, 1997; *Vaughan et al.*, this volume] owing largely to oceanic processes. Until better understanding is available, we cannot rule out the possibility that broad regions beneath ice shelves could switch toward or even beyond the extreme melting rates observed today.

SUMMARY

Recently, we have seen a revolution in our ability to observe the West Antarctic ice sheet. Parallel to this have been significant advances in the modeling of the ice sheet, and in characterization of its atmospheric, oceanic, and geologic environment today and in the past. Large gaps remain at the interfaces of these efforts--the new insights to "fast" physics, for example, have not yet been reduced to reliable governing equations that can be used in models.

As the interfaces between data and models are filled in by projects such as FRISP and WAIS, we can hope for a much clearer view of the West Antarctic ice sheet and its likely future. The papers in this volume combine a general review of our present knowledge of the West Antarctic ice sheet with some of the newest, most exciting results.

Pending further advances, West Antarctic ice-sheet collapse triggering sea-level rise of meters over centuries remains a possible if improbable event. Prudence dictates concern and continued research, though not panic.

Acknowledgments. We thank Charles Bentley, Todd Dupont, Mark Fahnestock, David Vaughan, Ian Whillans and other colleagues in WAIS, FRISP, and related projects. We thank the National Science Foundation for funding; funding of FRISP by the European Science Foundation is also gratefully acknowledged. This volume grew out of an American Geophysical Union Chapman Conference originally proposed by Robert Bindschadler and Hal Borns, and hosted by Hal Borns and colleagues at the University of Maine at Orono in Sept., 1998.

REFERENCES

Ackert, R. P., Jr., D. J. Barclay, H. W. Borns, Jr., P. E. Calkin, M. D. Kurz, J. L. Fastook and E. J. Steig, Measurements of past ice sheet elevations in interior West Antarctica, *Science, 286*, 276-280, 1999.

Alley, R. B. and C. R. Bentley, Ice-core analysis on the Siple Coast of West Antarctica, *Ann. Glaciol., 11*, 1-7, 1988.

Alley, R. B. and P. U. Clark, The deglaciation of the northern hemisphere: a global perspective, *Ann. Rev. Earth Planet. Sci.*, 27, 149-182, 1999.

Alley, R. B., S. Anandakrishnan, C. R. Bentley and N. Lord, A water-piracy hypothesis for the stagnation of ice stream C, *Ann. Glaciol., 20*, 187-194, 1994.

Alley, R. B. and D. R. MacAyeal, Ice-rafted debris associated with binge/purge oscillations of the Laurentide ice sheet, *Paleoceanography, 9*, 503-511, 1994.

Alley, R. B. and I. M. Whillans, Changes in the West Antarctic ice sheet, *Science, 254*, 959-963, 1991.

Anandakrishnan, S. and R. B. Alley, Stagnation of ice stream C, West Antarctica by water piracy, *Geophys. Res. Lett., 24*, 265-268, 1997a.

Anandakrishnan, S. and R. B. Alley, Tidal forcing of basal seismicity of ice stream C, West Antarctica, observed far inland, *J. Geophys. Res., 102B*, 15183-15196, 1997b.

Anandakrishnan, S., R. B. Alley, R. W. Jacobel and H. Conway, The flow regime of ice stream C and hypotheses concerning its recent stagnation, *This volume.*

Anderson, J. B. and S. S. Shipp, Evolution of the West Antarctic ice sheet, *This volume.*

Bentley, C. R, Rapid sea-level rise from a West Antarctic ice-sheet collapse: a short-term perspective, *J. Glaciol., 44*, 157-163, 1998.

Bentley, C. R. and J. M. Wahr, Satellite gravity and the mass balance of the Antarctic ice sheet, *J. Glaciol., 44*, 207-213, 1998.

Bentley, C. R., N. Lord and C. Liu, Radar reflections reveal a wet bed beneath stagnant ice stream C and a frozen bed beneath ridge BC, West Antarctica, *J. Glaciol., 44*, 157-164, 1998.

Berkman, P.A., J. T. Andrews, S. D. Emslie, I. D. Goodwin, B. L. Hall, C. P. Hart, K. Hirakawa, A. Igarashi, O. Ingolfsson., J. Lopez-Martinez, W. B. Lyons, M. C. G. Mabin, P. G. Quilty, M. Taviani and Y. Yoshida, Circum-Antarctic coastal environmental shifts during the Late Quaternary reflected by emerged marine deposits, *Antarctic Sci., 10*, 345-362, 1998.

Bindschadler, R. A., Future of the West Antarctic ice sheet, *Science, 282*, 428-429, 1998.

Bindschadler, R. A, and P. L. Vornberger, Detailed elevation map of ice stream C using satellite imagery and airborne radar, *Ann. Glaciol., 20*, 327-335, 1994.

Bindschadler, R., J. Bamber, S. Anandakrishnan, Onset of streaming flow in the Siple Coast region, West Antarctica, *This volume.*

Bindschadler, R. A., K. C. Jezek and J. Crawford, Glaciological Investigations using the Synthetic Aperture Radar imaging system, *Ann. Glaciol., 9*, 11-19, 1987.

Bindschadler, R. and P. Vornberger, Changes in the West Antarctica ice sheet since 1963 from declassified satellite photography, *Science, 279*, 689-692, 1998.

Bindschadler, R. A., E. P. Roberts and A. Iken, Age of Crary Ice Rise, Antarctica determined from temperature-depth profiles, *Ann. Glaciol., 14*, 13-16, 1990.

Bindschadler, R., P. Vornberger, D. Blankenship, T. Scambos and R. Jacobel, Surface velocity and mass balance of ice streams D and E, West Antarctica, *J. Glaciol., 42*, 461-475, 1996.

Blankenship, D. D., C. R. Bentley, S. T. Rooney and R. B. Alley, Seismic measurements reveal a saturated, porous layer beneath an active Antarctic ice stream, *Nature, 322*, 54-57, 1986.

Blankenship, D. D., R. E. Bell, S. M. Hodge, J. M. Behrendt J. C. Brozena, and C. A. Finn, Active volcanism beneath the West Antarctic ice sheet and implications for ice-sheet stability, *Nature, 361*, 526-529, 1993.

Blankenship, D. D., D. L. Morse, C. A. Finn, R. E. Bell, M. E. Peters, S. D. Kempf, S. M. Hodge, M. Studinger, J. C. Behrendt and J. M. Brozena, Geologic controls on the initiation of rapid basal motion for West Antarctic ice streams, *This volume.*

Bond, G., H. Heinrich, W. Broecker, L. Labeyrie, J. McManus, J. Andrews, S. Huon, R. Jantschik, S. Clasen, C. Simet, K. Tedesco, M. Klas, G. Bonani and S. Ivy, Evidence for massive discharges of icebergs into the North Atlantic ocean during the last glacial period, *Nature, 360*, 245-249, 1992.

Borns, H. W., Jr., The glacial geologic terrestrial record from West Antarctica with emphasis on the last glacial cycle, *This volume.*

Broecker, W. S., Paleocean circulation during the last deglaciation; a bipolar seesaw? *Paleoceanography, 13*, 119-121, 1998.

Broecker, W. S., Massive iceberg discharges as triggers for global climate change, *Nature, 372*, 421-424, 1994.

Bromwich, D. H. and A. N. Rogers, The El Niño-Southern Oscillation modulation of West Antarctic precipitation, *This volume*.

Burckle, L. H. and N. Potter, Jr., Pliocene-Pleistocene diatoms in Paleozoic and Mesozoic sedimentary and igneous rocks from Antarctica; a Sirius problem solved, *Geology, 24*, 235-238, 1996.

Casassa, G., K. C. Jezek, J. Turner and I. M. Whillans, Relict flow stripes on the Ross Ice Shelf, *Ann. Glaciol., 15*, 132-138, 1991.

Clarke, T. S. and C. R. Bentley, Evidence for a recently abandoned ice stream shear margin, *Eos (Transactions of the American Geophysical Union) 76*(46), F194 (abstract), 1995.

Comiso, J. C., Surface temperatures in the polar regions from Nimbus & temperature humidity infrared radiometer, *J. Geophys. Res., 99C*, 5181-5200, 1994.

Conway, H., B. L. Hall, G. H. Denton, A. M. Gades and E. D. Waddington, Past and future grounding-line retreat of the West Antarctic ice sheet, *Science, 286*, 280-283, 1999.

Cuffey, K.M. and S.J. Marshall, Substantial contribution to sea-level rise during the last interglacial from the Greenland ice sheet, *Nature, 404*, 591-594, 2000.

Dalziel, I. W. D. and L. A. Lawver, The lithospheric setting of the West Antarctic ice sheet, *This volume*.

Doake, C.S. M., H. F. J. Corr, A. Jenkins, K. Makinson, K. W. Nicholls, C. Nath, A. M. Smith and D. G. Vaughan, Rutford Ice Stream, Antarctica, *This volume*.

Drewry, D. J., ed., Antarctica: Glaciological and Geophysical Folio, Scott Polar Research Institute, University of Cambridge, Cambridge, UK, 1983.

Droxler, A. W., R. Poor and L. Burckle, Data on past climate warmth may lead to better model of warm future, *Eos (Transactions of the American Geophysical Union) 80*, 289-290, 1999.

Echelmeyer, K. A. and W. D. Harrison, Ongoing margin migration of ice stream B, Antarctica, *J. Glaciol., 45*, 361-369, 1999.

Fahnestock, M. and J. Bamber, Morphology and surface characteristics of the West Antarctic ice sheet, *This volume*.

Fahnestock, M. F., T. A. Scambos, R. A. Bindschadler and G. Kvaran, A millennium of variable ice flow recorded by the Ross Ice Shelf, Antarctica, *J. Glaciol.*, in press.

Gerdes, R., J. Determann and K. Grosfeld, Ocean circulation beneath Filchner-Ronne Ice Shelf from three-dimensional model results, *J. Geophys. Res., 104C*, 15827-15842, 1999.

Gordon, A. L., B. Barnier, K. Speer and L. Stramma, Introduction to special section: World Ocean Circulation Experiment: South Atlantic results, *J. Geophys. Res., 104C*, 20,859-20,861, 1999.

Hamilton, G. S., I. M. Whillans and P. J. Morgan, First point measurements of ice-sheet thickness change in Antarctica, *Ann. Glaciol., 27*, 125-129, 1998.

Heinrich, H., Origin and consequences of cyclic ice rafting in the northeast Atlantic Ocean during the past 130,000 years, *Quaternary Res., 29*, 143-152, 1988.

Hindmarsh, R. C. A., Stability of ice rises and uncoupled marine ice sheets, *Ann. Glaciol., 23*, 105-115, 1996.

Houghton, J. T., L. G. Meira Filho, B. A. Callander, N. Harris, A. Kattenberg and K. Maskell, Eds., *Climate change 1995: the science of climate change*, Cambridge University Press, 572 pp., 1996.

Hughes, T. J., Is the West Antarctic ice sheet disintegrating? Scientific Justification, *Ice Streamline Cooperative Antarctic Project (ISCAP) Bulletin No. 1*, Institute of Polar Studies, The Ohio State University, Columbus, OH, USA, 1972.

Hughes, T. J., Is the West Antarctic ice sheet disintegrating? Science Plan, *Ice Stability Coordinated Antarctic Program (ISCAP) Bulletin No. 2*, Institute of Polar Studies, The Ohio State University, Columbus, OH, USA, 1973.

Hulbe, C. L. and A. J. Payne, The contribution of numerical modelling to our understanding of the West Antarctic Ice Sheet, *This volume*.

Huybrechts, P., A 3-D model for the Antarctic ice sheet; a sensitivity study on the glacial-interglacial contrast, *Climate Dynamics, 5*, 79-92, 1990.

Jacobel, R. W., T. A. Scambos, N. A. Nereson and C. F. Raymond, Changes in the margin of ice stream C, Antarctica, *J. Glaciol.*, in press.

Jenkins, A. and A. Bombosch, Modeling the effects of frazil ice crystals on the dynamics and thermodynamics of ice shelf water plumes, *J. Geophys. Res., 100C*, 6967-6981, 1995.

Jenkins, A., D. G. Vaughan, S. S. Jacobs, H. H. Hellmer and J. R. Keys, Glaciological and oceanographic evidence of high melt rates beneath Pine Island Glacier, West Antarctica, *J. Glaciol. 43*, 114-121, 1997.

Jezek, K. C., R. B. Alley and R. H. Thomas, Rheology of glacier ice, *Science, 227*, 1335-1337, 1985.

Joughin I., M. Fahnestock, R. H. Thomas, and R. Kwok, Ice flow in Northeast Greenland Ice Stream derived using balance velocities as control, *IGARSS '98*, Seattle, 1998.

Joughin, I., L. Gray, R. Bindschadler, S. Price, D. Morse, C. Hulbe, K. Mattar and C. Werner, Tributaries of West Antarctic ice streams revealed by RADARSAT interferometry, *Science, 286*, 283-286, 1999.

Kamb, W. B., Basal zone of the West Antarctic ice streams and its role in their streaming motions, *This volume*.

MacAyeal, D. R., Irregular oscillations of the West Antarctic ice sheet, *Nature, 359*, 29-32, 1992.

MacAyeal, D. R., Binge/purge oscillations of the Laurentide Ice Sheet as a cause of the North Atlantic's Heinrich events, *Paleoceanography, 8*, 775-784, 1993.

MacAyeal, D. R., R. A. Bindschadler, S. Shabtaie, S. Stephenson and C. R. Bentley, Force, mass and energy budgets of the Crary Ice Rise complex, Antarctica, *J. Glaciol. 33*, 218-230, 1987.

MacAyeal, D. R., R. A. Bindschadler, S. Shabtaie, S. Stephenson and C. R. Bentley, Correction to: Force, mass and energy budgets of the Crary Ice Rise complex, Antarctica, *J. Glaciol. 35*, 151-152, 1989.

Marshall, S. J. and G. K. C. Clarke, A continuum mixture model of ice stream thermomechanics in the Laurentide ice sheet; 2, Application to the Hudson strait ice stream, *J. Geophys. Res., 102B,* 20,615-20,637, 1997.

Mercer, J. H., Antarctic ice and Sangamon sea level, *International Association of Hydrological Sciences Publication No. 179,* 217-225, 1968.

Mercer, J. H., West Antarctic ice sheet and CO_2 greenhouse effect; a threat of disaster, *Nature, 271,* 321-325, 1978.

Nereson, N. A., C. F. Raymond, E. D. Waddington and R. W. Jacobel, Migration of the Siple Dome ice divide, West Antarctica, *J. Glaciol., 44,* 632-652, 1998.

Oppenheimer, M., Global warming and the stability of the West Antarctic ice sheet, *Nature, 393,* 325-332, 1998.

Payne, A. J., Limit cycles in the basal thermal regime of ice sheets, *J. Geophys. Res., 100B,* 4249-4263, 1995.

Raymond, C. F., K. A. Echelmeyer, I. M. Whillans and C. S. M. Doake, Ice stream shear margins, *This volume.*

Retzlaff, R. and C. R. Bentley, Timing of stagnation of ice stream C, West Antarctica, from short-pulse radar studies of buried surface crevasses, *J. Glaciol., 39,* 553-561, 1993.

Rignot, E. J., Fast recession of a West Antarctic glacier, *Science 281,* 549-551, 1998.

Rott, H., P. Skvarca and T. Naegler, Rapid collapse of northern Larsen ice shelf, Antarctica, *Science, 271,* 788-792, 1996.

Scambos, T. A. and M. A. Fahnestock, Improving digital elevation models over ice sheets using AVHRR-based photoclinometry, *J. Glaciol., 44,* 97-103, 1998.

Scherer, R. P., A. Aldahan, S. Tulaczyk, G. Possnert, H. Engelhardt and B. Kamb, Pleistocene collapse of the West Antarctic ice sheet, *Science, 281,* 82-85, 1998.

Shabtaie, S. and C. R. Bentley, West Antarctic ice streams draining into the Ross Ice Shelf: configuration and mass balance, *J. Geophys. Res., 92B,* 1311-1336, 1987.

Shabtaie, S., C. R. Bentley, R. A. Bindschadler and D. R. MacAyeal, Mass-balance studies of ice streams A, B, and C, West Antarctica, and possible surging behavior of ice stream B, *Ann. Glaciol., 11,* 137-149, 1988.

Shabtaie, S., I. M. Whillans and C. R. Bentley, The morphology of ice streams A, B, and C, West Antarctica, and their environs, *J. Geophys. Res., 92B,* 8865-8883, 1987.

Shepherd, A., D. Wingham and J. Mansley, Inland thinning of Pine Island Glacier (abstract), *Eos (Transactions of the American Geophysical Union), 80,* supplement, F369, 1999.

Shuman, C. A., R. B. Alley and S. Anandakrishnan, Characterization of a hoar-development episode using SSM/I brightness temperatures in the vicinity of the GISP2 site, Greenland, *Ann. Glaciol., 17,* 183-188, 1993.

Smith, A. M., Basal conditions on Rutford Ice Stream, West Antarctica, from seismic observations, *J. Geophys. Res., 102B,* 543-552, 1997.

Steig, E. J., E. J. Brook, J. W. C. White, C. M. Sucher, M. L. Bender, S. J. Lehman D. L. Morse, E. D. Waddington, and G. D. Clow, Synchronous climate changes in Antarctica and the North Atlantic, *Science, 281,* 92-95, 1998.

Steig, E. J., J. W. Fastook, C. Zweck, R. P. Ackert, J. W. C. White, K. J. Licht and I. D. Goodwin, West Antarctic ice sheet elevation changes, *This volume.*

Taylor, K, Rapid climate change, *Am. Scientist, 87,* 320-327, 1999.

Thomas, R. H. and C. R. Bentley, A model for Holocene retreat of the West Antarctic ice sheet, *Quaternary Res., 10,* 150-170, 1978.

Vaughan, D. G., H. F. J. Corr, A. M. Smith, A. Jenkins, C. R. Bentley, M. D. Stenoien, S. S. Jacobs, T. B. Kellogg, E. Rignot and B. K. Lucchita, A review of ice-sheet dynamics in the Pine Island Glacier basin, West Antarctica: hypotheses of instability vs. observations of change, *This volume.*

Vaughan, D. V. and C. S. Doake, Recent atmospheric warming and retreat of ice shelves on the Antarctic Peninsula, *Nature, 379,* 328-331, 1996.

Wang, Z., P. H. Stone and J. Marotzke, Global thermohaline circulation. Part I: Sensitivity to atmospheric moisture transport, *J. Climate, 12,* 71-82, 1999.

Weertman, J., Stability of the junction of an ice sheet and an ice shelf, *J. Glaciol., 13,* 3-11, 1974.

Whillans, I. M., Radio-echo layers and the recent stability of the West Antarctic ice sheet, *Nature, 264,* 152-155, 1976.

Whillans, I. M., Inland ice sheet thinning due to Holocene warmth, *Science, 201,* 1014-1016, 1978.

Whillans, I. M., C. R. Bentley and C. J. van der Veen, Ice Streams B and C, *This volume.*

Wingham, D. J., A. J. Ridout, R. Scharroo, R. J. Arthern and C. K. Shum, Antarctic elevation change from 1992 to 1996, *Science, 282,* 456-458, 1998.

Richard B. Alley, Environment Institute and Department of Geosciences, The Pennsylvania State University, Deike Building, University Park, PA 16802

Robert A. Bindschadler, Oceans and Ice Branch, NASA/Goddard Space Flight Center, Greenbelt, MD, 20771

MORPHOLOGY AND SURFACE CHARACTERISTICS OF THE WEST ANTARCTIC ICE SHEET

Mark Fahnestock

Earth System Science Interdisciplinary Center, University of Maryland, College Park, Maryland

Jonathan Bamber

Bristol Glaciology Centre, School of Geographical Sciences, University of Bristol, UK

Over the last twenty five years a large body of knowledge about the characteristics of the West Antarctic Ice Sheet has been assembled from satellite and airborne remote sensing observations. Presented here is an overview of a number of these observations, covering surface morphology, ice thickness, surface characteristics, and ice motion. The value of these data sets for improving our understanding of the behavior of the West Antarctic Ice Sheet is considered with reference to other papers in this volume and previously published glaciological investigations of the ice sheet.

INTRODUCTION

Antarctica is divided into "eastern" and "western" ice sheets by the Transantarctic Mountains. Ice from the East Antarctic Ice Sheet (EAIS) does make its way through the mountains, but this flow represents a relatively small component of the total input to the West Antarctic Ice Sheet (WAIS) and, for most purposes, the WAIS can be treated as an autonomous flow unit.

The WAIS has a number of attributes that differentiate it from the East Antarctic and Greenland Ice Sheets. Ice discharge from the WAIS is dominated by ice streams that feed the two largest ice shelves on the planet: the Ross and Filchner-Ronne. While there are ice streams in East Antarctica and Greenland, they do not appear to control as large a fraction of the discharge from the interior. The catchments drained by the WAIS ice streams reach well into the interior, causing most topographic profiles along flow lines to depart from the typical parabolic shape of profiles found on the other ice sheets. Several of the ice streams that drain the WAIS show a variability in discharge patterns on a few-hundred year time scale (something that has not been observed on the other ice sheets) that significantly complicates attempts to understand long-term balance. Perhaps most importantly, the WAIS is a marine ice sheet with a bed that would still be largely below sea level if the ice were removed, even after isostatic rebound.

Knowledge of the morphology and surface character of the WAIS has been accumulated over the last four decades from numerous field and remote sensing sources. The first comprehensive overview of the continent was published in 1983 [*Drewry*, 1983] and contained maps of surface and bed topography, bedrock characteristics, internal layering in the ice, ice driving stress, ice thickness and aero-magnetics. These data were collected primarily from airborne and oversnow-seismic surveys during the 1970's. As a consequence, the coverage was highly variable, with the WAIS being better served than the EAIS. Due to rapid developments in technology since then, there is now a suite of data sets covering many aspects of the surface characteristics and ice flow, derived, primarily, from remote sensing instruments. We present an overview of some of these data sets, describing, briefly, what they can tell us about the WAIS. Applications of these data sets in studies of WAIS dynamics are described by other papers in this volume.

GENERAL MORPHOLOGY

A comprehensive representation of the WAIS, in a visual sense, is provided by the mosaic produced by the USGS from Advanced Very High Resolution Radiometer (AVHRR) instruments on NOAA polar-orbiting weather satellites [USGS Map I-2560 by *Ferrigno et al.*, 1996]. The part of this mosaic covering the WAIS is shown in Figure 1. The complex morphology of the ice sheet is immediately apparent from this mosaic, which shows an apparent symmetry, with the subtle undulating topography of the interior of the ice sheet giving way, via ice streams at either side, to the extremely flat surfaces of the large ice shelves.

The eastern side of the WAIS (right in Figure 1) contains a number of ice streams and outlet glaciers that discharge into the Ronne Ice Shelf (see Figures 2 and 3). The large outlets from the WAIS into the Ronne IS include Rutford, Evans, Institute, and Foundation Ice Streams, separated by inter-stream ridges, and Carlson Inlet, which has an ice-stream-like configuration but is not moving rapidly [*Frolich and Doake*, 1998; *Rignot*, 1998a].

The pattern is similar for ice streams flowing west into the Ross Ice Shelf. From south to north along the west side of the WAIS, the ice streams are named simply A through F (Figures 2 and 3). The initial overview of the ice streams came from aircraft-based radio echo sounding, aerial photography, and early Landsat imagery [*Rose*, 1979]. Ice stream A has the largest part of its discharge contributed by Reedy Glacier, which begins in East Antarctica and flows through the Transantarctic Mountains. Ice stream B, which converges with A into a stream more than 100 km wide near the grounding line, has shown a recent deceleration in its lower part [*Bindschadler and Vornberger*, 1998]. The lower half of ice stream C ceased flowing rapidly about 140 years ago [*Rose*, 1979; *Retzlaf and Bentley*, 1993]. The slow moving lower half is noticeably smoother in the mosaic than the upper part, which is still moving at ice-stream-like speeds [*Anandakrishnan and Alley*, 1997]. Ice stream D has several tributary branches. The southern-most tributary has been the site of onset studies in recent years [*Bindschadler et al.*, this volume]. Ice stream E has a short broad trunk, while ice stream F is smaller and of a different character than the other five streams. The trunks of ice streams A, B, D, and E are characterised by low surface slopes, high flow speeds, and intense marginal shear. Ice stream C appears to have had similar characteristics when it was flowing rapidly. A detailed description of feature interpretation from AVHRR imagery of the western WAIS is given by *Bindschadler and Vornberger* [1990].

The WAIS ice streams flow rapidly in spite of low driving stress, due to weak basal resistance to flow. Research into the causes of rapid motion is extensive; there are several summaries of work on these ice streams [*Alley and Whillans*, 1991; *Bentley*, 1987; *Bindschadler* 1993; papers in this volume]. The numerous streams draining the interior of the WAIS have a series of catchments that control the shape of the ice sheet's surface. The large outlet glaciers that drain the northern WAIS are similar to glaciers that drain the northern rim of the EAIS and much of Greenland. In these systems the rapidly moving ice is subject to a relatively high driving stress and, therefore, a significant amount of internal deformation, in addition to basal sliding [*Bindschadler et al.*, this volume].

The large ice shelves flanking the WAIS cover an area that is half as large as the sheet itself. These shelves are fed primarily by ice stream discharge from the WAIS, with a smaller component coming from East Antarctica [*Shabtaie and Bentley*, 1987], and by snow accumulation on their upper surfaces. In contrast, the smaller fringing ice shelves on the northern coast are fed by outlet glaciers that have relatively small catchments. A summary of surface work on the Ross Ice Shelf is provided by *Thomas et al.* [1984].

TOPOGRAPHY AND ICE THICKNESS

The surface topography of the ice sheet is portrayed in Figures 4a (contours) and 4b (shaded isometric). The data used to generate these diagrams were derived from radar altimetry [*Bamber and Huybrechts*, 1996; *Bamber and Bindschadler*, 1997]. The radar altimeter is most accurate when ranging to a flat surface; because of the slopes present on the ice sheet the accuracy of the resultant digital elevation model (DEM) ranges between about 1 and 10 m. The coverage of the satellite altimetry data only extends to 81.5 south (shown by the solid black line in Fig 4b). South of this limit sparse terrestrially-derived data have to be used, with a substantial reduction in detail and accuracy (to hundreds of metres) [*Bamber*, 1994]. None the less, the surface topography, particularly north of 81.5, is probably the most well defined boundary condition for the WAIS.

Accurate topography is required for determining the magnitude of the gravitational driving force (which

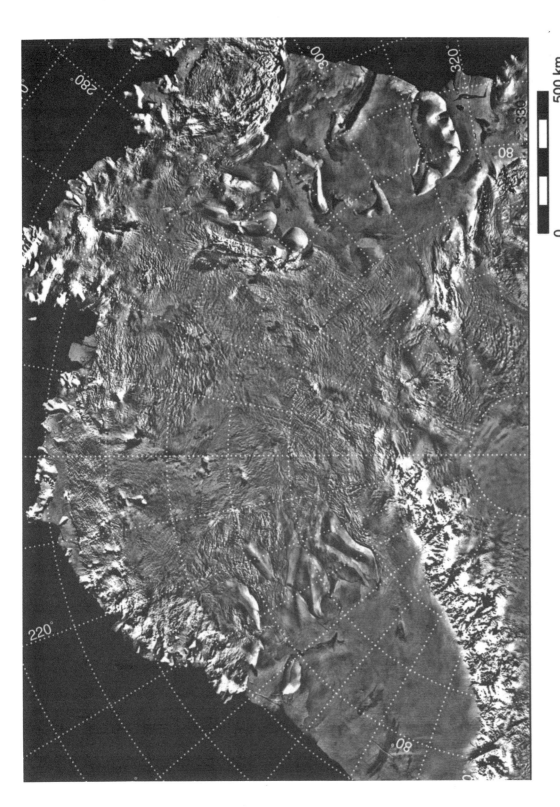

Fig. 1. AVHRR image mosaic of the West Antarctic Ice Sheet (after Ferrigno, et al., 1996). The undulating topography of the interior of the ice sheet is distinct from the domes, ridges, and ice streams which border the flat ice shelves. The features visible in this mosaic are identified in Figure 2. The image used for this figure can be found at the USGS Terra Web site [http://terraweb.wr.usgs.gov/TRS/projects/Antarctica/AVHRR.html].

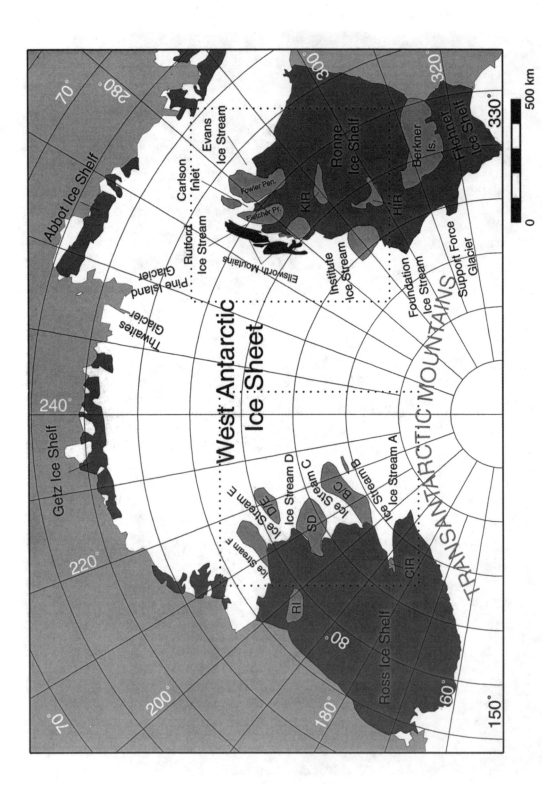

Fig. 2. Features visible in Figure 1. The dark grey areas the flat surfaces of ice shelves, which the lighter grey areas on the ice sheet represent the slow-moving ice of the domes and ridges which separate the various ice streams. The two squares outlined with dots show the locations of the more detailed views of the mosaic shown in Figure 3. RI - Roosevelt Island; D/E - Ridge D/E; SD - Siple Dome; B/C - Ridge B/C; CIR - Crary Ice Rise; KIR - Korff Ice Rise; HIR - Henry Ice Rise.

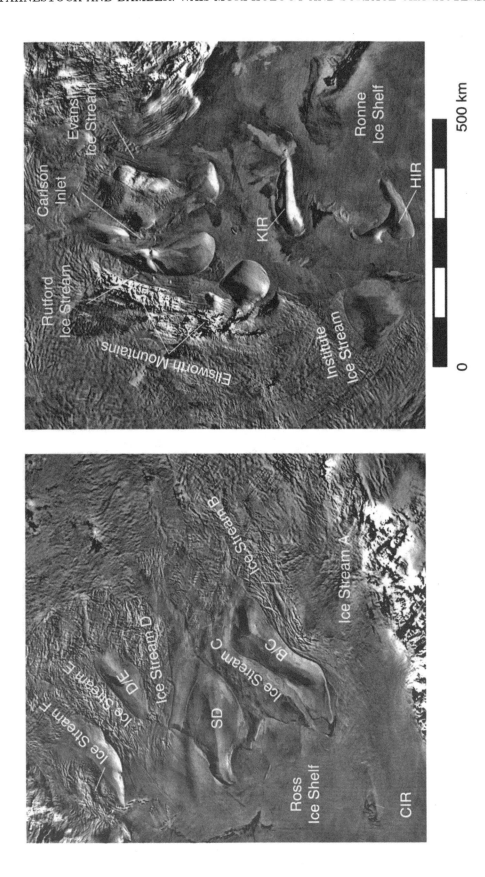

Fig. 3. Two detailed views of the mosaic in Figure 1. The features and locations are identified in Figure 2.

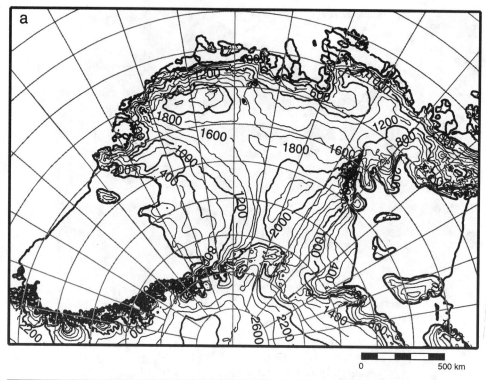

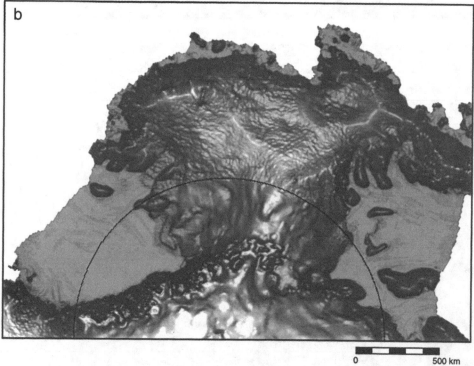

Fig. 4. a) Surface topography of the WAIS. Topography derived from ERS-1 radar altimetry, except south of 81.5 S, where the elevations are based on more limited surface data. Contour interval 200m. b) Shaded isometric view of the elevation data. Illumination is from above. Note the change in detail across the black line at 81.5 S, which is the southern limit of the radar altimeter coverage.

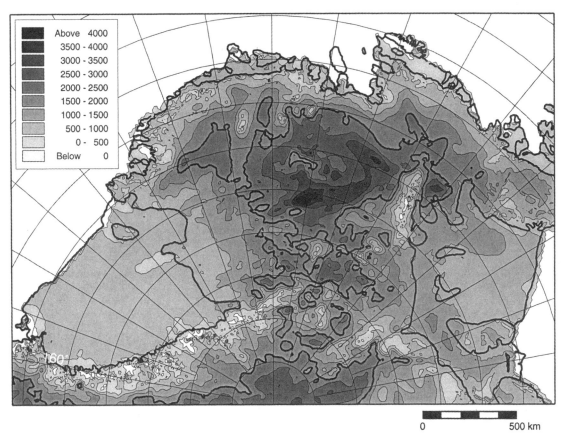

Fig. 5. Ice thickness derived from airborne and surface radio echo sounding and seismic measurements. Contour interval 500 m. The flight lines for much of the data collection are shown in Figure 6.

is proportional to the local slope) and is a valuable input boundary condition and validation data set for ice sheet numerical modelling [Hulbe and Payne, this volume; Bindschadler et al., this volume]. The DEM can be used to delimit ice divides, allowing identification of separate drainage systems. The illumination from directly overhead in Figure 4b highlights the low-slope areas found at the divides and makes the higher slopes near the ice margin darker. Elevation models can also be used to determine flow lines (surface particle paths) and catchment areas for outlet glaciers and ice streams [Vaughan et al., 1999]. Related to this application, an elevation model has been used to calculate balance fluxes over the ice sheet by combining flow lines with accumulation data [Budd and Warner, 1996; Bamber et al., 2000].

The ice shelves are remarkably flat with only a few meters of variation over the central 400 km of the Ross Ice Shelf. The variation across the floating part of the Ronne Ice Shelf is larger (25 m - equivalent to about 250 m in ice thickness) due to substantial basal melting and freezing affecting the central portion of the shelf [Jenkins and Doake, 1991].

Ice Thickness

The thickness of the ice in West Antarctica is shown in Figure 5. This data set was derived from airborne and over-snow radio echo sounding and seismic surveys. The coverage of the flight lines and traverses is shown in Figure 6. There is relatively good radio-echo sounding coverage over the Ross Ice Shelf and southern sector of the WAIS but the Peninsula and northern sector have relatively sparse coverage by comparison. New ice thickness data sets have been collected since the compilation of the Antarctic folio [Drewry, 1983] and these are now being incorporated into a new database for the whole of Antarctica with the objective of improving the bed elevation data set generally available. This Scientific Committee on Antarctic Research (SCAR) sponsored project, known as BEDMAP, is being co-

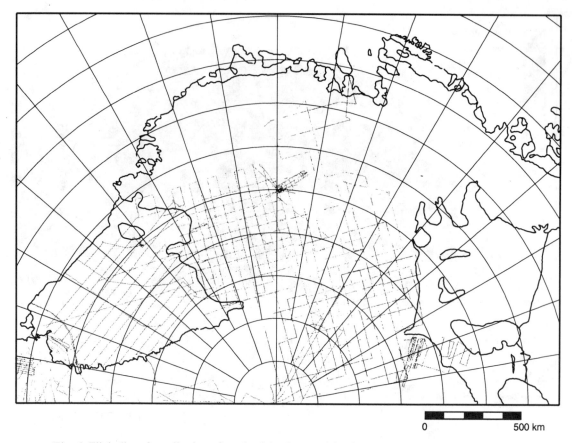

Fig. 6. Flight lines for collection of much of the data used for the production of Figure 5. Note the dense coverage over the Ross Ice Shelf and southern WAIS. Updates on the data to be included in the project can be found at the BEDMAP project home page [http://www.nerc-bas.ac.uk/public/aedc/bedmap/bedmap.html].

ordinated by the British Antarctic Survey, Cambridge, England. Most of the data used to produce Figure 5 was obtained as part of an NSF/SPRI collaboration during the 1970s. The quoted accuracy for ice thickness is 10 percent although this does not take into account geolocation errors which, due to the navigation systems available at that time, could be several kilometers [Drewry et al., 1982]. Recent additions to this data set from the National Science Foundation's Support Office for Aerogeophysical Research are on grids as fine as 5 km.

At a distance of approximately ten ice thicknesses downstream from the grounding line, it is reasonable to assume that the ice shelves are in hydrostatic equilibrium. Thus, if the seawater and ice densities and geoid are known with sufficient accuracy it is possible to use the altimeter-derived elevations to determine ice thickness over the shelves [Bamber and Bentley, 1994].

Bed Topography

Bed elevations are derived by subtracting the ice thickness from surface elevation obtained either directly from the auxiliary data collected during the flights or from the previously discussed satellite radar altimeter measurements. A contour plot of bed elevations is shown in Figure 7.

One of the most striking features of the bed topography is the amount of the ice sheet that is substantially below sea level. A histogram of the bed elevation for the grounded sector of the WAIS (i.e. excluding the ice shelves) is shown in Figure 8. This figure shows that about 75 percent of the grounded ice sheet rests on bedrock below sea level and about 21 percent of this is more than 1000 m b.s.l. It is this characteristic of the ice sheet, and the possible influence of the ice shelves on ice discharge rates, that has led various authors to speculate about the stability

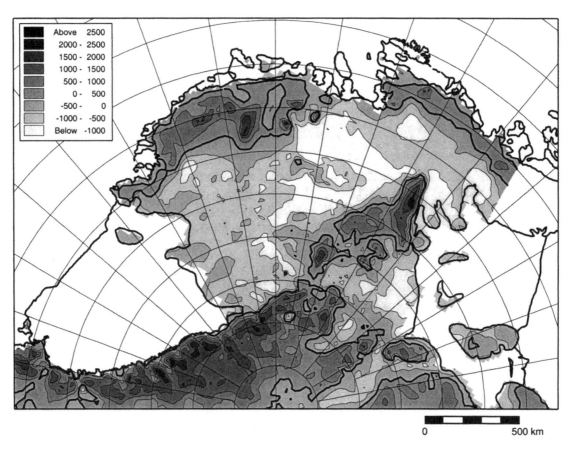

Fig. 7. Bed elevations for the grounded parts of the WAIS. Note the large areas which are below sea level. This data set was derived by subtracting the ice thickness data of Figure 5 from the surface elevations of Figure 4.

of the WAIS under conditions of oceanic warming [*Hughes*, 1975; *Mercer*, 1978; *Denton and Hughes*, 1981]. One notable feature in the bed topography is the Bentley subglacial trench, which reaches a depth of 2555 m b.s.l. Due to the sparse radio echo sounding coverage it is likely that there are other short wavelength features such as this, that are, as yet, undiscovered. There are very few airborne flight lines for the northern half of the ice sheet (between about 75 and 80 degrees south, Figure 6); the limited information on ice thickness in this region has come mainly from sparse over-snow seismic profiles.

SURFACE CHARACTERISTICS

Due to improvements in satellite technology and in the coverage of the polar regions since the 1970s, it has become possible to remotely measure or infer some surface characteristics over the ice sheet. Presented here are a few observations of surface variables of glaciological relevance.

Melt

Surface melt can be detected from satellites by measuring changes in the amount of microwave energy emitted from the firn. The presence of small amounts of water in the surface snow radically alters the conditions at the interface between the firn and the atmosphere, producing a spike in energy emission. Detection of this spike allowed *Zwally and Fiegles* [1994] to map the extent and duration of surface melt in Antarctica . It is clear from this work that most of the melt occurs at the coast, predominantly on the ice shelves on the northern coast and on the Antarctic Peninsula. Figure 9 shows an unusual melt episode on the Ross Ice Shelf which extended inland to a point upstream of Siple Dome. Similar events have occurred only a few times during

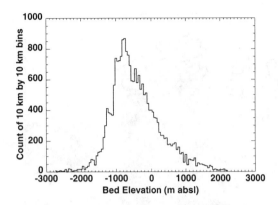

Fig. 8. A histogram of bed elevations averaged over 10 km by 10 km squares, showing the preponderance of bed elevations below sea level.

and the planned coring at several inland sites, which will provide a high-resolution record (temporally) comparable to the cores from the GISP 2 and GRIP sites on the summit of the Greenland Ice Sheet.

Traditional surface balance estimates over the ice sheet have been obtained from ice core and shallow pit measurements at roughly ten year intervals [*Kotlyakov*, 1961; *Bull*, 1971; *Giovinetto and Bentley*, 1985]. More recently, however, satellite-derived passive microwave data have been incorporated in a new representation of the pattern of accumulation [*Vaughan et al.*, 1999; initial work by *Zwally*, 1977]. This new compilation incorporates an updated database of in-situ measurements, the use of the 19 years covered by passive microwave measurements from satellites [*Christopher Shuman*, personal communication, *Wilson and Jezek*, 1993].

Temperature

Surface temperature or annual average temperature can be measured or estimated from satellites using two different techniques. Figure 10 shows contours of the average temperature for 1978 derived from thermal infrared imagery from the THIR sensor [*Comiso*, 1994]. Because the thermal infrared signal is affected by clouds, the data used to produce the figure reflect cloud-free conditions, and are, therefore, likely to be biased cold. The data do, however, provide a good indication of the spatial distribution of temperatures.

An estimate of the temperature in the firn, at a depth below most of the annual variation, can be made from microwave emission at 6 GHz [*Winebrenner*, 1994]. This method is not sensitive to cloud cover and gives an estimate of temperature that is directly related to the temperature measured in shallow boreholes. In addition, microwave emission from the firn at higher frequencies can be used to estimate near-surface temperatures, with uncertainties introduced by variations in grain size, density stratification, and surface snow conditions [*Shuman et al.*, 1995].

Accumulation

Accumulation rates on the WAIS are significantly higher than those over most of East Antarctica. This fact has motivated the coring program at Siple Dome

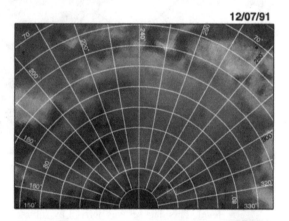
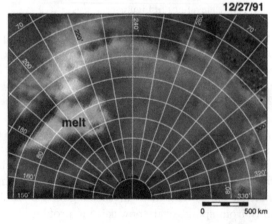

Fig. 9. Surface melt in December, 1991, detected as a peak in microwave emission (grey level proportional to microwave brightness temperature). Images from the Special Sensor Microwave/Imager (SSM/I), 37 GHz, vertical polarization. Top: Image showing background brightness temperatures. Bottom: Image from 20 days later showing melt event on Ross Ice Shelf. Images from Christopher Shuman, ESSIC, University of Maryland.

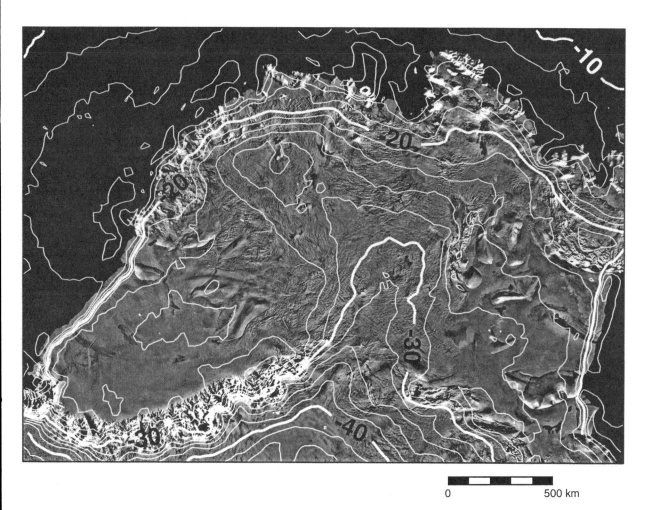

Fig. 10. Annual mean surface temperature during 1978, measured by the Temperature Humidity Infrared Radiometer on Nimbus 7. Contour interval 2C. (After Comiso, 1994). Clouds can affect the measurement of temperature in the infrared, so masking must be used. As measurements are only made during cloud-free periods, the temperatures would be biased low. The data do not cover a long enough period to indicate true mean temperatures, but they give a good indication of the spatial variation of surface temperature on the WAIS.

SMMR data as a background field for interpolation between the sparse ground measurements, and improved basin-wide integration using a geographical information system. Atmospheric forecast models used for reanalysis of meteorological observations are beginning to providing an alternative to field measurements when assessing the spatial and temporal patterns of accumulation [eg. *Cullather et al.*, 1998].

Flow-Related Features

Satellite imagery reveals a number of subtle features that are generated by the flow of ice. AVHRR imagery (Figure 1 and Figure 3) shows topographic undulations in the surface of the ice sheet, reflecting the interaction of the moving ice with the bed. On the inter-stream ridges the surface is much smoother due to lower ice flow speeds and low driving stress near the ridge crests. There are features on the flanks of these ridges that are related to past ice flow configurations, such as the linear feature on the northeast side of Siple Dome. This "scar" is the remnant of an ice stream shear margin from a previous configuration of ice stream D [*Jacobel et al.*, in press; *Raymond et al.*, this volume]. The upstream end of this scar is truncated by later flow in ice stream C.

On the ice shelves there are a number of features that record the discharge from the ice streams, and allow some account to be made of ice flow history. Flow past ice rises, such as Crary Ice Rise on the Ross

Ice Shelf or Korff or Henry ice rises on the Ronne Ice Shelf, produces downstream tracks in the ice that provide a record of past ice flow. In the case of the FRIS, this record shows relatively constant or stable behaviour; in the case of the ice streams flowing into the Ross Ice Shelf, the record shows a pattern of variability [Casassa and Turner, 1991; Jezek, 1984]. The ice takes about one thousand years to cross these shelves, and so preserves a fairly long record of flow patterns.

Detailed representations of flow-related features can be found in imagery from Landsat, SPOT, and other high-resolution imagers [Bindschadler el al., 1988; Merry and Whillans, 1993; Sievers, 1990]; in addition, some buried features can be seen in high resolution radar imagery from Synthetic Aperture Radar (SAR) sensors.

ICE FLOW

The general pattern of flow can be inferred from the shaded isometric view of the surface topography in Figure 4b. The ice streams act as the major channels for discharge, flowing directly into the Ross and Filchner-Ronne ice shelves. Three ice divides at the southern end of the Peninsula form the northern boundary of the very active Thwaites-Pine Island glacier system [Doake et al., this volume]. These divides have saddles due to the convergence of different drainages - this is distinct from all but a few places in Greenland and East Antarctica.

Quantitative observations of surface velocities have been carried out on the ground using GPS and its predecessors, and from space using two techniques: feature tracking and interferometry. These studies have focussed on the ice streams [Bindschadler and Scambos, 1991] and ice shelves [Thomas et al., 1984; Vaughan and Jonas, 1996]. An example of a velocity field measured by feature tracking in two sequential Landsat Thematic Mapper images is shown in Figure 11 [Bindschadler et al., 1996]. This map shows motion vectors for a selection of matched features and contours for the entire data set. This velocity field for ice stream E is characteristic of the motion found on most of the ice streams which drain the WAIS into the large ice shelves. Note that shear strain is concentrated at the margins, and that the rapid flow in the interior of the stream shows relatively small gradients in velocity. Also note in this high-resolution image the difference between the smooth surfaces of the inter-stream ridges and the more undulating, creavassed surface of the stream. The feature-tracking technique [Scambos and Bindschadler, 1993] requires the presence of crevasses or other resolvable surface features which move with the ice, limiting its application to rapidly flowing areas. In contrast, interferometry using imaging radar data is able to track the motion of essentially featureless areas. Several areas of the WAIS have been studied with interferometry. Goldstien et al. [1993] and Frolich and Doake [1998] used interferometry to look at the motion and grounding line position of the Rutford Ice Stream. Recently, Rignot has shown a sizeable change in grounding line position on Pine Island Glacier using multiple interferograms [Rignot, 1998b]. The uses of interferometry continue to expand, promising at some point to provide detailed ice motion from almost any area, providing that SAR satellites with appropriate capabilities and orbits are flown.

On the north coast of the WAIS, feature tracking has been used to measure ice flow in Pine Island and Thwaites glaciers. These glaciers drain relatively small basins, but their discharges are high due to the high accumulation along the coast. The measured velocities on these glaciers are among the highest in West Antarctica [Ferrigno et al., 1993; Lucchitta et al., 1993; Lucchitta et al., 1995].

CONCLUSIONS

There has been, over the last twenty years, a rapid growth in the breadth of observational data sets that can be applied to studying the characteristics and behaviour of the WAIS. These data sets have helped define the flow patterns of the ice sheet and document changes in flow. Direct measurement of motion and surface elevation is making more accurate estimates of mass balance possible. Of particular relevance for this application are recent advances in satellite SAR interferometry and altimetry. Interferometry is able to provide detailed two-dimensional surface velocity measurements, and is being used to identify rapid changes. Satellite radar altimetry is being used to investigate volume changes [e.g. Wingham, et al.,1998]. Satellite altimetry will enter a new era with the launch of the Geoscience Laser Altimeter System (GLAS) in 2001. One of the primary objectives of this instrument is to improve estimates of volume changes in the ice sheets.

These improvements in observational data sets have improved boundary conditions for numerical mod-

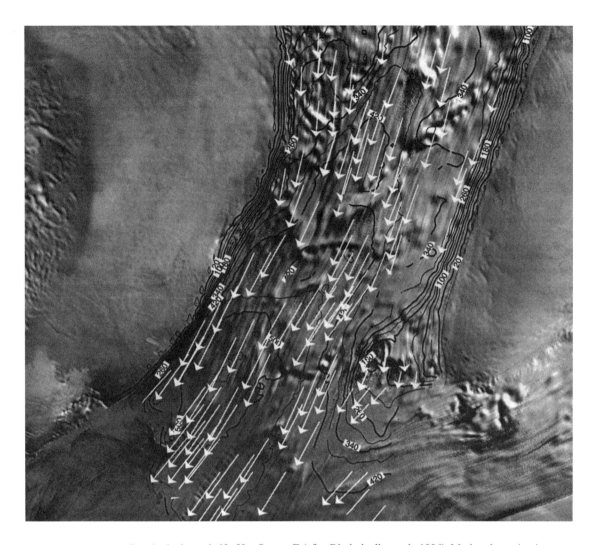

Fig. 11. Ice flow in the lower half of Ice Stream E (after Bindschadler et al., 1996). Motion determined from two Landsat TM images of the area taken one year apart. One out of thirty velocity vectors determined from feature tracking is shown. The entire dataset was used to draw the contours. Velocity map obtained from the National Snow and Ice Data Center web archive of Antarctic ice velocities [http://www-nsidc.colorado.edu].

elling of ice sheet behaviour [*Hulbe and Payne*, this volume]. This should help our understanding of the interactions between the ice streams and the surrounding sheet. The question of the stability of the WAIS, in the face of a warming climate, is still open; however, improving satellite, airborne, and surface observations, when combined with evolving numerical models, promise continued advances in our understanding.

Acknowledgments. We are grateful to J. Ferrigno, R. Bindschadler, J. Comiso, and T. Scambos for providing imagery for this article.

REFERENCES

Alley, R.B. and I.M. Whillans, Changes in the West Antarctic ice sheet. *Science*, 254, p. 959-963, 1991.

Anandakrishnan, S.; Alley, R.B. Stagnation of ice stream C, West Antarctica by water piracy. *Geophysical Research Letters*, 24(3), p. 265-268, 1997.

Bamber, J.L., Vaughan, D.G. Joughin, I, Widespread complex flow in the interior of the Antarctic Ice Sheet, *Science*, in press.

Bamber, J. L. and R. A. Bindschadler. An improved elevation data set for climate and ice sheet modelling: validation with satellite imagery. *Annals of Glaciology*, 25, p. 439 - 444, 1997.

Bamber, J.L. Digital elevation model of the Antarctic ice sheet derived from ERS-1 altimeter data and comparison with terrestrial measurements., *Annals of Glaciology*, 20, p.48-54, 1994.

Bamber, J.L., and P. Huybrechts, Geometric boundary conditions for modelling the velocity field of the antarctic ice sheet., *Annals of Glaciology*, 23, p. 364-373, 1996.

Bamber, J.L., and C.R. Bentley, Comparison of satellite-altimetry and ice -thickness measurements of the Ross Ice Shelf, Antarctica, *Annals of Glaciology*, 20, p.357-364, 1994.

Bentley, Charles R. Antarctic ice streams: a review. *Journal of Geophysical Research*; 92(B9), p. 8843-8858, 1987.

Bindschadler, R.A., Siple Coast Project research of Crary Ice Rise and the mouths of ice streams B and C, west Antarctica: review and new perspectives. *Journal of Glaciology*, 39(133), p. 538-552, 1993.

Bindschadler, R.A., J. Bamber, and S. Anandakrishnan, Onset of Streaming Flow in the Siple Coast Region, West Antarctica, this volume.

Bindschadler, R.; and P. Vornberger, Changes in the West Antarctic Ice Sheet since 1963 from declassified satellite photography, *Science*, 279(5351), p.689-692, 1998.

Bindschadler, R.A., P. Vornberger, D. Blankenship, T. Scambos and R. Jacobel.., Surface velocity and mass balance of Ice Streams D and E, West Antarctica, *Journal of Glaciology*, 42(142), p. 461-475, 1996.

Bindschadler, R. A., and T. A. Scambos, Satellite-image-derived velocity field of an Antarctic ice stream, *Science*, 252(5003), p. 242-246, 1991.

Bindschadler, Robert A.; Vornberger, Patricia L. AVHRR imagery reveals Antarctic ice dynamics.B *Eos. Transactions*, American Geophysical Union; 71(23), p. 741-742, 1990.

Bindschadler, R.A., F.S. Brownworth, and S.N. Stephenson, S.N.,Landsat Thematic Mapper imagery of the Siple Coast, Antarctica, *Antarctic Journal of the United States*, 23(5), p. 214-215, 1988.

Budd, W.F., and R.C. Warner, Computer scheme for rapid calculations of balance-flux distributions, *Annals of Glaciology*, 23, p. 21-27, 1996.

Bull, C., Snow accumulation in Antarctica., in Research in the Antarctic Washington, D.C., American Association for the Advancement of Science, p.367-421, 1971.

Casassa, G. and J. Turner, Dynamics of the Ross Ice Shelf, *Eos. Transactions*, American Geophysical Union, 72(44), 473-475, 1991.

Comiso, J.C., Surface temperatures in the polar regions from Nimbus 7 temperature humidity infrared radiometer, *Journal of Geophysical Research*, 99(C3), p. 5181-5200, 1994.

Cullather, R.I., D.H. Bromwich, and M.L. Van Woert, Spatial and temporal variability of Antarctic precipitation from atmospheric methods, *Journal of Climate*, 11, (3) 334-367, 1998.

Denton, G.H. and T. J. Hughes (Editors), The last great ice sheets New York : John Wiley, 1981.

Drewry, D.J. S.R. Jordan and E Jankowski, Measured properties of the Antarctic ice sheet: surface configuration, ice thickness and bedrock characteristics, *Annals of Glaciology*, 3, p. 83-91. 1982

Drewry, D. J., ed. Antarctica: glaciological and geophysical folio. Scott Polar Research Institute; Cambridge; 1983.

Doake, C.S.M., H. F. J. Corr, A. Jenkins, K. Makinson, K. W. Nicholls, C. Nath, A. M. Smith, and D. G. Vaughan, Rutford Ice Stream, Antarctica, this volume.

Ferrigno, J.G., J.L. Mullins, J. Stapleton, P.S. Chavez, Jr., M.G. Velasco, R.S. Williams, Jr., G.F. Delinski, Jr., and D. Lear, Satellite Image Map of Antarctica, USGS Miscellaneous Investigations Series Map I-2560, United States Geological Survey, Washington, D.C., 1996.

Ferrigno, Jane G.; Lucchitta, B.K.; Mullins, K.F.; Allison, A.L.; Allen, R.J.; Gould, W.G.A, Velocity measurements and changes in position of Thwaites Glacier/iceberg tongue from aerial photography, Landsat images and NOAA AVHRR data, *Annals of Glaciology*, No. 17, p. 239-244, 1993.

Frolich, R.M., and C.S.M. Doake, Synthetic Aperture Radar Interferometry over Rutford Ice Stream and Carlson Inlet, Antarctica, *Journal of Glaciology*, 44(146), p. 77-92, 1998.

Giovinetto, M.B. and C.R. Bentley, Surface balance in ice drainage systems of Antarctica, *Antarctic Journal of the United States*; 20(4), p. 6-13, 1985.

Goldstein, R.M.,H. Engelhardt, W.B. Kamb, R.M. Frolich, Satellite radar interferometry for monitoring ice sheet motion: application to an Antarctic ice stream, *Science*; 262(5139), p. 1525-1530, 1993.

Hughes, T. West Antarctic ice sheet: instability, disintegration, and initiation of ice ages, *Reviews Of Geophysics And Space Physics*, 13(4), p.502-526, 1975.

Hulbe, C. L. and A. J. Payne, The contribution of numerical modelling to our understanding of the West Antarctic Ice Sheet, this volume.

Jacobel, R. W., T.A. Scambos, N.A. Nereson, and C.F. Raymond, Changes in the margin of Ice Stream C, Antarctica, *Journal of Glaciology* in press.

Jenkins, A., and C.S.M. Doake, Ice-ocean interaction on the Ronne Ice Shelf, Antarctica. *Journal of Geophysical Research*, 96(C1), 791-813, 1991.

Jezek, K.C.. Recent changes in the dynamic condition of the Ross Ice Shelf, Antarctica, *Journal of Geophysical Research*, 89(B1), 409-416, 1984

Kotlyakov, V.M., The intensity of nourishment of the Antarctic ice sheet, International Association of Scientific Hydrology, Committee Snow & Ice, IASH Publ. No. 55, p. 100-110, 1961.

Lucchitta, B.K., C.E. Rosanova, and K.F. Mullins, Velocities of Pine Island Glacier, West Antarctica, from ERS-1 SAR images, International Symposium on the Role of the Cryosphere in Global Change, Columbus, OH, Edited by D.A. Rothrock, *Annals of Glaciology*, 21, p. 277-283, 1995.

Lucchitta, B.K.; Mullins, K.F.; Allison, A.L.; Ferrigno, Jane G. Antarctic glacier-tongue velocities from Landsat images: first results. *Annals of Glaciology*, 17, p. 356-366, 1993.

Mercer, J. H. West Antarctic ice sheet and carbon dioxide greenhouse effect : a threat of disaster, *Nature*, 271(5643), p.321-25, 1978.

Merry, C.J., and I.M. Whillans, Ice-flow features on Ice Stream B, Antarctica, revealed by SPOT HRV imagery, *Journal of Glaciology*, 39(133), p. 515-527, 1993.

Raymond, C.F., K. A. Echelmeyer, I. M. Whillans, and C. S. M. Doake, Ice Stream Shear Margins, this volume.

Retzlaff, R.; and C.R. Bentley; Timing of stagnation of Ice Streams C, West Antarctica, from short-pulse radar studies of buried surface crevasses. *Journal of Glaciology*, 133(pt.3), p.553-561, 1993.

Rignot, E., Radar interferometry detection of hinge-line migration on Rutford Ice Stream and Carlson Inlet, Antarctica. *Annals of Glaciology*, 27, 25-32, 1998a.

Rignot, E.J., Fast Recession of a West Antarctic Glacier, *Science*, 281(5376), p. 549-550, 1998b.

Rose, K.E., Characteristics of ice flow in Marie Byrd Land, Antarctica. *Journal of Glaciology*, 24 (90), p. 63-75, 1979.

Scambos, T.A., and R.A. Bindschadler, Complex ice stream flow revealed by sequential satellite imagery, *Annals of Glaciology*, 17, p. 177-182, 1993.

Shabtaie, Sion; Bentley, Charles R. West Antarctic ice streams draining into the Ross Ice Shelf: configuration and mass balance, *Journal of Geophysical Research*; 92(B2) :1311-1336, 1987.

Shuman, C.A., R.B. Alley, S. Anandakrishnan, and C.R. Stearns, An empirical technique for estimating near-surface air temperature trends in central Greenland from SSM/I brightness temperatures, *Remote Sensing of Environment*, 51(2), p. 245-252, 1995.

Sievers, J., Two thematic satellite image maps of Filchnerschelfeis at 1:250,000 scale Filchner-Ronne Ice Shelf Programme Report 4, edited by H. Miller: Bremerhaven, Alfred Wegener Institute for Polar and Marine Research, p. 86-87, 1990

Thomas, R.H., MacAyeal, D.R., Eilers, D.H; Gaylord, D.R.A Glaciological studies on the Ross Ice Shelf, Antarctica, 1973 –1978, *Antarctic Research Series* vol. 42, p. 21-53, AGU, Washington, D.C., 1984.

Vaughan, D.G.; Jonas, M. Measurements of velocity of Filchner-Ronne Ice Shelf. In Filchner-Ronne Ice Shelf Programme (FRISP) Report No.10, compiled by H. Oerter: Bremerhaven, Alfred-Wegener-Institute for Polar and Marine Research, p.111-116, 1996.

Vaughan, D.G., J.L. Bamber, M. Giovinetto, J. Russell, and A.P.R. Cooper, Reassessment of net surface mass balance in Antarctica. *Journal of Climate*, 12, (4) 933-946, 1999.

Wilson, J.D., and K.C. Jezek, Co-registration of an Antarctic digital elevation model with SSM/I brightness temperatures. *Annals of Glaciology*, 17, p. 93-97, 1993.

Winebrenner, D.P., R.D. West, and L. Tsang, Estimating Antarctic firn temperatures from dual-polarization microwave emission observations at 6 cm wavelength, abstract in *Eos. Transactions*, American Geophysical Union; 75(44) , p. 213; 1994.

Wingham, D. J., A. J. Ridout, R. Scharroo, R. J. Arthern, C. K. Shum, Antarctic elevation change from 1992 to 1996, *Science*, 282 (5388), p. 456-458, 1998.

Zwally, H. Jay; and S. Fiegles, Extent and duration of Antarctic surface melting, *Journal of Glaciology*, 40(136), p. 463-476, 1994.

Zwally, H.J., Microwave emissivity and accumulation rate of polar firn, *Journal of Glaciology*, 18(79), p.195-215, 1977.

M. Fahnestock, Earth System Science Interdisciplinary Center, University of Maryland, College Park, Maryland, 20742 USA

J. Bamber, Bristol Glaciology Centre, School of Geographical Sciences, University of Bristol, BS8 1SS UK

THE LITHOSPHERIC SETTING OF THE WEST ANTARCTIC ICE SHEET

I. W. D. Dalziel[1,2] and L. A. Lawver

Institute for Geophysics, University of Texas at Austin, Austin, Texas

Antarctica consists of two geologically distinct provinces, a Precambrian craton in the eastern hemisphere, and a younger series of mobile belts south of the Pacific Ocean. Unlike its land-based counterpart covering the East Antarctic craton, the base of the West Antarctic ice sheet (WAIS) is largely below sea level in the Ross and Weddell embayments. The East Antarctic craton separated from the other southern continents during the Mesozoic fragmentation of the Gondwanaland supercontinent. Since separation, East Antarctica has been near the South Pole and during the immediate past ~40 million year history of Cenozoic continental glaciation, Antarctica has remained close to its present position. During the breakup of Gondwanaland, the four major crustal units that comprise the exposed rocks of West Antarctica—the Antarctic Peninsula, Thurston Island, the Ellsworth-Whitmore mountains, and Marie Byrd Land—rotated outward from the convergent Pacific margin of the East Antarctic craton as rigid blocks. The driving forces for this relative motion appear to have been a major mantle plume in the case of the Antarctic Peninsula, Ellsworth-Whitmore mountains, and Thurston Island blocks, and ridge-crest subduction in the case of Marie Byrd Land. Both processes resulted in the generation of unusually large areas of extended continental crust, modified by underplating and the intrusion of mafic material between the rigid crustal blocks and the craton margin. This extended and modified crust forms the floors of the Weddell and Ross embayments at elevations permitting the very existence of the WAIS. Cenozoic impingement of a mantle plume beneath Marie Byrd Land and the Ross embayment further altered this crust, initiating formation of the West Antarctic rift system. Ongoing fracturing, volcanic activity, and bordering uplift associated with the development of the rift system combine to modify the predominantly sub-sea level, lithospheric "cradle" of the unique marine-based WAIS, possibly influencing its present and future behavior.

INTRODUCTION

Antarctica's name is derived from its present location, in the south polar region antipodal to that beneath the northern Constellation Arctus, the Bear. Of the two geologically distinct parts that make up the Antarctic continent, the larger portion is located in the eastern hemisphere, the smaller portion is to the south of the Pacific Ocean in the western hemisphere. Sometimes referred to as Greater and Lesser Antarctica, they are known more commonly as East and West Antarctica. East Antarctica was a substantial portion of Gondwanaland, the southern part of Pangea, and itself a long-lived supercontinent amalgamated from preexisting continental entities during global plate reorganization at the end of Precambrian times, ~550 Ma. Six of the seven present major lithospheric plates comprising the cold upper thermal boundary layer of the planet, including the Antarctic plate, contain embedded remnants of Pangea. Pangea fragmented after the Triassic Period that ended at

---

[1] Also at Department of Geological Sciences, University of Texas at Austin, Austin, Texas
[2] Also at Tectonics Special Research Centre, Department of Geology and Geophysics, University of Western Australia, Nedlands, WA, Australia

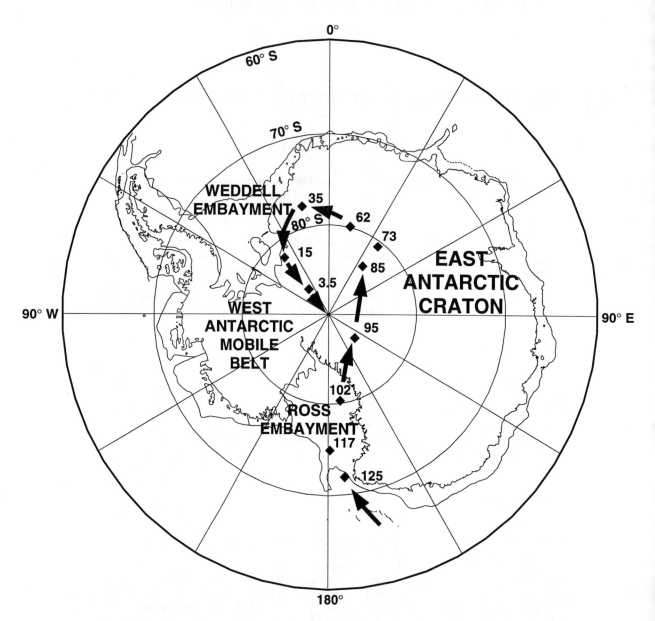

Fig. 1. Antarctic continent with a synthetic apparent polar wander path (Arrows, APWP, derived in part from rotation of data from other continents) for 125 Ma to Present [*Di Venere*, 1994]. Diamond symbols represent the position along the APWP of the South Pole relative to the continent at the time indicated.

~200 Ma. During the next 100 m.y. period, the Antarctic continent moved from high temperate latitudes to its nearly polar position (Figure 1). Between 185 and 95 Ma, four continents broke away from the present East Antarctic margin during Gondwana fragmentation, as illustrated by the geologically based reconstruction drawn by South African geologist Alex du Toit during the 1930's [*DuToit*, 1937, Figure 2]. Antarctica is classically known, therefore, as the "keystone" of Gondwanaland and for the past 100 m.y. it has been located over the South Pole.

Earth has undergone at least six major glaciations since the birth of the Solar System at ~4.55 Ga. Planetary "ice house" conditions, to use Fischer's term [1984], have extended over the planet for comparatively short time intervals of a few million to a few tens of million years during the Paleoproterozoic at ~2.1 Ga, during the Neoproterozoic at ~700 Ma and again at ~600 Ma, during the Paleozoic at ~460 and ~250 Ma, and most recently during the past ~40 m.y. of the Cenozoic Era. In all probability, the two existing remnants of more extensive Cenozoic glaciation, which cover Greenland in

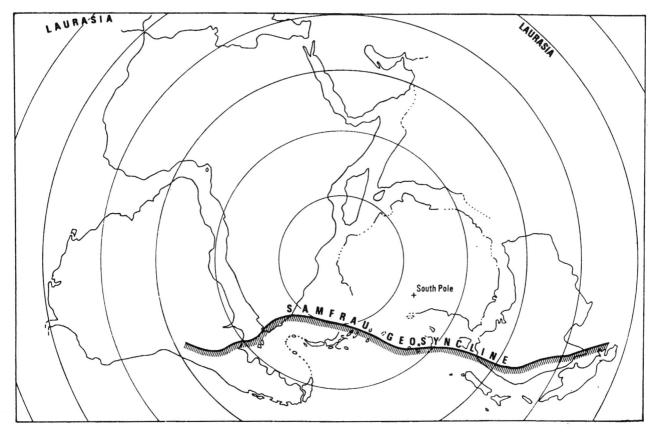

Fig. 2. Simplified version of Du Toit's geologically-based reconstruction of Gondwanaland [*Du Toit*, 1937].

the north and the Antarctic continent in the south, are ephemeral features of the Earth's surface. Even though the Weddell Sea margin of Antarctica lay very close to the South Pole at the onset of the Cenozoic glaciation, a paleoseaway had yet to open between the Antarctic Peninsula and South America (Figure 3). Thus the present Antarctic ice sheet may have begun to form prior to the establishment of a vigorous circum-polar current, though there was probably marine circulation through West Antarctica at the time of initial ice build-up [*Lawver et al.*, 1992; 1998].

THE WEST ANTARCTIC ICE SHEET IN TIME AND SPACE

East Antarctica was part of the tectonically stable Precambrian craton of the Gondwanaland supercontinent from 550 Ma to 200 Ma. It has apparently been a relatively stable area of Precambrian crust since the supercontinent began to break up at that time, though the presence of the small Gaussberg volcano on the coast, the Gamburtzev Subglacial Mountains, and depressions beneath the Lambert-Amery glacier and Lake Vostok have led to speculation concerning widespread recent volcanism and rifting. In contrast, West Antarctica has been part of the tectonically active mobile belt generally known as the circum-Pacific "ring of fire," the zone of plate convergence along ancient continental margins fringing the Earth's largest ocean basin, throughout the Phanerozoic. These margins were formed during the Neoproterozoic (~800 Ma) during the breakup of the Rodinia supercontinent that gave birth to the Pacific Ocean basin. The boundary between East and West Antarctica, now marked by the Pacific side of the Transantarctic Mountains, originated during that event (see *Dalziel*, 1997, for review). Stable East Antarctica has remained predominantly above sea level and has a land based ice sheet. The West Antarctic Ice Sheet (WAIS) is largely marine-based. Understanding of the history of the WAIS, and consideration of its present and future behavior, must start with an appreciation of the lithospheric setting that permits its existence.

Reconstruction of Gondwana using the seafloor spreading data obtained from marine geophysical surveys confirmed Du Toit's reconstruction, but resulted in a geologically unacceptable overlap between an in-place Antarctic Peninsula and the Falkland Plateau [*Norton and Sclater*, 1979, Figure 4]. Together with the earlier

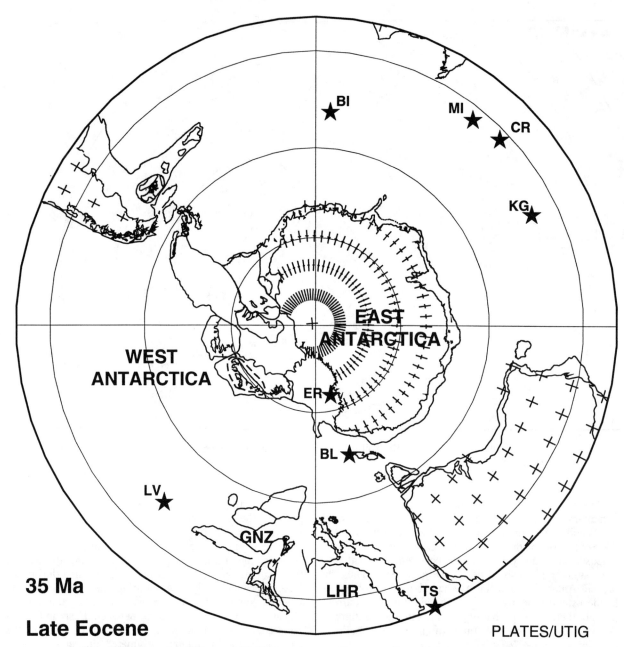

Fig. 3. Southern hemishere at ~35 Ma based on seafloor spreading data. Cross - present South Pole; Stars - selected present-day hot spot; BI - Bouvet Island; BL - Balleny Islands; CR - Crozet Island; ER - Mount Erebus; GNZ - New Zealand microcontinent; KG Kerghuelen; LHR - Lord Howe Rise; LV - Louisville Island; MI - Marion Island; TS - Tasman seamounts. Dashed line - extent of plume related to dome and volcanism [*LeMasurier and Landis*, 1996]. (PLATES project, Institute for Geophysics, University of Texas at Austin)

recognition that the Ellsworth Mountains of West Antarctica, the highest on the continent, were geologically part of the margin of the East Antarctic craton [*Schopf*, 1969], this led to the suggestion that West Antarctica consists of at least four major crustal blocks that have moved relative to each other, and also relative to the East Antarctic craton during the Mesozoic and Cenozoic fragmentation of Gondwanaland [*Dalziel and Elliot*, 1982, Figure 5]. Geological and paleomagnetic studies over the past decade and a half have confirmed this hypothesis, and resulted in a generally consistent model for the tectonic evolution of the

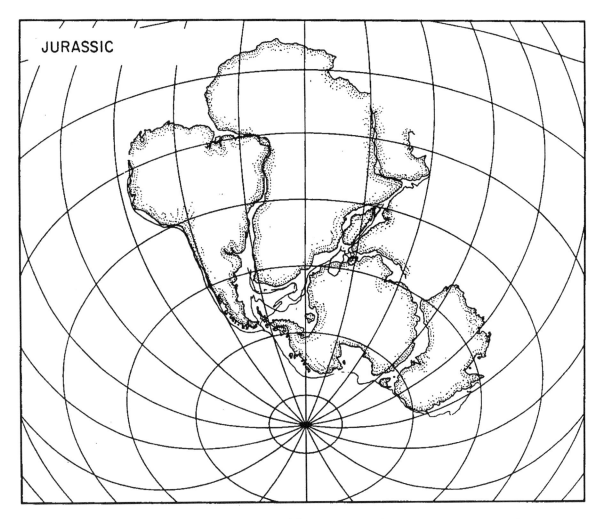

Fig. 4. Norton and Sclater's reconstruction of Gondwanaland using present-day geography of the Falkland Plateau and West Antarctica and seafloor spreading data [*Norton and Sclater*, 1979]. Note overlap of the Falkland Plateau and Antarctic Peninsula.

Antarctic continent since it separated from the other southern continents [for review, see *Dalziel*, 1992]. Hence, the WAIS is located on crust developed in association with microplate movement during the Mesozoic fragmentation of the Gondwanaland supercontinent.

If the crustal blocks of West Antarctica delineated in Figure 5 had rotated strictly as rigid continental microplates as a result of seafloor spreading, however, they would now be separated by Mesozoic ocean basins with depths of perhaps 3–4000 m or more. As in the Arctic, there would be multi-year sea ice, or possibly an ice shelf between isolated islands fringing the Pacific margin of the ice covered East Antarctic craton, but there would be no grounded WAIS. We believe that it is in new consideration of the processes that resulted in the movement of the West Antarctic crustal blocks presented in this contribution, that understanding of the lithospheric setting of the WAIS emerges. An unusual combination of tectonic processes generated the crustal environment in which a marine based ice sheet extending across an entire continent was able to develop. In contrast, the marine-based ice sheets of Hudson Bay and the Gulf of Bothnia formed in embayments of the North American and Eurasian continents and had single restricted outflows to the world ocean.

ORIGIN OF THE WEDDELL SEA EMBAYMENT

The Weddell Sea embayment developed at the junction of three continents, Antarctica (ANT), Africa (AFR), and South America (SAM) in a late Precambrian embayment of the Gondwanaland margin [*Dalziel*, 1982; *Dalziel*, 1997]. The growth of oceanic lithosphere between these three major continents during the

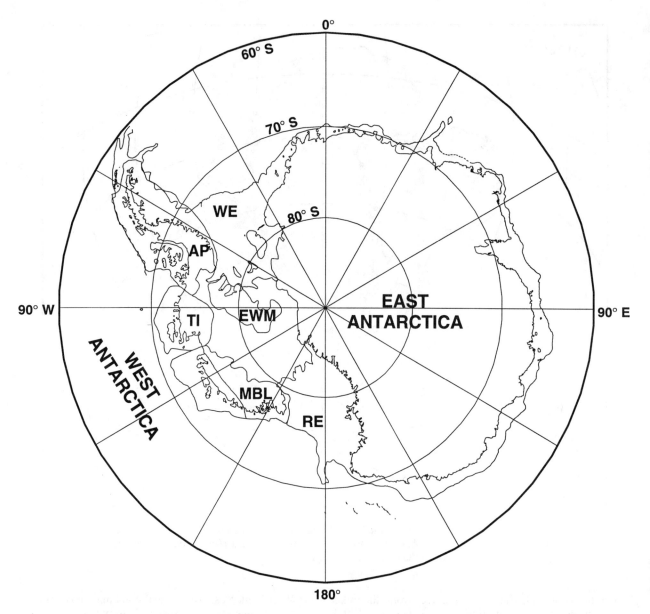

Fig. 5. Antarctica showing East Antarctic craton and displaced rigid crustal blocks of West Antarctica [after *Dalziel and Elliot*, 1982]. AP - Antarctic Peninsula crustal block; EWM - Ellsworth-Whitmore mountains crustal block; RE - Ross embayment; MBL; Marie Byrd Land crustal block (line across the block represents boundary of "west" and "east" Marie Byrd Land [*Di Venere*, 1995, see Fig. 8]; TI - Thurston Island crustal block; WE - Weddell embayment.

fragmentation of Gondwanaland has established their relative motion over the past 165 m.y. (AFR-ANT) or 130 m.y. (SAM-AFR) [*Lawver et al.*, 1985; *Lawver and Scotese*, 1987; *Norton and Sclater*, 1979]. Geologic and paleomagnetic evidence from the Falkland Islands and the Ellsworth Mountains and adjacent nunataks, however, indicate that independent motion of smaller rigid fragments of continental lithosphere relative to the three major continents commenced prior to the development of the present major ocean basins.

Adie [1952] first pointed out on geologic grounds that the Falkland Islands appear to have originated as the missing southeastern corner of the Karoo basin off the Natal margin. Based solely on seafloor spreading data from the South Atlantic, the Falkland Islands would have been located only off the Agulhas Plateau of southern Africa in a Mesozoic reconstruction, over 1000 km from their geologically probable original position off the corner of the Karoo basin. Adie [1952] pointed out that restoration of the structures in the rocks of the islands to

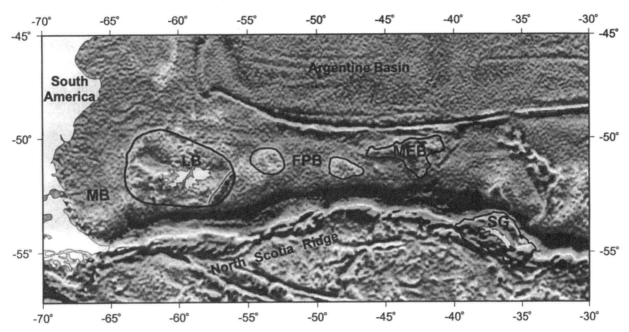

Fig. 6. Tectonic elements of the Falkland Plateau (GTOPO topography and predicted bathymetry is gray-scale version of *Smith and Sandwell*, 1997). LB - Lafonian (Falkland Island) crustal block; FPB - central Falkland Plateau basin; MB - south Malvinas basin; MEB - Maurice Ewing Bank crustal block; SG - South Georgia microcontinent.

their original orientation required ~180° of rotation relative to Africa. Rotation of the Falkland Plateau relative to Africa by closing the South Atlantic Ocean basin accounts for only ~30° of that rotation. Moreover, rotation of Antarctica relative to Africa by closure of the southwest Indian Ocean basin accounts for none of the rotation that Schopf [1969] realized is required to restore the structures in the rocks of the Ellsworth Mountains to a geologically reasonable position aligned with the contemporaneous Gondwanide fold belt in the Cape Mountains of southern Africa and the Pensacola Mountains on the margin of the East Antarctic craton. Paleomagnetic data support the geologic arguments of both Adie [1952] and Schopf [1969]. As noted in a review of the evidence by Dalziel and Grunow [1992], the temporal intervals in which both the clockwise motion of the Falkland Islands and the anticlockwise motion of the Ellsworth Mountains occurred are the same. Rotation could only have taken place between the cessation of the Gondwanide folding that affects rocks as young as Permian and rift-drift transition in the South Atlantic Ocean (Early Cretaceous) and Southwest Indian Ocean (Middle Jurassic).

The stumbling block in the acceptance of these geologically and paleomagnetically reasonable movements has been the absence of an obvious mechanism. There is limited space for rotation of the Ellsworth-Whitmore crustal block between the East Antarctic craton and the Antarctic Peninsula in geometrically precise paleotectonic reconstructions (Figure 5). Crust at the head of the Weddell Sea has been regarded as comprising an additional continental block, the Ronne, Filchner, or Weddell Sea embayment block [*Dalziel and Elliot*, 1982; *de Wit et al.*, 1988]. Seismic reflection data obtained from both the Falkland Plateau and the Weddell Sea and Weddell embayment show no sign of disturbance of the reflectors that could be attributed to such lithospheric rotations [*Jokat et al.*, 1997; *Richards et al.*, 1996].

The Falkland Plateau is a marginal plateau of the South American continent (Figure 6). It consists of two main "nodes" of Precambrian continental crust, one of which is exposed in the Falkland Islands while the other was recovered by drilling Maurice Ewing Bank at the eastern tip of the plateau [*Barker et al.*, 1977]. The limit of rigid continental crust is marked by edge anomalies in the satellite altimetry-derived gravity field (reflected in the "predicted bathymetry" for the plateau [*Smith and Sandwell*, 1997]. The islands occupy the central area of the larger continental Lafonian block or microplate. These proven continental nodes, and other possible smaller ones, are separated by rift-bounded Mesozoic sedimentary basins, the largest of which, the central Falkland Plateau basin, may be floored by oceanic crust [*Barker*, 1999; *Lorenzo and Mutter*, 1988, Figure 6]. Despite extensive seismic reflection profiling of the basins, no notable evidence of shortening has been found, even in the south Malvinas basin which separates the Lafonian block from South America (see *Dalziel*, 1997, Figure 10B). Thus the rotation of the Lafonian block indicated by the geologic and paleomagnetic data, took

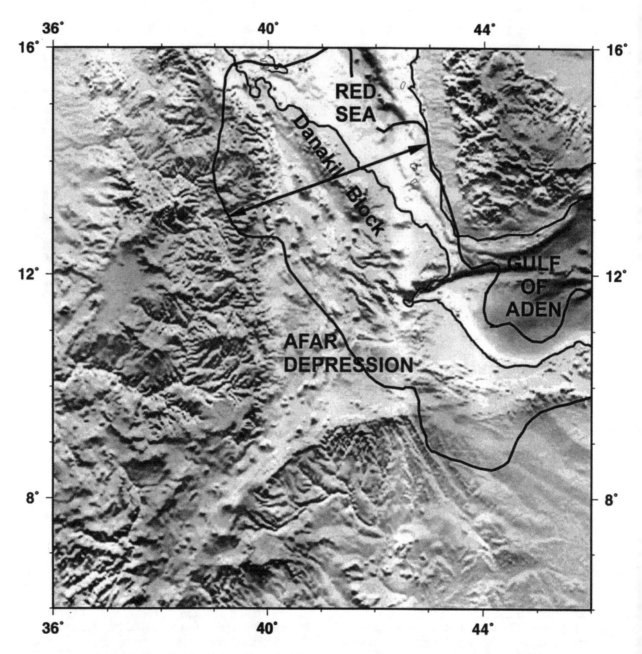

Fig. 7. Topography of the Afar triangle and surrounding regions of the northest Africa, the Red Sea, Gulf of Aden, and Arabian Peninsula (GTOPO topography and predicted bathymetry is gray-scale version of *Smith and Sandwell*, 1997). Black line shows the extent of the Ellsworth-Whitmore mountains crustal block of Antarctica on the same scale for comparison (see Fig. 5), the line with double arrows showing the position and length of the Ellsworth Mountains within that block.

place purely by extension of continental crust, possibly involving development of an incipient ocean basin beneath the central Falkland Plateau basin, and involving intrusion and extrusion of mafic igneous material. A large percentage of the area of the Falkland Plateau may have been generated by extension in the Mesozoic. We suggest that the crust beneath the head of the Weddell Sea and the Filchner-Ronne Ice Shelf is likewise an areal addition to the extent of the Antarctic continent generated by contemporaneous extension and magmatism associated with the rotation of the Ellsworth-Whitmore mountains crustal block or continental node.

The non-rigid rheologic behavior of the crust of Gondwanaland in the area of the developing Weddell Sea

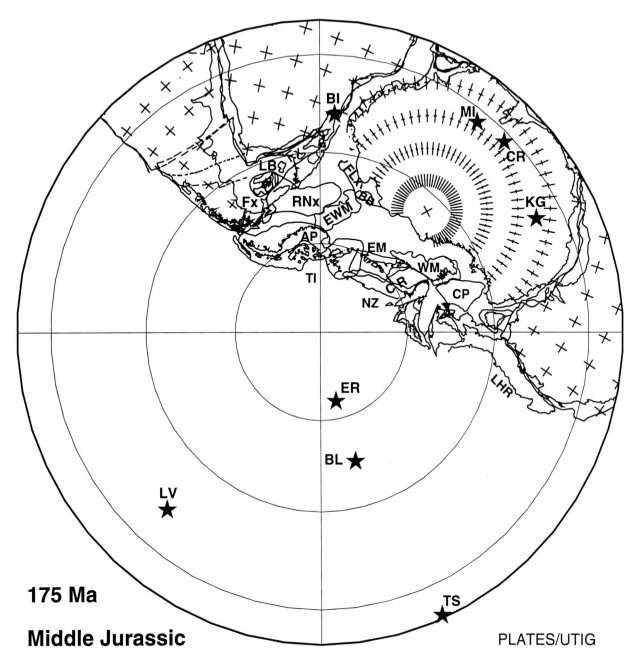

Fig. 8. Reconstruction of the region of Gondwanaland surrounding the incipient Weddell embayment at 175 Ma. Hot spots (stars) and abbreviations as in previous figures with additions: BB - Berkner Island block; CP - Campbell Plateau; CR - Chatham Rise; EM - East Marie Byrd Land block; FLx - Filchner extended lithosphere; Fx - Falkland Plateau extended lithosphere; RNx - Ronne extended lithosphere; WM - West Marie Byrd Land block.
(PLATES)

during Jurassic times is associated with the rapid emplacement of the dominantly mafic Karroo-Ferrar large igneous province (LIP) that extended from southern Africa over 4000 km to southern Australia and New Zealand and has been widely attributed to the impingement on the base of the lithosphere of the "head" of a large hot mantle plume [*Dalziel*, 1992; *Storey*, 1995; *White and McKenzie*, 1989; *Dalziel et al.*, 2000]. High precision U-Pb dating of minerals from the Karoo and Ferrar rocks suggests very rapid emplacement at ~182

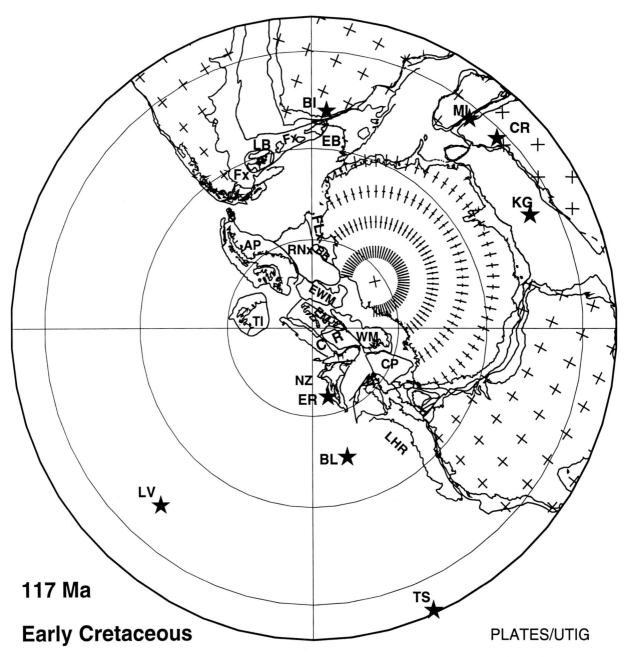

Fig. 9. West Antarctica within Gondwanaland at 117 Ma. Hot spots (stars) and abbreviations as in previous figures with addition: NZ - New Zealand blocks. (PLATES)

Ma [*Encarnación et al.*, 1996]. We envisage the process of formation of the new continental crust of South America and the Weddell Sea embayment as having been very rapid. Paleomagnetic data from granitic rocks associated with the Karoo-Ferrar LIP intruded into the Ellsworth-Whitmore continental block indicate significant rotation involving a paleolatitudinal shift had occurred by ~175 Ma [*Grunow et al.*, 1987a]. The enlargement of the area of the Falkland Plateau of South America and the Weddell embayment of Antarctica occurred by stretching of continental lithosphere and opening of rifts partly floored by oceanic crust [*Barker*, 1999] and infilled with sedimentary detritus in a rising and emergent dome above the impinging plume head. Development of the consequent drainage system on the dome has been documented by Cox [1989]. It is noteworthy that the area of non-rigid behavior and augmentation of the area of crust within Gondwanaland corresponds to a part of the supercontinent formed in Mesoproterozoic times rather than the Archean-Paleoproterozoic Kalahari nucleus.

A present-day analogue for the formation and rotation of the rigid blocks of the Falkland Plateau and the Weddell Sea embayment is the Afar region of northeast Africa. There the Danakil block that is approximately the same length as the Ellsworth Mountains within the EWM block, has rotated counterclockwise by ~10° as a result of extension and the propagation of rifts in the Afar triangle during only the last few million years [*Souriot and Brun*, 1992, Figure 7]. This process has effectively generated "continental" lithosphere. Even if it includes sediment filled, small, failed, ocean basins presently below sea level, the Afar depression has been added to the area of Africa [*Mohr*, 1989]. The rotation of the Danakil horst is occurring seemingly independent of true seafloor spreading. Thus the rotated, rigid Lafonian and Ellsworth-Whitmore mountains crustal blocks, are part of the South American and Antarctic continents together with the oceanic central basin of the Falkland Plateau and the extended continental lithosphere beneath the Ronne and Filchner ice shelves at the head of the Weddell Sea (Figures 8 and 9).

The subsequent tectonic history of the Weddell Sea basin was one of seafloor spreading with "Weddellia" (the Antarctic Peninsula and Thurston Island blocks of West Antarctica—the Pacific margin magmatic arc, together with the Ellsworth-Whitmore continental node and the stretched crust of the Weddell Sea embayment) rotating counterclockwise as the oceanic lithosphere beneath the Weddell Sea formed between 165 and 130 Ma *Grunow et al.*, 1987a, 1987b, 1991]. Weddellia may also include part of eastern and coastal Marie Byrd Land [*DiVenere et al.*, 1995, 1996; *Mukasa and Dalziel*, 2000].

ORIGIN OF THE ROSS EMBAYMENT

The sparsely exposed basement rocks of western and interior Marie Byrd Land are similar to those of the margin of the East Antarctic craton on the Transantarctic Mountains side of the Ross Sea [*Bradshaw et al.*, 1997]. Paleomagnetic data from volcanic and plutonic complexes in Marie Byrd Land, moreover, indicate that the Ross Sea embayment had not formed by the mid-Cretaceous [*DiVenere et al.*, 1994]. Sea floor spreading data on the other hand, clearly show that the continental margin of the Ross embayment had essentially its present geography by the time the Campbell Plateau of the New Zealand microcontinent separated from the West Antarctic margin at ~85 Ma [*Lawver and Gahagan*, 1994, Figure 10]. Separation of the rigid block of western Marie Byrd Land from the East Antarctic craton and development of the crust of the Ross embayment accompanied a marked change in the magmatism along the Pacific margin. Prior to ~100 Ma this margin was an active subduction zone with calc-alkaline arc activity. At that time it changed abruptly to bimodal rift-related magmatism [*Weaver et al.*, 1994; *Mukasa and Dalziel*, 2000].

The change in magmatism along the Marie Byrd Land margin has been ascribed to the Cretaceous impingement of a plume head on the lithosphere [*Storey et al.*, 1999; *Weaver et al.*, 1994]. As discussed in Mukasa and Dalziel [2000] however, there is little evidence for such an event, indeed, impingement of a plume in the region of a subduction zone more likely results in uplift and the cessation of continental margin magmatism [*Murphy et al.*, 1998; *Dalziel et al.*, 2000] rather than an immediate switch to rift related magmatism as happened in Marie Byrd Land in the mid-Cretaceous. On the contrary, an active spreading ridge, the Pacific-Phoenix ridge, was subducted beneath the New Zealand margin of Gondwanaland at that time, [*Lawver and Gahagan*, 1994]. In our view, this is the most likely cause of both the extension in the Ross Sea embayment and the separation of the New Zealand microcontinent (Plate 1).

Cenozoic alkaline volcanism in Marie Byrd Land, which started at ~35 Ma (Figure 3), included several active volcanoes [*LeMasurier*, 1990] and was preceded by uplift of the peneplained surface of the Marie Byrd Land block [*LeMasurier*, 1990; *LeMasurier and Landis*, 1996; Figure 3]. Geochemical data indicate that these are plume-related volcanics, hence the uplift of the Marie Byrd Land block may be largely or wholly the result of doming above a plume [*Hole and LeMasurier*, 1994]. The rest of the Ross embayment and most of the New Zealand microcontinent appear to represent the same sort of stretched continental crust. Indeed, the islands of New Zealand exist solely as a result of tectonism and volcanism along the Australian-Pacific plate boundary that traverses the microcontinent.

DEVELOPMENT OF THE WEST ANTARCTIC RIFT SYSTEM

There may be active volcanoes beneath the WAIS itself [*Blankenship et al.*, 1993]. Active volcanoes along the western side of the Ross embayment and in Marie Byrd Land indicate possible present-day rifting of the Antarctic continent. The young escarpment of the Transantarctic Mountains and the uplifted marine peneplain in Marie Byrd Land are taken to bound what is known as the West Antarctic rift system—'WARS' [*Behrendt and Cooper*, 1991; *Behrendt et al.*, 1994; 1996, Figure 10]. The geochemistry of the bimodal volcanic rocks indicates impingement of a mantle plume beneath the nearly stationary Antarctic continental lithosphere as the likely cause of both uplift and volcanism [*Behrendt et al.*, 1992; *Hole and LeMasurier*, 1994]. The WARS is believed to extend from North Victoria Land to the base of the Antarctic Peninsula [*Behrendt and Cooper*, 1991, Plate 1] with volcanicity that dates back to ~35 Ma. Uplift of the Transantarctic Mountains, widely believed to be related to the inception of WARS, dates from the early Cenozoic [*Fitzgerald*, 1992].

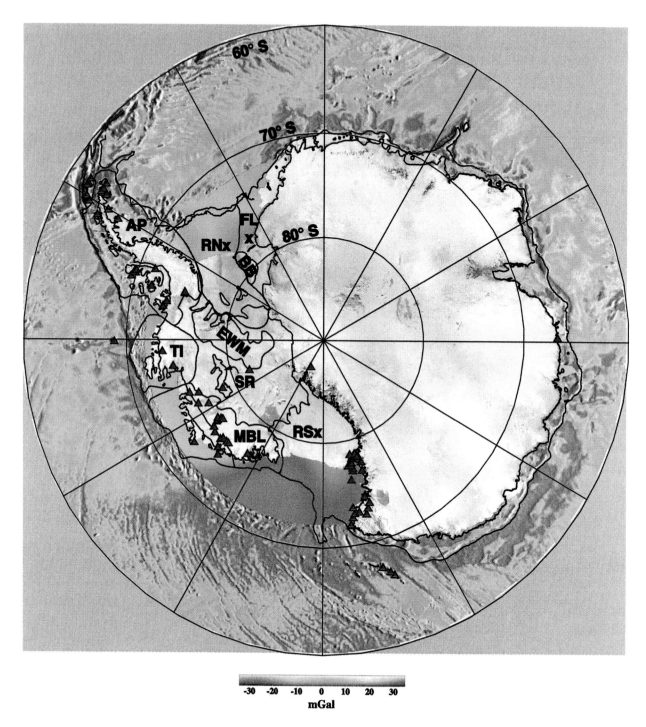

Plate 1. Combined AVHRR-satellite gravity image of present-day Antarctica [*Lawver et al.*, 1993], with active and recently active volcanoes [*LeMasurier*, 1990; *Blankenship et al.*, 1993; *Behrendt et al.*, 1999], and sub-ice lithospheric provinces of this paper. Abbreviations as in previous figures with addition: SR - Sinuous Ridge (indicated with arrow).

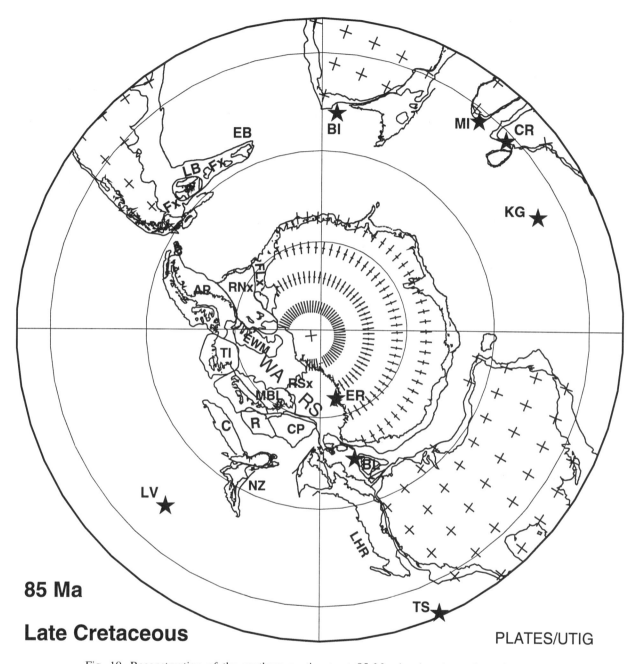

Fig. 10. Reconstruction of the southern continents at 85 Ma showing the setting of the Ross embayment. Hot spots (stars) and abbreviations as in previous figures. (PLATES)

Behrendt and Cooper [1991] suggested that the Cenozoic rift system resulted from the "capture" by a mantle plume of a rift propagating clockwise around the Antarctic continent during the break-up of Gondwanaland (see also *Behrendt et al., 1992*). As discussed above, we distinguish several phases in this so-called propagation, ascribing them to different causes. Thus, we see the volcanically active West Antarctic rift system of the present day, including the sinuous ridge that may be volcanically active [*Behrendt et al.*, 1998], as the result of plume-driven uplift, extension and of fracturing along pre-existing lines of weakness. Those were most notably: the Neoproterozoic margin of the East Antarctic craton; continental lithosphere of the Ross Sea embayment between Marie Byrd Land and East Antarctica stretched during mid-Cretaceous separation of the New Zealand microcontinent from West Antarctica; and continental lithosphere between Marie Byrd Land and the EWM block. The situation appears analogous to that of the 'capture' of the spreading center in the northwest Indian Ocean by the impingement of the Afar plume in the Neogene to form the Gulf of Aden and the Red Sea [*Omar and Steckler*, 1995, Figure 7].

CONCLUSIONS: THE LITHOSPHERIC "CRADLE" OF THE WAIS

We suggest that in a coincidence of time and space, several major tectonic events have combined to create the environment in which the unique marine-based WAIS that straddles an entire continent could form and endure for several millions of years. First, a major mantle plume impinged on the base of the lithosphere in the Middle Jurassic and generated the extended and magmatically modified crust of the Weddell Sea embayment. It also resulted in the uplift of the Ellsworth-Whitmore mountains crustal block and in the rotation of this block into a position impeding drainage of ice into the Weddell Sea. Second, subduction of the Pacific-Phoenix ridge in the mid-Cretaceous shut off subduction along the South Pacific margin of the continent and resulted in extension to form the crust of the Ross embayment. Third, impingement of another plume during the Cenozoic resulted in the uplift of Marie Byrd Land and possibly of the Transantarctic Mountains bordering the WAIS, as well as the volcanically active region which most of West Antarctica occupies today. The relatively shallow depth (<1000 m) of the Ross and Weddell embayments allows the formation of ice to the sea floor. The fact that both embayments have open access to the world's oceans allows escape of cold, dense seawater. Otherwise a dense unfrozen brine layer would be trapped beneath floating ice in the West Antarctic region. Hence, without the abnormal area of stretched continental crust in the Weddell and Ross embayments, the uplifted Ellsworth-Whitmore mountains crustal block, and the flanking uplifts of the Transantarctic Mountains and Marie Byrd Land plateau, there probably would not now be a WAIS.

REFERENCES

Adie, R.J., The position of the Falkland Islands in a construction of Gondwanaland, *Geological Magazine*, 89, 401-410, 1952.

Barker, P.F., Evidence for a volcanic rifted margin and oceanic crustal structure for the Falkland Plateau basin, *Journal of the Geological Society of London*, 156, Part 5, 889-900, 1999.

Barker, P.F., I.W.D. Dalziel, et al., The Evolution of the southwestern Atlantic Ocean Basin: Leg 36 data, in *Initial Rep. Deep Sea Drill. Proj.*, edited by S. Wise, pp. 1079, U.S. Government Printing Office, Washington, D.C., 1977.

Behrendt, J.C., D.D. Blankenship, C.A. Finn, R.E. Bell, R.E. Sweeney, S.M. Hodge, and J.M. Brozena, CASERTZ aeromagnetic data reveal late Cenozoic flood basalts(?) in the West Antarctic rift system, *Geology (Boulder)*, 22 (6), 527-530, 1994.

Behrendt, J.C., and A. Cooper, Evidence of rapid Cenozoic uplift of the shoulder escarpment of the West Antarctic: rift system and a speculation on possible climate forcing, *Geology*, 19 (4), 315-319, 1991.

Behrendt, J.C., C.A. Finn, D.D. Blankenship, and R.E. Bell, Aeromagnetic evidence for a volcanic caldera(?) complex beneath the divide of the West Antarctic Ice Sheet, *Geophysical Research Letters*, 25 (23), 4385-4388, 1998.

Behrendt, J.C., W. LeMasurier, and A.K. Cooper, The West Antarctic rift system: a propagating rift "captured" by a mantle plume?, in *Recent Progress in Antarctic Earth Science*, edited by Y. Yoshida, K. Kaminuma, and K. Shiraishi, pp. 315-322, Terra Scientific Publishing Company (TERRAPUB), Tokyo, 1992.

Behrendt, J.C., R. Saltus, D. Damaske, A.E. McCafferty, C.A. Finn, D. Blankenship, and R.E. Bell, Patterns of late Cenozoic volcanic and tectonic activity in the West Antarctic rift system revealed by aeromagnetic surveys, *Tectonics*, 15 (3), 660-676, 1996.

Blankenship, D.D., R.E. Bell, S.M. Hodges, J.M. Brozena, J.C. Behrendt, and C.A. Finn, Active volcanism beneath the West Antarctic ice sheet and implications for ice-sheet stability, *Nature*, 361, 526-529, 1993.

Bradshaw, J.D., R.J. Pankhurst, S.D. Weaver, B.C. Storey, R.J. Muir, and T.R. Ireland, New Zealand superterranes recognized in Marie Byrd Land and Thurston Island, in *The Antarctic region; geological evolution and processes; proceedings of the VII international symposium on Antarctic Earth sciences*, edited by C.A. Ricci, pp. 429-436, Siena, Italy, 1997.

Cox, K.G., The role of mantle plumes in the development of continental drainage patterns, *Nature*, 342 (6252), 873-877, 1989.

Dalziel, I.W.D., Pre-Jurassic history of the Scotia Arc region, in *Antarctic Geoscience*, edited by C. Craddock, pp. 111-126, University of Wisconsin Press, Madison, 1982.

Dalziel, I.W.D., Antarctica: a tale of two supercontinents?, *Annual Review of Earth and Planetary Sciences*, 20, 501-526, 1992.

Dalziel, I.W.D., Neoproterozoic-Paleozoic geography and tectonics: Review, hypothesis, environmental speculation, *GSA Bulletin*, 108 (1), 16-42, 1997.

Dalziel, I.W.D., and D.H. Elliot, West Antarctica: problem child of Gondwanaland, *Tectonics*, 1, 3-19, 1982.

Dalziel, I.W.D., and A.M. Grunow, Late Gondwanide tectonic rotations within Gondwanaland: Causes and consequences, *Tectonics*, 11 (3), 603-606, 1992.

Dalziel, I.W.D., L.A. Lawver, and J.B. Murphy, Plumes, orogenesis, and supercontinental fragmentation, Earth and Planetary Science Letters, 178 (1-2), 1-11, 2000.

de Wit, M.J., M. Jeffery, H. Bergh, and L. Nicolaysen, Geological Map of sectors of Gondwana, American Association of Petroleum Geologists, Tulsa, Oklahoma, 1988.

DiVenere, V.J., D.V. Kent, and I.W.D. Dalziel, Mid-Cretaceous paleomagnetic results from Marie Byrd Land, West Antarctica; a test of post-100 Ma relative motion between East and West Antarctica, *Journal of Geophysical Research*, *99* (B8), 15,115-15,139, 1994.

DiVenere, V.J., D.V. Kent, and I.W.D. Dalziel, Early Cretaceous paleomagnetic results from Marie Byrd Land, West Antarctica: implications for the Weddellia collage of crustal blocks, *Journal of Geophysical Research*, *100* (B5), 8133-8152, 1995.

DiVenere, V., D.V. Kent, and I.W.D. Dalziel, Summary of paleomagnetic results from West Antarctica: implications for the tectonic evolution of the Pacific margin of Gondwana during the Mesozoic, in *Weddell Sea Tectonics and Gondwana Break-up*, edited by B.C. Storey, E.C. King, and R.A. Livermore, pp. 31-43, The Geological Society, London, 1996.

DuToit, A.L., *Our Wandering Continents*, 355 pp., Oliver and Boyd, Edinburgh, Scotland, 1937.

Encarnación, J., T.H. Fleming, D.H. Elliot, and H.V. Eales, Synchronous emplacement of Ferrar and Karoo dolerites and the early breakup of Gondwana, *Geology*, *24* (6), 535-538, 1996.

Fischer, A.G., The two Phanerozoic supercycles, in *Catastrophes and Earth History*, edited by W.A. Berggren, and J.A. Van Couvering, pp. 129-150, Princeton University Press, Princeton, 1984.

Fitzgerald, P.G., The Transantarctic Mountains of Southern Victoria Land: the application of apatite fission track analysis to a rift shoulder uplift, *Tectonics*, *11* (3), 634-662, 1992.

Grunow, A.M., I.W.D. Dalziel, and D.V. Kent, Ellsworth-Whitmore mountains crustal block, Western Antarctica: New paleomagnetic results and their tectonic significance, in *Gondwana Six: Structure, Tectonics and Geophysics*, edited by G.D. McKenzie, pp. 161-172, American Geophysical Union, 1987a.

Grunow, A.M., D.V. Kent, and I.W.D. Dalziel, Mesozoic evolution of West Antarctica and the Weddell Sea Basin: New paleomagnetic constraints, *Earth Planet. Sci. Lett.*, *86*, 16-26, 1987b.

Grunow, A.M., D.V. Kent, and I.W.D. Dalziel, New paleomagnetic data from Thurston Island: Implications for the tectonics of West Antarctica and Weddell Sea opening, *Journal of Geophysical Research*, *96*, 17937-17954, 1991.

Hole, M.J., and W.E. LeMasurier, Tectonic controls on the geochemical composition of Cenozoic, mafic alkaline volcanic rocks from West Antarctica, *Contributions to Mineralogy and Petrology*, *117* (2), 187-202, 1994.

Jokat, W., N. Fechner, and M. Studinger, Geodynamic Models of the Weddell Sea Embayment in view of New Geophysical Data, in *The Antarctic Region: Geological Evolution and Processes*, edited by C.A. Ricci, pp. 453-460, Terra Antartica Publishers, Siena, 1997.

Lawver, L., and C.R. Scotese, A revised reconstruction of Gondwanaland, in *Gondwana Six: Structure, Tectonics and Geophysics*, edited by G.D. McKenzie, pp. 17-23, Am. Geophysical Union, Geophysical Monograph 40, 1987.

Lawver, L.A., I.W.D. Dalziel, and D.T. Sandwell, Antarctic plate: Tectonics from a gravity anomaly and infrared satellite image, *GSA Today*, *3*, 117-122, 1993.

Lawver, L.A., and L.M. Gahagan, Constraints on timing of extension in the Ross Sea region, *Terra Antartica*, *1* (3), 545-552, 1994.

Lawver, L.A., L.M. Gahagan, and M.F. Coffin, The development of paleoseaways around Antarctica, in *The Antarctic Paleoenvironment: A Perspective on Global Change, Part I*, edited by J.P. Kennett, and D.A. Warnke, pp. 7-30, American Geophysical Union, Washington, DC, 1992.

Lawver, L.A., L.M. Gahagan, and I.W.D. Dalziel, A tight fit-Early Mesozoic Gondwana, a plate reconstruction perspective, in *Memoirs of the National Institute of Polar Research: International Symposium on the Origin and Evolution of Continents*, pp. 214-229, National Institute of Polar Research, Tokyo, 1998.

Lawver, L.A., J.G. Sclater, and L. Meinke, Mesozoic and Cenozoic reconstructions of the South Atlantic, *Tectonophysics*, *114*, 233-254, 1985.

LeMasurier, W.E., Late Cenozoic volcanism on the Antarctic plate–An overview, in *Volcanoes of the Antarctic Plate and Southern Oceans, Antarctic Research Series*, edited by W.E. Le Masurier, and J.W. Thomson, pp. 1-19, American Geophysical Union, Washington, DC, 1990.

LeMasurier, W.E., and C.A. Landis, Mantle-plume activity recorded by low-relief erosion surfaces in West Antarctica and New Zealand, *GSA Bulletin*, *108* (11), 1450-1466, 1996.

Lorenzo, J.M., and J.C. Mutter, Seismic stratigraphy and tectonic evolution of the Falkland/Malvinas Plateau, *Revista Brasileira de Geociências*, *18*, 191-200, 1988.

Mohr, P., Nature of the crust under Afar; new igneous, not thinned continental, *Tectonophysics*, *167* (1), 1-11, 1989.

Mukasa, S.B., and I.W.D. Dalziel, Marie Byrd Land, West Antarctica; evolution of Gondwana's Pacific margin constrained by zircon U-Pb geochronology and feldspar common-Pb isotopic compositions, *GSA Bulletin*, *112* (4), 611-627, 2000..

Murphy, J.B., G.L. Oppliger, G.H. Brimhall, Jr., and A. Hynes, Plume-modified orogeny; an example from the Western United States, *Geology*, *26* (8), 731-734, 1998.

Norton, I.O., and J.G. Sclater, A model for the evolution of the Indian Ocean and the breakup of Gondwanaland, *Journal of Geophysical Research*, *84*, 6803-6830, 1979.

Omar, G.I., and M.S. Steckler, Fission track evidence on the initial rifting of the Red Sea: Two pulses, no propagation, *Science*, *270*, 1341-1344, 1995.

Richards, P.C., R.W. Gatliff, M.F. Quinn, J.P. Williamson, and N.G.T. Fannin, The gelogical evolution of the Falkland Islands continental shelf, in *Weddell Sea Tectonics and Gondwana Break-up: Geological Society Special Publication No. 108*, edited by B.C. Storey, E.C. King, and R.A. Livermore, pp. 105-128, The Geological Sociey, London, 1996.

Schopf, J.M., Ellsworth Mountains: position in West Antarctica due to sea-floor spreading, *Science*, *164* (3875), 63-66, 1969.

Smith, W.H.F., and D.T. Sandwell, Global sea floor topography from satellite altimetry and ship depth soundings, *Science*, *277*, 1956-1962, 1997.

Souriot, T., and J.P. Brun, Faulting and block rotation in the Afar triangle, East Africa; the Danakil "crank-arm" model,

Geology, *20* (10), 911-914, 1992.

Storey, B.C., The role of mantle plumes in continental breakup; case histories from Gondwanaland, *Nature*, *377* (6547), 301-308, 1995.

Storey, B.C., P.T. Leat, S.D. Weaver, R.J. Pankhurst, J.D. Bradshaw, and S. Kelley, Mantle plumes and Antarctica-New Zealand rifting; evidence from mid-Cretaceous mafic dykes, *Journal of the Geological Society of London*, *156, Part 4*, 659-671, 1999.

Weaver, S.D., B.C. Storey, R.J. Pankhurst, S.B. Mukasa, V.J. DiVenere, and J.D. Bradshaw, Antarctica-New Zealand rifting and Marie Byrd Land lithospheric magmatism linked to ridge subduction and mantle plume activity, *Geology*, *22* (9), 811-814, 1994.

White, R., and D. McKenzie, Magmatism at rift zones: the generation of volcanic continental margins and flood basalts, *Journal of Geophysical Research*, *94* (B6), 7685-7729, 1989.

The University of Texas Institute for Geophysics, 4412 Spicewood Springs Road, Bldg. 600, Austin TX 78759-8500 U.S.A.

[1]Department of Geological Sciences, The University of Texas at Austin, Austin TX 78712 U.S.A.

[2]Tectonics Special Research Centre, Department of Geology & Geophysics, The University of Western Australia, Nedlands, WA 6907 Australia

The University of Texas Institute for Geophysics, contribution no. 1489.

Tectonics Special Research Centre, University of Western Australia, contribution no. 113.

EVOLUTION OF THE WEST ANTARCTIC ICE SHEET

John B. Anderson and Stephanie S. Shipp

Department of Geology and Geophysics, Rice University Houston, Texas

During the Oligocene through early Miocene, the West Antarctic Ice Sheet (WAIS) consisted of a number of isolated ice caps centered over islands and continental blocks. The ice caps coalesced into an ice sheet that advanced onto the continental shelf on several occasions throughout the late Miocene through Pleistocene. During the LGM, the WAIS extended to the shelf break in eastern and central Ross Sea and to within 100 km of the shelf break in the western Ross Sea. Mega-scale glacial lineations, deformation till and grounding zone wedges within troughs indicate streaming ice flowing over a deforming bed.

Retreat of the WAIS from the shelf began shortly after the LGM and continued into the late Holocene. In western Ross Sea, grounding-zone-proximal geomorphic features were formed by the retreating ice streams and indicate that collapse of the ice sheet did not occur, at least not on the exposed shelf. Differences in the number and locations of grounding-zone wedges and smaller grounding-zone features from trough to trough imply that the ice streams acted independently during their retreat.

Glacial troughs on the Amundsen Sea and Bellingshausen Sea continental shelves contain subglacial bedforms that support an expanded, grounded WAIS. Within Pine Island Bay, an extensive network of subglacial channels and a virtual absence of recessional moraines indicate that subglacial meltwater may have contributed to collapse of the expanded ice sheet.

INTRODUCTION

Examination of the evolution of the West Antarctic Ice Sheet (WAIS) permits addressing of fundamental questions concerning ice-sheet stability and mechanisms that regulate stability over centuries to hundreds of thousands of years: 1) what factors contributed to the evolution of the WAIS?; 2) how large is the ice sheet during a glacial maximum?; 3) what is the minimum configuration?; 4) does the ice sheet have a history of catastrophic collapse?; 5) has the ice sheet advanced and retreated in concert with Northern Hemisphere ice sheets, suggesting a strong climatic and/or eustatic control?; 6) to what degree have volume changes in the East Antarctic Ice Sheet (EAIS) influenced the behavior of the WAIS?; 7) has deformation of the bed beneath the ice sheet resulted in increased ice flow and thinning and contributed to collapse?; and 8) what role did subglacial meltwater play in ice sheet behavior?

This paper reviews the history of the WAIS. Initial discussion focuses on the long-term evolution, which provides an assessment of factors contributing to ice-sheet growth and development over long periods of geological time (e.g., tectonic changes) and that are, essentially, irreversible. Figure 1 summarizes the evidence for WAIS evolution during the Tertiary. *Barrett* [1997], *Abreu* and *Anderson* [1998] and *Anderson* [1999] provide recent summaries of the evidence used to establish the ice sheet's history. An examination of the more recent (Late Quaternary) record follows, focusing on the marine geological evidence for the maximum configuration and retreat history of the WAIS. *Denton et al.* [1991], *Bentley* and *Anderson* [1998], *Licht et al.* [1996, 1999], *Anderson* [1999], *Shipp et al.* [1999], and *Domack et al.* [1999] provide reviews of the criteria used to reconstruct the ice sheet and discussions of chronostratigraphic data. The present drainage for the WAIS is depicted in Figure 2. Ross and Weddell seas, which together receive more than half of the drainage of the WAIS, also receive a

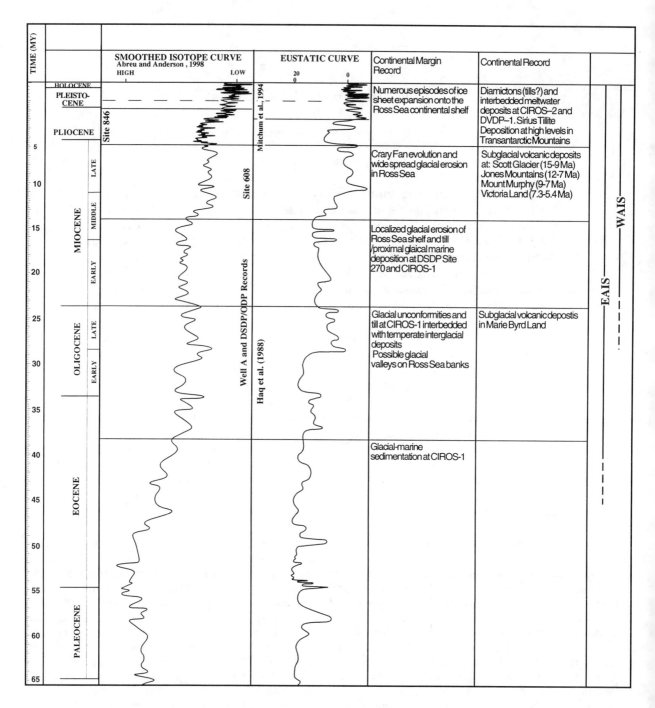

Fig. 1. Summary of evidence for West Antarctic Ice Sheet evolution. See text for sources of information. Modified from *Abreu and Anderson* [1998].

significant component of EAIS drainage. These two embayments present the opportunity to assess the relative fluctuations of the EAIS and WAIS during the LGM. The record preserved on the continental shelves of central and eastern Ross Sea, central and western Weddell (non-Antarctic Peninsula), and Bellingshausen and Amundsen seas, offers insights into the nature and timing of WAIS fluctuations. At present, the WAIS ice sheet contains enough water equivalent to raise sea level by 6 m. Reliable estimates of the LGM water-equivalent volume of the WAIS await better glacial reconstructions for that time interval.

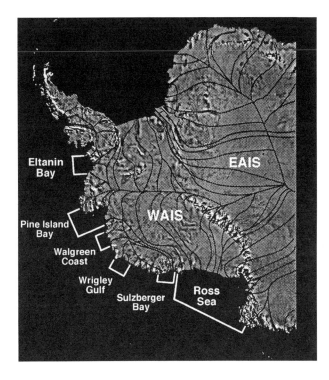

Fig. 2. Most drainage from the West Antarctic Ice Sheet is into the Ross Sea and Weddell Sea, with the remaining drainage through outlets all along the Marie Byrd Land and Ellsworth Land coasts. The boxes are areas where detailed marine geological surveys have been conducted, which includes all the main drainage outlets except those in the southwestern Weddell Sea.

TERTIARY RECORD OF WAIS EVOLUTION

There is significant evidence for mountain glaciation in West Antarctica throughout much of the Tertiary [*Barrett*, 1997; *Abreu and Anderson*, 1998, 1999], but when did the first ice sheets evolve? The oxygen isotope and eustatic records for the Cenozoic indicate that measurable growth of the Antarctic Ice Sheet began in the Middle Eocene (Figure 1). The early ice sheet is believed to have been confined to East Antarctica. Conventional theory holds that the EAIS initially formed by the coalescence of temperate glaciers flowing from centers on the Gamburtsev Subglacial Mountains [*Denton et al.*, 1971] and the ancestral Transantarctic Mountains [*TAM; Calkin*, 1964; *Drewry*, 1975; *Mercer*, 1978]. The ancestral TAM may have been a minor obstacle to the spread of the EAIS into West Antarctica during its early evolution [*Webb and Harwood*, 1991]. Bentley and Ostenso [1961] argued that the WAIS formed when ice shelves thickened and grounded on the continental shelf. These ice shelves were believed to have created a protective barrier behind which the ice sheet formed [*Mercer*, 1978].

Evidence for initial glaciation in West Antarctica occurs in the form of hyaloclastites (subglacial volcanic deposits), deposited 25 Ma, in Marie Byrd Land [*LeMasurier and Rex*, 1983]. Wilch [1997] reported glaciovolcanic deposits at Mt. Petras in West Antarctica that are Oligocene (29 to 27 Ma) in age. He concludes that these deposits reflect the presence of thin, local ice caps.

The CIROS-1 drill site in the western Ross Sea recovered abundant evidence for glaciation in the area during the Oligocene [*Barrett et al.*, 1989]. Early Oligocene deposits drilled at this site indicate mountain glaciation in the region. The Late Oligocene section at CIROS-1 consists of alternating till and glacial-marine deposits and interglacial units that include fluvial to deep-water mudstone facies that contain a temperate pollen-spore assemblage [*Mildenhall*, 1989]. These combined data reflect cool temperate interglacial conditions during the Late Oligocene [*Hambrey*, 1993], but provide little direct evidence for existence of the WAIS.

The continental record of glaciation in West Antarctica during the Miocene includes several examples of subglacially erupted volcanics that span the time interval from 15 Ma to 7 Ma [*Rutford et al.*, 1972; *Stump et al.*, 1980; *Wilch*, 1997]. Volcanic ashes, interbedded with tills, yielded 40Ar/39Ar radiometric dates from feldspars that indicate the EAIS overrode the Transantarctic Mountains and flowed into the Ross Sea prior to the Late Miocene [*Mayewski*, 1975; *Denton et al.*, 1984; *Marchant and Denton*, 1996].

During the Oligocene and Early Miocene, the Ross Sea continental shelf was characterized by banks and islands, separated by relatively deep basins formed during rifting of the Ross Embayment [*De Santis et al.*, in press]. Initially, these basins limited the growth of small ice caps centered over the islands [*De Santis et al.*, 1995]. As the basins filled with sediment, the ice caps grew and coalesced to form the WAIS. Several advances of the WAIS onto the continental shelf during the Late Miocene carved widespread glacial unconformities [*Cooper et al.*, 1991; *Anderson and Bartek*, 1992; *De Santis et al.*, 1995, in press]. DSDP Site 270 sampled Miocene ice-proximal deposits that contain mineral and rock compositions indicative of source areas in the interior portions of West Antarctica [*Barrett*, 1975; *Hambrey*, 1993]. The Miocene glacial deposits recovered at DSDP sites on the Ross Sea shelf and at CIROS-1 are interbedded with meltwater deposits and diatomaceous oozes, indicating strong climatic shifts.

The record of glaciation in the Weddell Sea region is not as well documented as for Ross Sea; extensive sea-ice cover often has prevented the acquisition of seismic data and drill cores on the shelf. Crary Fan, a large trough-mouth/submarine-fan complex in southeastern Weddell Sea, contains one of the best records of glaciation in the region. The fan comprises material excavated from Crary Trough during repeated expansion of the ice sheet. Detailed sequence stratigraphic analyses of Crary Fan show construction of several channel–levee units, each representing a period of significant fan development [Moons et al., 1992; Kaul, 1992; Bart et al., 1994, in press, a; De Batist et al., 1994]. De Batist [1994] argue that episodes of fan development correlate with ice-sheet grounding events on the continental shelf. Based on correlation to ODP Leg 113 Site 693 on the Weddell Sea continental rise, they conclude that at least five long-term glacial expansions, and no fewer than 14 smaller-scale expansions, occurred in the Weddell Sea region since the Middle Miocene. Hence, the record of glaciation on the Weddell Sea continental shelf appears to mirror that of the Ross Sea. Both regions experienced widespread glacial erosion of the shelf during the Miocene, implying the existence of a WAIS that was as large as and, at times, larger than at present.

The Pliocene global oxygen isotope and eustatic records indicate significant ice volume fluctuations [Prentice and Fastook, 1990; Miller et al., 1991; Mitchum et al., 1994 Abru and Anderson, 1998] (Figure 1). Micropaleontological investigations of deep sea sediments from around the Southern Ocean record significant shifts in oceanic fronts during this period. These shifts are believed to be associated with major climatic events in and around Antarctica [e.g., Ciesielski and Weaver, 1974; Burckle and Cirilli, 1986; Abelmann et al., 1990; Barron, 1996], but considerable debate has centered over the impact of these climatic events on the ice sheet. One school argues that the ice sheets responded radically to these climatic changes [Webb and Harwood, 1991; Barrett et al., 1992], while another school contends that these events had a relatively minor impact on the ice sheet [Denton et al., 1991; Sugden et al., 1993; Marchant and Denton, 1996]. When did the shift from temperate/subpolar to polar conditions occur? Webb and Harwood [1991] suggest that the occurrence of *Nothofagus* plant fossils in the Pliocene Sirius Group in the TAM indicates that temperate interglacial conditions occurred well into the Pliocene. CIROS-2 (Ferrar Fjord) and DVDP-11 (Taylor Valley) drill sites in the McMurdo Sound region support this climate interpretation. Diamictons at these sites are interbedded with mud, sand, and gravel facies interpreted to be meltwater deposits [Barrett et al., 1992; Hambrey, 1993]. However, Denton et al. [1991], Sugden et al., [1993] and Marchant and Denton, [1996] argue that the Dry Valleys region experienced subpolar to polar conditions throughout the Pliocene. This includes desert pavements and thermal expansion cracks covered by isotopically dated Pliocene ash beds, and an absence of landforms associated with temperate glaciers. The debate continues over the stability of the ice sheet during the Pliocene. What is known is that the WAIS advanced to the outer continental shelf several times during the Pliocene [Alonso et al., 1992], but this does not necessarily mean that a polar climate existed throughout the Pliocene.

The record of Pleistocene glaciation in West Antarctica is fragmentary. Hyaloclastites of Pleistocene age (1.18 Ma) on Ross Island indicate that the ice sheet overrode the island and was hundreds of meters thicker than at present [Kyle, 1981]. Glaciovolcanic tuffs at Sechrist Peak, Mount Murphy, record an ice-sheet highstand of greater than 550 m during the Middle Pleistocene [590 ± 15 ka; Wilch, 1997]. Seismic records from the Ross Sea show glacial unconformities and subglacial seismic facies within the Pleistocene section. These record several episodes of ice-sheet grounding on the shelf [Alonso et al., 1992; Bart et al., in press, b]. The exact timing of these grounding events remains unknown due to a lack of drill core or chronostratigraphic control on the sites drilled to date. Even the number of glacial/interglacial cycles during the Pleistocene remains problematic. More data are necessary to unravel the Pleistocene glacial history of Antarctica. Without this information, the relative influence of eustasy and climate on ice-sheet stability or the interactions of Northern and Southern hemisphere ice sheets will remain unclear. At present, the late Quaternary glacial history must be used to assess these forcing mechanisms; the last glacial/interglacial cycle offers the most complete record of WAIS glacial conditions.

LATE QUATERNARY RECORD OF ICE-SHEET EVOLUTION

Pre-LGM Ice-sheet Configuration

OI Stage 6. The isotopic record from Vostok provides strong evidence that the Antarctic Ice Sheet experienced the same 120 ka glacial/interglacial cyclicity that so strongly influenced Northern Hemisphere ice sheets during the late Pleistocene [Petit et al., 1997]. However, little is known about the configuration of the Antarctic Ice Sheet during the previous glacial maximum (Oxygen Isotope Stage 6) or interglacial (Oxygen Isotope Stage 5e). Denton et al. [1991] review the evidence for higher than present ice sheet elevations in the McMurdo Sound region during Stage 6. This includes glacial drifts that have been dated using uranium-series methods as between 130,000 and 185,000 years old.

OI Stage 5e. During the last interglacial (Oxygen Isotope Stage 5e), sea level was approximately five meters higher than present [Emiliani, 1969; Bloom et al., 1974]. A five meter rise could result from nearly complete melting of the WAIS [*Mercer*, 1968], complete melting of the Greenland Ice Sheet [*Emiliani*, 1969], or partial melting of both ice sheets [*Denton et al.*, 1971; *Robin*, 1983; *Hughes*, 1987]. *Denton et al.* [1991] point out that CLIMAP sea surface temperature reconstructions [*Stuiver et al.*, 1981] and ice-core $\delta^{18}O$ records from Vostok [*Petit et al.*, 1997] indicate that summer temperatures were approximately 2°C warmer during Stage 5e than present. *Scherer* [1998] interpreted marine diatoms in sediment sampled beneath Ice Steam B as supporting evidence for a WAIS grounding line positioned inland of its present location sometime during the Late Pleistocene.

OI Stage 3. Data for Stage 3 indicate a much reduced ice sheet, possibly similar in extent to today. Piston cores from the eastern continental shelf of the Weddell Sea recovered tills overlain by glacial-marine deposits [*Anderson et al.*, 1980] with an East Antarctic provenance [*Anderson et al.*, 1991]. Shell material from glacial-marine deposits yielded radiocarbon ages that extend back to 31,000 yr BP, indicating that the EAIS retreated from the continental shelf prior to this time [*Elverhoi*, 1981; *Bentley and Anderson*, 1998]. Anderson and Andrews [1999] found that ice rafting in the Weddell Sea was significantly greater than at present during and perhaps prior to Stage 3. They suggested that the sediment-laden basal portion of the ice sheet was exposed at that time, implying less extensive ice shelves than exist today. Similar observations have been made in western Ross Sea. There, total organic carbon dates from glacial-marine deposits above a shelf-wide unconformity yield radiocarbon ages in the range of >35,000 to 22,000 yr B.P. [*Anderson et al.*, 1992; *Hilfinger*, 1995, *Licht et al.*, 1996]. In-situ bioclastic carbonates on outer shelf banks have ages ranging from >34,000 to 22,700 yr B.P. [*Taviani et al.*, 1993]. These observations are consistent with results from coastal studies in other regions of East Antarctica that indicate retreat of the ice sheet prior to the LGM [*Yoshida*, 1983; *Orombelli et al.*, 1991; *Colhoun*, 1991; *Domack et al.*, 1998]. In the Ross and Weddell seas, earlier retreat of the EAIS from the shelf created embayments in the ice margin that may have contributed to retreat of the WAIS [*Kellogg et al.*, 1996; *Shipp et al.*, 1999].

LGM [Oxygen Isotope Stage 2] Ice-sheet Configuration

The configuration of the Antarctic Ice Sheet during the Last Glacial Maximum (LGM) has been modeled by several investigators, with widely different results. *Drewry's* [1979] model for the Ross Sea requires only minimum expansion of the WAIS during the LGM. The model predicts a vast ice shelf, pinned on shelf banks, floating above much of the continental shelf. *Stuiver et al.* [1981; CLIMAP reconstruction] provide an LGM ice-sheet reconstruction that places the ice sheet at the continental shelf edge all around Antarctica. This model was later revised by *Denton et al.* [1991] who show little interior surface elevation change during the LGM but considerable thickening of peripheral ice. *Huybrechts* [1990] constructed a three-dimensional, thermo-mechanical ice-sheet model that shows a glacial-maximum configuration similar to that of *Denton et al.* [1991] and a glacial-minimum configuration in which the WAIS is smaller than at present. *Kellogg et al.* [1996] developed a glacial reconstruction for the western Ross Sea that shows a deep embayment in the grounding line in western Ross Sea shortly after the LGM.

Detailed marine geological investigations, including swath bathymetry mapping, have been conducted in Ross Sea to reconstruct the ice-sheet configuration during the LGM [*Kellogg et al.*, 1979; *Anderson et al.* 1980; 1984; 1992; *Shipp et al.*, 1999; *Domack et al.*, 1999; *Licht et al.*, 1996, 1999] and offshore of all of the major drainage outlets of Marie Byrd Land and Ellsworth Land (Figure 2). Only preliminary results of the offshore Marie Byrd Land and Ellsworth Land investigations are presented in this paper. The evidence for ice sheets grounding on these continental shelves during the late Pleistocene is irrefutable. The most compelling evidence consists of subglacial geomorphic features including striations, rouche mountaines, flutes, drumlins, and mega-scale glacial lineations (Figure 3). Studies in Weddell Sea have focused on scattered high-resolution seismic profiles and analyses of sediment cores [*Anderson et al.*, 1980; 1991; *Elverhoi*, 1981; *Futterer and Melles*, 1990; *Bentley and Anderson*, 1998].

Ross Sea

Perched erratics on Beaufort and Franklin Islands in the southwestern Ross Sea occur at elevations of up to 320 m above sea level [*Stuiver et al.*, 1981]. *Hall and Denton* [in press] mapped moraines of the Ross Sea drift in the Taylor Valley region 350 m above sea level and concluded that the ice sheet grounded across the shelf during the LGM. These are only two examples of onshore data that indicate higher than present ice-sheet elevations in the Ross Sea region during the recent past. *Denton et al.* [1991] provide a more thorough review of the onshore record of ice-sheet configuration.

Piston cores from the Ross Sea continental shelf sampled a general stratigraphy of diatomaceous glacial-marine sediment, transitional glacial-marine sediment, and till [*Kellogg et al.*, 1979; *Anderson et al.*, 1980; 1992; *Licht et al*, 1996, 1999; *Domack et al.*, 1999]. Two types of till are recognized, stiff lodgement till and soft, water-saturated deformation till [*Anderson*, 1999; *Domack et al.*, 1999; *Shipp et al.*, 1999]. Both types of till exhibit compositional trends that have been

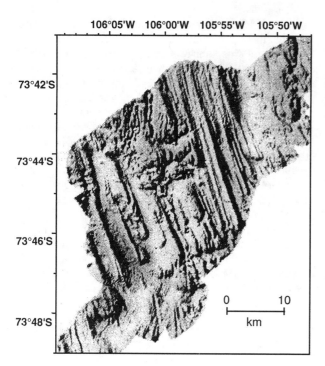

Fig. 3a. Examples of subglacial geomorphic features that document the presence of grounded ice on the continental shelf. Multibeam mosaics of striations, furrows and rouche mountainees that occur on basement rocks in Pine Island Bay.

used to reconstruct ice sheet paleodrainage [*Myers*, 1982; *Anderson et al.*, 1984; 1992; *Jahns*, 1994].

The distribution of geomorphic features within troughs implies changes in the interaction of the ice sheet and the bed as the ice sheet advanced across the shelf. The landward portions of troughs are characterized by exposed, striated bedrock. In central Ross Sea, highly attenuated drumlins and flutes occur at the transition between exposed bedrock and the offlapping wedge of sedimentary strata that covers the outer shelf (Figure 3b). The most prominent geomorphic features on the continental shelf are mega-scale glacial lineations (MSGL's; Figure 3c). These extend for tens of kilometers along trough axes and exhibit a unidirectional flow pattern, except where they diverge around bathymetric highs. MSGL's are associated with a soft till unit, interpreted to be deformation till, that thickens in an offshore direction [*Shipp et al.*, 1999]. These combined features suggest that the ice sheet was coupled to the sea floor on the inner shelf and on banks and that streaming ice, sliding across a mobile bed, occupied troughs on the outer shelf [*Alley et al.*, 1989; *Anderson*, 1999; *Shipp et al.*, 1999; *Stokes and Clark*, in press; *Clark*, in press]. The flanks of the troughs are marked by either abrupt escarpments or ice-stream boundary ridges with laterally accreting strata [*Anderson et al.*, 1992; *Alonso et al.*, 1992; *Shipp et al.*, 1999]. In addition, superimposed bidirectional MSGL's may indicate lateral shifts in paleo-ice streams. The regional paleodrainage patterns shown by the lineations generally are consistent with those derived from earlier till provenance studies [*Anderson et al.*, 1984, 1992; *Myers*, 1984; *Jahns*, 1994].

The preliminary grounding line positions and paleodrainage map derived from the combined sedimentologic and geophysical data are shown in Figure 4. These data show that the WAIS grounded at the shelf break in the central and eastern Ross Sea and that the grounding line in the western Ross Sea, which is associated with that portion of the ice sheet that was nourished by the EAIS, was located approximately 100 km inland of the shelf break. The large embayment in the grounding line in the western Ross Sea is consistent with the 16,000 yr B.P. glaciological model of *Kellogg et al.* [1996].

Retreat of the ice sheet from the Ross Sea continental shelf is marked by a variety of geomorphic features, including morainal ridges and wedges, corrugated moraines, and iceberg furrows in water depths to 700 m, well below modern iceberg furrows (Figure 5) [*Shipp*, 1999]. The ridges, wedges and corrugated moraines indicate that the ice sheet did not lose contact with the sea floor during its retreat. Thus, rising sea level did not cause the ice sheet to collapse [*Shipp*, 1999]. This indicates that the ice sheet was too thick to float off the bed. However, the deep iceberg furrows reflect later episodes when the retreating ice

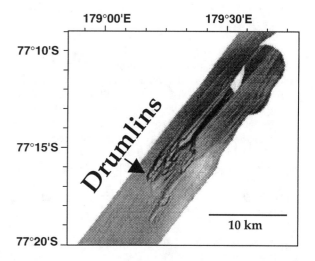

Fig. 3b. Examples of subglacial geomorphic features that document the presence of grounded ice on the continental shelf. Multibeam mosaics of drumlins at the transition between basement rocks and sedimentary strata in central Ross Sea.

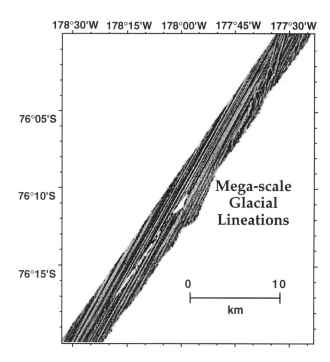

Fig. 3c. Examples of subglacial geomorphic features that document the presence of grounded ice on the continental shelf. Multibeam mosaics of mega-scale glacial lineations formed as the ice sheet advanced across progressively thicker sedimentary strata (central Ross Sea).

margin was an ice cliff (no ice shelf existed), and thus, a late phase of rapid retreat. But, when did these different stages of ice sheet retreat occur?

Studies of late Quaternary deposits exposed along the seaward side of the TAM, south of Coulman Island, have yielded a good chronostratigraphic record of ice-sheet grounding and retreat in this region [*Stuiver et al.*, 1981; *Denton et al.*, 1991; *Baroni and Orombelli*, 1991; *Hall and Denton*, in press]. These studies indicated that the ice sheet was grounded on this portion of the shelf between 21,200 and 17,000 yr B.P. [*Stuiver et al.*, 1981; *Denton et al.*, 1991; *Hall and Denton*, in press]. Glaciovolcanic deposits at ~ 350 m above the present ice surface at Mount Takahe yielded 40Ar/39Ar ages of 29,000 to 12,000 yr B.P. and record the maximum glaciation in this region. A parasitic cinder cone dated at 34,000 ± 8,000 yr B.P. on the west flank of Mount Frakes records ice elevations ~ 100 to 150 m above the present ice sheet [*Wilch*, 1997].

Cores collected south of the grounding-zone wedge (near Coulman Island) in the western Ross Sea yielded ages that range from ~18,000 to ~11,000 yr B.P. [*Anderson et al.*, 1992; *Licht et al.*, 1996; *Domack et al.*, 1999]. Cores collected north of the grounding-zone wedge penetrated glacial-marine deposits and bioclastic carbonates (on banks) that are older than 22,000 yr B.P. [*Anderson et al.*, 1992; *Taviani et al.*, 1993; *Licht et al.*, 1996].

By 11,500 yr B.P. the grounding line in the western Ross Sea retreated south to the vicinity of the Drygalski Ice Tongue [*Baroni and Orombelli*, 1991]. *Shipp* [1999] interpreted corrugated moraines on the western shelf (Figure 5b) as marking annual retreat positions of the grounding line, which yielded an annual retreat rate of 50 m/yr. This rate is consistent with that derived from radiocarbon dates. By ~7,000 yr B.P. the grounding line reached a position near Ross Island [*Stuiver et al.*, 1981; *Denton et al.*, 1991]. *Denton et al.* [1991] argued for completion of the Holocene grounding line recession in the western Ross Sea region between ~ 7,000 and 5,000 yr B.P. The grounding line retreated approximately 1000 km during this final stage of retreat. This is likely the time when the deep iceberg furrows that overprint other recessional geomorphic features where found.

Radiocarbon ages from reworked organic matter in tills on the central Ross Sea shelf have yielded uncorrected ages ranging from 30,000 to 34,000 yr B.P. [*Domack et al.*, 1999]. A large hiatus in radiocarbon ages between 23,000 and 29,000 yr B.P. (uncorrected ages) is believed to mark ice-sheet grounding on the continental shelf (excluding the northwestern Ross Sea), hence excluding formation of

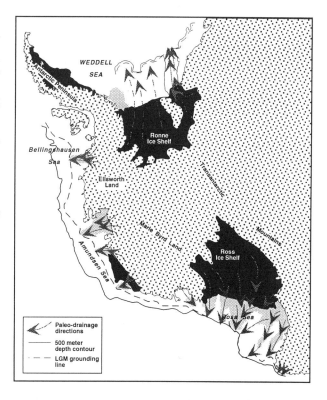

Fig. 4. Ice Sheet reconstructions for the Last Glacial Maximum. Large glacial troughs (shaded) occur offshore of major drainage outlets (Figure 2) and mark the locations of paleo-ice streams.

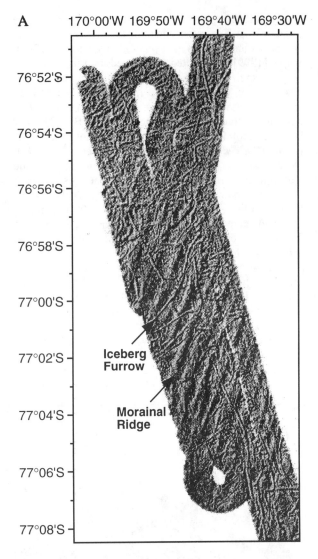

any organic carbon during this time [*Domack et al.*, 1999].

The retreat of the ice sheet from the central portion of the Ross Sea continental shelf, which receives drainage exclusively from West Antarctica, appears to have lagged behind the retreat in the western Ross Sea [*Domack et al.*, 1999; *Shipp et al.*, 1999]. Seismic records show morainal ridges that prograded from Iselin Bank (the paleodrainage divide between the EAIS and WAIS in the Ross Sea) toward the west and downlapped onto older subglacial and glacial-marine deposits to the west [*Shipp et al.*, 1999]. This is further evidence that the EAIS-nourished portion of the western Ross Sea ice sheet retreated first from the shelf, creating an embayment in the ice front, although the timing of this retreat could have been during an earlier glaciation. Differences in the number and spacing of grounding-zone wedges within each of the major troughs indicates that each of the major drainage outlets and associated ice streams behaved independently during the retreat of the ice sheet.

During the final stage of ice-sheet retreat, an extensive ice shelf grounded on shallow banks in the Ross Sea. This is indicated by the flat-topped nature of the banks and pinch-out of glacial units onto the banks [*Shipp et al.*, 1999], an apparent hiatus of ~15,000 yr B.P. in radiocarbon dates from the outer shelf [*Licht et al.*, 1996], and by sedimentary facies in piston cores collected from the outer shelf [*Domack et al.*, 1999a,b]. Radiocarbon ages on sub-ice-shelf deposits and the diatomaceous glacial-marine sediments that overly them indicate that retreat of the ice shelf in the western Ross Sea occurred at an average rate of 100m/yr. *Domack* and colleagues [1999b] recall that corrugation moraine in western Ross Sea indicate grounding line retreat rates of 50 m/yr. Hence, the ice shelf and grounding line appear to have retreated at approximately the same rate, at least during the early stages of retreat. However, deep iceberg furrows superimposed on backstepping moraines and wedges (Figure 5a) imply that the ice shelf eventually disintegrated into substantial blocks that gouged the sea floor.

Amundsen-Bellingshausen Seas

During the 1999 austral summer, marine geological surveys focused on major drainage outlets of the WAIS

Fig. 5. Examples of geomorphic features formed by the retreating ice sheet: (a) multibeam mosaic showing iceberg furrows superimposed on morainal ridges; and (b) side-scan sonar record showing backstepping, corrugated moraines. Sediment cores from both sets of features sampled till overlain by a thin layer of diatomaceous glacial marine sediments.

between the Ross Sea and the Antarctic Peninsula (Figure 2). These investigations included a regional bathymetric and swath bathymetry survey along the outer shelf of the Amundsen-Bellingshausen margin to map troughs. The modern drainage outlets of Marie Byrd Land and Ellsworth Land have large glacial troughs that extend across the inner shelf (Figure 4). No other large troughs were identified along the Marie Byrd Land/Amundsen-Bellingshausen margin. Swath bathymetry records from these troughs show striations, flutes, MSGL's and drumlins that provide the basis for the preliminary LGM reconstruction shown in Figure 4. Ongoing research focuses on more accurately defining the maximum grounding line position and examining the ice-sheet retreat history from the individual drainage basins. It is anticipated that the different paleo-ice streams have varying retreat histories, given their diversity of sizes and locations.

Pine Island Bay is unique among all the drainage systems in that the sea floor is riddled with subglacial channels (Figure 6). The presence of these channels indicate that meltwater flowed beneath the ice sheet that once occupied the bay. Indeed, meltwater may have contributed to the retreat of the expanded ice sheet. An apparent absence of morainal ridges and wedges in the bay implies rapid retreat. *Kellogg and Kellogg* [1987] examined seismic records and piston cores from Pine Island Bay and the adjacent shelf and suggested that the ice sheet grounded on the outer shelf during the LGM. Subglacial geomorphic features occur only to the mouth of the bay, so the exact grounding line position remains problematic. To date, a paucity of calcareous and organic material in sediment cores has prevented radiocarbon age dating of Pine Island Bay deposits.

Weddell Sea

Carrara [1981] conducted field studies along the Orville Coast of the Antarctic Peninsula region and presented evidence that the ice sheet once stood 450 m higher than at present, possibly during the LGM. In the northern Pensacola Mountains, glacial erratics occur 400 m above the present ice-sheet surface. *Waitt* [1983] examined nunataks in the Lassiter Coast of Southern Palmer Land and noted striations and erratics up to 500 m higher than the present ice elevation. He projected the ice-sheet grounding line nearly 800 km north of its present position in the Weddell embayment. *Denton et al.* [1992] observed a former trimline in the Ellsworth Mountains about 400 to 650 m above the present ice surface. The age of these features remains problematic. *Bentley and Anderson* [1998] reviewed the onshore and offshore data from the Weddell Sea region and generated an LGM ice-sheet reconstruction.

The extent of the grounded ice sheet on the southern Weddell Sea continental shelf, where the WAIS would

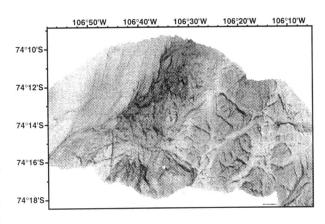

Fig. 6. Multibeam mosaic showing subglacial meltwater channels on the Pine Island Bay seafloor.

have grounded, is poorly constrained. Piston cores from Crary Trough sampled till, indicating that an ice sheet grounded in the trough [*Anderson et al.*, 1980; 1991]. Short sediment cores collected on the shelf west of Crary Trough during a 1968 *Glacier* cruise did not penetrate tills. Cores acquired during several expeditions of the R/V *Polarstern* penetrated "stiff" diamictons, interpreted as till, along the western front of the Ronne Ice Shelf [*Fütterer and Melles*, 1990]. No detailed sedimentological information has been published on these cores. A 3.5-kHz subbottom profiler record collected along the front of the ice shelf showed hummocky seafloor topography, typical of glaciated seafloor and similar to seafloor morphology in the Ross Sea.

Attempts to radiocarbon date glacial-marine sediments on the western Weddell Sea shelf have been hampered by a lack of carbonate material [*Bentley and Anderson*, 1998], hence the ice-sheet retreat history remains virtually unknown. The only radiocarbon dates that help constrain the timing of retreat of the WAIS from the shelf are from core 2-19-1, collected on the outer shelf, just west of Crary Trough. This core sampled disturbed (iceberg disturbed) glacial-marine sediment that indicates the ice sheet retreated from the outer shelf prior to 14,000 yr B.P [*Bentley and Anderson*, 1998].

CONCLUSIONS

1. Geological evidence from onshore and offshore indicates that mountain glaciers and ice caps existed in West Antarctica during the Oligocene. In the Ross Sea, a number of different ice caps rested on small continental blocks and islands, separated by seaways that occupied rift basins. With time, these seaways filled with sediment, allowing the WAIS to spread across the West Antarctic continent and adjacent

continental shelf. By the late Miocene, the WAIS reached present proportions. At times it advanced across the Ross Sea and Weddell Sea continental shelves. The ice sheet continued to advance and retreat across the continental shelf during the Plio-Pleistocene, but it is not known whether these oscillations were caused by climatic events in Antarctica or were a response to rising and falling sea level caused by Northern Hemisphere glaciations.

2. There is evidence that the EAIS advanced onto the continental shelf prior to the LGM. Indeed, it appears to have retreated from the shelf during Oxygen Isotope Stage 3. Retreat of the EAIS from the shelf in western Ross Sea and eastern Weddell Sea may have contributed to instability of the WAIS by creating embayments in the ice sheet margin.

3. During the LGM, the WAIS advanced onto the continental shelf around West Antarctica. The LGM configuration for the Ross Sea is well constrained, but the maximum extent of grounded ice elsewhere on the shelf remains problematic. The expanded ice sheet included large ice streams.

4. In Ross Sea the ice sheet is interpreted to have rested on a relatively thick deforming bed, at least during the late stages of advance. The deforming bed is much thicker on the outer shelf where the ice flowed over unconsolidated sedimentary strata. It is thin to absent on the inner shelf where sedimentary strata are thin to absent. Ice stream boundary ridges, backstepping moraines on the flanks of banks, and superimposed bi-directional glacial lineations suggest the laterally shifting nature of ice streams in central and eastern Ross Sea.

5. Initial deglaciation of the central and eastern Ross Sea shelf began shortly after the LGM. Ice-sheet and ice-shelf retreat from the exposed continental shelf (seaward of the present ice shelf) was relatively continuous with average rates of 50 and 100 m/yr, respectively. There is no evidence in western Ross Sea for ice-sheet decoupling and collapse during this phase of retreat.

6. Differences in the number and locations of recessional grounding line features from trough to trough indicates that each of the major ice streams behaved independently during their retreat from the shelf.

7. The combined onshore and offshore data from Ross Sea indicate that the WAIS grounding line retreated from a position near the present ice-shelf edge to its present location, a distance of up to 1000 km, since about 7,000 yr B.P. Large iceberg furrows on the inner shelf extend to water depths of just over 700 meters and display consistent offshore directions, further indicating that a late episode of retreat occurred when the calving margin of the ice sheet had little or no ice shelf. If the WAIS experienced collapse, it was during this phase of retreat.

8. Pine Island Bay is unique to other areas that were studied in that there is evidence for significant subglacial meltwater beneath the grounded ice sheet during the LGM. There is also an apparent absence of recessional geomorphic features. This implies rapid retreat of the ice sheet from the bay.

Acknowledgments. The research presented in this manuscript is a summary of the efforts of many individuals. The authors extend their appreciation for several pleasant and productive cruises to the Antarctic Support Associates personnel, the crew of the RV/IB *Nathaniel B. Palmer*, and the many students who have assisted in data collection and processing. Dale Chayes, Rob Haag, and Suzanne O'Hara of Lamont Doherty Geological Observatory and Michelle Fassell of Rice University assisted in the processing of multibeam data. Gratitude is extended to Tom Janacek and Matt Curren of the Florida State University Antarctic Research Facility for curating the 1994, 1995, 1998, and 1999 core databases. Thank you to Breezy Long, Julia Smith Wellner, and Ashley Lowe for assistance with the figure presentation. This research was funded by the National Science Foundation, Office of Polar Programs grant numbers DPP 9119683 and OPP-9527876 to John Anderson.

REFERENCES

Abelmann, A., R. Gersonde, and V. Spiess, Pliocene-Pleistocene paleoceanography in the Weddell Sea–siliceous microfossil evidence, in *Geological History of the Polar Oceans: Arctic Versus Antarctic*, edited by U. Bleil and J. Thiede, pp. 729–759, Kluwer Academic Publishers, Boston, 1990.

Abreu, V. S., and J. B. Anderson, Glacial eustasy during the Cenozoic: sequence stratigraphic implications, *Amer. Assoc. Petroleum Geologists Bull.*, 82, 1385-1400, 1998.

Alley, R. B., D. D. Blankenship, S. T. Rooney, and C. R. Bentley, Sedimentation beneath ice shelves: the view from ice stream B, *Marine Geology*, 85, 101-120, 1989.

Alonso, B., J. B. Anderson, J. I. Díaz, and L. R. Bartek, Pliocene-Pleistocene seismic stratigraphy of the Ross Sea: evidence for multiple ice sheet grounding episodes, in *Contributions to Antarctic Research III: Antarctic Res. Ser.*, vol. 57, edited by D. H. Elliot, pp. 93–103, AGU, Washington, D.C., 1992.

Anderson, J. B., *Antarctic Marine Geology*, Cambridge University Press, London, 289 pp., 1999.

Anderson, J. B., and J. T. Andrews, Radiocarbon constraints on ice sheet advance and retreat in the Weddell Sea, Antarctica, *Geology*, 27, 179-182, 1999.

Anderson, J. B., B. A. Andrews, L. Bartek, and E. M. Truswell, Petrology and palynology of Weddell Sea glacial sediments: implications for subglacial geology, in *Geological Evolution of Antarctica*, edited by M. R. A. Thomson, J. A. Crame, and J. W. Thomson, pp. 231–235, Cambridge University Press, New York, 1991.

Anderson, J. B., and L. R. Bartek, Cenozoic glacial history of the Ross Sea revealed by intermediate resolution seismic reflection data combined with drill site information, in *The Antarctic Paleoenvironment: A Perspective on Global Change I. Antarctic Res. Ser.*, vol. 56, edited by J. P. Kennett and D. A. Warnke, pp. 231–263, AGU, Washington, D.C., 1992.

Anderson, J. B., C. F. Brake, and N. C. Myers, Sedimentation on the Ross Sea continental shelf, Antarctica, *Mar. Geol.*, 57, 295–333, 1984.

Anderson, J. B., D. D. Kurtz, E. W. Domack, and K. M. Balshaw, Glacial and glacial marine sediments of the Antarctic continental shelves, *J. Geol.*, 88, 399–414, 1980.

Anderson, J. B., S. S. Shipp, L. R. Bartek, and D. E. Reid, Evidence for a grounded ice sheet on the Ross Sea contiental shelf during the late Pleistocene and preliminary paleodrainage reconstruction, in *Contributions to Antarctic Res. Ser. III*, vol. 57, edited by D. H. Elliot, pp. 39-42, AGU, Washington, D.C., 1992.

Baroni, C., and G. Orombelli, 1981, Holocene raised beaches at Terra Nova Bay, Victoria Land, Antarctica, *Quat. Res.*, 36, 157–177, 1991.

Barrett, P. J., Textural characteristics of Cenozoic preglacial and glacial sediments at Site 270, Ross Sea, Antarctica, in *Initial Reports of the Deep Sea Drilling Project*, vol. 28, edited by D. E. Hayes, L. A. Frakes, pp. 757–767, U.S. Government Printing Office, Washington D.C., 1975.

Barrett, P. J., Antarctic paleoenvironment through Cenozoic times: a review, *Terra Antartica*, 3, 103–119, 1997.

Barrett, P. J., C. J. Adams, W. C. McIntosh, C. C. Swisher III, and G. S. Wilson, Geochronological evidence supporting Antarctic deglaciation three million years ago, *Nature*, 359, 816–818, 1992.

Barrett, P. J., M. J. Hambrey, D. M. Harwood, A. R. Pyne, and P. N. Webb, Synthesis, in Antarctic Cenozoic History from the CIROS-1 Drillhole, McMurdo Sound, *DSIR Bull.* 245, edited by P. J. Barrett, pp. 241-251, DSIR Publishing, Wellington, New Zealand.

Barron, J. A., Diatom constraints on the position of the Antarctic Polar Front in the middle part of the Pliocene, *Mar. Micropaleontol.*, 27, 195–213, 1996.

Bart, P. J., M. De Batist, and H. Miller, Neogene collapse of glacially-deposited, shelf-edge deltas in the Weddell Sea: relationships between deposition during glacial periods and sub-marine fan development, in *The Antarctic Continental Margin. Geophysical and Geological Stratigraphic Records of Cenozoic Glaciation, Paleoenvironments and Sea-Level Change*, A. K. Cooper, P. F. Barker, P. N. Webb, and G. Brancolini, pp. 317–318, *Terra Antartica*, vol. 1, 1994.

Bart, P. J., M. De Batist, and W. Jokat, Interglacial collapse of Crary Trough Mouth Fan, Weddell Sea, Antarctica: implications for Antarctic glacial history analysis, *J. Sediment. Res.*, in press, a.

Bart, P.J., J.B. Anderson, F. Trincardi, and S.S. Shipp, Manifestation of the East Antarctic Ice Sheet Late Neogene grounding events with Victoria Trough Mouth Fan, northwestern Ross Sea, *Marine Geology*, in press, b.

Bentley, M. J., and J. B. Anderson, Glacial and marine geological evidence for the ice sheet configuration in the Weddell Sea-Antarctic Peninsula region during the last glacial maximum, *Ant. Sci.*, 10, 309–325, 1998.

Bentley, C. R., and N. A. Ostenso, Glacial and subglacial topography of West Antarctica, *J. Glaciol.*, 3, 882–911, 1961.

Bloom, A. L, W. S. Broecker, J. Chappell, R. K. Matthews, and K. J. Mesolella, Quaternary sea level fluctuations on a tectonic coast: new $^{230}Th/^{234}U$ dates from the Huon Peninsula, New Guinea, *Quat. Res.*, 4, 185–205, 1974.

Burckle, L. H., and J. Cirilli, Origin of diatom ooze belt in the Southern Ocean; implications for late Quaternary paleoceanography, Micropaleontology, 33, 82-86, 1987.

Calkin, P. E., Glacial geology of the Mount Gran area, southern Victoria Land, Antarctica, *Geolog. Soc. Am. Bull.*, 75, 1031–1036, 1964.

Carrara, P., Evidence for a former large ice sheet in the Orville Coast-Ronne Ice Shelf area, Antarctica, *J. Glaciol.*, 27, 487–491, 1981.

Ciesielski, P. F., and F. M. Weaver, Early Pliocene temperature changes in the Antarctic seas, *Geology*, 2, 511–515, 1974.

Clark, C.D., Glaciaodynamic context of subglacial bedform generation and preservation, *Annals of Glaciol.*, 28, in press.

Colhoun, E. A., Geological evidence for changes in the East Antarctic Ice Sheet [60 degrees–120 degrees E] during the last glaciation, *Polar Rec.*, 27, 345–355, 1991.

Cooper, A. K., P. J. Barrett, K. Hinz, V. Traube, G. Leitchenkov, and H. M. J. Stagg, Cenozoic prograding sequences of the Antarctic continental margin: a record of glacio-eustatic and tectonic events, *Mar. Geol.*, 102, 175–213, 1991.

Denton, G. H., R. P. Ackert, M. L. Prentice, and N. Potter Jr., Ice-sheet overriding of the ice-free valleys of southern Victoria Land, *Ant. J. U.S.*, 19, 47–48, 1984.

Denton, G. H., R. L. Armstrong, and M. Suiver, The late Cenozoic glacial history of Antarctica: *The Late Cenozoic Gglacial Ages*, Yale University Press, New Haven-London, pp. 267-306, 1971.

Denton, G. H., Bockheim, J. G., Rutford, R. H., and B. G. Anderson, Glacial history of the Ellsworth Mountains, West Antarctica, in *Geology and Paleontology of the Ellsworth Mountains, West Antarctica*, edited by G. F. Webers, C. Craddock, and J. F. Splettstoesser, pp. 403–432, 1992, Geological Society of America Memoir 170, Boulder, Colorado.

Denton, G. H., M. L. Prentice, and L. H. Burckle, Cainozoic history of the Antarctic Ice Sheet, in *The Geology of Antarctica*, edited by R.J. Tingey, pp. 365-433, Clarendon Press, Oxford, England, 1991.

De Batist, M., P. Bart, B. Kuvaas, A. Moons, and H. Miller, Detailed seismic stratigraphy of the Crary Fan, southeastern Weddell Sea, in The Antarctic Continental Margin: Geophysical and Geological Stratigraphic Records of Cenozoic Glaciation,

Paleoenvironments and Sea-Level Change, edited by A. K. Cooper, P. F. Barker, P. N. Webb, and G. Brancolini, pp. 321–323, *Terra Antarctica*, vol. 1, 1994.

De Santis, L., J. B. Anderson, G. Brancolini, and I. Zayatz, Seismic record of late Oligocene through Miocene glaciation on the central and eastern continental shelf of the Ross Sea, in *Geology and Seismic Stratigraphy of the Antarctic Margin, Antarctic Res. Ser.* vol. 68. edited by A. K. Cooper, P. F. Barker, and G. Brancolini, pp. 235–260, AGU, Washington D.C., 1995.

De Santis, L., S. Prato, G. Brancolini, M. Lovo, and T. Luigi, The eastern Ross Sea continental shelf during the Cenozoic: implications for the West Antarctic Ice Sheet development, in *Global and Planetary Change, Spec. Issue, Lithosphere dynamics and environmental change of the Cenozoic West Antarctic Rift System*, edited by F. M. Van der Wateren and S.A.P.L. Cloetingh, in press.

Domack, E. W., P. O'Brien, P. Harris, F. Taylor, P. G.Quilty, L. De Santis, and B. Raker, Late Quaternary sediment facies in Prydz Bay, East Antarctica and their relationship to glacial advance onto the continental shelf, *Antarctic Sci.*, 10, 236-246, 1999a.

Domack, E. W., E. A., Jacobson, S. S., Shipp, and J. B. Anderson, Late Pleistocene/Holocene retreat of the West Antarctic ice sheet in the Ross Sea: Part 2 - sedimentologic and stratigraphic signature, Geological Society of America Bulletin, 1999b.

Drewry, D. J., Radio echo sounding map of Antarctica ~90 E - 180, *Polar Rec.*, 17, 359-374, 1975.

Drewry, D. J., Late Wisconsin reconstruction for the Ross Sea region, Antarctica, *J. Glaciol.*, 18, 231-244.

Elverhoi, A., Evidence for a late Wisconsin glaciation of the Weddell Sea, *Nature*, 293, 641-642, 1981.

Emiliani, C., Interglacial high sea levels and the control of Greenland ice by the precession of the equinoxes, *Science*, 166, 1503–1504, 1969.

Fütterer, D. K., and M. Melles, Sediment patterns in the southern Weddell Sea: Filchner Shelf and Filchner Depression, in *Geologic History of the Polar Oceans: Arctic versus Antarctic*, edited by U. Bleil and J. Thiede, Kluwer Academic Publishers, Boston, pp. 381–401, 1990.

Hambrey, M. J., Cenozoic sedimentary and climatic record, Ross Sea region, Antarctica, in *The Antarctic Paleoenvironment: A Perspective on Global ChangeII, Antarctic Research Series*, vol. 60, edited by J. P. Kennett and D. A. Warnke, pp. 91–124, AGU, Washington, D.C., 1993.

Hilfinger, M., J., Franceshini, and E.W. Domack, Chronology of glacial marine lithofacies related to the recession of the West Antarctic ice sheet in the Ross Sea, *Ant. J. U.S.*, 30, 82-84, 1995.

Hughes, T., Deluge II and the continent of doom: rising sea level and collapsing Antarctic ice, *Boreas*, 16, 89–100, 1987.

Huybrechts, P., Antarctic ice sheet during the last Glacial-interglacial cycle: a three-dimensional experiment, *Annals of Glacialogy*, 14, 115-119, 1990.

Jahns, E., Evidence for a fluidized till deposit on the Ross Sea continental shelf, *Ant. J. U.S.*, 29, 139–141, 1994.

Kaul, N., High resolution seismics and stratigraphy off Kapp Norvegia, Antarctica, *Zeitschrift für Geomorphologie*, pp. 105–112, 1992.

Kellogg, D. E., and T. B. Kellogg, Microfossil distributions in modern Amundsen Sea sediments, *Marine Micropaleontol.*, 12, 203–222, 1987.

Kellogg, T. B., R. S. Truesdale, and L. E. Osterman, Late Quaternary extent of the West Antarctic Ice Sheet: new evidence from Ross Sea cores, *Geology*, 7, 249–253, 1979.

Kyle, P. R., Glacial history of the McMurdo Sound area as indicated by the distribution and nature of McMurdo Volcanic Group rocks, in *Dry Valley Drilling Project, Antarctic Research Series*, vol. 33, edited by L. D. McGinnis, pp. 403–412, AGU, Washington D.C., 1981.

LeMasurier, W. E., and D. C. Rex, Rates of uplift and the scale of ice level instabilities recorded by volcanic rocks in Marie Byrd Land, West Antarctica, in *Antarctic Earth Science*, edited by R. L. Oliver, P. R. James, and J. B. Jago, pp. 663–670, Cambridge University Press, New York, 1983.

Licht, K. J., N. W. Dunbar, J. T. Andrews, and A. E. Jennings, Distinguishing subglacial till and glacial marine diamictons in the western Ross Sea, Antarctica: implications for last glacial maximum grounding line, *Geo. Soc. Am. Bull.*, 91-103, 1999

Licht, K. J., A. E. Jennings, J. T. Andrews, and K. M. Williams, Chronology of late Wisconsin ice retreat from the western Ross Sea, Antarctica, *Geology*, 24, 223–226, 1996.

Marchant, D. R. and G. H. Denton, Miocene and Pliocene paleoclimate of the Dry Valleys region, southern Victoria Land; a geomorphological approach, *Marine Micropaleontol.*, 27, 253–271, 1996.

Mayewski, P. A., Glacial geology and late Cenozoic history of the Transantarctic Mountains, Antarctica, Ohio State University Institute of Polar Studies Report, 168 pp, 1975.

Mercer, J. H., Antarctic ice and Sangamon sea level. *Commission of Snow and Ice Reports and Discussions*, Publication, no. 79, De L'Association Internationale D'Hydrologie Scientifique, Gentbrugge, Belgium, pp. 217–225, 1968.

Mercer, J. H., Glacial development and temperature trends in the Antarctic and in South America, in *Antarctic Glacial History and World Paleoenvironments*, edited by E. M. van Zinderen Bakker, pp. 73–93, A. A. Balkema, Rotterdam, Netherlands, 1978.

Mildenhall, D., Terrestrial palynology, in *Antarctic Cenozoic History from the CIROS-1 Drillhole, McMurdo Sound*, vol. 245, edited by P. Barrett, pp. 119–127, Deptartment of Scientific and Industrial Research Bulletin, Wellington, 1989.

Miller, K. G., J. D. Wright, and R. G. Fairbanks, Unlocking the ice house: Oligocene-Miocene oxygen isotopes, eustasy, and margin erosion, *J. Geophys. Res.*, 96, 6829–6848, 1991.

Mitchum, R. M. Jr., J. B. Sangree, P. R. Vail, and W. W. Wornardt, Recognizing sequences and systems tracts from well logs, seismic data, and biostratigraphy; examples from the late Cenozoic of the Gulf of

Mexico, *Amer. Assoc. Petroleum Geologists Mem.* 58, 163–197, 1994.

Moons, A., M. De Batist, J. P. Henriet, and H. Miller, Sequence stratigraphy of the Crary Fan, southeastern Weddell Sea, in *Recent Progress in Antarctic Earth Science*, edited by Y. Yoshida, K. Kaminuma, and K. Shiraishi, pp. 613–618, Terra Scientific Publishing Co., Tokyo, Japan, 1992.

Myers, N. C., Petrology of Ross Sea basal tills: implications for Antarctic glacial history, *Ant. J. U.S.*, 17, 123–124, 1982.

Petit, J. R., I. Basile, A. Leruyuet, D. Raynaud, C. Lorius, J. Jouzel, M. Stievenard, V. Y. Lipenkov, N. I. Barkov, B. B. Kudryashov, M. Davis, E. Saltzman, V. Kotlyakov, Four climate cycles in Vostok ice core, *Nature*, 387, 359, 1997.

Prentice, M. L. and J. L. Fastook, Late Neogene Antarctic ice sheet dynamics; ice model-data convergence. EOS [Transactions of the American Geophysical Union], 71, 1378, 1990.

Rutford, R. H., C. Craddock, C. M. White, and R. L. Armstrong. Tertiary glaciation in the Jones Mountains. In *Antarctic Geology and Geophysics*, edited by R. J. Adie, International Union of Geological Sciences, Series B, vol. 1. Universitetsforlaget, Oslo, Norway, pp. 239–243, 1972.

Scherer, R. P., A. Aldahan, S. Tulaczyk, G. Possnert, H. Engelhardt, and B. Kamb, Pleistocene collapse of the West Antarctic Ice Sheet. *Science*, 281, 82–85, 1998.

Shipp, S. S. Retreat History of the Last Glacial Maximum Ice Sheet in Ross Sea, Antarctica, Unpublished Ph.D. Dissertation, Rice University, Houston, Texas, 296 p, 1999.

Shipp, S. S., J. B. Anderson, and E. W. Domack, Seismic signature of the late Pleistocene fluctuations of the West Antarctic ice sheet system in Ross Sea: A new perspective, Part 1. *Geol. Soc. Amer. Bull.* 1999.

Stokes, C. R., and Clark, C. D., Geomorphological criteria for identifying Pleistocene ice streams, in press.

Stuiver, M., G. H. Denton, T. J. Hughes, and J. L. Fastook. History of the marine ice sheet in West Antarctica during the last glaciation: a working hypothesis, in *The Last Great Ice Sheets, edited by* G. H. Denton and T. J. Hughes, pp. 319–436 John Wiley & Sons, New York, 1981.

Stump, E., M. F. Sheridan, S. G. Borg, and J. F. Sutter. Early Miocene subglacial basalts, the East Antarctic ice sheet, and uplift of the Transantarctic Mountains, *Science*, 207, 757–759, 1980.

Sugden, D. E., D. R. Marchant, and G. H. Denton, The case for a stable East Antarctic ice sheet; the background. *Geografiska Annaler*, 75A, 151–155, 1993.

Taviani, M., D. E. Reid, and J. B. Anderson, Skeletal and isotopic composition and paleoclimatic significance of Late Pleistocene carbonates, Ross Sea, Antarctica. *J. Sediment. Petrol.*, vol. 63, pp. 84–90, 1993.

Waitt, T. B., Thicker West Antarctic ice sheet and peninsula icecap in the late–Wisconsin time-sparse evidence from northern Lassiter Coast. *Antarctic J. of the U. S.*, 18, 91-93, 1983.

Webb, P. N., and D. M. Harwood, Late Cenozoic glacial history of the Ross embayment, Antarctica. *Quat. Sci. Rev.*, 10, 215–223, 1991.

Wilch, T. I., Volcanic record of the West Antarctic Ice Sheet in Marie Byrd Land, Unpublished PhD Dissertation. New Mexico Institute of Mining and Technology, Socorro, New Mexico, 1997.

Yoshida, Y., Physiography of the Prince Olav and Prince Harald Coasts, East Antarctica. *Memoirs of the NIPR Research*, Series C, Earth Sciences, vol. 13, 83 pp., 1983.

J. B. Anderson and S. S. Shipp, Department of Geology and Geophysics, Rice University, 6100 South Main, Post Office Box 1892, Houston, TX 77251-1892

THE GLACIAL GEOLOGIC TERRESTRIAL RECORD FROM WEST ANTARCTICA WITH EMPHASIS ON THE LAST GLACIAL CYCLE

Harold W. Borns, Jr.

*Institute for Quaternary Studies and Department of Geological Sciences,
University of Maine, Orono, ME*

The terrestrial glacial geologic record of the last glacial cycle in West Antarctica is restricted to the foothills of the Transantarctic Mountains bordering the western Ross Sea, the coastal low areas and mountains and the interior nunataks. This record consists of glacially eroded bedrock outcrops, glacial deposits and glacial-volcanic hyaloclastites. By far the most complete record to date has been reported from foothills of the Transantarctic Mountains. The remainder of the record in West Antarctica is sparse and difficult to interpret because of its nature, wide distribution, lack of adequate chronological control and the regional effects of tectonism. However, available data indicates that at the last glacial maximum the level of the ice sheet surface was higher, at least as recorded at several locations in Marie Byrd Land. This evidence is consistent with the glaciomarine record from adjacent areas of the continental shelf which infers a more expansive ice sheet than at present.

INTRODUCTION

The terrestrial glacial geologic record in West Antarctica is preserved along the flanks of the Transantarctic Mountains abutting the western Ross Sea, on the mountains and coastal lowlands of Marie Byrd Land and Ellsworth Land, and on interior nunataks (Figure 1).

The earliest glacial geologic observations were made during the British Antarctic Expeditions of Scott and Shackleton near the turn of the century followed by renewed, but limited, activity during the Byrd expeditions of the late 1920s. After a lapse of about 40 years the significant expansion of research began during the International Geophysical Year (IGY) 1957-58 and has continued since then.

Renewed and increasing interest since the IGY has focused on West Antarctica since Mercer pointed out in 1978 that the history of the marine-based West Antarctic Ice Sheet [WAIS] suggests that the ice sheet may be unstable in the face of anthropogenic-induced global warming [Figure 1]. Following this in 1992, MacAyeal further suggested that WAIS may also experience irregular, internally driven collapse. Given these possibilities, and the potential impact on global sea level and climate, knowledge of the past configurations and dynamic behavior of WAIS is necessary to calibrate glaciological models which provide a method for predicting future responses of the ice sheet so important to civilization as we know it. Here the glacial geological record of West Antarctica as it is presently understood is reviewed.

BACKGROUND

Ross [1847] reported the results of his mapping of the edge of the Ross Ice Shelf, which he called the Great Icy Barrier, in 1841 and 1842. It was remapped in 1902 during Scott's National Antarctic Expedition and the edge of the shelf seemed to be 24 to 32 km south of the position previously mapped by Ross, while another remeasurement in 1911 showed the Barrier front to be close to its 1902 position [Scott, 1915]. It is not clear

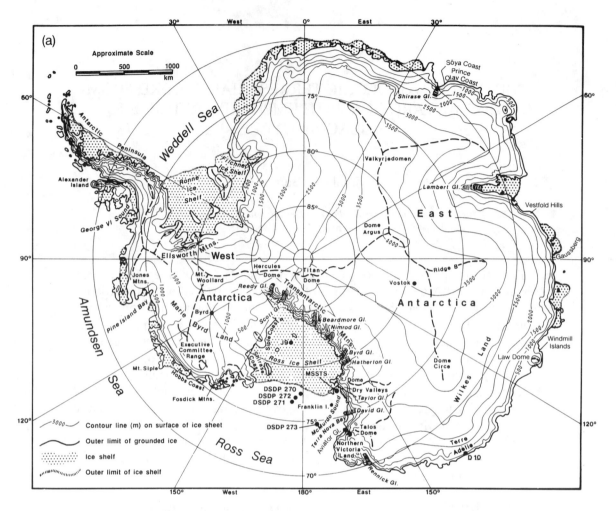

Fig. 1. Index map showing the present Antarctic Ice Sheet and its subglacial topography [Modified from Denton and others, 1991].

that Ross understood whether the barrier was afloat or grounded.

The earliest glacial geologic observations were made on Ross Island by members of the 1901-1903 National Antarctic Expedition and summarized by Scott [1905]. Among other things they determined that the Barrier was floating and that granite erratics were present high on the slopes of Mt. Erebus which Scott attributed to a more expansive and thicker "Barrier". In summarizing the former extent of the Antarctic Ice Sheet, represented by the changes in the position of the Barrier, he suggested the following:

> "when the Southern glaciation was at a maximum... The Great Barrier was a very different formation from what it is today... My opinion is that at or near the time of maximum glaciation the huge glacier, no longer able to float on a sea of 400 fathoms, spread out over the Ross Sea, completely filling it with an immense sheet of ice. At that time the edge of the sheet and the first place at which it could be water borne bordered on the ocean depths to the north of Cape Adare".

Furthermore Scott [1905] also postulated that the ice sheet in the Ross Sea became buoyant, broke away gradually and left the present remaining Barrier.

Geologists David and Priestley [1914] on Shackleton's 1907-09 British Antarctic Expedition, also hypothesized an expansion of the Ross Ice Barrier northward in the Ross Sea. They determined that the surface of the ice rose 305 m above present sea level based upon the distribution of erratics at Cape Royds foreign to Ross Island [Figure 2]. They also proposed that at least some of the "Ross Barrier" was composed of ice draining from outlet glaciers of the East Antarctic Ice Sheet along the west side of the Ross Sea which merged with the "Ross Barrier" and flowed northward.

In 1921 Debenham, geologist on Scott's British Antarctic Expedition of 1910-13, attributed the glacial deposits in McMurdo Sound area to local glaciation

Fig. 2. Granite boulder resting on basaltic volcanics at Cape Royds, Ross Island, Antarctica [Modified after Denton and others, 1981].

rather than to the widespread advance and grounding of the Great Ice Barrier proposed by David and Priestly [1914]. The difference in interpretation of the glacial deposits in McMurdo Sound area was not addressed until additional research was undertaken during and since the International Geophysical Year [IGY] program in Antarctica.

With the advent of the IGY, and continuing therefore, observations of glacial deposits have been made by many researchers. Much of this research has focussed primarily on the western Ross Sea area, including Ross and nearby islands, the Transantarctic Mountains, and to a lesser extent on the widely scattered nunataks and coastal mountains of West Antarctica (Figure 1).

Modern studies of glacial deposits, especially moraines, in the Transantarctic Mountains of South Victoria Land [Figure 1], demonstrate that outlet glaciers of the East Antarctic Ice Sheet flowed into the thick, grounded ice in the Ross Sea, where only the floating Ross Ice Shelf now exists [Mercer, 1968; Denton et al., 1975; Bockheim et al., 1989; Denton et al., 1989]. In the McMurdo Sound region, radiocarbon and cosmogenic surface exposure dates on drift deposited by grounded ice in the Ross Sea date to the last glacial maximum [LGM] [Denton et al., 1971; Hall and Denton, 1999; Shipp et al., 1998]. The distribution of tills and radiocarbon dates on marine organics in post-glacial sediments overlying the tills. [Licht et al., 1998; Licht et al., 1999] suggest that grounded ice extended to near the continental shelf about 74° S just south of Coulman Island at the LGM, over 1000 km seaward of the present grounding line of WAIS in the Ross Sea [Figure 3]. The work of Conway et al. [1999] based upon an interpretation of glacial, emerged marine deposits and radiocarbon dates from Terra Nova Bay to McMurdo Sound, indicates that the grounding line of WAIS has retreated nearly 1300 km since the LGM about 20,000 years before present, and retreated through McMurdo Sound between 8340-5730 yrs B.P. In the interior of West Antarctica glacial trimlines in the Ellsworth Mountains [Denton et al., 1991] indicate a thicker, more extensive WAIS at the LGM. Borns et al. [1996] and Ackert et al. [1999] report ice-sheet moraines 100 m above the ice sheet surface on Mt. Waesche in the Executive Committee Range [Figures 1, 9] that date to the LGM [Figure 1]. However, the first direct constraints on ice-surface elevations in interior West Antarctica were provided by total gas content of the Byrd Station ice core. This has been interpreted to indicate that ice-surface elevations at the LGM were 200-250 m lower than at present [Raynaud et al., 1979;

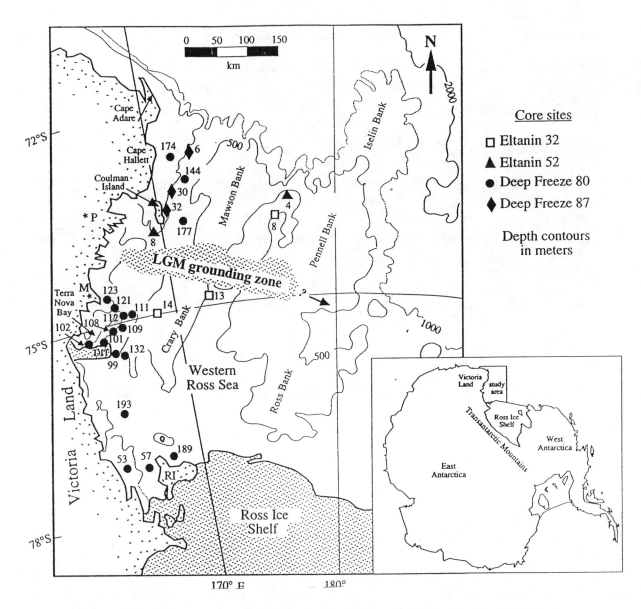

Fig. 3. Map of the western Ross Sea showing the location of 25 analyzed cores of glacial and glaciomarine sediments. The position of the grounding line of the WAIS at the last glacial maximum is shown at approximately 74° S latitude [Modified after Licht and others, 1999].

Raynaud *et al.*, 1982]. Resolution of all these seemingly ambivalent data sets has important implications for modeling the configuration and dynamics of WAIS at and since the LGM.

WESTERN ROSS SEA–TRANSANTARCTIC MOUNTAINS AREA

In the McMurdo Sound area Péwé, in 1960, attributed the glacial deposits to local outlet glaciers, in particular to the Koettlitz Glacier [Figure 4]. In doing so he agreed with Debenham's [1921] conclusions. Denton and others in 1981 concluded that during the latest glacial cycle ice, grounded in the Ross Sea deposited the Ross Sea Drift along the western coast of McMurdo Sound up to elevations ranging from 240 to 610 m a.s.l. in the foothills of the Transantarctic Mountains. The upper edge of the Ross Sea Drift in the foothills of the Royal Society Range is marked by a very prominent moraine throughout the area [Figures 5, 6]. This drift is bordered at many locations by older, more-weathered drift. The outer edge of the Ross Sea

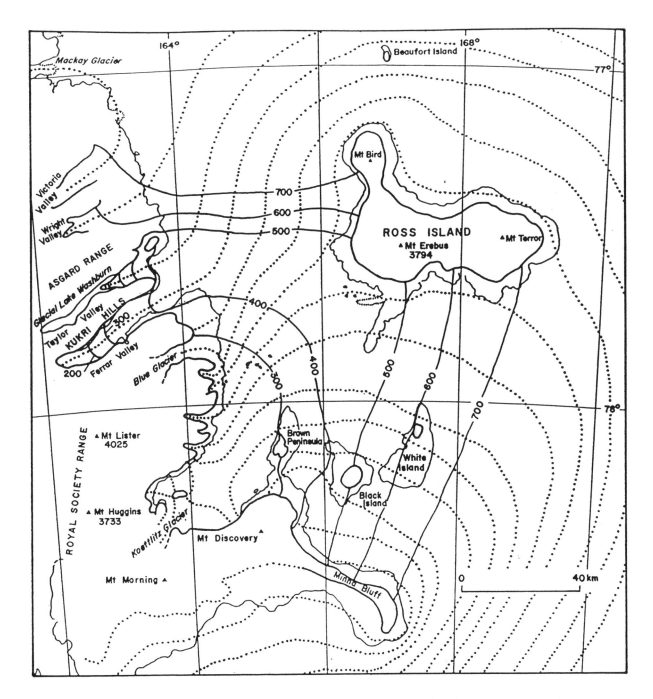

Fig. 4. Map of surface contours and flowlines of the ice sheet that deposited the Ross Sea drift [Modified after Denton and others, 1981].

Drift was deposited abut 17,000 to 21,200 ^{14}C yrs B.P. Denton et al. [1981] concluded the following. The lithologies and distribution of the Ross Sea Drift indicate that a grounded Ross Sea Ice Sheet occupied McMurdo Sound and that it received ice from two directions: 1. ice flowed into the sound from the southeast around Minna Bluff, 2. thick ice flowed around the northern end of Ross Island then eastward and southeastward into the sound [Figure 4]. At their maximum extent tongues of Ross Sea Ice extended westward into the ice-free valleys on the west coast of McMurdo Sound [Figure 4], and a maximum ice

Fig. 5. Aerial view of the upper extent of Ross Sea drift on the west coast of McMurdo Sound looking north [Modified after Denton and others, 1981].

thickness of 1325 m was achieved in McMurdo Sound. The fact that the ice flow was directed around Ross Island rather than northward implies that it was being buttressed by the grounded WAIS in the Ross Sea to the east.

In addition, the subsequent research of Hall and Denton [1998] examined the deglacial chronology of the McMurdo Sound area and reported that Ross Sea Ice blocked Taylor Valley between 8,340 and 23,800 ^{14}C yrs B.P. and that the ice maximum was achieved between 14,600 and 12,700 ^{14}C yrs B.P. and formed a moraine 500 m distance from its maximum position as late as 10,800 ^{14}C yrs B.P. This is based upon lacustrine algae that formed in ice-damned Glacial Lake Washburn.

COASTAL AREAS OF MARIE BYRD LAND AND ELLSWORTH LAND

The coastal mountain area of West Antarctica was discovered and examined by Admiral Byrd from the air in 1929, and he named the area Marie Byrd Land [Gould, 1931; Byrd, 1933]. During the second Byrd Antarctic Expedition of 1934 and 1935, the geography and glaciology were examined by two tractor parties and one sled party led by Paul Siple, in addition to four additional exploratory flights. Wade [1937] as a geologist on the sledding trip examined glaciology of the northeastern borderlands of the Ross Sea in King Edward VII Land and northwestern Marie Byrd Land. [Figure 7]. He reported many indications of former greater glacierization present in the Edsel Ford Range. This evidence consisted of striated and gouged rock surfaces on the heights of Mt. Grace McKinley and Garland Hershey Ridge well above the present glacier ice surface, and on the flat crest of Mt. Donald Woodward about 200 m above the present glacier surface [Byrd, 1935]. Presumably these surfaces were eroded by a higher WAIS. The most striking evidence of former glaciation is recorded on the Rea-Cooper Massif. Here remnants of alpine glacier valleys are present at many locations up to levels of 600 m above the present glacier surface. In addition, Wade also reports the presence of terraced moraines deposited at lower elevations presumably by the former alpine glaciers.

In 1978, Karlén and Melander reported patches of striated bedrock and roche mountnnees on nunataks in the Hobbs and Ruppert Coast areas [Figure 7]. In the

Fig. 6. Aerial view of dark, volcanic rich Ross Sea drift in the mouths of Garwood, Marshall and Meirs Valleys on the west coast of McMurdo Sound. The drift tongues were deposited by ice lobes of the grounded ice sheet in the Ross Sea flowing inland [Modified after Denton and others, 1981].

Hobbs coastal area the striae, which are above, but near the level of the present ice surface indicate former ice flow towards the north. On the Ruppert coast similar features indicate former ice flow towards the north and northwest. These data, coupled with the fresh appearance of the bedrock surfaces, suggest thicker ice during the most recent glacial cycle. They also reported that for the areas above the level of the exposed fresh appearing bedrock, the surfaces are badly weathered and striae are generally absent although some examples are presumed.

Richard and Luyendyk [1991] reported striae on the flat summit ridges between the Chester and Fosdick Mountains and in the eastern Fosdick Mountains in the Ford Ranges of Marie Byrd Land [Figure 8]. The striae in the Fosdick Mountains trend 150° - 160° and 115° - 120° on the nunataks between the Chester and Fosdick Mountains. They also recorded roches mountonnees on

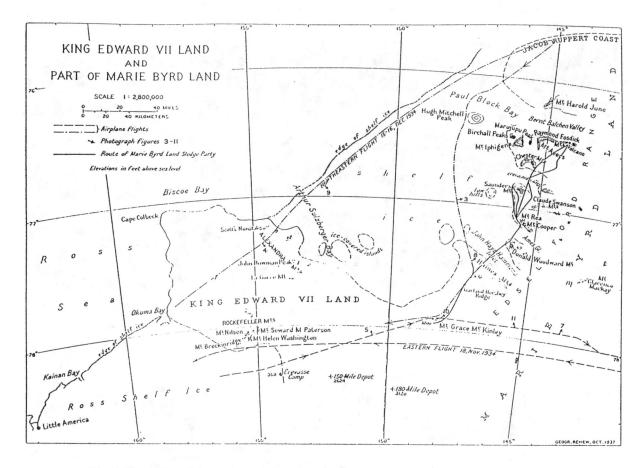

Fig. 7. Sketch map of the northeastern borderlands of the Ross Sea in King Edward VII Land and Marie Byrd Land [Modified after Wade, 1937].

the southern part of Bird Bluff indicating former ice flow towards the northwest. Modern alpine glaciers on the south flanks of the Fosdick Mountains now flow towards the south and southwest. The difference in flow directions between high and low elevations can be attributed to either a decrease in ice thickness or to a reoriented flow due to possible Neogene tectonic uplift of the ranges above the ice surface.

INTERIOR NUNATAKS

Introduction

The earliest observations of some of these areas was accomplished during the "over-the-snow traverses during the IGY emanating from Byrd Station in Marie Byrd Land [Doumani, 1970].

The terrestrial geology was observed on those traverses that passed interior nunataks and exposed bedrock areas in some of the coastal zone. In particular the Executive Committee Range Traverse [1959] and the Marie Byrd Land Traverse [1959-60; Antarctica – a map of the American Geog. Soc. of New York]. On these traverses considerable areas of exposed bedrock were examined. However, there are scattered observations relating to features that indicated possible higher ice sheet levels [Craddock and others, 1964]. More geological research was focused on this remote area following IGY by groups visiting the Ellsworth and Whitmore Mountains [Craddock and others, 1964]. Further major work in the Ellsworth Mountains was undertaken in 1979-80 seasons [Denton and others, 1992].

Since the IGY there has been ongoing research focussed upon determining the history of volcanism from the volcanoes in West Antarctica. During these studies observations related to higher levels of glaciation have been reported [e.g. LeMasurier and others, 1990]. More recently a reconnaissance research team visited the Executive Committee Range [Figs. 1, 9] in 1996 specifically to look for and document evidence of former high levels of the ice sheet (Borns and others, 1997; Ackert *et al.*, 1999).

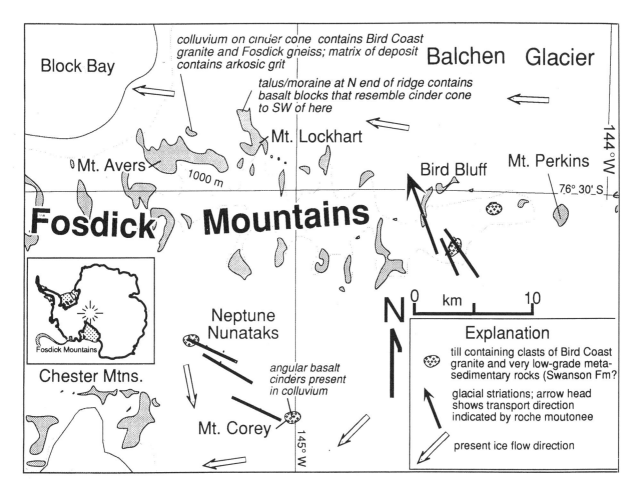

Fig. 8. Map of the locations and orientations of glacial triae and the locations and nature of some glacial deposits documented in the Fosdick Mountains area [Modified after Richard and others, 1991].

Interior Nunataks

Evidence for former glaciation above present levels is extensive in the Heritage and Sentinel Ranges of the Ellsworth Mountains located at the head of the Ronne Ice Shelf on the Weddell Sea (Figure 10). Because of their location close to the flow line passing through the Rutford Ice Stream into the Ronne Ice Shelf these ranges should preserve any evidence of a thicker grounded ice sheet in the Ronne Ice Shelf and Weddell Sea region similar to the history inferred for the Ross Sea. Early work by Rutford [1969] and Craddock and others [1964] reported planed, striated and polished bedrock surfaces as well as poorly developed roches mountonnees in the Heritage Range and foothills of the Sentinel Range. This evidence was ascribed to a higher level of the WAIS that overran most of the Heritage Range, but left the Sentinel Range as a nunatak. The ice-sheet surface at this time was estimated to have been 300-500 m above that of present [Craddock and others, 1964]. They also reported the presence of erratic boulders at least 500 m above the present surface as evidence of a former thicker ice sheet. Craddock and others [1964] also attributed the freshness of these glaciated bedrock surfaces to recent exposure during rapid deglaciation.

The most recent and largest geologic field camp in the Ellsworth Mountains was conducted in the 1979-80 austral summer. The resulting glacial geological work by Denton and others [1992] concluded that two types of major glacial erosion features characterize the Ellsworth Mountains. They reported classic features of alpine glacier erosion which are best developed in the Sentinel Range. These features include cirques arêtes, horns and spurs below the summit plateau of Vinson Massif [4,897 m] which remains as pre-alpine, undulating topography. In addition, a glacial trimline is present on the alpine ridges and spurs throughout the

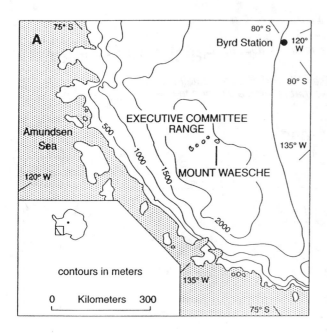

Fig. 9. Map showing the location of the Executive Committee Range and Mt. Waesche in Marie Byrd Land, West Antarctica [Modified after Borns and others, 1996].

Ellsworth Mountains [Figure 11]. This trimline on the ridges separates the serrated bedrock above from the smooth glacially polished and striated surfaces below. The elevations of the trimline range from about 2500-3000 m and generally diminish in the down glacier direction, north to south. Denton and others [1992] inferred from the trimline and striation evidence that the former surface of the WAIS stood 1,300 to 1,900 m above the present glacier surface along the eastern flanks of the Ellsworth Mountains. They also inferred that the alpine glaciation antedated the cutting of the trimline and that the trimline, with associated striae, documents a major thickening of the WAIS which flowed seaward around and through the Ellsworth Mountains either during late Wisconsin-Holocene time or pre Quaternary during the Tertiary Period.

LeMasurier and Rex [1982] noted that moraines and striated bedrock pavements are uncommon on the volcanoes of Marie Byrd Land and, when observed, are very difficult to date. They also identified possible moraines in several areas, such as in the Executive Committee Range and the Kohler Range, where moraines are present at the level of the ice sheet or outlet glacier surfaces. However, elevated moraines are more difficult to recognize as such except where they are composed of mixtures of volcanic and basement rocks. These mixtures are to heterogeneous to have been derived by local glaciers [Figure 12]. They also reported elevated moraines on Chang Peak, just north of Mt. Waesche, in the Executive Committee Range [Figure 12] resting upon 1.6 m y. old pumice at an elevation of 2900 m, approximately 800-1000 m above the present ice sheet level. Similar deposits rest on the upper Miocene-age lavas of near by Mt. Hartigan 500-700 m above ice level. On Toney Mountain [Figure 12] a less heterogeneous bouldery deposit is present just below the summit cone 3000 m above sea level and 1400 m above the present ice sheet surface. This deposit rests upon lava dated to 0.5 m y. LeMasurier noted a striated and polished bedrock surface on the summit of Dorrel Rock near Mt. Murphy [Figure 12] at an elevation of 400-500 m above the present ice-sheet surface. However, he suggests the possibility that the striae and polish resulted from aeolian rather than ice action. Doumani [1964] reported similar features on the southern flank of Mt. Sidley, southern most volcano, in the Executive Committee Range [Figure 12] and he also suggested that the striae may have originated as glacial striae and grooves later modified by aeolian action on the crenulated surface of a lava flow. Borns and Calkin [personal communication] studied similarly eroded surfaces on the lava of Mt. Waesche at several elevations well above the present ice surface and concluded that the striae were not caused by glaciation but by wind driven ice crystals or lithic fragments. One of these surfaces was dated at 430 ka by the surface exposure age method [Borns et al., 1996, Ackert et al., 1999].

Doumani [1964] visited the Executive Committee Range on the IGY Executive Committee Range Traverse in 1959 and reported his observations and syntheses of the nature of the volcanoes comprising the range in 1964. At that time the age of the volcanoes was estimated at 6.2 myr during middle Pliocene time. He also reported that morainic debris is spread over a wide area on and around the slopes and describes two main lines of lateral moraines extending from the south end of Mt. Waesche southward for 3-4 km in the ice. These are composed of angular clasts of diverse lithologies ranging in size from a few centimeters to 2m in diameter, all of which have been derived from the bedrock sequence exposed in the range. More recent analyses [Dunbar et al., 1998] suggest that Doumani's "lateral moraines" which are like down-stream tails from Mt. Waesche within the ice are composed of tephra that erupted onto the ice surface to be buried and folded into the deforming ice around and southward of the volcano. These are not iceward extensions of the true lateral ice sheet moraines that are present on the flanks of Mt Waesche [Borns et al., 1997; Ackert et al. 1999]. Doumani and others (1962) also reported narrow

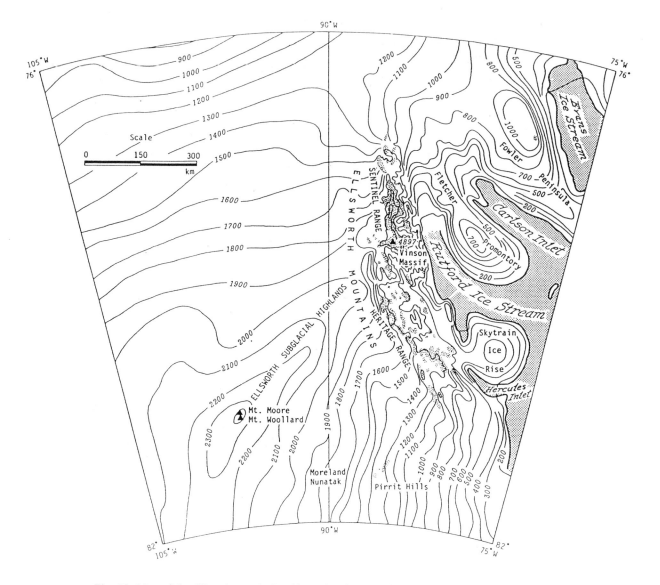

Fig. 10. Map of the West Antarctic Ice Sheet showing the Ellsworth Mountains and the surface topography, in meters, of the ice sheet [Modified after Denton and others, 1992].

symmetrical bedrock grooves 10-15 cm deep and trending north-south on the flanks of Mt. Sidley which they interpreted as true erosion features.

Recent work (Borns et al. 1996; Ackert et al., 1999] reported lateral ice-sheet moraines on the flanks of Mt. Waesche, a one million year old volcano, at the southern end of the Executive Committee Range [LeMasurier and Rex, 1982]. The present level of the ice sheet surface in the area of Mt. Waesche is approximately 2,000 m above sea level. Presently, the ice flows southward, around the volcano, from a major ice dome of the WAIS centered near the northern end of the Executive Committee Range. The southern flank of Mt. Waesche rises from 1,200 m from the present ice surface to the summit caldera and supports local glacier tongues. These tongues currently terminate 2,400 m above sea level, 400 m above the level of the surrounding ice sheet. An ice-cored lateral moraine is present at the present ice level above which is a band of lateral moraines extending up to about 80 m above this ice level [Figure 9]. The moraine band consists of many sub-parallel ridges of drift composed primarily of basalt with some gabbro/diorite clasts. The upper 20-30 m is a morainal terrace with subdued relief that parallels the present ice-sheet surface as its level falls towards the south [Borns et al., 1996; Ackert et al. 1999].

Fig. 11. Photo of the spur at the foot of Mount Epperly in the Sentinel Range of the Ellsworth Mountains. The glacial trimline is clearly shown at an elevation of 2,870 m where it separates the higher serrated ridge crest from the lower smooth ridge crest [Modified after Denton and others, 1992].

Preliminary cosmogenic surface exposure dates from four samples indicate that the portion of the moraine belt below the subdued terrace form a tight cluster whose mean is 12.5 ± 6 ka. The subdued terrace not only appears older, but the three dates cluster around 105 ka [Ackert et al., 1999]. These data indicate that the surface of the ice sheet was up to 80 m above its present level prior to 12.5 ka. The last high stand of ice occurred during the Younger Dryas Chronozone. This is in direct contrast with the evidence inferred from the total gas content of the nearby Byrd ice core which suggests that the ice sheet surface was 200-250 m lower at the LGM at that location [Raynaud et al., 1979; 1982]. The Byrd data are on a different flowline where the ice may have behaved differently from that of the Mt. Waesche flowline. However, Grootes et al., [1986] and Setig et al., [1997] concluded that the level of the WAIS stood about 500 m higher than the present level at the LGM. The evidence of higher ice-sheet surface levels, recorded at Mt. Waesche infer increased snow accumulation due to warming temperatures and an ice sheet grounding line in the Ross Sea well seaward of its present position on the Siple Coast. The data are attributed to a combination of increased snow accumulation due to warming temperatures and a grounding-line position seaward of its present position.

Extensive hyaloclastites have been reported to be present on the volcanoes of West Antarctica. These have been interpreted as products of volcanic eruptions beneath or adjacent to a thicker continental ice sheet present during each eruptive episode. For West Antarctica these have been reported from Marie Byrd Land [Figure 13] [LeMasurier, 1972; Wilch, 1997], Ross and Beaufort islands [Harrington, 1958; Luckman,

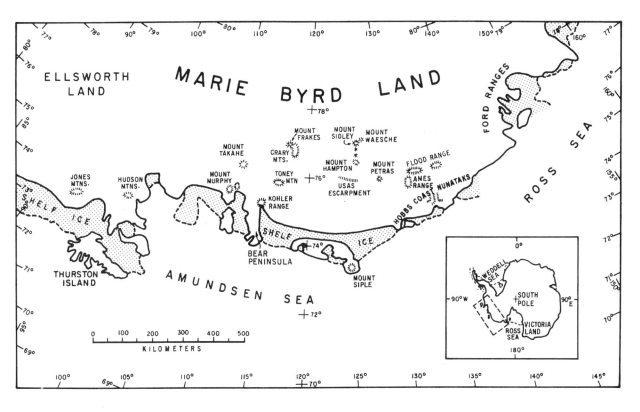

Fig. 12. Index map for Marie Byrd Land and western Ellsworth Land showing the location nunataks [Modified after LeMasurier and others, 1982].

1974; Lyon, 1974, Kyle and Treves, 1974; Treves, 1975; Wilch, 1997]. Hyaloclastite is a common rock type in the coastal volcanic belt of Marie Byrd Land and Ellsworth Land that extends from Fosdick Mountains to the Jones Mountains and inland to the Crary Mountains [Figure 13]. The ages of the hyaloclastites range from Oligocene [27-28 m y.] to early Quaternary [<0.2 m y.] [LeMasurier and Rex, 1982]. Wilch et al., [1997] indicate that the hyaloclastite record shows that the WAIS reached full-bodied configuration by late Miocene time (Ca 9 Ma). They also report a thickening of the WAIS, inferred from Mt. Takahe [about 150 km from the coast] of about 350 m above the present glacier surface level during late Pleistocene time about 29 ka.

Mercer [1975] attributed the Jones Mountains hyaloclastite to eruption beneath a Miocene-age mountain glacier. However, the thickness of the deposits in the Crary Mountains and Turtle Peak [Figure 13] imply an ice thickness of over 1000 m. This would be exceptional for a mountain glacier [LeMasurier, 19982]. It is more likely that the older hyaloclastites of western Ellsworth Land and Marie Byrd Land formed beneath terrestrially-based continental ice sheet [LeMasurier, 1990; Wilch et al., 1997].

In Marie Byrd Land the largest mountains are volcanoes formed after the formation of the older hyaloclastic sequences. The tectonism in West Antarctica played a role in determining the present elevation of the hyaloclastites. However, the history is inadequately known at this time to allow a significant evaluation of the history of level changes. Nevertheless, current evidence suggests that block faulting has displaced the flat pre-volcanic surface of Marie Byrd Land with vertical offsets of 3000 to 5000 m within the last 60 to 80 m y. LeMasurier [1982] feels that:

> "Geologic history in West Antarctica has evidently involved a complex interplay of volcanic, tectonic, and glacial phenomena, all of which must be viewed together in order to understand any one of the three".

SUMMARY AND CONCLUSIONS

The available terrestrial glacial geologic evidence from West Antarctica points to a greatly expanded West Antarctic Ice Sheet at the last glacial maximum. The most voluminous and convincing terrestrial evidence comes from the interpretation and chronology of the terrestrial glacial and emerged glacial marine deposits along the flanks of the Transantarctic Mountains

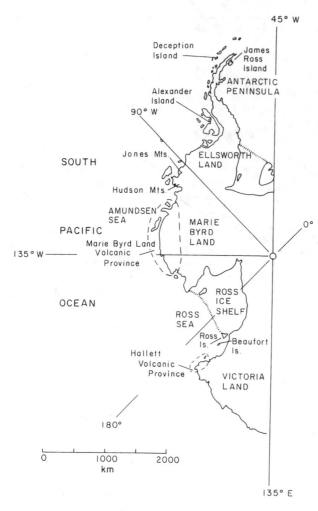

Fig. 13. Location map of West Antarctica showing locations of reported hyalocastite localities [Modified after LeMasurier and others, 1982].

bordering the western side of the Ross Sea. This record demonstrates that at the LGM grounded ice expanded in the Ross Sea north of Terra Nova Bay to the position just south of Coulman Island, along the Transantarctic Mountains. This ice invaded the mouths of many of the valleys of the foothills. The grounding line in the Ross Sea retreated from its extended position on the continental shelf at the LGM passing through McMurdo Sound in mid Holocene time, to its present position approximately 200 km to the south along the flowline of McMurdo Sound, a total distance of about 1300 km since the LGM. Furthermore, Hall and Denton [1998] pointed out that because of the timing of this recession through the McMurdo area in mid-to-late Holocene then rising sea level could not have directly driven ice recession to its present grounding line on the Siple coast.

In the coastal and interior areas of WAIS the paleo glacial record is primarily limited to exposed glaciated rock surfaces in Marie Byrd Land and Ellsworth Land. At various locations evidence is found at above the present level of the ice sheet on the faulted mountain blocks and Cenozoic volcanoes.

In the foothills of the Transantarctic Mountains, along the western border of the Ross Sea, the chronology of glacial events is tightly controlled by several hundred of radiocarbon dates primarily on terrestrial lacustrine algae and secondarily on marine shells in the emerged glacial marine deposits. In contrast, no terrestrial radiocarbon dates have been reported for the entire remainder of West Antarctica. However, there are other types of dates bearing on the past glacial record of WAIS. About 50 of these are K/Ar dates from the volcanoes which control the maximum age of the formation of volcanic glacio-hyaloclastite rocks, and about 20 cosmogenic surface exposure age dates from Mt. Waesche in the Executive Committee Range. In addition the oxygen-isotopic profiles from cores at Byrd Station and Siple dome also indicate an expanded WAILS in Pleistocene and Holocene times.

Many of the glacial erosion surfaces are reported to have a fresh appearance which has suggested that they may date from the LGM. However, their various elevations above the present ice surface may indeed document former higher elevations of the ice surface, upward motion of block faults in this tectonic zone, or both. Without better knowledge of the ages of both the erosion and of the tectonic action the age of glaciation and the former surface elevations of the ice sheet remain problematic.

In conclusion, the available, well-controlled and widely distributed glacial geologic record, the glacial volcanic record and the ice-core record from Marie Byrd Land, all indicate higher surface levels of the West Antarctic ice sheet at the LGM and early Holocene times. Clearly evidence of higher levels is preserved in the glacial geologic, glacial volcanic and ice-core records.

Future Research Directions

The numerical modeling of the configuration of the West Antarctic Ice Sheet at and since the last glacial maximum requires the documentation of former elevation and chronology of the surface levels. The terrestrial record must be than correlated with the glacial marine record in the Ross Sea and Weddell Sea embayments in order to determine the configuration of the WAIS at and since the LGM.

In order to accomplish these tasks the following are required. The terrestrial geologic record and chronology recorded on the coastal mountains and interior nunataks must be determined over as many areas of West Antarctica as possible. This record is in two forms, the glacial geologic record and the glacial-volcanic record. In the case of both records chronology is critical. The probability of the presence of organics suitable for radiocarbon dating is extremely low and therefore cosmogenic surface exposure-age dating offers the best hope of providing the chronology for the glacial geologic record, not only for times since the LGM, but also for earlier events. In the case of the glacial-volcanic record K/AR dating provides perhaps the only method if the proper stratigraphic sequence of deposits can be identified on the volcanoes. The terrestrial geologic records and ice-core records must be then calibrated with the glacial-marine record on the continental shelf to determine the positions of the grounding line since the LGM on the continental shelf. However, the chronologies of events between two records is presently inadequately calibrated due, in part due to the results of dating the total organic content of samples for the marine record which provide, at best maximum ages for events. In addition, the radiocarbon reservoir correction factor has yet to be clearly defined in time and place in order to determine real ages of marine shells.

Acknowledgements. The work from Mt. Waesche was supported by NSF grants OPP-93-18872 (Harold W. Borns, Jr. and Parker E. Calkin), and OPP-94-1833 and EAR 96-14561 (Mark D. Kurz). The author thanks Deborah Seymour for help in manuscript preparation. George A. Doumain, Bruce P. Luyendyk, Robert P. Ackert and other researchers of the West Antarctic glacial record.

REFERENCES

Ackert,, R. P., Jr., Barclay, D. J., Borns, H. W.; Jr., Calkin, P. E., Kurz, M. D., Fastook, J. L., and Steig, E. J., Measurements of past ice sheet elevations in interior West Antarctica, *Science 286*, pp. 276-280, 1999.

Bockheim, J. G., Soil development in the Taylor Valley and McMurdo Sound area, *Ant. Jour. of the US, 12,* (5), pp. 105-108, 1977.

Borns, H. W., Jr., Dorion, C., Calkin, P. E., Wiles, G., and Barclay, D., Evidence for thicker ice in interior West Antarctica, *Ant. Jour. of the US, 30,* (5), pp. 100-101, 1996.

Byrd, R. E., The flight to Marie Byrd Land, *Geogr. Rev. 23*, pp. 177-195, 1933.

Byrd, R. E., Discovery, New York, 1935.

Conway, H., Hall, B. L., Denton, G. H., Gades, A. M., and Waddington, E. D., Past and future grounding-line retreat of the West Antarctic ice sheet, *Science, 286*, pp. 280-283, 1999.

Craddock, C., Anderson, J. J., and Webbers, G. F., Geologic outline of the Ellsworth Mountains, in *Antarctic Geology Proceedings of the First International Symposium on Antarctic Geology, Cape Town 16-21 Sept. 1963*, edited by R. J. Adie, Amsterdam North-Holland, pp. 155-170, 1963.

David, T. W. E., and Priestley, H. E., Glaciology, physiography, stratigraphy, and tectonic geology of South Victoria Land: British Antarctic Expedition, 1907-1909, Reports on the Scientific Investigations, *Geology, 1,319*, 1914.

Debenham, F., A new mode of transportation by ice: The raised marine muds of South Victoria Land, *Quarterly Journal of the Geological Society of London*, 75, pp. 51-76, 1920.

Debenham, F., Recent and local deposits of McMurdo Sound region: London, British Museum, British Antarctic [Terra Nova] Expedition, *Natural History Report, Geology, 1,* (3), pp. 63-90, 1921.

Denton, G. H., and Armstrong, R. L., Glacial geology and chronology of the McMurdo Sound region, *Ant. Jour. of the US, 3*, p. 99-101, 1968.

Denton, G. H., Armstrong, R. L., and Stuiver, M., Late Cenozoic glaciation in Antarctica, *Ant. Jour. of the US, 5*, pp. 15-22, 1970.

Denton, G.H., Armstrong, R.L., and Stuiver, M., The Late Cenozoic glacial history of Antarctica, in *The Late Cenozoic Glacial Ages*, edited by K. K. Turekian, New Haven, CT, Yale University Press, pp. 267-306, 1971.

Denton, G. H., and Borns, H. W., Jr., Former grounded ice sheets in the Ross Sea, *Ant. Jour. of the US, 9*, p. 167, 1974.

Denton, G. H., Borns, H. W., Jr., Grosswald, M. G. Stuiver, M., and Nichols, R. L., Glacial history of the Ross Sea, *Ant. Jour. of the US, 10*, pp. 160-164, 1975.

Denton, G. H., Andersen, B. G., and Conway, H. B., Late Quaternary surface ice level fluctuations of the Beardmore Glacier, Antarctica, *Ant. Jour. of the US, 21* (5), pp. 90-92, 1986.

Denton, G. H., Bockheim, J. G., Wilson, S. C. and Stuiver, M., Late Wisconsin and Early Holocene Glacial History, Inner Ross Sea embayment, Antarctica, *Quaternary Research, 31*, pp. 151-182, 1989.

Denton, G. H., Bockheim, J. G., Rutford, R. H. and Andersen, B. G., Glacial history of the Ellsworth Mountains, West Antarctica, *Geol. Soc. Am., 170*, pp. 403-432, 1992.

Doumani, G. A., Geological observations in West Antarctica during recent oversnow traverses, IGY *Bull. Nat. Acad. Sci.*, 41, pp. 6-10, 1960.

Doumani, G. A., Volcanoes of the Executive Committee Range, Byrd Land, in *Antarctic Geology*, edited by R. J. Adie, Amsterdam, North-Holland, pp. 666-675, 1964.

Dunbar, N. W., McIntosh, W. C., Esser, R. P., Wilch, T. L., and Zielinski, G. A., Direct dating and geochemical correlations of englacial tephra layers at two sites in West Antarctica, The West Antarctic Ice Sheet abstract book, AGU Chapman Conference, University of Maine, Orono, pp. 18-21, 1998.

Gould, L. M., Some geographical results of the Byrd Antarctic Expedition, *Geogr. Rev., 21*, p. 177-200, 1931.

Grootes, P. and Stuiver, M., Ross Ice Shelf oxygen isotopes and West Antarctic climate history, *Quaternary Research*, 26, pp. 49-67, 1986.

Hall, B. L. and Denton, G. H., Deglacial chronology of the western Ross Sea from terrestrial data, abst., in The West Antarctic Ice Sheet – Abstract Volume, edited by R. Bindschadler, and H. W. Borns, Jr., AGU Chapman Conference, University of Maine, Orono, 140 p., 1998.

Hall, B. L., and Denton, G. H., New relative sea-level curves for the southern Scott Coast, Antarctica: Evidence for Holocene deglaciation of the western Ross Sea, *Journal of Quaternary Science,* in press.

Hamilton, W., The Hallett volcanic province, Antarctica *Prof. Pap. U. S. Geol. Surv., 456-C*, 62 pp., 1972.

Harrington, H. J., Beaufort Island, remnant Quaternary volcano in Ross Sea, Antarctica. *N.Z. J. Geol. Geophys., 1*, pp. 595-603, 1958.

Karlén, W. and Melander, O., Reconnaissance of the glacial geology of Hobbs Coast and Ruppert Coast, Marie Byrd Land, *Ant. Jour. of the U. S., 13,* (4), pp. 46-47, 1978.

Kyle, P. R., and Treves, S. B., Geology of DVDP 3 Hut Point Peninsula, Ross Island, Antarctica, *Bull. Dry Valley Drilling Project, 3*, pp. 13-48, 1974.

LeMasurier, W. E., Volcanic record of Cenozoic glacial history of Marie Byrd Land, in *Antarctic Geology and Geophysics,* ed. by R. J. Adie, Oslo, Universitetsforlaget, pp. 251-60, 1972.

LeMasurier, W. E., and D. C. Rex, Volcanic record of Cenozoic glacial history in Marie Byrd Land and western Ellsworth Land: Revised chronology and evaluation of tectonic factors, in *Antarctic Geoscience,* edited by C. Craddock, University of Wisconsin Press, Madison, pp. 725-734, 1982.

LeMasurier, W. E., Marie Byrd Land: Summary, in Volcanoes of the Antarctic plate and southern oceans, edited by W. E. LeMasurier and J. W. Thompson, Antarctic Research Series, 48, pp. 147-174, 1997.

Licht, K. J. and Fastook, J. L., Constraining a numerical ice sheet model with geologic data over one ice-sheet advance/retreat cycle in the Ross Sea, Abstract Vol. AGU Chapman Conf, Univ. of Maine, 1998.

Licht, K. J., Dunbar, N. W., Andrews, J. T., and Jennings, A. E., Distinguishing subglacial till and glacial marine diamictons in the western Ross Sea, Antarctica: Implications for a last glacial maximum grounding line, *Geol. Soc. Am. Bull., 111*, pp. 91-103, 1999.

Luckman, P. G., Products of Submarine and Subglacial Volcanism in the McMurdo Sound region, Ross Island, Antarctica, *B. Sc. Hons. Proj.*, Victoria University of Wellington, 115 pp, 1974.

Lyon, G. L., Stable isotope analyses of ice from DVDP 3. *Bull. Dry Valley Drilling Project, 3*, pp. 160-70, 1974.

MacAyeal, D. R., Transient temperature-depth profiles of the Ross Ice Shelf [M. S. thesis]: University of Maine at Orono, 1979.

MacAyeal, D. R., Irregular oscillations of the West Antarctic ice sheet: *Nature, 359,* pp. 29-32, 1992.

Mayewski, P. A., Glacial geology and late Cenozoic history of the Transantarctic Mountains, Antarctica: Columbus, Ohio State University Institute of Polar Studies Report 56, 168 p., 1975.

Mercer, J. H., Glacial geology of the Reedy Glacier area, Antarctica, *Geol. Soc. Amer. Bull., 79*, pp. 471-486, 1968.

Péwé, T. L., Multiple glaciation in the McMurdo Sound region, Antarctica, a progressive report, *J. Geol., 68*, p. 489-514, 1960.

Raynaud, D. and Lebel, B., Total gas content and surface elevation of polar ice sheets. *Nature, 281*, pp. 289-291, 1979.

Raynaud, D. and Whillans, I. M., Air content of the Byrd core and past changes in the West Antarctic Ice Sheet, *Annals of Glaciology, 3*, pp. 269-273, 1982.

Richard, S. M. and B. P. Luyendyk, Glacial flow reorientation in the southwestern Fosdick Mountains, Ford Range, Marie Byrd Land, *Ant. Jour. of the United States*, 26, (5), pp. 67-69, 1991.

Ross, Captain Sir James Clark, *A Voyage of Discovery and Research in the Southern and Antarctic Regions During the Years 1839-1843*, London, John Murray, 2 vols., pp. 366 and 447, 1847.

Rutford, R. H., The glacial geology and geomorphology of the Ellsworth Mountains, West Antarctica [Ph.D. dissertation], Minneapolis, University of Minnesota, 289 p., 1969.

Scott, R. F., The Voyage of the Discovery: New York, Charles Scribner's Sons, 2 vols., pp. 556 and 508, 1905.

Steig, E. J. and White, W. D., Elevation change in West Antarctica from stable isotope profiles, Abstract, Fourth Annual Workshop [WAIS] Agenda and Abstracts, 1997.

Stuiver, M., Denton, G. H., Hughes, T. J., and Fastook, J. L., History of the marine ice sheet in West Antarctica during the last glaciation: A working hypothesis, *in The Last Great Ice Sheets*, edited by G. H. Denton and T. J. Hughes, New York, Wiley-Interscience, p. 319-436, 1981.

Teves, S. B., Hyaloclastite of DVDP 3, Hut Point Peninsular, Ross Island, Antarctica. *Bull. Dry Valley Drilling Project, 6*, (32), 1975.

Vella, P., Surficial geological sequence, Black Island and Brown peninsula, McMurdo Sound, Antarctica, *New Zealand Journal of Geology and Geophysics, 12*, pp. 761-770, 1969.

Wade, F. A., Northeastern borderlands of the Ross Sea: Glaciological studies in King Edward VII Land and northeastern Marie Byrd Land, *Geog. Rev, 27*, pp. 584-597, 1937.

Wilch, T. I., Volcanic record of the West Antarctic ice sheet in Marie Byrd Land, [Ph.D dissertation], New Mexico Institute of Mining and Technology, Socorro, 1997.

Wilch, T. I. and McIntosh, W. C., Paleo-ice-level reconstructions at Marie Byrd Land volcanoes, West Antarctica: Abstract, Fourth Annual Workshop [WAIS] Agenda and Abstracts, 1997.

Harold W. Borns, Jr., Institute for Quaternary Studies and Department of Geological Sciences, University of Maine, Orono, ME 04469.

WEST ANTARCTIC ICE SHEET ELEVATION CHANGES

Eric J. Steig,[1] James L. Fastook,[2] Christopher Zweck,[3,4]
Ian D. Goodwin,[3] Kathy J. Licht,[5] James W. C. White,[5] Robert P. Ackert, Jr.[6]

Changes in climate and sea level have caused significant fluctuations of the margins of the West Antarctic Ice Sheet (WAIS). Geologic evidence shows that changes in the elevation of the WAIS accompanied these ice-margin fluctuations. Results from two numerical modeling experiments, one forced by prescribed climate and sea level variations, the other by prescribed ice-margin retreat, suggest that the WAIS interior was at a maximum elevation during the early Holocene and has declined by tens to hundreds of meters in the last 10,000 years. Cosmogenic ^{3}He exposure ages on moraines at Mount Waesche, north Marie Byrd Land, provide direct evidence for an early Holocene maximum. Estimated elevation changes from ice core stable isotope anomalies at Byrd Station show an overall decline of ~200 m in the last 8000 years, also confirming higher elevations in the early Holocene. Together, the model results and empirical evidence indicate that the WAIS has largely adjusted to the accumulation increase and sea level rise associated with the last glacial-interglacial transition. It may still be responding to more recent climate and sea level changes.

1. INTRODUCTION

The West Antarctic ice sheet (WAIS, Figure 1) has varied considerably in volume and areal extent. As the only large ice sheet grounded below sea level at its margins, the WAIS is believed to be sensitive to changes in sea level and may be subject to significant retreat or collapse in response to future climate change [*Bindschadler*, 1998]. Terrestrial and marine geologic data together document changes in the position of the WAIS margin since the end of the last glacial period [*Anderson et al.*, 1980; *Stuiver et al.*, 1981; *Denton et al.*, 1989a; *Licht et al.*, 1996; 1999; *Licht*, 1999; *Hall and Denton*, 1999; *Higgins et al.*, 1999; *Shipp et al.*, 1999]. Considerably less is known about changes in elevation that must have accompanied ice-margin fluctuations [*Calkin and Bull*, 1972; *Stuiver et al.*, 1981; *Bockheim et al.*, 1989; *Denton et al.*, 1989b]. Because even small changes in elevation may represent large changes in volume, knowledge of past WAIS elevation is important for evaluating the contribution of the ice sheet to sea level variations. It is also essential for understanding the present-day behavior of the ice sheet, which is largely independent of current climate forcing but represents a lagged response to past changes in sea level, temperature and snow accumulation [*Alley and Whillans*, 1991].

Several methods have been used to determine past WAIS elevations, including total gas content measurements [*Raynaud and Lebel*, 1979; *Raynaud and Whillans*, 1982; *Jenssen*, 1983] and stable isotope ratios [*Grootes and Stuiver*, 1986; 1987] from ice cores, mapping of ice-sheet trimlines and other erosional features [*Denton et al.*, 1992], and back-calculation of ice-sheet surface profiles from documented ice-margin positions and outlet glacier surface profiles [*Anderson et al.*, 1980; *Stuiver et al.*, 1981; *Bentley and Anderson*, 1998].

---

[1] Department of Earth and Environmental Science, University of Pennsylvania, Philadelphia, PA 19104.
[2] Institute for Quaternary Studies and Department of Computer Science, University of Maine, Orono, ME 04469.
[3] Antarctic Cooperative Research Centre and SCAR Global Change Program, Hobart 7001, Tasmania, Australia.
[4] Now at Geophysical Institute, University of Alaska, Fairbanks, AK 99775.
[5] Institute of Arctic and Alpine Research and Department of Geological Sciences, University of Colorado, Boulder, CO 80309.
[6] Massachusetts Institute of Technology/Woods Hole Oceanographic Institution, Woods Hole, MA 02543.

This empirical evidence has been largely inconclusive because of a lack of good chronological control. Most numerical modeling studies [*Budd and Jenssen*, 1989; *Huybrechts*, 1990a; 1990b; 1990c; 1993; *MacAyeal*, 1992; *Fastook and Prentice*, 1994; *Verbitsky and Saltzman*, 1995; *Verbitsky and Oglesby*, 1995; *Warner and Budd*, 1998] have focussed on the sensitivity of the ice sheet to different internal conditions (i.e. ice dynamics, basal boundary conditions) and external forcings (i.e. climate, sea level) rather than attempting to calculate ice sheet configurations at specific times in the past. Some recent studies [*Budd et al.*, 1994; 1998; *Verbitsky and Saltzman*, 1997] have combined ice sheet models with paleoclimate data from ice and sediment cores to develop time series of ice sheet configuration through the last glacial-interglacial cycle. However, none of the published work has specifically examined WAIS elevation changes nor made direct comparisons with empirical data.

In this paper, we review current understanding of WAIS elevation change since the early Holocene, taking advantage of new, independent empirical constraints on former ice elevations: 1) cosmogenic isotope exposure-age data from moraines marking the most recent WAIS high-stand at Mount Waesche, a volcanic nunatak near the Executive Committee Range in north Marie Byrd Land and 2) a revised estimate of elevation change at Byrd Station, central West Antarctica, using stable-isotope anomalies for the Byrd ice core. As context for interpretation of the empirical data, we first present model calculations of former ice sheet elevations using 1) the model of *Budd and Jenssen* [1989] under prescribed climate and sea level forcing histories and 2) the model of *Fastook and Prentice* [1994], tuned to the empirically-documented Ross Sea deglaciation chronology. Comparisons of empirical and model-derived results provide the basis for discussion of elevation history over the last 10,000 years and its implications for earlier, present and future changes in the WAIS.

2. VARIATIONS IN ICE SHEET ELEVATION

Prior to the presentation of new data and modeling results, it is worth reviewing the basic physics that control the response of interior ice sheet elevations to changing boundary conditions. The best treatment of this subject is probably that of *Alley and Whillans* [1984] who developed a two-dimensional model and used it to estimate elevation changes at Dome C, East Antarctica. *Alley and Whillans* cautioned against applying their model to West Antarctica because of the complexity of basal boundary conditions; however, their model is general enough that it provides a useful framework for discussion. Here, we give a simplified one-dimensional treatment and include some comments on the specific problem for West Antarctica.

Given an initial ice-sheet configuration, the response of elevation to small changes in boundary conditions can be reduced to a straightforward advection-diffusion problem, which we can write [*Nye*, 1960] in the form of a perturbation equation

$$\frac{\partial \Delta h}{\partial t} = \Delta \dot{b} - \frac{\partial c}{\partial x} \Delta h - \left(c - \frac{\partial D}{\partial x}\right)\frac{\partial (\Delta h)}{\partial x} + D\frac{\partial^2 (\Delta h)}{\partial x^2} \quad (1)$$

where t is time since the initial perturbation, x is horizontal distance (positive in the direction from the ice divide towards the margin), c is the kinematic wave speed, D is the local diffusivity, $\dot{b}$ is accumulation rate, and h is ice sheet thickness. The perturbation terms $\Delta \dot{b}$ and Δh allow for instantaneous changes in key external variables: for the timescales of primary interest (the last 10^3 to 10^5 years) these are snow accumulation rate, temperature, and relative sea level. We ignore the influence of temperature change for the moment, noting that for central WAIS the time required for surface temperate to begin to significantly influence ice deformation near the bed is on the order of the vertical ice advection time, $>10^4$ years for most of the WAIS. The effects of sea level change and accumulation rate change, however, will be immediate. The fundamental internal influences on ice sheet response may be grouped under the general term "ice dynamics" which are incorporated in the values c and D; of particular importance is whether the ice sheet flows entirely by internal deformation or with a significant contribution from basal sliding. For East Antarctica, basal sliding can be neglected, as in *Alley and Whillans* [1984]. For the WAIS, basal sliding is known to be important, and considerable work has gone into developing a quantitative description of basal sliding processes [*Weertman*, 1957; *Kamb*, 1970; *Budd et al.*, 1979]. Even more important than sliding may be subglacial till deformation, which is believed to be responsible for the fast flow of ice streams [*Alley et al.*, 1986; *Bindschadler et al.*, 1987; *Alley and Whillans*, 1991]. In current-generation numerical models, subglacial till deformation is usually parameterized within the basal sliding terms.

For a localized perturbation in ice sheet thickness – as corresponds to a rapid retreat at the ice sheet margin induced by sea level rise – the spatial derivatives of Δh in (1) are large. Because velocity is very sensitive to

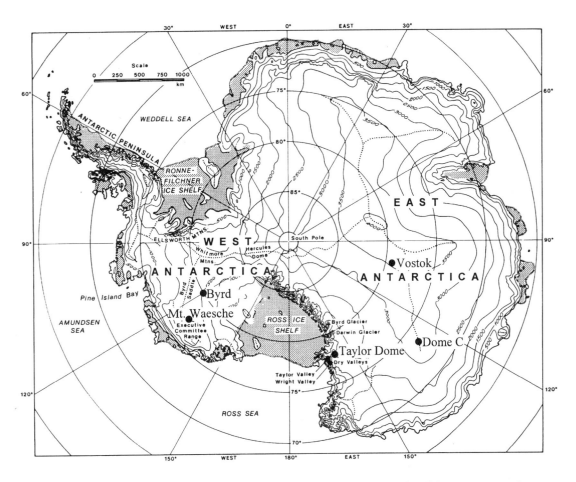

Fig. 1. Map of Antarctica, showing locations referred to in the text (modified from *Denton et al.* [1992]).

surface slope, a sudden thinning at the ice sheet margin will lead to increased velocity, resulting in a thinning wave of initial height Δh that propagates upglacier. Under these conditions, the diffusive term tends to dominate, suggesting that we may write simply

$$\frac{\partial \Delta h}{\partial t} \cong D \frac{\partial^2 (\Delta h)}{\partial x^2} \qquad (2)$$

for the case where $\Delta \dot{b} = 0$. For the highest point on the ice sheet (i.e. the ice divide, $x = 0$) we can write

$$\Delta h(t) = \Delta h(\infty) 2 \sum_{j=0}^{\infty} \frac{(-1)^j}{\omega_j \ell} e^{-Dw_j^2 t}$$
$$\omega_j = \frac{(2j+1)\pi}{2\ell} \qquad (3)$$

which is the Fourier series solution to (2), subject to the boundary condition $\partial h(t)/\partial x \leq 0$ (the position of the divide does not change laterally). The distance from ice divide to margin is given by ℓ. A function $h(0,x)$ defines the ice sheet profile and determines the magnitude of the final response $\Delta h(\infty)$; rather than specifying a particular form for $h(0,x)$ it is useful, following *Cuffey and Clow* [1997], to generalize (3) as a non-dimensional filter

$$\Phi_m(t) = 2 \sum_{j=0}^{\infty} \frac{(-1)^j}{\omega_j \ell} e^{-D\omega_j^2 t} \qquad (4)$$

which gives the time-dependent response at the ice divide to a unit change in ℓ.

For a *non-localized* perturbation in ice sheet thickness – corresponding to an increase in accumulation rate – we have $\partial \Delta \dot{b} / \partial x = \partial \Delta h / \partial x = 0$ and (1) reduces to

$$\frac{\partial \Delta h}{\partial t} = \Delta \dot{b} - \frac{\partial c}{\partial x} \Delta h \qquad (5)$$

with the solution:

$$\Delta h(t) = \frac{\Delta \dot{b}}{(\partial c/\partial x)}\left(1 - e^{-(\partial c/\partial x)t}\right) \quad (6)$$

for $\partial c/\partial t = 0$. As above, (6) can be expressed more usefully in non-dimensional form, noting that $\Delta h(\infty) = \Delta \dot{b}/(\partial c/\partial x)$; that is,

$$\Phi_{\dot{b}}(t) = 1 - e^{-(\partial c/\partial x)t} \quad (7)$$

which is a filter giving the response of the ice sheet height to a step change in accumulation rate across the entire ice sheet.

How do the response filters $\Phi_m(t)$ and $\Phi_{\dot{b}}(t)$ apply to a real ice sheet? The assumptions made in deriving equations (4) and (7) are reasonable for West Antarctica for the transition from the last glacial period to the Holocene. From the glacial geologic record we know that the WAIS began retreating from its maximum position in the Ross Sea some time after ~15 ka, largely in response to the sea level rise that marks the end of the last glacial period [Stuiver et al., 1981; Denton et al., 1986; 1989a; 1989b; 1991; Licht et al., 1996; Licht, 1999; Domack et al., 1999]. We also know from ice core data that accumulation rates over the WAIS increased across the glacial-interglacial transition, beginning perhaps as early as 24 ka [Raisbeck et al., 1987; Blunier et al., 1998; Steig et al., 1998a]. This leaves as unknowns the diffusivity D and the spatial derivative $\partial c/\partial x$ of the kinematic wave speed. We can make reasonable estimates of the magnitude of these quantities by recognizing that they are primarily functions of the ice velocity. For an ice sheet that is moving entirely by internal deformation, $\partial c/\partial x$ can be shown to equal $(n+2)(\partial \bar{u}/\partial x)$ where n is the familiar exponent from the glacier flow law and $\bar{u}$ is the depth-averaged ice velocity. The term $\partial \bar{u}/\partial x$ is equivalent to the longitudinal strain rate $\dot{\varepsilon}_x$; thus

$$\frac{\partial c}{\partial x} = (n+2)\dot{\varepsilon}_x \quad (8)$$

Similarly, an average value for the diffusivity may be obtained from

$$D = \frac{nh\bar{u}}{\sin \theta} \quad (9)$$

where θ is the surface slope.

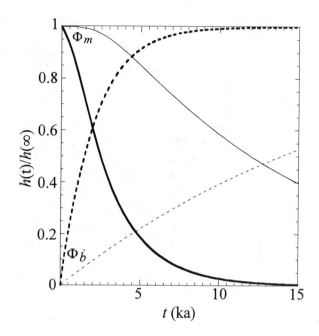

Fig. 2. Theoretical response filters for perturbations in accumulation ($\Phi_{\dot{b}}$, dashed lines) and sea level (Φ_m, solid lines) for a one dimensional ice sheet comparable to East Antarctica (thin lines) and West Antarctica (bold lines).

For purposes of illustration, Figure 2 (thin lines) shows the two response filters Φ_m and $\Phi_{\dot{b}}$, given $\ell = 10^6$ m, $D = 2$ m^{2}s^{-1}, and $\dot{\varepsilon}_x = 10^{-5}$ a^{-1}. These values are reasonable estimates for an ice sheet having simple geometry and undergoing flow entirely by internal deformation. This is appropriate for East Antarctica and as a consequence our calculated response filters agree well with the results of Alley and Whillans [1984]. Following a sudden retreat of the margin, the elevation at the ice divide at first responds slowly, then more quickly as the crest of the thinning wave approaches the divide, then more slowly again as equilibrium conditions are restored. Following a sudden change in accumulation, the initial response is a rapid thickening; the rate of thickening gradually decreases as ice velocities increase. When both a change at the margin and a change in accumulation occurs, the response filter will be some nonlinear combination of Φ_m and $\Phi_{\dot{b}}$, the exact shape depending on the relative magnitude of the individual perturbations.

If we include the influences of basal sliding and subglacial till deformation, but hold the total velocity constant, $\partial c/\partial x$ and D will be slightly lower than implied by (8) and (9). That this is the case is readily seen if one considers that both $\dot{\varepsilon}_x$ and h must decrease as the total velocity approaches the sliding velocity. For the real WAIS, however, the distribution of sliding is non-uniform, and in general we expect that $\partial c/\partial x$ and D will be significantly *higher* because surface slopes are lower, average velocities are higher, and there are very large longitudinal strain rates between the slow-moving interior ice and fast moving ice streams. As an example, consider the flow line from Byrd Station (velocity ~5 m^{-1}) towards the Ross Ice Sheet through ice stream D, which is currently moving at a velocity of about 300 m a^{-1} [*Bindschadler et al.*, 1996]. For this flow line, we have average $\theta < 0.1°$, $h \sim 10^3$ m, $\ell \sim 10^6$ m, $\dot\varepsilon_x \sim 10^{-4}$ a^{-1} and therefore $D \sim 5$ m^{2}s^{-1}, $\partial c/\partial x \sim 10^{-3}$ a^{-1}. The resulting response filters (Figure 2, bold lines) suggest that the WAIS should adjust to changes in boundary conditions much more rapidly – by perhaps an order of magnitude – than East Antarctica. According to these calculations, the WAIS today should already be fully adjusted to the increase in accumulation associated with the last glacial-interglacial transition at ~20 to 10 ka. It may still be adjusting to sea level change because the fastest sea level rise occurred more recently, between about 11 ka and 6 ka [*Fairbanks*, 1989; *Bard et al.*, 1990].

3. ELEVATION HISTORY FROM ICE SHEET MODELS

3.1 Climate and Sea Level Forcing

We now consider the response of the WAIS to climate and sea level using a more realistic ice sheet model forced by empirical accumulation and sea level histories obtained from the paleoclimate record. The model ("BJM") is that of *Budd and Jenssen* [1994], with refinements discussed by *Budd et al.* [1994; 1998] and *Warner and Budd* [1998]. Briefly, BJM is a finite-difference thermomechanical model with 31 vertical layers. The model has a horizontal resolution of 100 × 100 km. Internal deformation and heating are calculated as functions of horizontal shear stress; the model also includes an explicit formulation for ice shelves that takes both transverse and longitudinal stresses into account [*Mavrakis*, 1993]. A simple calving law allows loss of ice from the ice shelves whenever the thickness is lower than 250 m. Sliding occurs whenever there is melting at the bed, as determined by frictional heating, effective normal stress, and prescribed geothermal heat flux (0.05 Wm^{-2}). The sliding velocity is taken to be inversely proportional to the effective normal stress, from the relationship of *Budd et al.* [1979].

We initialize BJM using the elevation data of *Drewry* [1983] and present-day climate and sea level conditions. We then run the model for 100,000 model years; this long spin-up time ensures both that numerical errors are minimized and that the simulated ice sheet is initially in near equilibrium. After 100,000 years, changes in external boundary conditions are permitted, and the model is run for another 160,000 years to simulate ice sheet dynamics since the penultimate glacial period. This length of time is great enough that memory of ice sheet configuration achieved during spin-up is largely lost by the time of the last glacial-interglacial transition; the ice sheet is not necessarily in steady-state equilibrium with climate at the time of the LGM.

For climate forcing, we use the ice core records from Vostok as a reference, using the original *Lorius et al.* [1985] timescale; more recent improvements to the Vostok timescale are expected to be negligible relative to other uncertainties in our calculations. For temperature, we use the δD data of *Jouzel et al.* [1987] where $\partial \delta D/\partial T = 6$ ‰/°C; uncertainty in this value is insignificant relative to other uncertainties (especially accumulation rate) in the climate forcing. For accumulation, we use the ^{10}Be data of *Raisbeck et al.* [1987] assuming that the air-snow flux of ^{10}Be is constant (see *Raisbeck and Yiou* [1985], *Jouzel et al.* [1989] and *Steig et al.* [1998a; 1998b] for discussions of this calculation). Because there are few data points in *Raisbeck et al.* [1987] for the Holocene, we assume constant accumulation rates after 10 ka. Temperature and accumulation at any given point on the ice sheet are scaled according to today's climatic gradients as functions of elevation and surface slope (see *Budd et al.* [1994; 1998] for details). For sea level forcing we use the eustatic sea level curve of *Chappell and Shackleton* [1986].

We ran BJM under three different sets of boundary conditions: 1) sea level and climate varying as described above, 2) accumulation rate held constant at today's value, and 3) sea level held constant at today's value. Figure 3 shows the calculated elevation history for each of these scenarios, for a point (79° 10' S, 113° 00' W, 1800 m) on the present-day ice divide that separates the Pine Island Bay and Ross Sea drainages (labeled "Byrd Saddle" in Figure 1). Considering the effect of sea level and temperature alone, ice sheet elevation decreases significantly since an early Holocene maximum. With sea level held constant, ice sheet elevation increases since the LGM. When both sea level and climate change are considered together, they tend to partially counteract one another. The effects are not purely additive; the non-linear interaction between climate and sea

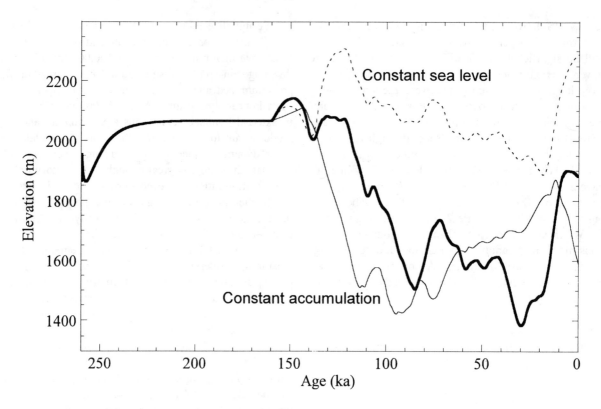

Fig. 3. Calculated elevation at present-day ice divide from the Budd and Jenssen model, 160 ka to present, subject to forcing by climate and sea level (bold line) and with sea level (dashed line) and accumulation rate (thin line) held constant.

level forcing is particularly noticeable for the LGM, where the model predicts lower ice sheet elevation than either the sea-level-constant or accumulation-constant simulations. This results from the delayed response to low LGM sea level – which causes thickening – compared with the relatively fast response to lower LGM accumulation rate. Similarly, when accumulation begins to increase at about 18 ka, the ice sheet initially thickens even though sea level has begun to rise.

A key result from BJM is that WAIS elevations are generally higher during interglacial than glacial periods. Elevations in the central WAIS are higher in the early to mid Holocene than at any other time in the past 100 ka. This result is obtained even when accumulation rates are held constant, which essentially shows that the ice sheet does not have time to reach equilibrium with low sea level during the LGM: it still continues to thicken temporarily after 20 ka, when sea level has already begun to rise. The greatest elevation is predicted to occur at about 7 ka, followed by a slight decline towards the present as the thinning wave from sea level rise reaches the interior. Because the modeled elevation history is nearly flat through the mid-to-late Holocene, small variations in accumulation rate – which we have assumed to be constant after 10 ka – could change the predicted timing of maximum ice elevation by perhaps a few thousand years.

3.2 Direct Forcing by Ice Margin Changes

An alternative to forcing an ice sheet model with climate and sea level changes is to force the model more directly, using a prescribed history of ice margin changes. By prescribing the position of the ice margin, we are effectively determining the mean mass balance: if accumulation rates are too high in the ice sheet interior, for example, ablation rates can be adjusted accordingly to give the required advance and retreat history. For these calculations, we use the *Fastook and Prentice* [1994] ice sheet model ("FM"). FM is a finite element model with a 111 X 102 node grid and spatial resolution of 20 x 20 km.

For our simulation, the primary inputs are accumulation rates, bedrock topography and bathymetry from *Drewry* [1983]. Accumulation rates are allowed to change as a function of surface slope and elevation

[*Fastook and Prentice*, 1994]; for simplicity, there are no other prescribed climate changes. Sliding occurs where basal ice temperatures are above the pressure melting point, with sliding velocities calculated for Weertman-type sliding [*Weertman*, 1964] with a linear dependence on basal shear stress; this sliding law is less conservative than that used in BJM, where the dependence on normal stress tends to produce less sliding for a given ice thickness. We assume a geothermal heat flux of 0.05 Wm^{-2} as in BJM.

To force changes in the position of the ice margin, we use an ablation term that represents the mass removed by flux of ice across a grounding line into a floating ice shelf from which calving occurs. As in BJM, the magnitude (i.e. mass per time) of ablation is proportional to the average ice column thickness but in this case is controlled by a prescribed calving rate. Calving occurs within any model elements that include the grounding line; these elements are identified as having one or more nodal ice surface elevations that lie at or below the flotation line (the minimum thickness at which an ice column would be grounded).

We initialize FM by setting the calving rate to zero and, starting with present-day ice sheet configuration, allowing the ice sheet to advance for 4900 model years, the amount of time required to reach a maximum position close to the continental shelf edge. Total ice sheet thickness in the western Ross Sea is somewhat higher in the maximum ice sheet reconstructions of *Denton et al.* [1991] and lower in *Kellogg et al.* [1996] than in our model ice sheet. Following the advance phase, calving is turned on and the ice sheet is allowed to retreat for 10,000 years. The advance and retreat simulation is repeated twice to eliminate model instabilities. Calving is held constant at a rate that allows retreat from the maximum position to Ross Island in 3000 years. This value was chosen on the basis of retreat chronologies from terrestrial geologic data [*Stuiver et al.*, 1981; *Denton et al.*, 1989a; *Orombelli et al.*, 1991] and Ross Sea sediment cores [*Licht et al.*, 1996]. With the calving rate held constant, the rate of ice-margin retreat decreases as ice thicknesses decreases, and a nearly stable ice sheet margin results after an additional 2000 to 3000 years. This result – conveniently but perhaps fortuitously – is in good correspondence with sea-level data [*Fairbanks*, 1989; *Bard et al.*, 1990] showing that close-to-modern sea level was reached by about 6 ka. Finally, modern ice sheet configuration is reached at 9000 – 10,000 model years.

We calculated ice sheet elevations throughout the simulated retreat for several locations on the ice sheet. Shown in Figure 4 are results for the Ross Sea/Pine Island Bay ice divide (79° 10' S, 113° 00' W), for Byrd Station (80°01' S, 199°31' W), and for the area of Mount Waesche (77°15' S, 127°00' W), a volcanic nunatak in north Marie Byrd Land. Both the ice divide and Byrd are centrally located on the WAIS, roughly midway between its two main domes. Mt. Waesche is on the flank of the northern dome that is centered on the Executive Committee Range (Figure 1). Also presented in Figure 4 are the results from BJM for the same locations, assuming varying sea level and climate as described in the previous section.

FM predicts maximum elevations in the WAIS interior in the early Holocene, delayed relative to the timing of maximum ice-margin position in the Ross Sea. Results for Mt. Waesche are nearly identical to those from BJM; both models show maximum ice elevations ~50 m above present at ~8 ka. For the location of the present-day ice divide and for Byrd Station, however, the magnitude of thinning following the Holocene maximum is considerably greater, nearly 200 m between ~8 ka and the present. The large magnitude of predicted thinning in the central WAIS, compared with the relatively modest amount of thinning at higher elevations in north Marie Byrd Land, is a key result of the FM simulation. This differential thinning is not a consequence of the assumption that accumulation rates remain constant in FM: when accumulation is held constant in BJM, total Holocene thinning increases by about the same amount at both Byrd and Mt. Waesche. Rather, the important difference between the models appears to be the less conservative sliding law used in FM; we return to this point in section 5.

4. EMPIRICAL CONSTRAINTS ON FORMER ICE ELEVATIONS

4.1 Glacial-geologic Data

Empirical evidence for former ice sheet high stands is apparent in glacial erosional and depositional features on the flanks of mountains and nunataks that penetrate the WAIS [e.g. *Denton et al.*, 1992]. The ages of most such features, however, are unknown. Recently, *Ackert et al.* [1999] obtained cosmogenic isotope exposure ages that provide chronological control for former WAIS high stands on the flank of Mt. Waesche, for which modeling results were presented in the previous section. Two moraines extend up to ~45 m and ~100 m above the present-day ice surface. Additional details on the geologic setting are given in *Ackert et al.* [1999] and are not elaborated on here. The concentrations of *in situ*-produced ³He in boulders collected from the moraine surfaces appear in Table 1. We calculated mean exposure ages for these samples using the ³He production rates of *Cerling and Craig* [1994] and altitude scaling of J. Stone (personal communication, 1999). (See *Lal* [1991], *Nishizuumi et al.* [1993] and references therein for general discussion of cosmogenic isotope exposure age methods; *Kurz* [1986] and *Brook and Kurz* [1993] for specific discussion of analytical methods used for the ³He analyses.) The mean exposure age of samples from the lower moraine is 9.1 ± 1.6 ka. The oldest of these samples (~50 ka) was not included in the calculation of the mean under the assumption that this anomalous date reflects prior exposure history. The data indicate that the most recent high stand occurred during the early Holocene and was delayed by several thousand years relative to the maximum extent of grounded ice in the Ross Sea. The exposure ages on the higher moraine are more scattered but with one exception, significantly older. Such a scattered exposure age distribution precludes a simple mean age calculation. However the data suggest that the WAIS (at Mt. Waesche) has been not higher than 100 m above its present elevation during the last 80 ka or more.

4.2 Ice-core Data

As a complement to the glacial-geologic data, constraints on former WAIS elevations may be obtained from ice core analyses. *Raynaud and Lebel* [1979], *Raynaud and Whillans* [1982], and *Jenssen* [1983] reconstructed past elevation changes for WAIS from gas measurements in the Byrd Station ice core. Total gas content (the gas fraction air/ice) was assumed to be a reliable proxy for ambient atmospheric pressure at the time of bubble close-off. *Jenssen's* [1983] results show elevations 400 – 600 m higher than present during the late Wisconsinan and early Holocene, and a few tens of meters lower than present during the LGM. A significant portion of the Holocene lowering almost certainly reflects lateral advection of ice from higher elevations, rather than lowering of the ice surface at the Byrd drilling site itself [*Budd and Young*, 1983; *Jenssen*, 1983]. An added complication is that total gas content in ice cores is affected not only by ambient air-pressure changes, but also by site-specific factors that influence the volume of open pore space [*Martinerie et al.*, 1992].

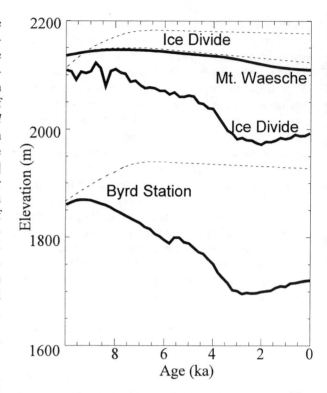

Fig. 4. Calculated elevation at Mount Waesche, the present-day ice divide, and Byrd Station, from the Fastook model (FM), subject to prescribed ice margin retreat over the last 10,000 years (bold lines). Dashed lines show the same from the Budd and Jenssen model (BJM) under combined climate and sea level forcing as in Fig. 3. To simplify comparison the BJM elevations have been shifted vertically so that they coincide with FM at 10 ka. Note that in general absolute model elevations correspond only approximately with modern ice sheet elevations.

An alternative to total gas measurements is to exploit the strong elevation-dependence of stable isotope concentrations on elevation. The nearly linear relationship between $\delta^{18}O$ (or δD) and elevation has been demonstrated empirically both in measurements of precipitation [*Dansgaard*, 1964; *Dansgaard et al.*, 1973] and in atmospheric water vapor [*Benson and White*, 1994]. In the Antarctic, the correlation is strong because air masses are always close to saturation with respect to atmospheric water vapor [*Robin*, 1977]. Adiabatic cooling in rising air masses therefore tends to cause clear air precipitation ("diamond dust") and preferential loss of the heavier isotope [*Jouzel and Merlivat*, 1984]. Subtraction of a given $\delta^{18}O$ profile δ_1 from another δ_2 produces an anomaly record that may be interpreted as a

TABLE 1. MOUNT WAESCHE COSMOGENIC ^{3}HE DATA

Sample	Altitude (m)	R/Ra (melt)	± 1σ	^{4}He (10^{-9} cc/g)	± 1σ	R/Ra (crsh)	± 1σ	Age (ka)	± 1σ
(L) WA-4D-1	2010	11.7	0.1	44.7	0.30	6.27	0.08	10.6	0.2
(L) WA-4D-3^{cpx}	2010	44.0	0.5	3.70	0.09	5.62	0.41	6.2	0.2
(L) WA-4C-2	2015	13.0	0.1	32.8	0.30	6.01	0.12	10.1	0.2
(L) WA-4C-2$^{cpx, r}$	2015	63.5	0.5	3.79	0.02	6.01	0.12	9.6	0.1
(L) WA-4C-1	2015	60.6	0.7	4.24	0.04	6.13	0.10	10.2	0.2
(L) WA-3E-1	1945	485	4.0	2.27	0.05	6.29	0.12	49.5	1.3
(L) WA-3D-1	1985	120	1.0	1.81	0.02	5.30	0.16	9.2	0.2
(L) WA-3D-2	1985	104	1.0	1.70	0.01	5.91	0.15	7.4	0.1
(U) WA-4B-1^{cpx}	2035	66.0	0.3	31.8	0.30	6.10	0.05	83.2	1.1
(U) WA-4B-1	2035	104	1.0	19.4	0.20	6.10	0.05	83.2	1.0
(U) WA-4B-1$^{cpx, r}$	2035	226	1.0	9.13	0.09	6.10	0.05	87.8	1.0
(U) WA-4B-3	2035	2201	30	0.93	0.01	6.05	0.09	87.8	1.5
(U) WA-4A-1	2040	40.9	0.2	40.9	0.40	5.42	0.13	62.1	0.9
(U) WA-4A-1^r	2040	328	2.0	4.49	0.09	5.42	0.13	62.1	1.3
(U) WA-3B-1	2025	174	2.0	2.23	0.03	6.03	0.08	16.1	0.3
(U) WA-3B-2	2025	198	1.0	5.45	0.04	5.63	0.06	47.0	0.5
(U) WA-3B-2^{cpx}	2025	169	1.0	6.37	0.07	6.08	0.19	46.5	0.7
(U) WA-3A-1	2035	1217	5.0	4.49	0.04	7.22	0.18	233.4	2.4
(U) WA-3A-2^{cpx}	2035	80.4	0.6	3.41	0.03	5.90	0.05	10.9	0.1
(U) WA-3A-2	2035	28.1	0.3	10.3	0.10	5.89	0.12	9.8	0.2

Table 1. Cosmogenic isotope data from Mount Waesche. Dotted line separates lower moraine (L) and upper moraine (U) samples. R/Ra (melt) is the ^{3}He/^{4}He ratio relative to the atmospheric ratio (1.384 x 10^{-6}). ^{4}He is the gas released on melting *in vacuo*. R/Ra (crsh) is the ratio of the inherited (magmatic) helium component. The mean crush R/Ra (6.11±.46, n=21) characterizes the mantle source. Samples with superscript "r" are replicates; typically the initial sample was not crushed, the R/Ra crush value of the replicate was used for both samples. All samples were olivine separates unless denoted by superscript "cpx" (clinopyroxene). Additional details are given in *Ackert et al.* [1999].

relative elevation time series, h_δ under the assumption that both sites have experienced the same climate changes; that is:

$$h_\delta(t) = \frac{(\delta_2(t) - \delta_1(t))}{\alpha \Gamma} - (h_1(0) - h_2(0)) \quad (10)$$
$$\alpha = \partial \delta^{18}O / \partial T$$

where T is the surface temperature, Γ is the atmospheric lapse rate and h_2-h_1 is the difference between the absolute elevations of the two sites. *Grootes and Stuiver* [1986; 1987] employed this method, using Dome C as a climate reference point, assumed to have experienced no change in elevation. *Grootes and Stuiver's* results for Byrd indicate higher elevations during the early Holocene. Their results are in reasonable agreement with the gas data cited above; like the gas data, the anomaly time series includes both site-specific elevation changes and lateral advection of ice originally at a higher location. Other possible complications are that elevation changes at Dome C have occurred, that Dome C and Byrd have experienced different climate changes over the period of interest, and that the anomaly time series contains significant noise attributable to inaccuracies in the timescales used for the two sites.

We make a new estimate of elevation changes at Byrd, taking advantage of the stable isotope records recently obtained from the East Antarctic Taylor Dome ice core [*Steig et al.*, 1998c; 1999] (Figure 5). For several reasons, the Taylor Dome core provides a firmer basis for comparison with Byrd than Dome C. First, the timescales available for Byrd and Taylor Dome are significantly more accurate than for Dome C, which at the time of the work cited above [*Grootes and Stuiver*,

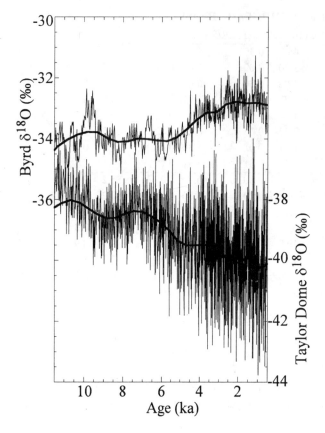

Fig. 5. Time series of stable isotopes ($\delta^{18}O$) for Byrd (top) and Taylor Dome (bottom). Bold lines show low-pass filtered time series used in the calculation of elevation changes.

1986; 1987], was based on a 1-D flow model with a constant vertical strain rate [*Lorius et al.*, 1979]. Second, there is independent evidence from the Dry Valleys glacial geologic record that Taylor Dome has experienced relatively small (<50 m) changes in elevation [*Denton et al.*, 1989a; *Marchant et al.*, 1994]. Numerical flow modeling results for Taylor Dome support this conclusion: because Taylor Dome is relatively shallow, the response time to thermal forcing is rapid, and the tendency for thinning due to lower accumulation rates was approximately balanced by the reduced flow rates of cold LGM ice [*Morse*, 1997]. Third, the location of Taylor Dome relative to Byrd suggests *a priori* that both sites may have experienced comparable climate changes during the Holocene. Although Taylor Dome is situated in East Antarctica, not West Antarctica, it is directly across the Ross Sea from Byrd, and both sites (in today's climate) are influenced by the same cyclonic systems that develop over the Ross and Amundsen Seas [*Carrasco and Bromwich*, 1993; *Waddington and Morse*, 1994]. While Taylor Dome and Byrd experienced demonstrably *different* climate changes during the glacial-interglacial transition [*Steig et al.*, 1998a], climate boundary conditions were clearly very different than today; changes in ice-sheet and/or sea-ice configuration may have altered atmospheric circulation patterns at Taylor Dome during the LGM [*Morse et al.*, 1998]. For this reason we restrict our analysis to the Holocene.

To estimate elevation change at Byrd with Taylor Dome as our reference, we use the most recent published Holocene timescales for each core. For Taylor Dome, this is *st9810* [*Steig et al.*, 1999], which is based on a 2-D finite element flow model [*Morse*, 1997] guided by layer thickness and accumulation estimates from ^{10}Be [*Steig et al.*, 1998b]. Timescale *st9810* is consistent with the pre-Holocene chronology of *Steig et al.* [1998a], derived by trace-gas correlation with Greenland ice cores. For Byrd, we use the layer-counted timescale of *Hammer et al.* [1994], which is compatible with the trace-gas chronology of *Blunier et al.* [1998]. Prior to differencing the two records, we smoothed each using a low-pass filter with a square 0.002 year cut-off frequency to remove high-frequency variations shorter than the precision of each independent timescale. The difference $\delta^{18}O_{Byrd} - \delta^{18}O_{Taylor\ Dome}$ yields an anomaly time series, which we convert to relative elevation using equation (10). We assume that $\alpha = 0.75$ ‰/°C from modern-day spatial measurements [*Morgan*, 1982] and that $\Gamma = 8$°C/km, the appropriate pseudoadiabatic lapse rate [*Jacobson*, 1999]. Various factors, including changes in moisture source [*Charles et al.*, 1994] and seasonality of precipitation [*Steig et al.*, 1994] may influence the effective value of α; for the Holocene, a conservative estimate for the precision of α is ±0.2‰/°C [e.g. *Cuffey et al.*, 1994; *Krinner et al.*, 1997], which suggests a precision of our elevation calculation of about ±25% (ignoring other sources of error).

Figure 6 shows the Byrd-Taylor Dome $\delta^{18}O$ anomaly series and calculated elevation changes for Byrd. The overall decrease of close to 600 m since the early Holocene is in good agreement with the data of *Jenssen* [1983]. As with the gas data, part of this trend is due to the advection of ice from higher elevations rather than to elevation change at the Byrd drilling site itself. To correct for this, we use the results of *Budd and Young* [1983], who calculated ice-flow paths between the present-day ice divide and Byrd using a 2-D flow model. From the ice-flow paths, we can determine the elevation – relative to Byrd – at which precipitation originally accumulated. Although *Budd and Young's* calculations

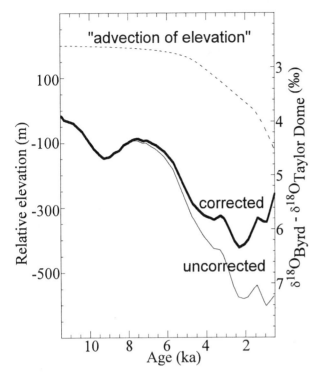

Fig. 6. Relative elevations at Byrd Station over the last 10 ka as estimated from the difference $\delta^{18}O_{Byrd} - \delta^{18}O_{Taylor\ Dome}$. Thin solid line shows elevation change uncorrected for lateral advection of ice from higher elevations. Bold line shows advection-corrected elevation change, using the steady-state flow model results of *Budd and Young* [1983], shown as a dashed line.

assumed a steady-state ice sheet, for each of our model runs the difference between Byrd and ice-divide elevations changes by less than 50 m, or about 0.025 ice thicknesses, between 10 ka and the present. Also, the position of the ice divide does not change significantly during the last 10 ka. Finally, ice of age 10 ka at Byrd is less than 1000 m (0.5 ice-thicknesses) in depth [*Hammer et al.*, 1994], whereas borehole data [*Garfield and Ueda*, 1976] indicate that most of the deformation occurs below a depth of ~0.75 ice thicknesses (i.e. horizontal ice velocities are roughly constant to a depth of ~1500 m [*Whillans*, 1983]). These observations suggest that the relevant Holocene flow paths between the ice divide and Byrd did not vary greatly from steady-state as ice-sheet elevations changed. When the correction for advection of ice from higher elevations is applied to the isotope anomaly series, the results suggest more modest elevation change at Byrd – less than 300 m between 10 ka and 2 ka, followed by a recent rise of up to 100 m (Figure 6).

5. DISCUSSION

5.1 Model/Data Comparison

Figure 7 summarizes data and model results. Calculated elevation changes from both the FM and BJM simulations are in good agreement with the cosmogenic exposure-age data for Mt. Waesche. All three independent estimates show an early Holocene maximum a few tens of meters above present. (Given elevations of moraines at Mt. Waesche are relative to the bedrock surface; if maximum isostatic compensation is taken into account, then ice elevations relative to modern sea level will be somewhat lower; this correction is small relative to other uncertainties.) Model/data agreement for Mt. Waesche occurs for somewhat different reasons in each

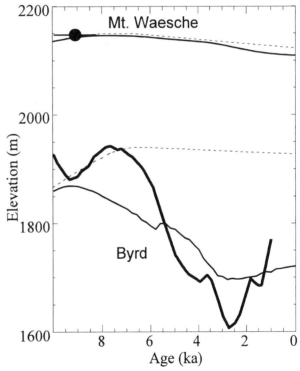

Fig. 7. Comparison of model results and empirical data. Circle shows former ice sheet elevation at Mt. Waesche from the cosmogenic isotope data, with age uncertainties (horizontal bar). Bold line shows relative elevations at Byrd station, corrected for advection as described in the text, from the stable isotope anomaly series from the Byrd ice core. Thin solid lines show results from the Fastook model (FM). Dashed lines show results from the Budd and Jenssen model; these have been shifted to coincide in elevation with the FM results at 10 ka.

model. In the FM results, the early-Holocene maximum is a consequence of our decision to advance the WAIS for only 4900 model years: when retreat begins, the ice sheet is out of equilibrium with its extended grounding line position and is still thickening in the interior. In BJM, the early Holocene maximum is partly a response to increased accumulation rates. If accumulation is held constant in BJM, significantly more thinning occurs, with a total thinning of >100 m since 8 ka.

For Byrd, the BJM and FM results differ considerably. Comparison with the isotope-anomaly results for Byrd clearly supports the latter, which shows net lowering at Byrd of about 200 m since the early Holocene. Indeed both the magnitude and timing of thinning in the FM results are in good agreement with the data, with the most rapid lowering at Byrd occurring between about 7 ka and 3 ka. That the FM results are consistent with both the glacial geologic data at Mt. Waesche and ice-core data from Byrd suggests that FM produces realistic topographic changes. However, while this is encouraging, it does not provide validation of FM: we may be right for the wrong reasons. Obvious criticisms include 1) the specific retreat history that we use to force the model, 2) our simplification that accumulation rates remain constant and 3) the geothermal heat flux we use in both models (0.05 W/m^2) may be both too uniform (see *Blankenship et al.* [1999]) and too low (*Alley and Bentley* [1988], give 0.07 W/m^2 at Byrd and 0.08 W/m^2 towards Siple Coast).

We believe that the Ross Sea maximum configuration and deglaciation estimates are unlikely to be significantly in error. Terrestrial data show ice advancing into the mouth of the Dry Valleys at ~25 ka with retreat from Taylor Valley beginning at ~10 ka, followed by recession from the McMurdo Sound region slightly before 7 ka [*Hall and Denton*, 1999]. Independent of the terrestrial data, sediment-core data from the Ross Sea show ice retreating from Ross Island prior to 7 ka [*Licht et al.*, 1996, *Licht*, 1999]. The maximum grounding line position for the LGM is also supported by sediment cores [*Licht et al.*, 1996; 1999] and seismic imagery [*Shipp et al.*, 1999].

The assumption that accumulation rates during the simulated advance were the same as today may be reasonable because the real ice sheet probably began to advance under relatively warm, high accumulation rate conditions (i.e. during marine isotope stage 3 [*Morse et al.*, 1998]). As noted earlier, the greater predicted thinning at Byrd than at Mt. Waesche is not simply a consequence of the assumption that accumulation rates remain constant in FM; allowing accumulation rates to change, as in the BJM simulation, would cause underprediction of thinning at Mt. Waesche. On the other hand, the spatial distribution of accumulation in FM does change as model elevations change. The parameterization used [*Fastook and Prentice*, 1994] may not be particularly realistic because it assumes that today's climatic gradients applied in the past [*Steig*, 1997]. The importance of this assumption is difficult to address because insufficient empirical data currently exist on which to base rigorous parameterization of the spatial and temporal distribution of accumulation rates for the WAIS.

The main consequence of using a uniform geothermal heat flux of 0.05 W/m^2 in our models may be an underestimation of the area of the ice sheet experiencing sliding. Consequently, we may have generally overestimated the ice-sheet response time, particularly along the Siple Coast. These caveats notwithstanding, we reiterate that a key result from the FM simulation is the *contrast* in the amount of Holocene thinning between central West Antarctica and north Marie Byrd Land. In this respect, the FM simulation is in remarkably good agreement with the empirical data. This suggests that the most important difference between the FM and BJM results is in the parameterization of basal sliding, which should affect ice velocities significantly at Byrd (which is melting at the base) but only indirectly at Mt. Waesche (where the ice is grounded above sea level and is below the pressure melting point). As noted earlier, FM uses a sliding law that tends to produce faster sliding velocities for a given driving stress.

5.2 Conclusions

Several conclusions can be drawn from the model/data comparisons presented in this paper. First, all of the evidence strongly points to the conclusion that interior WAIS elevations have decreased since the early Holocene. This conclusion is consistent with a WAIS component to post-LGM sea level rise [*Peltier*, 1998]. In this context, it is interesting to note that an *increase* in elevation since about 2 ka is observed both in the isotope anomaly series for Byrd and in the Fastook model results. In the model, the late Holocene increase apparently reflects the stabilization of the ice sheet at approximately its current position. A late Holocene increase in WAIS elevation may have contributed to the 1 to 2 m eustatic sea level lowering since ~4 ka documented on mid-oceanic islands (*Goodwin*, [1998] attributed this sea level change to East Antarctic elevation change).

Second, it is clear that West Antarctica has largely adjusted to the accumulation-rate change associated with the last glacial-interglacial transition. It has probably also adjusted fully to glacial-interglacial sea level rise, as indicated by the reduced rate of lowering in both the empirical data and the model results for Byrd. A corrollary is that any changes occurring now – including documented thinning of the inland ice and thickening of the ice streams [*Alley and Whillans*, 1991] – must reflect climate and sea level forcings that have occurred much more recently than the glacial-interglacial transition, or may reflect the complex dynamics of ice streams that are largely independent of, and may force sea level and climate [*Whillans and van der Veen*, 1993].

Third, the early Holocene maximum that appears in the model simulations, and is clearly supported by the empirical evidence, provides support for the concept of a non-equilibrium LGM ice sheet [e.g. *Kellogg et al.*, 1996]. Because the WAIS did not apparently have sufficient time to equilibrate with its extended grounding line position during the LGM, interior elevations continued to thicken during initial sea level rise and grounding line retreat. Indeed, had the WAIS reached an equilibrium configuration with lower sea level, it would have been significantly thicker during the LGM than it was during the early Holocene, which is difficult to reconcile with either the Byrd or the Mt. Waesche data.

Finally, our results illustrate the sensitivity of ice sheet models to the parameterization of basal sliding. The next stage of model development will need to include a more explicit treatment of spatial variations in sliding and in particular, of the physics of ice streams. Although both BJM and FM do produce fast-flowing ice stream-like features, neither allows for the complexity of basal hydrologic gradients, subglacial till deformation, or other features suspected as important controlling factors in ice stream behavior [*Jackson and Kamb*, 1997; *Anandakrishnan et al.*, 1998].

Acknowledgments. We thank R. Alley, J. Crider, and two anonymous reviewers for helpful comments on the manuscript. This work was supported by the National Science Foundation's Office of Polar Programs under grants 9318872, 9418333, 9526348, 9526979, and 9614287. We are grateful to W. Budd and R. Warner of the Antarctic CRC for discussions of the BJM model.

REFERENCES

Ackert, R. P., Jr., D. J. Barclay, H. W. Borns, P. E. Calkin, M. D. Kurz, E. J. Steig and J. L. Fastook, Measurement of past ice sheet elevations in interior West Antarctica, *Science*, in press, 1999.

Alley, R. B. and C. R. Bentley, Ice-core analysis on the Siple Coast of West Antarctica, *Annal. Glaciol.*, 11, 1-7, 1988.

Alley, R. B. and I. M. Whillans, Changes in the West Antarctic Ice Sheet, *Science*, 254, 959-963, 1991.

Alley, R. B. and I. M. Whillans, Response of the East Antarctica ice sheet to sea-level rise, *J. Geophys. Res.*, 89, 6487-6493, 1984.

Alley, R. B., D. D. Blankenship, C. R. Bentley and S. T. Rooney, Deformation of till beneath ice stream B, West Antarctica, *Nature*, 322, 57-59, 1986.

Anandakrishnan, S., D. D. Blankenship, R. B. Alley and P. L. Stoffa, Influence of subglacial geology on the position of a West Antarctic ice stream from seismic observations, *Nature*, 394, 62-65, 1998.

Anderson, J. B., D. D. Kurtz, E. W. Domack and K. M. Balshaw, Glacial and glacial marine sediments of the Antarctic continental shelf, *J. Geol.*, 88, 399-414, 1980.

Bard, E., B. Hamelin and R. G. Fairbanks, U-Th ages obtained by mass spectrometry in corals from Barbados: sea level during the past 130,000 years, *Nature*, 346, 456-458, 1990.

Benson, L. V. and J. W. C. White, Behavior of the stable isotopes of oxygen and hydrogen in the Truckee River - Pyramid Lake surface water system, Part 3: Source of water vapor over Pyramid Lake, Nevada, *Limnol. Oceanog.*, 39, 1945-1958, 1994.

Bentley, M. J. and J. B. Anderson, Glacial and marine geological evidence for the ice sheet configuration in the Weddell Sea-Antarctic Peninsula region during the Last Glacial Maximum, *Antarc. Sci.*, 10, 309-325, 1998.

Bindschadler, R., Future of the West Antarctic Ice Sheet, *Science*, 282, 428-429, 1998.

Bindschadler, R. A., S. N. Stephenson, D. R. Macayeal and S. Shabtaie, Ice dynamics at the mouth of Ice Stream B, Antarctica, *J. Geophys. Res.*, 92, 8885-8894, 1987.

Bindschadler, R. A., P. Vornberger, D. Blankenship, T. Scambos and R. Jacobel., Surface velocity and mass balance of Ice Streams D and E, West Antarctica, *J. Glaciol.*, 42, 461-475, 1996.

Blankenship, D. D., D. L. Morse, C. A. Finn, R. E. Bell, M. E. Peters, S. D. Kempf, S. M. Hodge, M. Studinger, J. C. Behrendt and J. M. Brozena. Geologic controls on the initiation of rapid basal motion for West Antarctic ice streams: A geophysical perspective including new airborne radar sounding and laser altimetry results. *Ant. Res. Ser.*, this volume, 1999.

Blunier, T., J. Chappellaz, J. Schwander, A. Dallenbach, B. Stauffer, T. F. Stocker, D. Raynaud, J. Jouzel, H. B. Clausen, C. U. Hammer and S. J. Johnson, Asynchrony of Antarctic and Greenland climate change during the last glacial period, *Nature*, 394, 739-743, 1998.

Bockheim, J. G., S. C. Wilson, G. H. Denton, B. G. Andersen

and M. Stuiver, Late Quaternary ice-surface fluctuations of Hatherton Glacier, Transantarctic Mountains, *Quat. Res.*, *31*, 229-245, 1989.

Brook, E. J. and M. D. Kurz, Surface-exposure chronology using in Situ cosmogenic ^{3}He in Antarctic quartz sandstone boulders., *Quat. Res.*, *39*, 1-10, 1993.

Budd, W. F. and N. Young, Application of modeling techniques to measured profiles of temperature and isotopes., in *The Climatic Record in Polar Ice Sheets*, edited by G. de Q. Robin, pp. 150-177, Cambridge University Press, Cambridge, 1983.

Budd, W. F. and D. Jennsen, The dynamics of the Antarctic ice sheet, *Annal. Glaciol.*, *12*, 16-22, 1989.

Budd, W. F., P. L. Keage and N. A. Blundy, Empirical studies of ice sliding, *J. Glaciol.*, *23*, 157-170, 1979.

Budd, W. F., D. Jenssen, E. Mavrakis and B. Coutts, Modelling the Antarctic ice-sheet changes through time, *Annal. Glaciol.*, *20*, 291-297, 1994.

Budd, W. F., B. Coutts and R. C. Warner, Modelling the Antarctic and Northern Hemisphere ice-sheet changes with global climate through the glacial cycle, *Annal. Glaciol.*, *27*, 153-160, 1998.

Calkin, P. E. and C. Bull, Interaction of the East Antarctic ice sheet, alpine glaciations and sea level in the Wright Valley areas, Southern Victoria Land, in *Antarctic Geology and Geophysics*, edited by R. J. Adie, pp. 435-440, Universitetsforlaget, Oslo, 1972.

Carrasco, J. F. and D. H. Bromwich, Satellite and automatic weather station analyses of katabatic surges across the Ross Ice Shelf, *Ant. Res. Ser.*, *61*, 93-108, 1993.

Cerling, T. E. and H. Craig, Cosmogenic ^{3}He production rates from 39°N to 46°N latitude, western USA and France, *Geochim. Cosmochim. Acta*, *58*, 349-255, 1994.

Chappell, J. and N. J. Shackleton, Oxygen isotopes and sea level, *Nature*, *324*, 137-140, 1986.

Charles, C. D., D. Rind, J. Jouzel, R. D. Koster and R. G. Fairbanks, Glacial-interglacial changes in moisture sources for Greenland: influences on the ice core record of climate, *Science*, *263*, 508-511, 1994.

Cuffey, K. M., R. B. Alley, P. G. Grootes, J. M. Bolzan and S. Anandakrishan, Calibration of the δ^{18}O isotopic paleothermometer for central Greenland, using borehole temperatures, *J. Glaciol.*, *40*, 341-349, 1994.

Cuffey, K. M. and G. D. Clow, Temperature, accumulation and ice sheet elevation in central Greenland through the last deglacial transition, *J. Geophys. Res.*, *102*, 26383-26396, 1997.

Dansgaard, Stable isotopes in precipitation, *Tellus*, *16*, 436-468, 1964.

Dansgaard, W., S. J. Johnsen, H. B. Clausen and N. Gundestrup, Stable isotope glaciology, *Medd. Groenl.*, *197*, 1-53, 1973.

Denton, G. H., T. J. Hughes and W. Karlen, Global ice-sheet system interlocked by sea level., *Quat. Res.*, *26*, 3-26, 1986.

Denton, G. H., J. G. Bockheim, S. C. Wilson and M. Stuiver, Late Wisconsin and Early Holocene Glacial History, Inner Ross Embayment, Antarctica, *Quat. Res.*, *31*, 151-182, 1989a.

Denton, G. H., J. G. Bockheim, S. C. Wilson, J. E. Leide and B. G. Anderson, Late Quaternary ice-surface fluctuations of Beardmore Glacier, Transantarctic Mountains, *Quat. Res.*, *31*, 183-209, 1989b.

Denton, G. H., M. L. Prentice and L. H. Burckle, Cainozoic history of the Antarctic ice sheet, in *The Geology of Antarctica*, edited by R. J. Tingey, pp. 365-433, Clarendon Press, Oxford, 1991.

Denton, G. H., J. G. Bockheim and R. H. Rutford, Glacial history of the Ellsworth Mountains, West Antarctica, *Geol. Soc. Amer. Mem.*, *170*, 403-432, 1992.

Domack, E. W., E. A. Jacobsen, S. S. Shipp and J. B. Anderson, Sedimentologic and stratigraphic signature of the Late Pleistocene/Holocene fluctuation of the West Antarctic Ice Sheet in the Ross Sea: A new perspective, Part 2, *Geol. Soc. Amer. Bull.*, in press, 1999.

Drewry, D. J., *Antarctica: glaciological and geophysical folio*, Scott Polar Research Institute, Cambridge, 1983.

Fairbanks, R. D., A 17,000-year glacio-eustatic sea level record: influence of glacial melting rates on the Younger Dryas event and deep-ocean circulation, *Nature*, *342*, 637-642, 1989.

Fastook, J. L. and M. Prentice, A finite-element model of Antarctica: sensitivity test of meteorological mass-balance relationships, *J. Glaciol.*, *40*, 167-175, 1994.

Garfield, D. E. and H. T. Ueda, Resurvey of the Byrd station, Antarctica, drill hole, *J. Glaciol.*, *17*, 29-34, 1976.

Goodwin, I. D., Did changes in Antarctic ice volume influence late Holocene sea-level lowering?, *Quat. Sci. Rev.*, *17*, 319-332, 1998.

Grootes, P. M. and M. Stuiver, Ross Ice Shelf oxygen isotopes and West Antarctic climate history, *Quat. Res.*, *26*, 49-67, 1986.

Grootes, P. M. and M. Stuiver, Ice sheet elevation changes from isotope profiles, in *The physical basis of ice sheet modelling*, edited by E. D. Waddington and J. S. Walder, pp. 269-281, IAHS-AISH, 1987.

Hall, B. L. and G. H. Denton, New relative sea-level curves for the southern Scott Coast, Antarctica: Evidence for Holocene deglaciation of the western Ross Sea, *Journal of Quaternary Science*, in press, 1999.

Hammer, C. U., H. B. Clausen and C. C. Langway, Electrical conductivity method (ECM) stratigraphic dating of the Byrd Station ice core, Antarctica, *Annal. Glaciol.*, *20*, 115-120, 1994.

Higgins, S. M., C. H. Hendy and G. H. Denton, Geochronology of Bonney Drift, Taylor Valley, Antarctica: Evidence for interglacial expansions of Taylor Glacier, *Geogr. Ann.*, in press, 1999.

Huybrechts, P., A 3-D model for the Antarctic ice sheet; a sensitivity study on the glacial-interglacial contrast, *Clim. Dyn.*, *5*, 79-92, 1990a.

Huybrechts, P., Response of the Antarctic Ice Sheet to future greenhouse warming., *Clim. Dyn.*, *5*, 93-102, 1990b.

Huybrechts, P., The Antarctic ice sheet during the last glacial-interglacial cycle: a three dimensional experiment, *Annal. Glaciol.*, *5*, 115-119, 1990c.

Huybrechts, P., Glaciological modelling of the late Cenezoic

East Antarctic Ice Sheet: stability or dynamism?, *Geogr. Ann.*, *75A*, 221-238, 1993.

Jackson, M. and B. Kamb, The marginal shear stress of Ice Stream B, West Antarctica, *J. Glaciol.*, *43*, 415-426, 1997.

Jacobson, M. Z., *Fundamentals of Atmospheric Modeling*, Cambridge University Press, 1999.

Jenssen, D., Elevation and climate changes from total gas content and stable isotopic measurements, in *The climatic record in polar ice sheets*, edited by G. d. Robin, pp. 138-144, Cambridge University Press, London, 1983.

Jouzel, J. and L. Merlivat, Deuterium and oxygen-18 in precipitation: modeling of the isotopic effects during snow formation, *J. Geophys. Res.*, *89*, 11749 - 11757, 1984.

Jouzel, J., C. Lorius, J. R. Petit, C. Genthon, N. I. Barkov, V. M. Kotlyakov and V. M. Petrov, Vostok ice core: a continuous isotope temperature record over the last climatic cycle (160,000 years), *Nature*, *329*, 403-407, 1987.

Jouzel, J., G. Raisbeck, J. P. Benoist, F. Yiou, C. Lorius, D. Raynaud, J. R. Petit, N. I. Barkov, Y. S. Korotkevitch and V. M. Kotlyakov, A comparison of deep Antarctic ice cores and their implications for climate between 65,000 and 15,000 years ago, *Quat. Res.*, *31*, 135-150, 1989.

Kamb, B., Sliding motion of glaciers: Theory and observation, *Rev. Geophys. Space Phys.*, *8*, 673-728, 1970.

Kellogg, T. B., T. Hughes and D. E. Kellogg, Late Pleistocene interactions of East and West Antarctic ice-flow regimes: evidence from the McMurdo Ice Shelf, *J. Glaciol.*, *42*, 486-500, 1996.

Krinner, G., C. Genthon and J. Jouzel, GCM analysis of local influences on ice core δ signals, *Geophys. Res. Lett.*, *24*, 2825-2828, 1997.

Kurz, M. D., *In situ* production of terrestrial cosmogenic helium and some applications to geochronology, *Geochim. Cosmochim. Acta*, *50*, 2855-2862, 1986.

Lal, D., Cosmic ray labeling of erosion surfaces: *In situ* production rates and erosion models, *Earth Planet. Sci. Lett.*, *104*, 424-439, 1991.

Licht, K. J., Investigations into the Late Quaternary history of the Ross Sea, Antarctica, Ph.D. thesis, University of Colorado, Boulder, 1999.

Licht, K. J., A. E. Jennings, J. T. Andrews and K. M. Williams, Chronology of the late Wisconsin ice retreat from the western Ross Sea, Antarctica, *Geology*, *24*, 223-226, 1996.

Licht, K. J., N. W. Dunbar, A. J. T and A. E. Jennings, Distinguishing subglacial till and glacial marine diamictons in the western Ross Sea, Antarctica: Implications for last glacial maximum, *Geol. Soc. Amer. Bull.*, *111*, 91-103., 1999.

Lorius, C., L. Merlivat, J. Jouzel and M. Pourchet, A 30,000 yr isotope climatic record from Antarctic ice, *Nature*, *280*, 644-648, 1979.

Lorius, C., J. Jouzel, C. Ritz, L. Merlivat, N. E. Barkov and Y. S. Korotkevich, A 150,000-year climatic record from Antarctic ice, *Nature*, *316*, 591-595, 1985.

MacAyeal, D. R., Irregular oscillations of the West Antarctic ice sheet, *Nature*, *359*, 29-32, 1992.

Marchant, D. R., G. H. Denton, J. G. Bockheim, S. C. Wilson and A. R. Kerr, Quaternary changes in level of the upper Taylor Glacier, Antarctica: implications for paleoclimate and East Antarctic Ice sheet dynamics, *Boreas*, *23*, 29-43, 1994.

Martinerie, P., D. Raynaud, D. M. Etheridge, J.-M. Barnola and D. Mazaudier, Physical and climatic parameters which influence the air content in polar ice, *Earth Planet. Sci. Lett.*, *112*, 1-13, 1992.

Mavrakis, E., Time dependent, three-dimensional modelling of the dynamics of large ice masses, M.S. thesis, University of Melbourne, 1993.

Morgan, V. I., Antarctic ice sheet surface oxygen isotope values, *J. Glaciol.*, *28*, 315-323, 1982.

Morse, Glacier Geophysics at Taylor Dome, Antarctica, Ph.D. thesis, University of Washington, 1997.

Morse, D. L., E. D. Waddington and E. J. Steig, Ice age storm trajectories inferred from radar stratigraphy at Taylor Dome, Antarctica., *Geophys. Res. Lett.*, *25*, 3383-3386, 1998.

Nishiizumi, K., C. P. Kokl, J. R. Arnold, R. Dorn, J. Kelin, D. Fink, R. Middleton and D. Lal, Role of *in situ* cosmogenic nuclides ^{10}Be and ^{26}Al in the study of diverse geomorphic processes, *Earth Surf. Proc. Landform.*, *18*, 407-425, 1993.

Nye, J. F., The response of glaciers and ice sheets to seasonal and climatic changes, *Proc. Roy. Soc.*, *A256*, 559-584, 1960.

Orombelli, G., C. Baroni and G. H. Denton, Late Cenozoic glacial history of the Terra Nova Bay region, Northern Victoria Land, Antarctica, *Geograf. Fis. Dinam. Quat.*, *13*, , 1991.

Peltier, W. R., Postglacial variations in the level of the sea; implications for climate dynamics and solid-Earth geophysics, *Rev. Geophys.*, *36*, 603-689, 1998.

Raisbeck, G. M. and F. Yiou, ^{10}Be in polar ice and atmospheres, *Annal. Glaciol.*, *7*, 138-140, 1985.

Raisbeck, G. M., F. Yiou, D. Bourles, C. Lorius, J. Jouzel and N. I. Barkov, Evidence for two intervals of enhanced ^{10}Be deposition in Antarctic ice during the last glacial period, *Nature*, *326*, 273 - 277, 1987.

Raynaud, D. and B. Lebel, Total gas content and surface elevation of polar ice sheets, *Nature*, *281*, 289-291, 1979.

Raynaud, D. and I. M. Whillans, Air content of the Byrd core and past changes in the West Antarctic Ice Sheet, *Annal. Glaciol.*, *3*, 269-273, 1982.

Robin, G. de Q., Ice cores and climatic changes, *Phil. Trans. Roy. Soc.*, *B280*, 143-168, 1977.

Shipp, S. S., J. B. Anderson and E. W. Domack, Seismic signature of the Late Pleistocene fluctuation of the West Antarctic Ice Sheet system in the Ross Sea: A new perspective, Part 1, *Geol. Soc. Amer. Bull.*, in press, 1999.

Steig, E. J., How well can we parameterize past accumulation rates in polar ice sheets?, *Annal. Glaciol.*, *25*, 418-422, 1997.

Steig, E. J., P. M. Grootes and M. Stuiver, Seasonal precipitation timing and ice core records, *Science*, *266*, 1885-1886, 1994.

Steig, E. J., E. J. Brook, J. W. C. White, C. M. Sucher, M. L. Bender, S. J. Lehman, E. D. Waddington, D. L. Morse and G. D. Clow, Synchronous climate changes in Antarctica and the North Atlantic, *Science*, *282*, 92-95, 1998a.

Steig, E. J., D. L. Morse, E. D. Waddington and P. J. Polissar, Using the sunspot cycle to date ice cores, *Geophys. Res. Lett.*, *25*, 163-166, 1998b.

Steig, E. J., C. Hart, J. W. C. White, W. L. Cunningham, M. D. Davis and E. S. Saltzman, Changes in climate, ocean and ice-sheet conditions in the Ross Embayment, Antarctica, at 6 ka, *Annal. Glaciol.*, *27*, 305-310, 1998c.

Steig, E. J., D. L. Morse, E. D. Waddington, M. Stuiver, P. M. Grootes, P. A. Mayewski, M. S. Twickler and S. I. Whitlow, Wisconsinan and Holocene climate history from an ice core at Taylor Dome, western Ross Embayment, Antarctica, *Geogr. Ann., in press,* 1999.

Stuiver, M., G. H. Denton, T. J. Hughes and J. L. Fastook, History of the marine ice sheet in West Antarctica during the last glaciation, a working hypothesis., in *The Last Great Ice Sheets*, edited by G. H. Denton and T. H. Hughes, pp. 319-436, Wiley-Interscience, New York, 1981.

Verbitsky, M. Y. and R. J. Oglesby, The CO_2-induced thickening/thinning of the Greenland and Antarctic ice sheets as simulated by a GCM (CCM1) and an ice-sheet model., *Clim. Dyn.*, *11*, 247-253, 1995.

Verbitsky, M. and B. Saltzman, Behavior of the east Antarctic ice sheet as deduced from a coupled GCM/ice-sheet model, *Geophys. Res. Lett.*, *22*, 2913-2916, 1995.

Verbitsky, M. and M. Saltzman, Modeling the Antarctic ice sheet, *Annal. Glaciol.*, *25*, 259-268, 1997.

Waddington, E. D. and D. L. Morse, Spatial variations of local climate at Taylor Dome: implications for paleoclimate from ice cores, *Annal. Glaciol.*, *20*, 219-225, 1994.

Warner, R. C. and W. F. Budd, Modelling the long-term response of the Antarctic ice sheet to global warming, *Annal. Glaciol.*, *27*, 161-168, 1998.

Weertman, J., On sliding of glaciers, *J. Glaciol.*, *3*, 33-38, 1957.

Weertman, J., Glacier sliding, *J. Glaciol.*, *5*, 287-303, 1964.

Whillans, I. M., Ice movement, in *The Climatic Record in Polar Ice Sheets*, edited by G. de Q. Robin, pp. 70-76, Cambridge University Press, Cambridge, 1983.

Whillans, I. M. and C. J. van der Veen, Controls on changes in the West Antarctic Ice Sheet, in *Ice in the Climate System*, edited by W. R. Peltier, pp. 47-54, Springer-Verlag, Berlin, 1993.

Robert P. Ackert, Jr., MIT/WHOI Joint Program, MS 25, Clark 419, Woods Hole Oceanographic Institution, Woods Hole, MA 02543

James L. Fastook, Institute for Quaternary Sciences, 5790 Bryand Global Sciences Center, University of Maine, Orono, ME 04469.

Ian D. Goodwin, Antarctic CRC, University of Tasmania, GPO Box 252-80, Hobart, Tasmania, 7001, Australia.

Kathy J. Licht, INSTAAR, Campus Box 450, University of Colorado, Boulder, CO 80309.

Eric J. Steig, Department of Earth and Environmental Science, 251 Hayden Hall, 240 South 33rd Street, University of Pennsylvania, Philadelphia, PA 19104.

James W. C. White, INSTAAR, Campus Box 450, University of Colorado, Boulder, CO 80309.

Christopher Zweck, Geophysical Institute, University of Alaska Fairbanks, P.O. Box 757320, 903 Koyukuk Drive., Fairbanks, AK 99775.

THE EL NIÑO-SOUTHERN OSCILLATION MODULATION OF WEST ANTARCTIC PRECIPITATION

David H. Bromwich[1] and Aric N. Rogers

Polar Meteorology Group, Byrd Polar Research Center, The Ohio State University, Columbus, Ohio

Atmospheric numerical analyses from the European Centre for Medium-Range Weather Forecasts (ECMWF) are used to compute the moisture flux convergence, and thus the net flux of water to the Earth's surface, or P-E, for a sector of West Antarctica (120° W -180°, 75° S -90° S) for 1980-1998. P-E closely approximates snow accumulation for most parts of the ice sheet. As found in a previous study using the same data through 1994, a close relation between P-E and the Southern Oscillation Index (SOI) is present for the entire period, including the newly added years from 1995-1998. The variations are in phase for the early 1980s to 1990, and then abruptly switch to antiphase for 1990-1998. Ground-based accumulation and deuterium isotope concentrations provided by annually and sub-annually dated ice cores from the sector provide strong qualitative support for the derived accumulation variations. The atmospheric circulation is examined for the strong El Niño events in 1982/83 and 1997/98. These respectively occur before and after the switch in association between the SOI and P-E, and are characterized by low and high precipitation amounts. In the former case the local atmospheric circulation is organized to deliver limited amounts of moisture to the sector while during the 1997/98 event moisture flows into the sector directly from the ocean in a deep atmospheric layer. On the broadscale, the circulation characteristics are quite similar for both El Niño periods.

1. INTRODUCTION

The pronounced sea-surface temperature (SST) warming every few years in the central and eastern parts of the tropical Pacific Ocean with slight cooling in the far western tropical Pacific is known as the El Niño phenomenon. This has an associated atmospheric variation called the Southern Oscillation which is an exchange of atmospheric mass between the Indonesia area and the southeastern Pacific Ocean. It is monitored by the Southern Oscillation Index (SOI) which is the normalized sea-level pressure difference between Tahiti (18° S, 150° W) and Darwin, Australia (12° S, 131° E). In the normal situation, the warmest Pacific SSTs are located in the western Pacific Ocean with cooler SSTs in the eastern tropical Pacific Ocean. This phase can become enhanced and is known as a La Niña event. The combined ocean-atmosphere cycle is known as the El Niño-Southern Oscillation (ENSO) phenomenon. The ENSO cycle is known to exert global, although variable impacts, on the atmospheric circulation [e.g., *Philander*, 1990; *Trenberth*, 1991]. The frequency and intensity of El Niño events has increased markedly since the mid 1970s, leading some to infer that this change is related to the expected warming of the global atmosphere in conjunction with the major increases of greenhouse gases [e.g., *Trenberth and Hoar*, 1996; *Timmermann et al.*, 1999].

A marked impact of the ENSO cycle in higher latitudes of the South Pacific Ocean has been known for some time. This paper re-examines the finding that the net precipitation (or precipitation minus evaporation/ sublimation) for part of West Antarctica (75° to 90° S,

---

[1]Additional Affiliation: Atmospheric Sciences Program, Department of Geography, The Ohio State University, Columbus, Ohio.

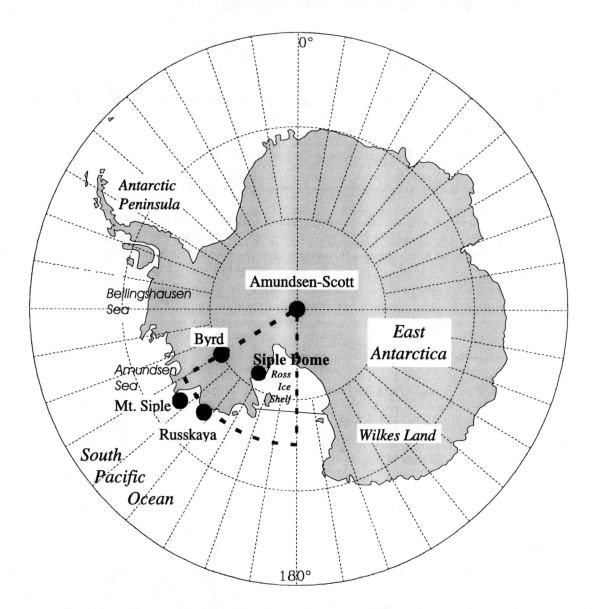

Fig. 1. Antarctic continent with the West Antarctic P-E sector indicated by dashed lines (adapted from *Cullather et al.* [1996]).

120° W to 180° - see Figure 1) is strongly correlated with the SOI from the early 1980s to 1990 and then abruptly switches to become anticorrelated after 1990 [*Cullather et al.*, 1996]. *van Loon and Shea* [1987] show that there is a close relationship between SST variations in the central tropical Pacific Ocean and the anomalies of sea-level pressure throughout the middle and higher latitude parts of the South Pacific Ocean. A related study by *Karoly* [1989] demonstrates that El Niño-anomalies of 500 hPa height arise because of Rossby wave propagation from the anomalous convective heating in the central Pacific associated with the SST warming. The chain of positive and negative anomalies stretching from the central tropical Pacific to southern South America is known as the Pacific-South American pattern (PSA), in analogy to the well-known and El Niño-related Pacific-North American (PNA) pattern. The 500 hPa anomalies support opposite variations in the South Pacific jet streams, with the subtropical jet (STJ) near 30° S strengthening and the polar front jet (PFJ) near 60° S weakening. *Chen et al.* [1996] find that the STJ weakens during La Niña events and the PFJ strengthens, although the latter association is not as pronounced as the former. Thus, there is a marked oscillation in the behavior of the split jet stream in the South Pacific Ocean as a result of tropical SST variations. Pronounced impacts on the atmospheric behavior over those parts of West Antarctica bordering the South Pacific Ocean are thus expected to result from the ENSO cycle. Confirmation of this expectation is provided by *Gloersen* [1995] who finds that the sea-ice areas in the Ross, Amundsen and Bellingshausen seas exhibit marked ENSO variability on time scales of 2.3 to 3.5 years. In addition, *Ledley and Huang* [1997] find that tropical Pacific SST variations precede much smaller but similar variations in Ross Sea SSTs by about three months. Similarly, sea-ice areas in the same region are reduced during ENSO warm events.

The marked warming of near-surface air temperatures on the west side of the Antarctic Peninsula [*King*, 1994] since the 1940s has been well documented, and is associated with increased winter precipitation amounts [*Turner et al.*, 1997] and retreat of ice shelves along both sides of the Peninsula [*Vaughn and Doake*, 1996]. *Jacobs and Comiso* [1997] find that the sea-ice cover in the Amundsen and Bellingshausen seas has decreased substantially in the two decades following 1973 with a subsequent increase. Superimposed on the warming trend is a definite 5-year cycle which is probably related to the ENSO cycle [*Stark*, 1994]. *Marshall and King* [1998] find that a large part of the winter temperature variations at Faraday (65° S, 64° W) is related to SST variations in the central tropical Pacific which affect the behavior of the split jet over the South Pacific Ocean. Local effects due to sea-ice variations in the Bellingshausen Sea also play a significant role in modulating the winter air temperatures.

Cullather et al. [1996] examine the snow accumulation variability over part of West Antarctica (Figure 1), as calculated from the atmospheric moisture budget (explained below) using operational atmospheric analyses from the European Centre for Medium-Range Weather Forecasts (ECMWF). They find that the derived snow accumulation is strongly correlated with the SOI from the early1980s to 1990 and then abruptly switches to become strongly anticorrelated for 1990-1998 (see Figure 2, which is discussed in greater detail later). This finding is surprising in view of the consistent relationship between central tropical SST variations and the circulation anomalies found by *van Loon and Shea* [1987]. Re-examination of this finding and some exploration of its consequences form the basis of this manuscript. Section 2 summarizes recent work directed toward validating this relationship. Section 3 compares the derived accumulation amounts with ground-based accumulation measurements from Siple Dome and Byrd Station area, and with a sub-annually resolved deuterium isotope record from Siple Dome. This is followed in Section 4 by an examination of the atmospheric circulation and derived accumulation

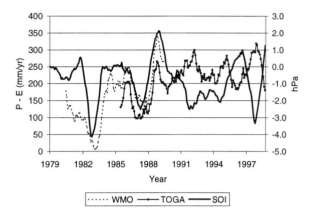

Fig. 2. Annual running mean of P-E (left scale) from monthly values for the West Antarctic sector calculated from ECMWF WMO and TOGA data (updated through March 1999) along with the SOI (right scale).

amounts for the very strong El Niño events of 1982/83 and 1997/98, which are characterized by very different accumulation amounts over the West Antarctic sector. Section 5 gives conclusions.

2. WEST ANTARCTIC ENSO VARIABILITY

2.1 Atmospheric Moisture Budget and Atmospheric Numerical Analyses

Before proceeding to the re-examination and extension of the work by *Cullather et al.* [1996], it is first necessary to provide a brief outline of the atmospheric moisture budget and its use to calculate accumulation amounts from ECMWF atmospheric analyses. An atmospheric column from the surface of the Earth to the top of the atmosphere is considered. The terms describing changes in the amount of water vapor in the column are the net flux of water vapor to the surface, or the precipitation minus evaporation/sublimation (P-E, term 1), the change in water vapor storage (term 2 below), and the net flux of water vapor across the vertical boundaries of the column (term 3). This is expressed by

$$\left\langle \overline{P} - \overline{E} \right\rangle = -\frac{\overline{\partial \langle W \rangle}}{\partial t} - \frac{1}{A} \oint \left\{ \int_0^{P_{sfc}} \frac{\overline{q\mathbf{V}}}{g} dp \right\} \cdot \mathbf{n} dl \quad (1)$$

where

$$W = \int_0^{P_{sfc}} \frac{q}{g} dp \quad (2)$$

Here, P is the precipitation rate, E is the rate of evaporation/sublimation, W is the column precipitable water vapor, A is the area over which P-E is calculated, g is gravity, q is specific humidity, P_{sfc} is the surface pressure, $\mathbf{V}$ is the horizontal wind field, and $\mathbf{n}$ is the outward normal to the area perimeter. Angle brackets indicate an areal average and overbars indicate a time average. The contributions from liquid and solid phases of water to the budget are very small in relation to water vapor [*Peixoto and Oort*, 1992]. This approach is suitable for areally averaged estimates of P-E (typically for 10^6 km^2 or more) over seasonal and longer time scales from atmospheric measurements of moisture content and winds, such as from radiosondes or from atmospheric analyses. It is often used because accurate areal averages of precipitation and evaporation are not available for most parts of the Earth. The ECMWF analyses are used for the present study because previous examinations have shown these analyses to be the most reliable of those which are widely available [*Bromwich et al.*, 1995; *Cullather et al.*, 1997]. For most parts of Antarctica, P-E closely approximates snow accumulation because the effects of drift snow redistribution are small overall, and except for right at the coast the impact of snow melting and runoff is negligible [*Bromwich*, 1990; *Giovinetto et al.*, 1992]. So to a good approximation the right-hand side of (1) (excluding storage term changes in precipitable water, which are found to be small on interannual time scales) can be approximated as areally- and temporally-averaged snow accumulation. This approximation is adopted here.

Operational analyses serve as the initial fields for global numerical weather prediction models, and these analyses are constantly being enhanced for the benefit of improved weather forecasts. Examples are increases in analysis resolution and improved parameterizations for key subgrid-scale processes. This has led to the concept of reanalyses where the atmospheric observations are re-evaluated with a fixed data assimilation scheme which is not subject to the artificial discontinuities (i.e., non-climatic signals) that characterize operational analyses. Reanalyses are completed by the National Centers for Environmental Prediction (NCEP) in conjunction with the National Center for Atmospheric Research (NCAR) for 1949-present [*Kalnay et al.*, 1996] and by ECMWF for 1979-1993 (known as ERA-15) [*Gibson et al.*, 1997]. The NCEP/NCAR reanalyses are not considered here because of known limitations over and around Antarctica [*Hines et al.*, 1999; *Bromwich et al.*, 1999; *Hines and Bromwich*, 2000].

The ECMWF Tropical Oceans Global Atmosphere (TOGA) operational analyses (1985-present) used here are available at 14 pressure levels twice a day. They are prepared with an assimilation model which in 1991 went to a horizontal resolution of T213 (varies before that date - see *Trenberth* [1992]) and most recently has gone to a horizontal resolution of T319 (or equivalently 56 km). The data are subsampled to a resolution of 2.5° by 2.5°, and were acquired from NCAR. The ERA are prepared 4 times a day with a T106 (or about 100 km) resolution and have 17 pressure levels. The data are also subsampled to 2.5° by 2.5° and were acquired from NCAR.

ERA-15 also produces forecast values of P and E; these derived quantities are quite dependent on the parameterizations used in the forecast model. The forecast values are not used here because *Genthon and Krinner* [1998] show that these values are much lower than P-E produced from the reanalyses by use of the atmospheric moisture budget. As the latter approach is much more closely linked to the observations, it is the preferred method. *Cullather et al.* [2000] demonstrate the same shortcomings in forecast P-E values from ERA-15 for the Arctic basin. Forecast values of P and E are not available from NCAR for the ECMWF operational analyses.

2.2 Re-examination of the Cullather et al. [1996] Results

Genthon and Krinner [1998] use ERA-15 to repeat the snow accumulation calculations of *Cullather et al.* [1996] for the West Antarctic sector. They find little ENSO modulation of their derived snow accumulation amounts. They attribute the major discrepancy to the expected deficiencies in the ECMWF operational analyses used by *Cullather et al.* [1996], but they do not perform any checks to verify this assumption. This outcome is quite surprising in view of the extensive Antarctic validation studies done on the ECMWF operational analyses for 1985 and later (earlier operational analyses had some significant shortcomings) [*Bromwich et al.*, 1995; *Cullather et al.*, 1997]. In view of this unsatisfactory situation *Bromwich et al.* [2000] perform a detailed comparative study of the moisture flux convergence into the West Antarctic sector as depicted by the ECMWF operational analyses and ERA-15. In collaboration with ECMWF they determine that an erroneously low elevation for Vostok station (78 °S, 106 °E) in East Antarctica was used throughout ERA-15 (1979-1993) to assimilate surface pressure observations from that manned location. This leads to a major distortion of the reanalyzed atmospheric circulation that suppresses not only overall accumulation in the West Antarctic sector but also the ENSO modulation of accumulation for this sector. The data sparse nature of central East Antarctica allows an error from one station to have a disproportionally large impact on ERA-15. This shortcoming is present in the ECMWF operational analyses for 1987-1990, but has a smaller impact on mean height and flow fields, which are found to have the greatest impact on moisture flux to the West Antarctic sector [*Bromwich et al.*, 2000]. Thus, the original *Cullather et al.* [1996] findings are still valid.

Figure 2 shows the time series of annual (centered) running mean West Antarctic sector accumulation from the ECMWF operational analyses (called WMO and TOGA) updated from *Cullather et al.* [1996] through March 1999 in comparison to the SOI (as shown in *Bromwich et al.* [2000]). Although the earlier periods of the WMO analyses are not as accurate as TOGA analyses, they are included to extend the time series back to 1980. As described above, the accumulation is highly correlated and in phase with the SOI from the early 1980s to 1990, and then abruptly switches to a high anticorrelation with the SOI from 1990 to 1998. In contrast, a time series of accumulation for 60° W to 120° W, 75° S to 90° S, the sector just east of the West Antarctic sector, does not indicate a similar strong relationship with the SOI (not shown). Although there are occasional hints at interannual variability that may be connected to ENSO cycles, the pattern of connection is in no way as clear as that seen in Figure 2. Interestingly, this sector just to the east of the West Antarctic sector generally has higher accumulation values in the ERA-15 depiction than in the depiction given by the ECMWF operational analyses for this sector, while ERA-15 values are low compared with operational values in the West Antarctic sector (again, not shown for brevity). This is likely attributable to the distortion of mean geopotential height fields due to the Vostok elevation error in ERA-15, which is shown to impact mean height fields even over West Antarctica [*Bromwich et al.*, 2000].

In order to assess the relative strength of the ENSO teleconnection with West Antarctic sector P-E, correlation analyses are performed in *Bromwich et al.* [2000] as follows for the period when both ERA-15 and TOGA operational analyses are available (1985-93). Each series of annual running mean data (at monthly intervals) is divided at the time of the correlation switch to create two new series (pre- and post-1990). The annual running mean values are used in order to place more emphasis on the interannual relationships that exist. Finally, each series is detrended to remove multi-annual trends in precipitation that obscure the correlation related to interannual variability. For 1985-1990, the TOGA-SOI correlation is 0.84, and the ERA-15-SOI correlation is 0.63. The value of 0.84 for TOGA-SOI series differs slightly from the value of 0.88 quoted in *Cullather et al.* [1996] due to the extra step of detrending that is done here. For 1990 through 1993, the correlation values are –0.75 and -0.74 for the TOGA-SOI and ERA-15-SOI series, respectively. The correlation values quoted above are impressive, yet it is important to determine the statistical significance due to the relatively short period for which the series overlap. This can only be determined once an effective number of independent variables is established. This number will be smaller than the number of data points in the series as a result of autocorrelation, which is common in meteorological time series and which is induced by using running means. An estimate of the effective number of independent observations is obtained using the technique outlined by *Angell* [1981] (see also *Quenouille* [1952], p. 168). It is given by

$$N/(1+2a_1b_1+2a_2b_2+\cdots) \quad (3)$$

where N is the original number of points in each of the two series, a and b. Series subscripts represent the lag-level autocorrelations. For example, a_2 is the correlation value of series a and its lag-two series. For 1985-1990, N is 56 (a centered running mean starts in the middle of 1985). The lag-one autocorrelations for the SOI, TOGA, and ERA-15 series for this period are 0.981, 0.953, and 0.886, respectively. For the lag-six autocorrelations the values are 0.492, 0.307, and 0.070, respectively. Although contributions to the denominator in equation (3) are small at this point for any combination of two series, we carried this process out to lag-eight autocorrelations. By this point, no product of any two eight-lagged autocorrelations was greater than 0.025, resulting in less than a 0.05 contribution to the denominator in equation (3). The resulting effective number of independent observations is 7 for the TOGA/SOI series and 9 for the ERA-15/SOI series for the period 1985-1990. Based on these and the correlation values quoted above, it is found that the TOGA-SOI correlation for 1985-90 is significant at nearly the 99% confidence level. The ERA-15-SOI correlation for the same period is significant at slightly greater than the 90% confidence level (typically, a value of 95% or greater is considered statistically significant using a two-tailed t-test). The post-1990 overlap period between ERA-15 and TOGA operational analyses is characterized by similar P-E-SOI correlation coefficients and by shorter series. Thus, analyses of correlation and statistical significance for TOGA-SOI variability using annual running mean data from 1990 through 1998 are performed to test the correlation stability seen in the 1990-93 series. Again, the series are detrended. The effective number of independent observations is found to be 15. The TOGA-SOI correlation value is –0.72, significant at greater than the 99% confidence level.

In Figure 2, it is observed that sector P-E lags behind the SOI for El Niño events (SOI < 0), which suggests forcing from the tropics. For the La Niña events (SOI > 0) during our period of study, it is observed that sector P-E leads the SOI, making the origin of the forcing less obvious. Particularly noticeable are the strong accumulation *decrease* associated with the major El Niño event of 1982/83 and the strong accumulation *increase* in association with the other major El Niño event of 1997/98. In view of the variable quality of the ECMWF operational analyses over the entire period it is desirable to seek confirmation against ground-based observations, particularly for the period prior to 1985.

3. CONFIRMATION OF CULLATHER ET AL. [1996] RESULTS BY SURFACE-BASED OBSERVATIONS

3.1 Accumulation Observations From the Siple Dome and Byrd Station Vicinity

As in *Bromwich et al.* [2000], ice cores taken from near Byrd station and Siple Dome (see Figure 1 for relative locations) are used to look at variability and trends in West Antarctic sector precipitation in comparison to annual values of moisture flux-based estimates of P-E from ERA-15 and TOGA operational analyses. Three of the six ice cores used in the evaluation of ice accumulation are from near Byrd station (locations 78.73° S 116.33° W, 79.46°S 118.05° W, and 80.01° S 119.56° W) along a 150 km north-northeast to south-southwest line with Byrd Station at the south-southwestern end. The other three ice cores are from the summit area of Siple Dome near the eastern edge of the Ross Ice Shelf in Figure 1 (locations 81.68° S 149.19° W, 81.64° S 148.77° W, and 81.71° S 148.60° W). All cores are dated with two independent techniques and yield annual estimates of accumulation rate [*Kreutz et al.,* 1999]. For each location the three

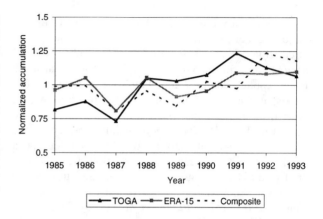

Fig. 3. Normalized values of annual TOGA and ERA-15 accumulation plotted with a normalized linear average (observational composite - labeled composite in the figure) of ice-core accumulation data from Byrd Station and Siple Dome.

records are averaged to mitigate the spatial accumulation variability that characterizes polar ice sheets. Although we eventually normalize the results for comparison with the numerical P-E values obtained for the sector, normalization is not carried out before the cores are averaged. This results in greater weight being given to higher accumulation cores. The reasoning for this is that, in general, higher accumulation ice cores will tend to provide a more reliable depiction of interannual variability than lower accumulation cores.

Upon calculating a linear average of the Byrd station and Siple Dome annual accumulation results, this series is normalized by dividing by the mean annual value and compared with similarly normalized ERA-15 and TOGA annual P-E values (Figure 3). The normalization is necessary in order to facilitate comparison of interannual variability and trends in P-E given how dry the ice core locations are relative to the sector averages obtained via the numerical analyses. From the late 1980s to the early 1990s, there is a significant increase in TOGA operational P-E (a total of 30% from the beginning to the end of the period shown). Although ERA-15 P-E is slightly higher in 1993 than in 1985, the increase is less than half that seen in the TOGA series. The tendency of accumulation from composited core data (up 20%) falls between those seen in the TOGA and ERA-15 series. Figure 3 also shows that there is general agreement in interannual variability among the series.

As alluded to above, the core data are obtained in areas that are relatively dry compared to the total average West Antarctic sector precipitation obtained from ECMWF numerical analyses. Much larger amounts of precipitation are known to fall over the marginal ice slopes of the West Antarctic sector near the Amundsen Sea [*Giovinetto and Bentley,* 1985; *Vaughn et al.,* 1997]. For instance, the spatial trend of low mean precipitation near Byrd to greater amounts farther north toward the coast is evidenced by the northern-most ice core near Byrd, which shows almost twice the mean accumulation as the other two ice cores farther inland. Without additional sampling from areas of the West Antarctic sector where precipitation values are higher, it is difficult to say with certainty that the magnitude of increase seen in the observational composite of normalized core data in Figure 3 is representative of the entire sector. Thus, from this result we can conclude that the core accumulation data tend to support the interannual variability seen in both of the numerical depictions presented. However, while both numerical P-E and ice accumulation results portray an upward trend in net precipitation, additional cores from the higher accumulation regions nearer the coast are needed to determine whether an observationally based upward trend lends more support to the larger positive trend obtained from TOGA analyses or to the smaller one from ERA-15.

3.2 Deuterium Isotope Variations in a Siple Dome ice Core

The chemical composition of an ice core can be used to determine a great deal of information about climate once the core has been properly dated. The ice core component employed in this study (as in *Bromwich et al.* [2000]) is deuterium isotope concentration (specifically, deuterium isotope depletion) from a core drilled at Siple Dome. Monthly variations in deuterium isotope levels in polar precipitation are primarily determined by the condensation temperature contrast between moisture source and precipitation location [*Dansgaard,* 1964]. In general, higher initial isotope concentrations are associated with warmer conditions at the precipitation location.

Given the importance of moisture source and route on regional precipitation, investigation of possible connections between deuterium isotope concentrations and temporal variability in West Antarctic sector P-E is warranted. In noting possible links between these variables, we acknowledge the importance of the position of the low pressure center in the vicinity of the Amundsen Sea and the distance of sector precipitation from the primary moisture source [*Bromwich and Weaver,* 1983; *Cole et al.,* 1999] as represented by mean flow characteristics. As demonstrated by Figure 5 of *Cullather et al.* [1996], the low pressure center during

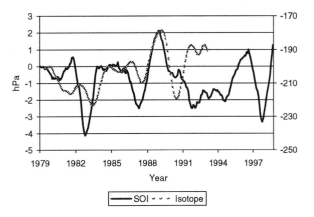

Fig. 4. Annual running mean of deuterium isotope depletions obtained from Siple Dome ice core analysis (right scale in per mil) and the SOI (left scale in hPa).

normal West Antarctic sector precipitation events (1980, for example) is located just north of the Ross Ice Shelf. During reduced precipitation events (1982, for example), the center of this low migrates to the far eastern side of the Amundsen Sea. The impact on vertically integrated moisture flux (also plotted in Figure 5 of *Cullather et al.* [1996]) shows that the mean flow during normal precipitation events is meridional and more direct than during reduced events, at which time the mean flow is from the east across the sector, suggestive of dryer and cooler conditions.

Based on the example from *Cullather et al.* [1996], we expect the less direct, cooler flow to the West Antarctic sector indicative of low precipitation events to be reflected in low deuterium isotope concentrations. Likewise, more direct, moister and warmer flow from the Ross Sea during higher, or more normal, precipitation events should be reflected by higher isotopic concentrations of deuterium. The fact that moisture budget P-E is linked to ENSO is germane here because of a previous case study by *Chen et al.* [1996] that showed how ENSO cycles are probably linked to the Amundsen Sea low via signal propagation along the South Pacific Convergence Zone. In Figure 4, the depletion values of deuterium isotopes obtained from the sub-annually dated Siple Dome B ice core (data provided by J.W.C. White) located at 81.5263° S 148.8139° W are shown plotted with the SOI (both as annual running means of monthly values). The plot indicates the same correlation pattern as is seen between P-E and the SOI in Figure 2. The correlation switch from positive to negative happens at the same time as with moisture flux convergence, circa 1990. The series in Figure 4 also show evidence of another possible switch occurring around the time of the 1982-83 ENSO event, from being negatively to positively correlated. This is reflected in a comparison between the SOI and the WMO series in Figure 2. However, at present there is not a long enough series available with which to see the antiphase relationship that appears to exist prior to 1982.

It should be noted that isotope concentrations obtained from ice cores are naturally smoothed by vapor diffusion in the snow pack. This process leads to more smoothing in the earlier part of the series than in the later part, so a Gaussian low-pass filter was applied to the more recent portion of the isotope series by J.W.C. White. Frequencies higher than one cycle per 18 months are smoothed away, simulating the vapor diffusion that has already removed higher frequencies in the earlier portion of the series.

The isotope results are from one ice core. Several additional cores will be drilled in the near future, sampling over a greater area, which will help determine the fidelity of this isotope-SOI relationship. However, the dependence of isotope variability on the mean circulation characteristics makes it a worthy proxy for sector-based P-E variability given that it is not as likely to suffer as much from the high spatial variability seen in snow accumulation. In addition, this isotope evaluation provides convincing glaciochemical support of the strong West Antarctic sector P-E/SOI correlation pattern seen in the numerical analyses.

4. CIRCULATION CHARACTERISTICS OF THE STRONG EL NIÑO EVENTS OF 1982/83 AND 1997/98

Now that the derived accumulation time series for the West Antarctic sector have garnered strong qualitative support even for the period prior to 1985 when the ECMWF WMO operational analyses have significant shortcomings, the following section contains new (original) material which contrasts the circulation characteristics for the very strong El Niño events in 1982/83 and 1997/98. These two time periods have very different accumulation rates in the West Antarctic sector, as seen in Figure 2. First, the evolution of the mean three-month sea-level pressure anomalies over the South Pacific Ocean is examined for the strong El Niño (warm) events of 1982/83 and 1997/98. We use the ECMWF TOGA operational analyses for 1997/98 (the only data source available) and ERA-15 for 1982/83; we infer that the ECMWF WMO operational analyses for this period are of lower quality than ERA-15 over the ocean areas of concern.

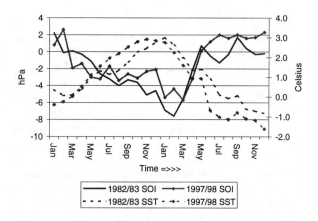

Fig. 5. Comparison of the onset of the El Niño events in 1982/83 and 1997/98.

A previous study by *van Loon and Shea* [1987, hereafter vLS87] looked at mean three-month anomalies of sea-level pressure on the Southern Hemisphere for the year before and year of an El Niño event. The study found that the largest anomaly response occurred in the Australia-South Pacific sector. While their study used the World Monthly Surface Station Climatology (WMSSC) data set and the Comprehensive Ocean-Atmosphere Data Set (COADS), this study uses the aforementioned numerical analysis and reanalysis data to calculate seasonal surface pressure fields. In addition, this study also uses the full range of years available to calculate the mean fields from which anomalies can be determined, whereas vLS97 do not incorporate El Niño or La Niña (cold event) years in their calculation of mean fields. For brevity, not all seasonal departures are shown here. Instead, it suffices to say that the broad scale circulation pattern anomalies as given by the departures in mean three-month pressures from climatology are very similar to those found by vLS87. The similarities get stronger the nearer the three-month period is to the El Niño event. During the year prior to the 1997/98 event, there are instances when mean three-month pressure anomalies appear to lead those shown by vLS87 by a season (or appear to share features of the current and immediately preceding season). This timing difference is consistent with the relatively early onset of the 1997/98 El Niño in comparison to that of the 1982/83 event. Figure 5 shows the El Niño sea-surface temperature anomaly (Niño region 3.4 - 120° W to 170° W, 5° N to 5° S) and SOI comparison for the two events (the values here are mean monthly values and not annual running means). It is interesting to note that the peak in sea-surface

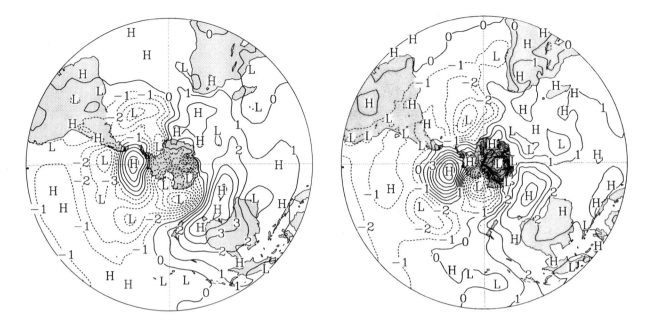

Fig. 6. August-September-October surface pressure anomaly field in hPa for a) 1982 and b) 1997.

temperature for the 1997/98 event clearly leads that of the 1982/83 event by 2 months; the timing contrast between events using the SOI alone is not nearly as distinct. Thus, it is important to look at tropical Pacific sea-surface temperatures in addition to the SOI to establish the timing of ENSO.

Although anomaly features using numerical data for the 1982/83 and 1997/98 events verify the observed pattern between the broad scale circulation anomalies in the South Pacific and the behavior of the SOI found by other investigators, event comparison of the mean sea-level pressure pattern just off the coastline bordering the West Antarctic sector indicates important differences in mean flow that impact P-E. These differences are illustrated by looking at the significant features for a sample three-month period in the Australia-South Pacific sector, namely for August-September-October of 1982 and 1997. Figure 6 shows the anomaly sea-level pressure fields for this three-month period in 1982 and 1997. In vLS87, (see their Figure 1g), there is a high anomaly just north of Thurston Island, a low anomaly in the mid-latitudes of the South Pacific sector, and another high anomaly centered near the southern coast of Australia for this season during the El Niño year. These patterns are also the prominent features of the Australia-South Pacific sector in 1982 and 1997 (Figure 6a,b). Another feature present in both 1982 and 1997 is relatively strong anomalous low pressure between the high anomaly north of Thurston Island and the high anomaly near the southern coast of Australia. In 1982, this low anomaly is located just north of the George V Coast, while in 1997 it is centered just north of the Ross Ice Shelf. There is some indication in vLS87 of a less prominent low anomaly over the Ross Ice Shelf, but it is weak and its extent does not appear to be much greater than that of the Ross Ice Shelf. The near-absence of this feature in the vLS87 composites is likely due to high variance of this feature (in magnitude and location) during ENSO events. In addition to the difference in the low anomaly spatial distribution between 1982 and 1997, the center of the high anomaly north of Thurston Island originally depicted in vLS87 is pushed just east of 90° W in 1982 (Figure 6a) and just west of 90° W in 1997 (Figure 6b, almost exactly the center placement of where it is in vLS87). The impact on the 1982 flow field is a less meridionally direct mean flow into the West Antarctic sector. In 1997, the proximity of the high and low anomalies just described results in a more direct oceanic flow into the West Antarctic sector as indicated by the anomaly field in Figure 6b. The August-September-October example illustrates the association of increased P-E when isobars are directed more perpendicular to the West Antarctic coast and the dryer conditions that prevail when the isobars are oriented in a more parallel fashion. In other words, it demonstrates the importance of the flow field immediately adjacent the coastline, which as it turns out, may only be reflected by subtle

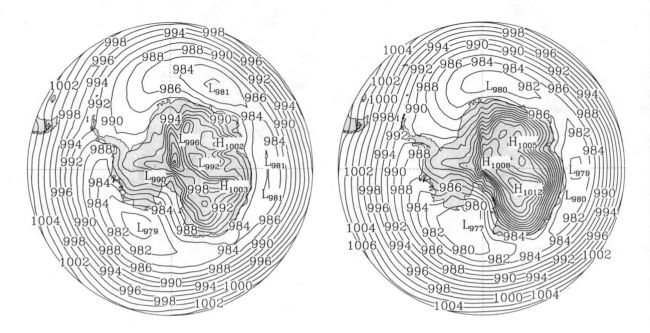

Fig. 7. Mean sea-level pressure in hPa for July through June of a) 1982/83 and b) 1997/98.

differences in the broad-scale circulation pattern anomalies. It now becomes important to demonstrate the persistence of the mean flow characteristics for the 1982/83 and 1997/98 events for which P-E to the West Antarctic sector varied so greatly.

Annual circulation fields centered in time on the 1982/83 and 1997/98 El Niño events help illustrate the contrasting circulation characteristics of these two periods. Figures 7 and 8 show the results for the sea-level pressure and 500 hPa geopotential height, respectively. During 1997/98 the "Amundsen Sea" low is centered in the Ross Sea, and the moisture enters the West Antarctic sector directly from the ocean on the eastern side of the low (Figure 7b). In 1982/83 the "Amundsen Sea" low is shifted farther to east (Figure 7a), and the geostrophic flow is nearly parallel to the coastline. Moisture flowing onto the continent further upstream (east) is depleted by orographic lifting over the ice sheet mostly to the east of 120° W (eastern edge of the West Antarctic sector), and thus less precipitation falls in the West Antarctic sector which is sheltered by the ice sheet topography. Although circulation characteristics for the year prior to the 1982/83 El Niño are not analyzed here, results show that there is an upward trend in precipitation from 1982 to 1983 for the sector just to the east of the West Antarctic sector. This picture is consistent with that presented by *Cullather et al.* [1996] using ECMWF WMO operational analyses for circulation variations associated with precipitation changes in the West Antarctic sector. The only difference is that the eastward shift of the Amundsen Sea low is less marked for 1982/83 in ERA-15 than the ECMWF WMO analyses. At 500 hPa in 1997/98, Figure 8b shows the trough centered just north of the Ross Ice Shelf. It position is only slightly south and west of the surface low (i.e., is almost equivalent barotropic), which indicates that moist air moves southward from the ocean directly into the West Antarctic sector over a deep layer. The 500 hPa ridge over Wilkes Land is highly amplified in relation to average conditions. *Cullather et al.* [1996] observe that this is a blocking situation which is particularly prominent in the early 1990s when West Antarctic sector precipitation is high, and the SOI was persistently negative. In 1982/83 the low at 500 hPa is shifted to the southwest (Figure 8a), which along with the eastward shift in the surface low, yields baroclinic conditions and weak cold air advection from East Antarctica. The zonally-elongated circulation means that there is no deep inflow of moisture into the West Antarctic sector, and thus precipitation is reduced. The ridge over Wilkes Land is almost completely suppressed, and the blocking conditions are absent. The ERA 500 hPa analysis for 1982/83 is negatively impacted by the suppressed 500 hPa heights over Vostok station due to the incorrect station elevation, and the ridge is probably present but situated to the west of its location in 1997/98. Therefore, there are coherent changes in the

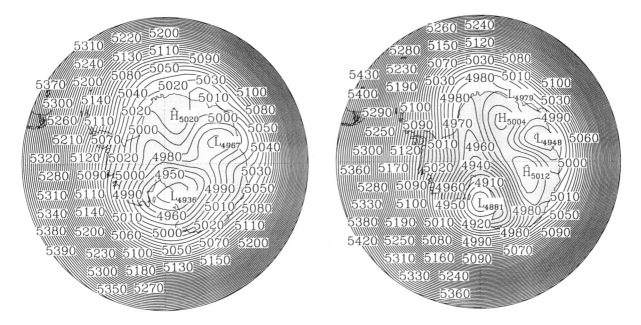

Fig. 8. Mean 500 hPa geopotential height in geopotential meters for July through June of a) 1982/83 and b) 1997/98.

atmospheric circulation over the Ross Sea and East Antarctica, which explain the great contrast in sector precipitation between 1997/98 and 1982/83.

5. DISCUSSION

The strong, but variable, association between West Antarctic precipitation/accumulation and the SOI has been confirmed by ground-based observations, and the less prominent SOI signal in ERA-15 calculated P-E has been traced to deficiencies in the ERA-15 analyses as a result of an erroneous elevation for Vostok station. Reliable numerical analyses for this area start in 1985, and thus a limited number of ENSO events can be evaluated at present. ECMWF will commence a new reanalysis in June 1999, and will span the period from 1957 to 2000, the so-called ERA-40 [*Gibson et al.*, 1999]. This effort is expected to produce a high quality product because of the large amount of input from the Antarctic research community (including the first author) in the form of accurate terrain information, prior validation studies, correction of station elevation errors, reliable sea ice data sets, etc. Prior to the modern satellite era (1979 and later), ERA-40 will inevitably be influenced to some extent by the varying quality and quantity of observations over the Southern Ocean. This data set will most likely form the basis for comprehensive exploration of the causality of the phenomenon described here.

The principal future task is to explain what causes the switch in correlation between West Antarctic sector precipitation and the SOI around 1990. Section 4 demonstrates how the mean circulation characteristics just off the coast of West Antarctica can lead to large differences in P-E even for years when the broad scale circulation in the South Pacific is generally the same. It is possible that it will be necessary to look to additional forcing mechanisms to understand the bimodal relationship seen between West Antarctic sector precipitation and ENSO. *Quintanar and Mechoso* [1995] point out that the southwestern corner of the South Pacific Ocean is subject to eddy energy which originates in the tropical Indian Ocean, propagates southward in those longitudes, and then turns northeastward just to the north of Marie Byrd Land. *Chen et al.* [1996] find that the strengthening of the PFJ in the 120° W to 180° sector in 1988/89 results from the meridional convergence of eddy momentum moving southward over the South Pacific Ocean and northward from Antarctica. The PFJ and the cyclonic activity which leads to West Antarctic precipitation are closely linked. On the 2-week time scale for winter 1988, *Bromwich et al.* [1993] observe that changes in the 500 hPa ridge over Wilkes Land precede those to the cyclonic activity in the eastern Ross Sea by two days. This sequence of events indicates a propagation of disturbance energy from the southern fringes of the Indian Ocean to the South Pacific Ocean. Collectively these studies suggest that the explanation for the correlation switch between the Marie Byrd Land precipitation and the SOI should be sought in the relative contributions of the tropical forcing from the Indian and Pacific Oceans on the atmospheric behavior in the eastern Ross Sea. This will be the subject of future investigations.

Local oceanographic conditions may also modulate the precipitation over Marie Byrd Land. *White and Peterson* [1996] describe the eastward propagation in the Southern Ocean of a coupled interaction between meridional wind stress anomalies and SST/sea-ice extent anomalies. The so-called Antarctic Circumpolar Wave, which has a period of 4-5 years, may impact Antarctic precipitation by analogy with that already demonstrated for New Zealand precipitation [*White and Cherry*, 1999]. Whether this phenomenon plays a significant role in the regime shift seen in 1990 deserves serious consideration.

The continuing ice-core studies in West Antarctica offer the opportunity for more comprehensive validation of the ENSO signal in West Antarctic accumulation. It is desirable to acquire many more accumulation time series with which to test the P-E time series produced from atmospheric analyses and reanalyses. With the elimination of the Vostok elevation problem, ERA-40 should be able to produce much higher resolution descriptions of P-E time series from the forecast values of P and E. According to *Genthon and Krinner* [1998], these forecast values exhibit the same temporal variations as the moisture budget derived values used here. The comparison shown in Figure 4 between deuterium isotope concentrations from a sub-annually dated ice core from Siple Dome and the SOI closely resembles the association between P-E and the SOI. It is desirable to confirm this association with more ice cores to ensure that the derived time scale is robust. In addition, the mechanisms governing the very similar behavior of deuterium isotope concentrations and P-E (both products of the atmospheric hydrologic cycle) should be explored.

Acknowledgments. This research was sponsored by National Science Foundation grant OPP-9725730. ECMWF data were obtained from NCAR. K. Kreutz supplied the accumulation data evaluated in Section 4 and J.W.C. White contributed the deuterium isotope analysis. All this assistance is gratefully acknowledged. Contribution 1149 of Byrd Polar Research Center.

REFERENCES

Angell, J.K., Comparison of variations in atmospheric quantities with sea surface temperature variations in the equatorial Eastern Pacific. *Mon. Wea. Rev., 109*, 230-243, 1981.

Bromwich, D.H., and C.J. Weaver, Latitudinal displacement from main moisture source controls $\delta^{18}O$ of snow in coastal Antarctica. *Nature, 301*, 145-147, 1983.

Bromwich, D.H., Estimates of Antarctic precipitation, *Nature, 343*, 627-629, 1990.

Bromwich, D.H., J.F. Carrasco, Z. Liu and R.-Y. Tzeng, Hemispheric atmospheric variations and oceanographic impacts associated with katabatic surges across the Ross Ice Shelf, Antarctica, *J. Geophys. Res., 98*, 13,045-13,062, 1993.

Bromwich, D.H., F.M. Robasky, R.I. Cullather, and M.L. Van Woert, The atmospheric hydrologic cycle over the Southern Ocean and Antarctica from operational numerical analyses, *Mon. Weather Rev., 123*, 3518-3538, 1995.

Bromwich, D.H., R.I. Cullather and R.W. Grumbine, An assessment of the NCEP operational global spectral model forecasts and analyses for Antarctica during FROST. *Weather and Forecasting*, in press, 1999.

Bromwich, D.H., A.N. Rogers, P. Kållberg, R.I. Cullather, J.W.C. White, and K. Kreutz, ECMWF analysis and reanalysis depiction of ENSO signal in Antarctic precipitation, *J. Clim.*, in press, 2000.

Chen, B., S.R. Smith, and D.H. Bromwich, Evolution of the tropospheric split jet over the South Pacific Ocean during the 1986-1989 ENSO cycle, *Mon. Weather Rev., 124*, 1711-1731, 1996.

Cole, J.E., D. Rind, R.S. Webb, J. Jouzel, and R. Healy, Climatic controls on interannual variability of precipitation $\delta^{18}O$: Simulated influence of temperature, precipitation amount and vapor source region, *J. Geophys. Res., 104*, 14,223-14,235, 1999.

Cullather, R.I., D.H. Bromwich, and M.L. Van Woert, Interannual variations in Antarctic precipitation related to El Niño-Southern Oscillation, *J. Geophys. Res., 101*, 19,109-19,118, 1996.

Cullather, R.I., D.H. Bromwich, and R.W. Grumbine, Validation of operational numerical analyses in Antarctic latitudes, *J. Geophys. Res, 102*, 13,761-13,784, 1997

Cullather, R.I., D.H. Bromwich, and M.C. Serreze, The atmospheric hydrologic cycle over the Arctic Basin from reanalyses Part 1. Comparison with observations and previous studies, *J. Clim.*, in press, 2000.

Dansgaard, W., Stable isotopes in precipitation. *Tellus, 16*, 436-468, 1964.

Genthon, C., and G. Krinner, Convergence and disposal of energy and moisture on the Antarctic polar cap from ECMWF reanalyses and forecasts, *J. Clim., 11*, 1703-1716, 1998.

Gibson, J.K., P. Kållberg, S. Uppala, A. Hernandez, A. Nomura, and E. Serrano, ECMWF re-analysis project report series. Part1. ERA description, ECMWF, Shinfield Park, Reading RG2 9AX, U.K., 1997.

Gibson, J.K., M. Fiorino, A. Hernandez, P. Kållberg, X. Li, K. Onogi, S. Saarinen, and S. Uppala, The ECMWF 40 year re-analysis (ERA-40) project - plans and current status, in *10th Symposium on Global Change Studies*, pp. 369-372, AMS, Boston, Massachusetts, 1999.

Giovinetto, M.B., and C.R. Bentley, Surface balance in ice drainage systems of Antarctica. *Antarct. J. U.S., 20*(4), 6-13, 1985.

Giovinetto, M.B., D.H. Bromwich, and G. Wendler,

Atmospheric net transport of water vapor and latent heat across 70° S. *J. Geophys. Res.*, 97, 917-930, 1992.

Gloersen, P., Modulation of hemispheric sea-ice cover by ENSO events, *Nature*, 373, 503-506, 1995.

Hines, K.M., R.W. Grumbine, D.H. Bromwich, and R. I. Cullather, Surface energy balance of the NCEP MRF and NCEP/NCAR Reanalysis in Antarctic latitudes during FROST. *Weather and Forecasting*, in press, 1999.

Hines, K.M. and D.H. Bromwich, Artificial surface pressure trends in the NCEP/NCAR reanalysis over the Southern Ocean, *J. Clim.*, in review, 2000.

Jacobs, S.S. and J.C. Comiso, Climate variability in the Amundsen and Bellingshausen Seas, *J. Clim.*, 10, 697-709, 1997.

Kalnay, E., et al., The NCEP/NCAR 40-year reanalysis project, *Bull. Am. Meteorol. Soc.*, 77, 437-471 plus CD-ROM, 1996.

Karoly, D.J., Southern Hemisphere circulation features associated with El Niño-Southern Oscillation events, *J. Clim.*, 2, 1239-1252, 1989.

King, J.C., Recent climate variability in the vicinity of the Antarctic Peninsula, *Int. J. Clim.*, 14, 357-369, 1994.

Kreutz, K.J., P.A. Mayewski, M.S. Twickler, and S.I. Whitlow, Spatial variability and relationships between glaciochemistry and accumulation rate at Siple Dome and Marie Byrd Land, West Antarctica. *J. Geophys. Res.*, in review, 1999.

Ledley, T.S., and Z. Huang, A possible ENSO signal in the Ross Sea, *Geophys. Res. Lett.*, 24, 3253-3256, 1997.

Marshall, G.J. and J.C. King, Southern Hemisphere circulation anomalies associated with extreme Antarctic Peninsula winter temperatures, *Geophys. Res. Lett.*, 25(13), 2437-2440, 1998.

Peixoto, J.P., and A.H. Oort, *Physics of Climate*, Amer. Instit. of Phys., 520 pp., 1992.

Philander, S.G., El Niño, La Niña, and the Southern Oscillation, *International Geophysics Series*, 46, edited by R. Dmowska and J.R. Holton, Academic Press, Inc., 1990.

Quenouille, M.H., Associated Measurements. Butterworths, 241pp, 1952.

Quintanar, A.I., and C.R. Mechoso, Quasi-stationary waves in the Southern Hemisphere. Part 1: observational data, *J. Clim.*, 8, 2659-2672, 1995.

Stark, P., Climate warming in the central Antarctic Peninsula area, *Weather*, 49, 215-220, 1994.

Timmermann, A., J. Oberhuber, A. Bacher, M. Esch, M. Latif and E. Roeckner, Increased El Niño frequency in a climate model forced by future greenhouse warming, *Nature*, 398, 694-696, 1999.

Trenberth, K.E., General characteristics of El Niño-Southern Oscillation, in *Teleconnections Linking Worldwide Climate Anomalies: Scientific Basis and Societal Impact*, edited by M.H. Glantz, R.W. Katz, and N. Nicholls, pp. 13-42, Cambridge Univ. Press, New York, 1991.

Trenberth, K.E., Global analyses from ECMWF and atlas of 1000 to 10 mb circulation statistics, NCAR Tech. Note NCAR/TN-0373+STR, 191 pp., plus 24 fiche, 1992.

Trenberth, K.E., and T.J. Hoar, The 1990-1995 El Niño-Southern oscillation event: Longest on record, *Geophys. Res. Lett.*, 23, 57-60, 1996.

Turner, J., S.R. Colwell, and S. Harangozo, Variability of precipitation over the coastal western Antarctic Peninsula from synoptic observations, *J. Geophys. Res.*, 102, 13,999-14,007, 1997.

van Loon, H., and D.J. Shea, The Southern Oscillation, VI, Anomalies of sea-level pressure on the Southern Hemisphere and Pacific sea-surface temperature during the development of a warm event, *Mon. Weather Rev.*, 115, 370-379, 1987.

Vaughan, D.G. and C.S. Doake, Recent atmospheric warming and retreat of ice shelves on the Antarctic Peninsula, *Nature*, 379, 328-331, 1996.

Vaughan, D.G., J.L Bamber, M. Giovinetto, J. Russell, A. Paul, and R. Cooper, Reassessment of net surface mass balance in Antarctica. *J. Clim.*, 12, 933-946, 1997.

White, W.B., and R.G. Peterson, An Antarctic circumpolar wave in surface pressure, wind, temperature and sea-ice extent, *Nature*, 380, 699-702, 1996.

White, W.B., and N.J. Cherry, Influence of the Antarctic circumpolar wave upon New Zealand temperature and precipitation during autumn-winter, *J. Clim.*, 12, 960-976, 1999.

David H. Bromwich, Polar Meteorology Group, Byrd Polar Research Center, The Ohio State University, 1090 Carmack Road, Columbus, OH 43210-1002, e-mail: bromwich.1@osu.edu

Aric N. Rogers, Polar Meteorology Group, Byrd Polar Research Center, The Ohio State University, 1090 Carmack Road, Columbus, OH 43210-1002, e-mail: aric@polarmet1.mps.ohio-state.edu

GEOLOGIC CONTROLS ON THE INITIATION OF RAPID BASAL MOTION FOR WEST ANTARCTIC ICE STREAMS: A GEOPHYSICAL PERSPECTIVE INCLUDING NEW AIRBORNE RADAR SOUNDING AND LASER ALTIMETRY RESULTS

D.D. Blankenship[1], D.L. Morse[1], C.A. Finn[2], R.E. Bell[3], M.E. Peters[1], S.D. Kempf[1], S.M. Hodge[4], M. Studinger[3], J.C. Behrendt[2,5], and J.M. Brozena[6]

The Ross Embayment of West Antarctica is characterized by the presence of the Siple Coast ice streams. The physical processes that enable their enhanced flow are subject to debate, but rapid basal motion (RBM) seems to be required. Basal conditions are indicative of the governing processes for overall ice stream behavior. As such, heterogenous boundary conditions implied by the analyses presented here are a crucial, but missing, component of models of the evolution of the ice stream system. We review existing geophysical data and examine new aerogeophysical results in an effort to identify geological controls on the initiation of RBM for the ice stream B and C system of the Southeastern Ross Embayment. Both seismic and aerogeophysical observations are compatible with a model of tectonic evolution for this region that implies Cretaceous rifting with a generally quiescent mid-Cenozoic punctuated by localized volcanism. This volcanism is associated with pre-established zones of crustal weakness. From the new aerogeophysical results we identify a band of transitional ice flow that separates the internally deforming interior ice and the fully developed ice stream system characterized by RBM. Four ice stream branches are clearly defined at the down-stream limit of this band. There exists a strong correspondence of this zone of RBM initiation with the transitional crust lying between the cold, thick and elevated crust of the Ellsworth/Whitmore crustal block and the warm, thin and depressed crust of the Ross Embayment. It appears that ice stream initiation relies on the availability of mid-Cenozoic sediments draped on the pre-existing rift-controlled topography of the embayment and up onto the flanks of the surrounding crustal blocks. Additionally, the gradient of regional geothermal flux across this zone cannot be ignored as potentially important. A third possible geologic control on the initiation of RBM is focused geothermal flux associated with localized volcanics both along the crustal block boundaries and within the Southeastern Ross Embayment.

INTRODUCTION

The ice streams of West Antarctica's Ross Embayment are differentiated from the inland ice of the West Antarctic ice sheet (WAIS) by their relative speed; however, the transition from slow flow to fast flow has been difficult to define. This transition is almost certainly controlled by the initiation of rapid basal motion (RBM) and the propagation kinematically of this initiation upstream into the inland ice to a point where the velocity becomes a function dominated by the driving stress [Alley and Whillans, 1991]. For the purposes of this paper we will define the onset of ice streaming to be the transition zone bounded downstream by the initiation of RBM and upstream by the limit of the kinematic response to this initiation as reflected in the driving stress. Our objective is to determine if the initiation of

[1] Institute for Geophysics, University of Texas at Austin, Austin, Texas
[2] U.S. Geological Survey, Denver, Colorado
[3] Lamont-Doherty Earth Observatory, Columbia University, Palisades, New York
[4] U.S. Geological Survey, Tacoma, Washington
[5] Institute of Arctic and Alpine Research, University of Colorado at Boulder, Boulder, Colorado
[6] Naval Research Laboratory, Washington, D.C.

RBM is controlled by geological characteristics of the subsurface other than subglacial topography.

The foundation for the hypothesis that subglacial geology controls the initiation of RBM is based on the general observation that some combination of saturated (and possibly mobile) sediments [e.g., *Blankenship et al.*, 1986; *Engelhardt et al.*, 1990] and thin (lubricating) water layers or gaps (possibly with canals) [*Engelhardt and Kamb*, 1997] is necessary but perhaps not sufficient for the RBM of ice stream B in West Antarctica. There is substantial debate over the relative contributions to RBM of distributed deformation within mobile saturated sediments verses sliding over any thin lubricating water or sediment layer [*Alley*, 1989; *Engelhardt and Kamb*, 1998]. The end-members of this debate imply different hypotheses for geological controls on the initiation of RBM. For the end-member where RBM is caused exclusively by deformation within saturated sediments [*Boulton and Jones*, 1979; *Alley et al.*, 1986, 1987a, 1987b] a substantial source of easily eroded sediment is a necessary condition. In this case, the expected geological controls on initiation of RBM would range from a strong correlation either with the tectonically controlled boundary of a sedimentary basin or an erosionally controlled boundary of a much thinner but regionally extensive sedimentary drape. It is important to note that for any combination of deformation and sliding, sediment availability is a necessary but not sufficient condition because water is also required. So, for the sliding end member or for any combination of deformation and sliding, geological controls on the initiation of RBM might also include variability in regional geothermal flux caused by thinner crust or contrasting thermal histories for adjacent crustal blocks [*Stern and ten Brink*, 1989] to a focusing of geothermal flux by volcanic activity [*Blankenship et al.*, 1993]. In all of these scenarios, increased geothermal flux may be a sufficient condition for the initiation of RBM but it is unlikely to be necessary because the simple concentration of water by gradients in the subglacial hydrological potential [*Anandakrishnan and Alley*, 1997; *Hindmarsh*, 1998] could provide an adequate supply for RBM even in the absence of heterogeneous geothermal flux [*Rose*, 1979].

What these glaciological hypotheses show is that knowledge of any correlations between subglacial geological character and the initiation of RBM is essential to our understanding of the mechanics of RBM. In addition, if subglacial geology provides a necessary condition (other than the topographic framework) for the initiation of RBM in the Ross Embayment, the inclusion of this heterogeneous and largely time-invariant boundary condition would be necessary for accurate models of WAIS evolution. Interestingly, this seems to be the case for the former ice sheets of the Northern Hemisphere [*Clark and Walder*, 1994].

THE ROSS EMBAYMENT AS A GEOLOGICAL FRAMEWORK FOR RBM

A general understanding of the crustal evolution of the Ross Embayment (see Dalziel and Lawver, this volume) has been obtained from the outcrops surrounding the WAIS combined with subglacial topography (Figure 1) from a comprehensive program of airborne radar sounding undertaken in the 1970's [*Drewry*, 1983; see also *Rose*, 1982 and *Jankowski and Drewry*, 1981]. The Southeastern portion of the Ross Embayment (SERE), which is dominated by the Siple Coast ice streams, is characterized by thin extended crust with subdued topography (a few hundred meters below sea level) bounded to the north by the Marie Byrd Land crustal block (MBL) and to the south by the Transantarctic Mountains (TAM). Downstream, this thin extended crust is thought to extend beneath the Ross Ice Shelf where it continues onto the Ross Sea continental shelf. Upstream of the southernmost Siple Coast ice streams (A, B and C of Figure 1) the ice sheet is underlain by the elevated and rougher topography of the Ellsworth/Whitmore Mountains crustal block (EWB). To the north of these ice streams and adjacent to the EWB lies the deep trough of the Bentley Subglacial Trench.

Any hypothesis regarding geological controls on RBM in the SERE is critically dependent on the nature of the "rifting" event that resulted in the thin crust that forms the cradle for the Siple Coast ice streams. Although the specific cause is debated, it is well documented that a period of rifting associated with Gondwana breakup in the Cretaceous is responsible for most of the crustal thinning in the Ross Embayment [*Lawver et al.*, 1991]. It has also been hypothesized that this rifting was reactivated in the late Cenozoic [*Behrendt and Cooper*, 1991; *Behrendt et al.*, 1992] by interaction with an extensive mantle plume (encompassing the entirety of the SERE and MBL) resulting in large-scale volcanism and possibly flood basalts. These two histories of crustal thinning and associated volcanism have contrasting implications for geologic controls on RBM in the SERE.

Because the East Antarctic ice sheet is likely to have been present throughout much of the latter half of the Cenozoic [see *Van der Wateren and Hindmarsh*, 1995, for a review] and by comparison the WAIS is viewed as relatively ephemeral at least until the Pliocene [*Scherer*, 1991], it is expected that mid-Cenozoic (Oligocene and Miocene) sedimentary deposition in the SERE would be dominated by high rates of glacial marine sedimentation from East Antarctic sources (except, for brief periods of deglaciation in West Antarctica through the Pleistocene [*Scherer*, 1998]). In the case of reactivated Cenozoic rifting in the Miocene [*Behrendt et al.*, 1996], much of

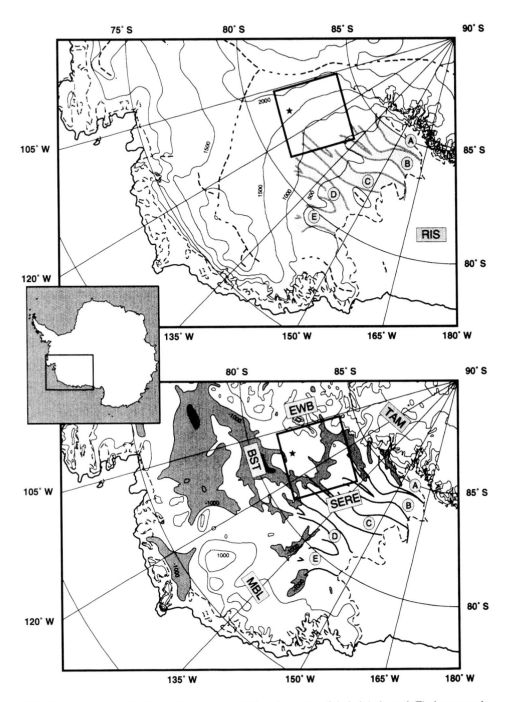

Fig. 1. Ice streams of the Ross Embayment of West Antarctica (labeled A through E) shown on the ice sheet surface (a) and subglacial topography (b) of Drewry (1983). For the surface topography the contour interval is 500 m. The subglacial topography is represented by a 1000 m contour interval with shaded topography below -1000 m a.s.l. The CASERTZ aerogeophysical survey over the Southeastern Ross Embayment (SERE) is outlined by the box overlying the catchments of ice streams B and C. EWB and BST, respectively, mark the locations of the Ellsworth/Whitmore crustal block and the Bentley Subglacial Trench; Marie Byrd Land (MBL), the Transantarctic Mountains (TAM) and the Ross Ice Shelf (RIS) are shown as well. The star indicates the location of possible active subglacial volcanism (Blankenship et al., 1993).

this mid-Cenozoic sediment would have been deposited syn-tectonically in structurally controlled sedimentary basins that may or may not be coincident with pre-existing Cretaceous grabens. For rifting limited to the Cretaceous, mid-Cenozoic marine sedimentation in the SERE would be primarily controlled by topography and therefore only subject to the indirect structural controls on this topography. In this case the mid-Cenozoic sediments would form a drape of variable thickness over the structurally controlled Cretaceous topography (and older sediments) with the greatest thickness in the topographic lows.

The expected distribution of geothermal flux would also be different for the two rift-timing scenarios in the Ross Embayment. For plume-reactivated Cenozoic rifting, the crust of the SERE would be expected to be characterized by high regional heat flow with large scale volcanism arising from zones of weakness associated with primary rift structures [Behrendt et al., 1994]. For rifting limited to the Cretaceous, the regional heat flux in the SERE should remain elevated relative to the surrounding crustal blocks because of the thinned crust. In this case however, any focused geothermal flux would be associated with volcanics probably derived from local mantle sources and emplaced along zones of weakness associated with the original rifting; it has been hypothesized that these zones of weakness were reactivated in the late Cenozoic as a result of plume related uplift limited to MBL (see Dalziel and Lawver, this volume).

What we can envision are different relationships between any Cenozoic volcanics and sediments from these two rift-timing scenarios for the SERE. For the case of plume-reactivated rifting in the Cenozoic we would expect syn-tectonic sedimentation with widespread volcanic capping and further structurally controlled sedimentation. For rifting limited to the Cretaceous, we would expect a mid-Cenozoic sediment drape of variable thickness over a rift controlled topography; the deposition of this drape could have been interrupted by volcanic emplacements from localized sources following possibly reactivated zones of weakness left from the Cretaceous rifting episode. Because Cretaceous rifting is not plume related, the volume of associated Cretaceous volcanics underlying these sediments would likely be smaller than for Cenozoic plume-reactivated rifting.

The impact of these contrasting rift-timing scenarios on the initiation of RBM in the SERE is clear. The stronger structural control of mid-Cenozoic sedimentation contemporaneous with plume-reactivated rifting would provide a reasonably rigid framework for the initiation of deformation-dominated RBM. This framework would be characterized by fault-bounded sedimentary basins and extensive interlayered (or capping) volcanics. On the other hand, a tectonically quiescent mid-Cenozoic and the resulting topographically controlled drape of sediments punctuated locally by volcanics would provide fewer spatial constraints on RBM except for the lack of long-term sediment sources represented by deeper structurally controlled sedimentary basins.

For both rift-timing scenarios we expect a higher regional geothermal flux for the SERE with respect to the colder thicker crust of the EWB; however, the plume-reactivated Cenozoic rifting scenario would presumably be associated with higher and more variable heat flux in the SERE accompanying widespread volcanism. For the case of rifting limited to the Cretaceous, any focused heat sources would be associated with locally concentrated volcanics (associated with zones of crustal weakness) that are less likely to cover substantial portions of the mid-Cenozoic sedimentary drape. In either of these cases, heterogeneous heat flux could cause enhanced concentrations of basal meltwater that would be available for the initiation of RBM.

SEISMIC EVIDENCE FOR POSSIBLE GEOLOGICAL CONTROL ON RBM IN THE SERE

Specific knowledge of the geological framework for RBM in the SERE has been limited until recently and, even now, seismic evidence relevant to resolving the critical issue of rift timing is sparse but not inconsequential. During the 1980's seismic reflection surveys over ice stream B confirmed the close association of rapid basal motion and low-velocity subglacial sediments [Blankenship et al., 1986; 1987; Atre and Bentley, 1994]. More extensive reflection surveys over ice stream B [Rooney et al., 1987a; 1991] showed that low-velocity and very likely mid-Cenozoic sediments do exist in the topographic low beneath this ice stream and that the thickness of these low-velocity sediments was generally less than one kilometer. (Paleontological studies of subglacial sediments recovered from the seismic site on ice stream B also indicate a mid-Cenozoic age for these sediments [Scherer, 1991]).

Rooney et al. [1987b; 1991] used seismic refraction and reflection techniques to show that, at least at one location on ice stream B, these sediments were underlain by either older sediments with a higher-velocity or crystalline basement; they also showed that a fault in this higher-velocity unit possibly formed a lateral boundary for the lower-velocity sediments which seemed to coincide with the northern boundary of ice stream B. On the other hand, Munson and Bentley [1992] interpreted seismic refraction work on ice stream C to show that a thin O(100 m) drape of low-velocity sediments rests on a higher-velocity unit of crystalline rock or older sediments without apparent structural control. Interestingly, they also infer a kilometers thick and presumably structurally controlled low-velocity unit under the ice ridge between ice streams B and C; however, the velocity for this unit is poorly constrained.

More recent seismic refraction work near the initiation of RBM for the ice stream B/C system (ice stream C1b of Plate 1b) has indicated a low-velocity sedimentary unit that is coincident with RBM [*Anandakrishnan et al.*, 1998] and that covers the base of a trough which appears from aerogeophysical interpretations [*Bell et al.*, 1998] to be fault bounded. However, Anandakrishnan et al. also show evidence that a thinner drape of these low-velocity sediments (that is not coincident with RBM) is present in an adjacent shallow trough which from the aerogeophysical results appears to lack structural control.

From these sparse seismic observations it appears that low-velocity and probably mid-Cenozoic sediments are uniquely associated with RBM in the SERE. In addition, up to one km of these sediments can reside in structurally controlled topographic lows. However, the frequent occurrence of a drape of low-velocity sediments with a thickness of up to a few hundred meters indicates that sedimentation probably did not occur in concert with large-scale Cenozoic rifting; the seismic results are more consistent with mid-Cenozoic sediments draping a preexisting topography resulting from Cretaceous rifting processes with thicker deposits residing in former topographic lows.

Reliable seismic observations of crustal thicknesses are also reasonably rare in the SERE [see *Bentley*, 1973]. Clarke et al. [1997] have presented results from a carefully controlled refraction study along a 235 km profile over the boundary between SERE and the EWB crust. They observed a 30-km crustal thickness nearest the EWB thinning to the west at the SERE end of the profile. They concluded, based on the crustal thickness and mantle velocity, that there was little evidence for recent rifting; this is in agreement with the magnetotellurics results of Wannamaker et al. [1996] from the same area which are inconsistent with rifting in the latest Cenozoic. Clarke et al. [1997] presented no substantive evidence for a low-velocity layer at the base of the ice along their profile, although the detectable thickness for their experiment was about 100 m. The context for these particular results will be presented in the next section.

AEROGEOPHYSICAL EVIDENCE FOR POSSIBLE GEOLOGICAL CONTROL ON RBM IN THE SERE

Until recently, aerogeophysical studies of the SERE lacked sufficient resolution to address directly the issue of geological control on the initiation of RBM. However, these early studies did contribute significantly to our knowledge of the geological framework for RBM in the SERE. Early aeromagnetics work [*Behrendt*, 1964] showed that short wavelength magnetic anomalies consistent with subglacial volcanic units were common throughout the region and that many of the magnetic units likely rested in close proximity to the base of the ice; however, the longer wavelength components of these data also indicated substantial intervening thicknesses of non-magnetic rocks [*Behrendt and Wold*, 1963]. In addition, the reconnaissance radar sounding of the 1970's [*Rose*, 1979, 1982; *Jankowski and Drewry*, 1981; *Jankowski, et al.*, 1983] provided knowledge of the subglacial topographic framework for understanding the mosaic of crustal blocks bounding the SERE (see Dalziel and Lawver, this volume) as well as a comprehensive description of its ice stream system [*Rose*, 1979; see also *Shabtaie et al.*, 1987]. This work included a brief program of aeromagnetic surveying and reanalysis of previous aeromagnetic results which were used to define the boundary of the EWB and the SERE (near the origins of ice streams B and C) and largely verified the earlier observations of Behrendt and Wold [1963]. These investigators showed that the EWB/SERE boundary beneath the catchments of ice streams B and C is defined by magnetic crust that is transitional between the topographically smooth and low lying crust of the SERE and the non-magnetic, rough and elevated crust of the EWB. This broad region of intervening "magnetic" transitional crust also possesses an elevation intermediate to that characterizing the EWB and SERE. We will use the term Whitmore Mountains/Ross Embayment transitional crust (WRT) to describe it.

These crustal boundary definitions are important because, as mentioned above, the thinner rifted crust of the SERE is likely to be characterized by a regional geothermal flux higher than the continental average. (A single calculation of geothermal flux from an ice coring site on an interstream ice ridge in the SERE [*Alley and Bentley*, 1988] supports this hypothesis.) Therefore, any transitional crust between the relatively warm SERE and the colder thicker crust of the EWB could imply a transition in regional geothermal flux. Similarly, these crustal boundaries could be characterized by zones of weakness that could provide conduits for volcanics and associated focused geothermal flux.

During the early to mid 1990's a series of integrated aerogeophysical experiments was undertaken over the SERE where it meets the EWB. These experiments, collectively known as Corridor Aerogeophysics of the Southeastern Ross Transect Zone (CASERTZ), represent approximately 50,000 km of aerogeophysical surveying (at a 5.3 km grid spacing) accomplished with a uniquely configured deHavilland Twin Otter (see Appendix).

These experiments were designed specifically to address the issue of geological controls at the onset of the ice stream B/C system; they included observations of ice-sheet thickness from radar sounding and surface elevation from laser altimetry coupled to coincident observations of the gravitational and geomagnetic fields. Individual results from these experiments have been previously presented [*Blankenship et al.*, 1993; *Brozena et al.*, 1993; *Behrendt et al.*, 1993, 1994, 1995, 1996, 1997; *Sweeney et al.*, 1994, 1999; *Bell et al.*, 1998; 1999]. In Plate 1 we present a compilation including new ice surface elevations and subglacial topography for the SERE region derived from the laser altimetry and radar sounding components of these experiments (described in the attached Appendix); we then use the subglacial topography to establish detailed crustal boundaries. In Plate 2 we use the ice surface and ice thickness observations to calculate the driving stress for the ice sheet [following *Paterson*, 1994] in order to establish both downstream and upstream limits on the initiation of RBM. We also correlate these limits with subglacial topography. In Plate 3 we present the CASERTZ potential field observations with the interpreted crustal boundaries and calculated driving stress limits for the SERE region. We then interpret these results in the context of possible geologic controls on the initiation of RBM as represented by tectonic verses topographic control on mid-Cenozoic sedimentation and regional versus focused geothermal flux.

We first use the CASERTZ aerogeophysical results to define the zone that should include the initiation of RBM for the ice streams in the SERE survey region (i.e., C2, C1a, C1b and B2 of Plate 1). Assuming that the existence of active lateral margins indicates a downstream limit for the initiation of RBM [*Alley and Whillans*, 1991], Plate 2a shows clearly that this limit coincides well with the 50 kPa driving stress contour. The upper bound for the possible initiation of RBM (also shown in Plate 2a) is chosen somewhat conservatively as the 100 kPa driving stress contour consistent with ice flow dominated by internal deformation [e.g., *Paterson*, 1994].

Using the CASERTZ subglacial topography, it can be seen that the downstream limits of RBM initiation (Plate 2b) for ice streams C2, C1b, C1a and possibly B2 are correlated with topographic lows and that in most cases these lows continue downstream of this limit. However, for ice stream B2, RBM downstream of the initiation zone is correlated with generally elevated topography. In addition, examples of topographic lows exist within the survey area (e.g., the long linear valley in the northwest corner) that do not fall anywhere within the zone we have defined for the possible initiation of RBM. It is also interesting that the troughs associated with the downstream limit on RBM initiation for ice streams C2 and C1b are truncated by elevated topography within a few ten's of kilometers upstream of this limit; while both ice streams C1a and B2 are characterized by depressed topography above this limit. In summary, we can conclude from these observations only that topographic lows characterize the downstream limit on the initiation of RBM and that either elevated or depressed topography can exist either upstream or downstream of this limit. This evidence generally supports the concept that factors other than topography are important in determining the initiation of RBM for the ice streams of the SERE.

The CASERTZ subglacial topography (Plate 1b) also supports the earlier interpretation of the relationship between the SERE and its bounding crustal blocks. The rugged and relatively high elevations (above -250 m a.s.l.) of the EWB are clearly discernable running from northeast to southwest across the survey area with smoother topography of intermediate elevation (above -500 m a.s.l.) outlining the WRT crust; the very smooth topography of the SERE crust lies primarily below -500 m a.s.l. The CASERTZ subglacial topography also shows an indentation in the WRT crust in the south-central portion of the survey area. Our downstream limit for the initiation of RBM lies entirely within what would be topographically defined as Ross Embayment crust (Plates 1b and 2b) and the EWB crustal block consistently lies inland of our upstream limit (Plate 3a). The WRT crust is often present between the two limits that we have defined for the possible initiation of RBM (Plate 3a) and in all but one case, ice stream C1a, the boundary of the WRT block is largely coincident with our upstream limit for the initiation of RBM.

The Bouguer gravity calculated for the CASERTZ survey [*Bell et al.*, 1999] generally supports the crustal block interpretation derived from the subglacial topography (Plate 3a). EWB crust, as defined by the topography, is characterized by a Bouguer low which is consistent with thick (40 km and presumably cold) crust predicted from early gravity observations in the region [see *Bentley*, 1983]. Similarly, much of the SERE crust is characterized by a positive Bouguer anomaly consistent with the predicted thin (about 20 km and presumably warm) rifted crust observed elsewhere in the southern Ross Embayment [e.g., *ten Brink et al.*, 1998]. The WRT crust, as expected, has a negative Bouguer anomaly intermediate to that for the EWB and the SERE which meshes well with the 30 km crustal thickness determined by the seismic work of Clarke et al. [1997] which sampled almost exclusively the northern unit of WRT crust. It is very likely that the regional geothermal flux expected for the WRT crust would also lie between that for the EWB and SERE; this could represent an important change of boundary conditions for the upstream limit on initiation of RBM which seem to be correlated with WRT crustal boundaries.

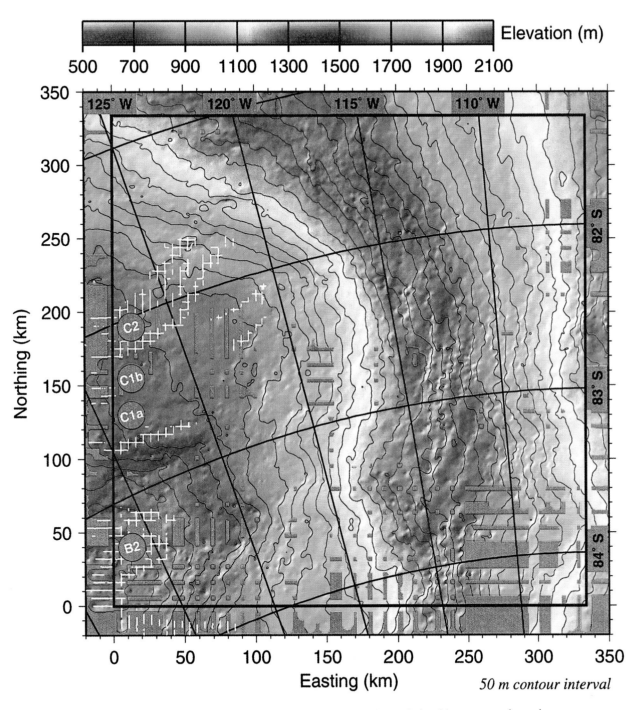

Plate 1a. Ice-sheet surface elevations from CASERTZ airborne laser altimetry over the region encompassing the initiation of rapid basal motion for the ice streams of the southeastern Ross Embayment. Crevassed shear margins for these ice streams as determined by airborne radar sounding are indicated by white shading along the tracklines. C2 and B2 indicate the northern limbs of ice streams C and B, respectively; C1b and C1a indicate the northern and southern tributaries of the southern limb of ice stream C. A brief description of the acquisition and interpretation of the CASERTZ laser altimetry data is presented in the attached Appendix.

112 THE WEST ANTARCTIC ICE SHEET: BEHAVIOR AND ENVIRONMENT

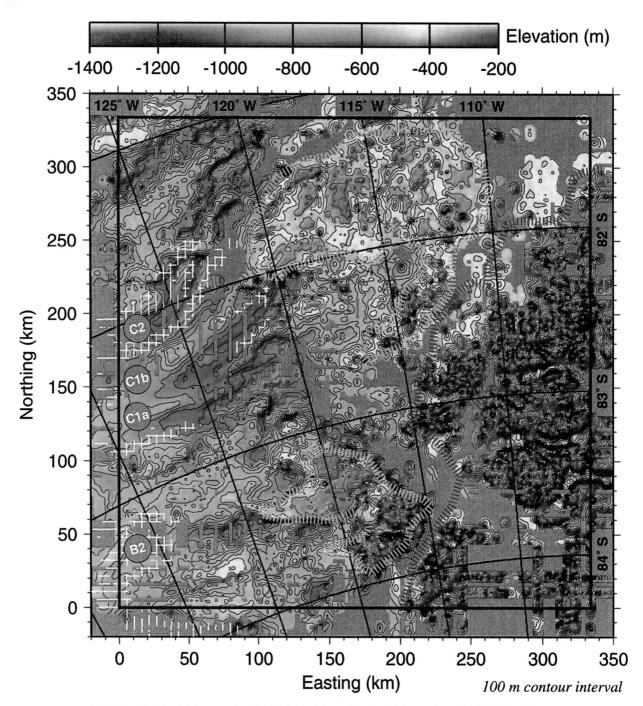

Plate 1b. Subglacial topography obtained by subtracting ice thickness from CASERTZ airborne radar sounding from the surface elevations of Plate 1a. The margins shown and naming scheme for the ice stream tributaries are the same as for Plate 1a. The assumed crustal boundary for the Ellsworth/Whitmore crustal block (EWB; -250m a.s.l.) is shown by the broken red line; the boundary for the Whitmore Mountains/Ross Embayment transitional crust (WRT; -500 m a.s.l.) is indicated by the broken orange line. A brief description of the acquisition and interpretation of the CASERTZ airborne radar sounding data is presented in the attached Appendix.

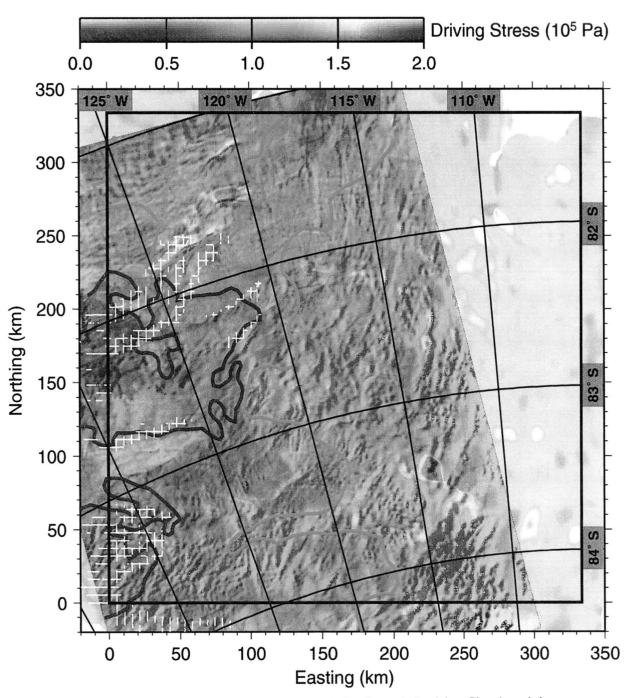

Plate 2a. The driving stress for the West Antarctic ice sheet calculated from Plate 1a and the CASERTZ ice thickness superimposed on the AVHRR image that includes this portion of the Southeastern Ross Embayment (T. Scambos, personal communication). Ice stream shear margins are as indicated in Plate 1a; note the agreement between these shear margins and many of the flow features apparent in the AVHRR image. Our interpretations of the 50 kPa and 100 kPa driving stress contours are indicated by the solid blue and green lines, respectively. These lines represent our estimates of the downstream (blue) and upstream (green) limits for the initiation of rapid basal motion for the ice streams of the region.

114 THE WEST ANTARCTIC ICE SHEET: BEHAVIOR AND ENVIRONMENT

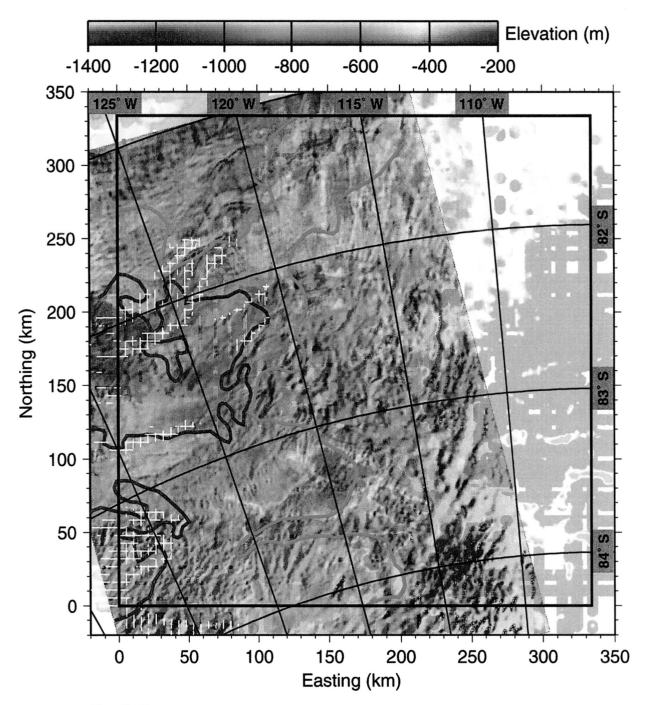

Plate 2b. The subglacial topography of Plate 1b superimposed on the AVHRR image of Plate 2a with our estimated downstream (blue line) and upstream (green line) limits for the initiation of rapid basal motion and active shear margins as defined for previous plates and figures. Interpretation of the relationship between subglacial topography and initiation of rapid basal motion is presented in the text.

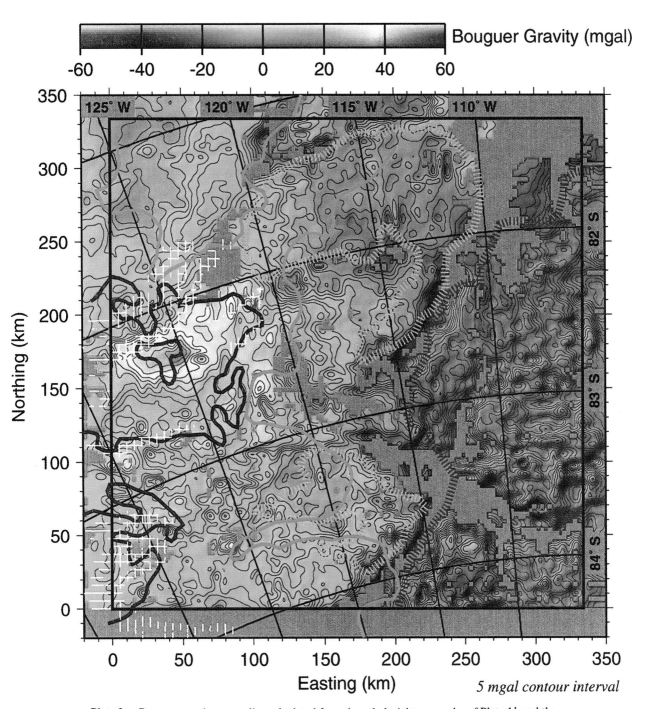

Plate 3a. Bouguer gravity anomalies calculated from the subglacial topography of Plate 1b and the CASERTZ airborne gravity observations (Bell et al., 1999).

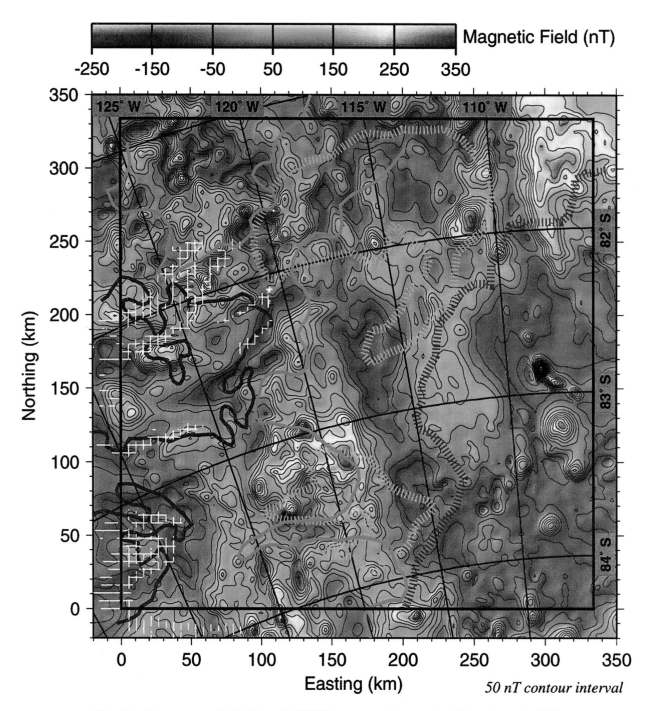

Plate 3b. The geomagnetic field from CASERTZ aeromagnetic surveying (Behrendt et al., 1994; Sweeney et al., 1994; 1999). For both figures, the ice stream shear margins, boundaries for the EWB and WRT crustal blocks and our limits for the initiation of rapid basal motion are as defined in Plates 1a, 1b and 2a, respectively.

A compilation of the CASERTZ aeromagnetic observations [*Behrendt et al.*, 1994; Sweeney et al., 1994, 1999] is presented in Plate 3b. In addition to the seismic work of Anandakrishnan et al. [1998], coupled modeling of coincident lows in CASERTZ free-air gravity and magnetics observations [*Bell et al.*, 1998] showed a structurally-controlled topographic low filled with about one km of sediment within the zone of expected RBM initiation for ice stream C1b. This work supported the hypothesis of strong structural control on sediments available for the initiation of RBM but was not able to explain the lack of RBM associated with a nearby topographically controlled drape of thinner sediments. It is interesting to note that this simple relationship between free-air gravity and magnetic lows exists only for ice stream C1b and to a lesser degree for ice stream C1a; it does not exist for ice streams C2 or B2. This apparent lack of correlation could imply either a general lack of strong structural control on these sediments or interference with the simple relationship by volcanics intermingled with or capping the sediments correlated with any gravity lows. At any rate, the CASERTZ potential field data present little evidence to counter the conclusion based on seismic evidence presented above that most of the mid-Cenozoic sediments in the SERE form a drape over a structurally controlled topography resulting from rifting in the Cretaceous. Using Drewry's [1983] value of isostatic rebound for the SERE would imply a present upper limit for these mid-Cenozoic sediments of between -250 and -500 m a.s.l. which is largely coincident with the SERE/WRT crustal boundary (Plates 1b, 2b and 3a) and very near our upstream limit for the initiation of RBM on ice streams C1b and B2 (e.g. Plate 3b). Conversely, this upstream limit for ice streams C2 and C1a correlates with a bedrock elevation below -500 m a.s.l. (Plate 2b) making it clear that access to mid-Cenozoic sediments may be a necessary condition for the initiation of RBM but in the case of these two ice streams it may not be sufficient.

Further evaluation of the CASERTZ aeromagnetic results for the SERE (Plate 3b) shows that both the EWB and the northern occurrence of WRT crust are largely non-magnetic although they are punctuated by high amplitude magnetic anomalies at their boundaries (recalling that an intense low like that at 260 km north and 100 km east can represent volcanics with reversed magnetization). In addition, the far northeastern corner of the survey area shows the abutment of an anomalous region of highly magnetic crust (possibly characteristic of the Bentley Subglacial Trench) against the intersection of the EWB and WRT crustal boundaries. It is the magnetic high (centered at north 250 km and east 260 km of Plate 3b) near this intersection that is the proposed site of active volcanism [*Blankenship et al.*, 1993]. In addition, the northern boundary of the southern occurrence of WRT crust is dominated by a magnetic high that could represent a volcanic center for flood basalt as proposed by Behrendt et al. [1994], although no studies have yet been undertaken to establish this.

Taken in their entirety, the CASERTZ magnetic observations are generally consistent with the picture described above of localized Cenozoic volcanism concentrated at zones of weakness (i.e., along crustal boundaries) that have possibly been reactivated in the late Cenozoic by plume related uplift centered on Marie Byrd Land. Note also from these observations that in some areas the short wavelength magnetic anomalies associated with these crustal boundaries lie in the zone we have specified for the probable initiation of RBM but that frequently they do not. In particular, the best-established site for prospective active volcanism lies well above our upper limit and seems to play no role in the initiation of RBM. The conclusion from this correlation is that the simple occurrence of focused geothermal flux associated with localized volcanism is not a sufficient condition for the initiation of RBM on the crustal blocks bounding the SERE.

Within the SERE proper, the CASERTZ aeromagnetic data (Plate 3b) indicate numerous very localized magnetic highs superimposed on a background of moderately magnetic crust. For ice stream C2 these short wavelength anomalies correlate closely with our zone of expected RBM initiation. In addition, the upper limit for RBM initiation on ice stream C1a correlates with the edge of one of the more extensive moderate amplitude magnetic units of the SERE crust. For both of these cases the limits on the zone of expected RBM initiation lie in part quite close together (where they are characterized by variations in subglacial topography) although this relationship could result from a kinematic response to the recent shut down of ice stream C. For ice streams C1b and B2 this correlation between narrowly separated limits for RBM initiation and magnetic units does not hold. It should also be noted that the moderate magnetic high along the ice stream C1/C2 boundary correlates with the highest Bouguer gravity anomaly shown in Plate 3a. The mass excess required for such a dramatic high could result from a shallow mantle which could be associated with focused geothermal flux.

In summary, the CASERTZ aeromagnetic observations are generally consistent with the expected geological framework for the SERE as represented by a "marble cake" of mid-Cenozoic sediments punctuated by localized volcanics. These observations also indicate that focused geothermal flux associated with short wavelength magnetic anomalies could play a role in the initiation of RBM when these anomalies occur at an elevation that is consistent with the presence of mid-Cenozoic sediments (i.e., for ice streams C2 and C1a). However, it is also possible that the locally elevated topography associated with these anomalies could represent volcanics capping (and therefore isolating) the mid-Cenozoic sediments necessary for the initiation of RBM.

CONCLUSIONS

From the evidence reviewed and presented here, it seems that the most likely geological control on the initiation of rapid basal motion (RBM) for the ice streams of the Southeastern Ross Embayment of West Antarctica is the availability of the mid-Cenozoic sediments deposited on the pre-existing rift controlled topography of the embayment and up onto the flanks of the surrounding crustal blocks. There is little geophysical evidence for the rigid structural controls on these sediments that would result from deposition that was syn-tectonic with reactivated rifting in the late Cenozoic. The CASERTZ aerogeophysical results indicate that in two cases, ice streams C1b and B2, the existence of sediments may be both necessary and sufficient for the initiation of RBM; for ice streams C2 and C1a these data indicate that sediment availability is probably a necessary but not sufficient condition for RBM initiation. A conclusive correlation of RBM with this drape of mid-Cenozoic sediments is probably not possible with aerogeophysical techniques [see *Bentley*, 1998] because the sediments could possibly be quite thin (O(100 m)); however, a conclusive verification should be straightforward with high resolution seismic reflection [e.g., *Rooney*, 1987b] or possibly refraction [e.g. *Anandakrishnan et al.*, 1998] techniques.

An additional known geological control based on the evidence reviewed above is certainly the gradient in geothermal flux from the cold thick crust characterizing the Whitmore Mountains (EWB) across the intermediate thickness of the transitional crust (WRT) to the thinner and warmer crust of the Southeastern Ross Embayment (SERE). Estimated values for this regional geothermal flux would range from near continental average for the EWB to the measured value of about twice continental average for SERE crust [*Alley* and *Bentley*, 1988] with the flux for the WRT crust expected to lie between these two end members. The aerogeophysical results presented here show that our observed zone of probable RBM initiation for ice streams C2, C1b, C1a and B2 lies entirely over crust with an expected regional geothermal flux that is probably higher than the continental average.

A final possible geological control on the initiation of RBM in the SERE is focused geothermal flux associated with localized volcanics (or alternatively a passive volcanic capping of the mid-Cenozoic sediments). The aerogeophysical evidence presented here is entirely consistent with the existence of these volcanics, particularly along the boundaries of the WRT crust as well as throughout the SERE. The concentration of these volcanics at crustal boundaries is thought to result from their migration along zones of weakness that were reactivated by the plume-related uplift of Marie Byrd Land in the late Cenozoic (see Dalziel and Lawver, this volume). This evidence for localized volcanics and possible focused geothermal flux is in particular associated with an anomalous narrowing of the zone we expect for the initiation of RBM on ice streams C2 and C1a although the kinematic response to ice stream C's shut down may play a role. Newly developed airborne radar sounding methods for discriminating water patches on subglacial rocks and sediments beneath ice streams [*Peters et al.*, submitted] may be sufficient for investigating the relationship between focused geothermal flux and the initiation of RBM for these ice streams.

APPENDIX

Acquisition and interpretation of CASERTZ airborne radar sounding and laser altimetry observations over the Southeastern Ross Embayment.

The CASERTZ surveys over the Southeastern Ross Embayment were accomplished with 116 flights in three seasons using a uniquely configured deHavilland Twin Otter aircraft (43 and 53 flights from the CASERTZ field camp in the 1991/92 and 1992/93 field seasons, respectively, combined with 20 flights from Byrd Surface Camp in the 1995/96 austral summer.) In general, these flights were approximately four hours in duration with a survey elevation of 350 to 1000 m above the ice surface and a survey air speed of 120 to 130 knots. The technical aspects of the airborne radar sounding and laser altimetry instrumentation and the processing techniques applied to the acquired data are detailed below.

The airborne radar sounding was accomplished with a radar constructed by the Technical University of Denmark (Skou and Sondergaard, 1976) and modified for digital data acquisition initially using a data acquisition system designed and constructed by the USGS (Wright et al., 1989) which was ultimately upgraded with a system designed and constructed by the University of Texas, Institute for Geophysics. In general, this radar transmitted a 250 ns pulse at 60 Mhz with a peak power of about 8 kW at a repetition rate of 12.5 kHz. The aircraft has been modified to support "flat-plate" dipole antennas (Skou and Teilgaard, 1978) suspended 1.25 m (i.e., approximately 1/4 wavelength) beneath each wing; a signal splitter/combiner allowed both of these antennas to transmit and receive simultaneously. After the received signals have been logarithmically detected, they were digitized at a sample rate of either 20 or 40 ns; 2048 subsequent transmissions were added to improve the signal-to-noise ratio. In the early surveys, the 2048 transmissions could be digitized and integrated about twice per second (i.e., every 30 m along track); in the later surveys using the upgraded system this rate was increased to approximately five times per second (i.e., every 12 m along track).

Data interpretation consisted of the display of differentiated radar-grams followed by an "interpretation" phase consisting of interactively bounding the transmitted pulse, surface return and bedrock return. The application of an auto tracker to these bounded returns identified the times of arrival by locating the maximum of the second derivative of the signal amplitude and extrapolating it to locate the end of the quiescent signal just before the return. This process has proved to be stable with a precision of better than a sample interval. To guarantee consistency across a survey region, each interpretation was verified by a second interpreter and line-to-line comparisons were made at survey crossover points as a final check of consistency. The radar wave velocity in ice was taken to be 168.4 m/μs and no firn correction was applied. The r.m.s. deviation of ice thickness at the crossover points varies with the ruggedness of the topography from +/- 82 m for the mountainous EWM block to +/- 11 m for the smooth topography of the SERE.

Laser altimetry was utilized to establish the best absolute surface morphology for the CASERTZ aerogeophysics experiments. The laser altimeter on the Twin Otter was a 1000W peak-power infrared unit (YAG) manufactured by Azimuth Corp. This unit was capable of 1000 pulses per second; in our configuration 64 of these pulses were integrated eight times per second for a range determination approximately every eight meters. In the absence of intervening clouds, range precisions of better than 0.1 m were readily achievable at ranges of up to 1500 m. These laser ranges were corrected for aircraft attitude as determined by an Inertial Navigation System (Litton LTN-92) and projected to a spot on the ice sheet surface using the absolute position of the aircraft.

Aircraft positions were established by solving kinematic differential carrier-phase observations of the constellation of GPS satellites visible at the aircraft and at the main base of operations (Brozena et al., 1993; Bell et al., 1999). Both Ashtech Z-12 and TurboRogue GPS receivers were used at various times (and often simultaneously) to make these observations. After adjusting for linear drift along a track line, the r.m.s. deviation of one half the observational discrepancy in laser-determined surface elevations at the crossover points was calculated for each survey; these range from +/-37 cm for the earliest survey, which was undertaken before the GPS constellation was complete, to +/-9 cm for the later surveys after the constellation was complete.

Acknowledgements. This paper is dedicated to the memory of Kathryn A. Moser who provided such valuable administrative assistance over the course of these experiments and analyses. We would also like to thank the members of the CASERTZ technical staff, from the University of Texas Institute for Geophysics (UTIG), the Byrd Polar Research Center at The Ohio State University, Lamont Doherty Earth Observatory at Columbia University, the U.S. Geological Survey and the Office of Naval Research for their efforts in the early surveys; in addition, we would like to thank the technical staff of the Support Office for Aerogeophysical Research at UTIG for the later surveys. The aerogeophysical experiments presented here also benefited from the dedicated efforts of personnel from both Kenn Borek Air Ltd. and Antarctic Support Associates. This work was supported by National Science Foundation grants OPP-9319369 and OPP-9319379.

REFERENCES

Alley, R.B., Water-pressure coupling of sliding and bed deformation: I. Water system, *Journal of Glaciology*, 35(119), 108-118, 1989.

Alley, R.B., and Bentley, C.R., Ice core analysis on the Siple Coast of West Antarctica, *Annals of Glaciology*, 11(1), 1-7, 1988.

Alley, R.B., and Whillans, I.M., Changes in the West Antarctic ice sheet, *Science*, 254(5034), 959-963, 1991.

Alley, R.B., Blankenship, D.D., Bentley, C.R., and Rooney, S.T., Deformation of till beneath ice stream B, West Antarctica, *Nature*, 322(6074), 57-59, 1986.

Alley, R.B., Blankenship, D.D., Bentley, C.R., and Rooney, S.T., Till beneath ice stream B. 3. Till deformation: Evidence and implications, *Journal of Geophysical Research*, 9(B9), 8921-8929, 1987.

Anandakrishnan, S., and Alley, R.B., Stagnation of ice stream C, West Antarctica by water piracy, *Geophysical Research Letters*, 24(3), 265-268, 1997.

Anandakrishnan, S., Blankenship, D.D., Alley, R.B., and Stoffa, P.L., Influence of subglacial geology on the position of a West Antarctic ice stream from seismic observations, *Nature*, 394(6688), 62-65, 1998.

Atre, S.R., and Bentley, C.R., Indication of a dilatant bed near downstream B camp, ice stream B, Antarctica, in: *Annals of Glaciology*, edited by E. M. Morris, vol. 20, pp. 177-182, International Glaciology Society, 1994.

Behrendt, J.C., Distribution of narrow width magnetic anomalies in Antarctica, *Science*, 144(3621), 993-995, 1964.

Behrendt, J.C., and Cooper, A., Evidence of rapid Cenozoic uplift of the shoulder escarpment of the Cenozoic West Antarctic rift system and a speculation on possible climate forcing, *Geology*, 19, 315-319, 1991.

Behrendt, J.C., and Wold, R.J., Depth to magnetic "basement" in West Antarctica, *Journal of Geophysical Research*, 68(4), 1145-1153, 1963.

Behrendt, J.C., LeMasurier, W.E., and Cooper, A.K., The West Antarctic rift system-A propagating rift "captured" by mantle plume?, in: *Recent Progress in Antarctic Earth*

Science, edited by Y. Yoshida et al., pp. 315-322, Tokyo: Terra Scientific Publishing Company, 1992.

Behrendt, J.C., Blankenship, D.D., Finn, C.A., Bell, R.E., Sweeney, R.E., Hodge, S.M., and Brozena, J.M., CASERTZ aeromagnetic data reveal late Cenozoic flood basalts(?) in the West Antarctic rift system, *Geology*, 22, 527-530, 1994.

Behrendt, J.C., Blankenship, D.D., Damaske, D., and Cooper, A.K., Glacial removal of late Cenozoic subglacially emplaced volcanic edifices by the West Antarctic ice sheet, *Geology*, 23(12), 1111-1114, 1995.

Behrendt, J.C., Saltus, R., Damaske, D., McCafferty, A., Finn, C.A., Blankenship, D.D., and Bell, R.E., Patterns of late Cenozoic volcanic and tectonic activity in the West Antarctic rift system revealed by aeromagnetic surveys, *Tectonics*, 15(2), 660-676, 1996.

Behrendt, J., Blankenship, D.D., Damaske, D., Cooper, A.K., Finn, C., and Bell, R.E., Geophysical evidence for late Cenozoic subglacial volcanism beneath the West Antarctic ice sheet and additional speculation as to its origin, in: *The Antarctic Region: Geological Evolution and Processes*, edited by C. A. Rice, pp. 539-546, Terra Antarctica Publication, 1997.

Bell, R.E., Blankenship, D.D., Finn, C.A., Morse, D.L., Scambos, T.A., Brozena, J.M., and Hodge, S.M., Influence of subglacial geology on the onset of a West Antarctic ice stream from aerogeophysical observations, *Nature*, 394, 58-62, 1998.

Bell, R.E., Childers, V.A., Arko, R.A., Blankenship, D.D., and Brozena, J.M., Airborne gravity and precise positioning for geologic applications, *Journal of Geophysical Research*, 104 (B7), 15281-15292, 1999.

Bentley, C.R., Crustal structure of Antarctica, *Tectonophysics*, 20, 229-240, 1973.

Bentley, C.R., Crustal structure of Antarctica from geophysical evidence - A review, in: *Antarctic Earth Science; Fourth International Symposium*, edited by R.L. Oliver, P.R. James, and J.B. Jago, pp. 491-497, Terra Scientific Publishing Company, 1983.

Bentley, C.R., Ice on the fast track, *Nature*, 394(6688), 21-22, 1998.

Blankenship, D.D., Bentley, C.R., Rooney, S.T., and Alley, R.B., Seismic measurements reveal a saturated porous layer beneath an active Antarctic ice stream, *Nature*, 322(6074), 54-57, 1986.

Blankenship, D.D., Bentley, C.R., Rooney, S.T., and Alley, R.B., Till beneath ice stream B. 1. Properties derived from seismic travel times, *Journal of Geophysical Research*, 92(B9), 8903-8911, 1987.

Blankenship, D.D., Bell, R.E., Hodge, S.M., Brozena, J.M., Behrendt, J.C., and Finn, C.A., Active volcanism beneath the West Antarctic ice sheet and implications for ice-sheet stability, *Nature*, 361, 526-529, 1993.

Brozena, J.M., Jarvis, J.L., Bell, R.E., Blankenship, D.D., Hodge, S.M., and Behrendt, J.C., CASERTZ 91-92: Airborne gravity, *Antarctic Journal of the United States*, 20(5), 1-3, 1993.

Boulton, G.S., and Jones, A.S., Stability of temperate ice caps and ice sheets resting on beds of deformable sediment, *Journal of Glaciology*, 24(90), 29-43, 1979.

Clark, P.U., and Walder, J.S., Subglacial drainage, eskers, and deforming beds beneath the Laurentide and Eurasian ice sheets, *Geological Society of America Bulletin*, 106(2), 304-314, 1994.

Clarke, T.S., Burkholder, P.D., Smithson, S.B., and Bentley, C.R., Optimum seismic shooting and recording parameters and a preliminary crustal model for the Byrd Subglacial Basin, Antarctica, in: *The Antarctic Region: Geological Evolution and Processes*, edited by C. A. Ricci, pp. 485-493, Terra Antarctica Publication, 1997.

Drewry, D.J., *Antarctica: Glaciological and Geophysical Folio*, Cambridge: Scott Polar Research Institute, 1983.

Engelhardt, H., and Kamb, B., Basal hydraulic system of a West Antarctic ice stream: Constraints from borehole observations, *Journal of Glaciology*, 43(144), 207-230, 1997.

Engelhardt, H., and Kamb, B., Basal sliding of ice stream B West Antarctica, *Journal of Glaciology*, 44(147), 223-230, 1998.

Engelhardt, H., Humphrey, N., Kamb, B., and Fahnestock, M., Physical conditions at the base of a fast moving Antarctic ice stream, *Science*, 248(4951), 57-59, 1990.

Hindmarsh, R.C.A., Ice stream surface texture, sticky spots, waves and breathers: The coupled flow of ice, till and water, *Journal of Glaciology*, 44(148), 589-614, 1998.

Jankowski, E.J., and Drewry, D.J., The structure of West Antarctica from geophysical studies, *Nature*, 291(5810), 17-21, 1981.

Jankowski, E.J., Drewry, D.J., and Behrendt, J.C., Magnetic studies of upper crustal structure in West Antarctica and the boundary with East Antarctica, in: *Antarctic Earth Science*, edited by R. L. Oliver and P. R. James, pp. 197-203, Australian Academy of Science, 1983.

Lawver, L.A., Royer, J.Y., Sandwell, D.T., and Scotese, C.R., *Evolution of the Antarctic Continental Margin,* Cambridge: Cambridge University Press, 1991.

Munson, C.G., and Bentley, C.R., The crustal structure beneath ice stream C and ridge BC, West Antarctica from seismic refraction and gravity measurements, in: *Recent Progress in Antarctic Earth Science*, edited by Y. Yoshida, et al., pp. 507-514, Terra Scientific Publishing Company, 1992.

Paterson, W.S.B., *The Physics of Glaciers 3rd ed.,* Elsevier Science Ltd., 1994.

Peters, M.E., Blankenship, D.D., and Morse, D.L., Scattering characteristics of ice streams B and C, West Antarctica, from airborne coherent ice-penetrating radar, submitted, *Journal of Glaciology*, 1999.

Rooney, S.T., Blankenship, D.D., Bentley, C.R., and Alley, R.B., Till beneath ice stream B. 2. Structure and continuity, *Journal of Geophysical Research*, 92(B9), 8913-8920, 1987a.

Rooney, S.T., Blankenship, D.D., and Bentley, C.R., Seismic refraction measurements of crustal structure in West Antarctica, in: *Gondwana Six: Structure, Tectonics, and Geophysics,* edited by G. D. McKenzie , vol. 40, pp. 1-7, American Geophysical Union, 1987b.

Rooney, S.T., Blankenship, D.D., Alley, R.B., and Bentley, C.R., Seismic reflection profiling of a sediment-filled graben beneath ice stream B, West Antarctica, in: *Geological Evolution of Antarctica*, edited by M. R. A. Thomson, J. A. Crame, and J. W. Thomson, pp. 261-265, Cambridge: Cambridge University Press, 1991.

Rose, K.E., Characteristics of ice flow in Marie Byrd Land, Antarctica, *Journal of Glaciology*, 24(90), 63-75, 1979.

Rose, K.E., Radio-echo studies of bedrock in southern Marie Byrd Land, West Antarctica, in: *Antarctic Geoscience*, edited by C. Craddock, pp. 985-992, Madison: The University of Wisconsin Press, 1982.

Scherer, R.P., Quaternary and Tertiary microfossils from beneath ice stream B: Evidence for a dynamic West Antarctic ice sheet history, *Palaeogeography, Palaeoclimatology, and Palaeoecology*, 90, 395-412, 1991.

Scherer, R.P., Aldahan, A., Tulaczyk, S., Possnert, G., Engelhardt, H., and Kamb, B., Pleistocene collapse of the West Antarctic ice sheet, *Science*, 281(5373), 82-85, 1998.

Shabtaie, S., and Bentley, C.R., West Antarctic ice streams draining into the Ross Ice Shelf: Configuration and mass balance, *Journal of Geophysical Research*, 92(B2), 1311-1336, 1987.

Shabtaie, S., Whillans, I.M., and Bentley, C.R., The morphology of ice streams A, B, and C, West Antarctica, and their environs, *Journal of Geophysical Research*, 92(B9), 8865-8883, 1987.

Skou, N., and Sondergaard, F., Radioglaciology: A 60 MHz ice sounder system, *Report R169*, Electromagnetics Institute, Technical University of Denmark, Lyngby, 124 pp., 1976.

Skou, N., and Teilgaard, J., Radioglaciology: Design and construction of ice sounding antennas for LC-130 Bureau No. 159131, *Report R200*, Electromagnetics Institute, Technical University of Denmark, Lyngby, 149 pp., 1978.

Stern, T.A., and ten Brink, U., Flexural uplift of the Transantarctic Mountains, *Journal of Geophysical Research*, 94(B8), 10315-10330, 1989.

Sweeney, R., and the CASERTZ group, Aeromagnetic maps of the Interior Ross Embayment, West Antarctica: Folio A, *Open-file Report No. 94-122*, U.S. Geological Survey, Denver, Colorado, 1994.

Sweeney, R., Finn, C.A., Blankenship, D.D., Bell, R.E., and Behrendt, J.C., Central West Antarctica aeromagnetic data: A web site for distribution of data and maps, *Open-file Report No. 99-0420*, 15 pp., U.S. Geological Survey, Denver, Colorado, 1999.

ten Brink, U.S., Bannister, S., Beaudoin, B.C., and Stern, T.A., Geophysical investigations of the tectonic boundary between East and West Antarctica, *Science*, 261(5117), 45-50, 1993.

van der Wateren, D., and Hindmarsh, R., Stabilists strike again, *Nature*, 376(6539), 389-391, 1995.

Wannamaker, P., Stodt, J.A., and Olsen, S.L., Dormant state of rifting below the Byrd Subglacial Basin, West Antarctica, implied by magnetotelluric (MT) profiling, *Geophysical Research Letters*, 23(21), 2983-2986, 1996.

Wright, D.L., Bradley, J.A., and Hodge, S.M., Use of new high-speed digital data acquisition system in airborne ice-sounding, *IEEE Transactions on Geoscience and Remote Sensing*, 27(5), 561-566, 1989.

J.C. Behrendt, U.S. Geological Survey, P.O. Box 25046, MS 964, Denver, CO 80225 U.S.A.

R.E. Bell, Lamont-Doherty Earth Observatory, P.O. Box 1000, 60 Route 9W, Palisades, NY 10964-1000 U.S.A.

D.D. Blankenship, The University of Texas Institute for Geophysics, 4412 Spicewood Springs Road, Bldg. 600, Austin TX 78759-8500 U.S.A.

J.M. Brozena, Naval Research Laboratory, Code 7420, Overlook Ave. S.W., Washington, D.C. 20375-5320 U.S.A.

C.A. Finn, U.S. Geological Survey, P.O. Box 25046, MS 964, Denver, CO 80225 U.S.A.

S.M. Hodge, U.S. Geological Survey, University of Puget Sound, Tacoma, WA 98416 U.S.A.

S.D. Kempf, The University of Texas Institute for Geophysics, 4412 Spicewood Springs Road, Bldg. 600, Austin TX 78759-8500 U.S.A.

D.L. Morse, The University of Texas Institute for Geophysics, 4412 Spicewood Springs Road, Bldg. 600, Austin TX 78759-8500 U.S.A.

M.E. Peters, The University of Texas Institute for Geophysics, 4412 Spicewood Springs Road, Bldg. 600, Austin TX 78759-8500 U.S.A.

M. Studinger, Lamont-Doherty Earth Observatory, P.O. Box 1000, 60 Route 9W, Palisades, NY 10964-1000 U.S.A.

ONSET OF STREAMING FLOW IN THE SIPLE COAST REGION, WEST ANTARCTICA

Robert Bindschadler

Oceans and Ice Branch, NASA Goddard Space Flight Center, Greenbelt, MD

Jonathan Bamber

Bristol Glaciology Centre, University of Bristol, Bristol, United Kingdom

Sridhar Anandakrishnan

Department of Geology, University of Alabama, Tuscaloosa, AL

Onsets of West Antarctic ice streams feeding the Ross Ice Shelf are reviewed. Inland flow and streaming flow are defined to clarify what is meant by the onset. Various means to locate the onset are discussed including: crevasses, flow-stripes, surface elevations, bed elevations and driving stress, with the last being the most reliable. A new map of driving stress in West Antarctica is presented which clearly shows the location of maximum driving stresses. Recent work is summarized and used to draw conclusions that onsets appear to share the properties of temperate basal conditions, location in a channel, and presence of a sedimentary bed. Consideration of previous analysis of shear margins suggests that the controlling process in the onset of streaming is basal dynamics rather than internal weaknesses such as strain softening or lateral shear margins. Migration of onsets is also concluded to be an inevitable consequence of their kinematics and may be an episodic process.

INTRODUCTION

Ice streams are responsible for the unique dynamic character of the West Antarctic ice sheet. These fast-moving conduits discharge ice rapidly from the slower inland ice and feed the large floating ice shelves. They also determine the shape of this marine-based ice sheet. Ice streams have lower surface slopes which, in turn, reduce the elevation of the ice-sheet interior and permit inland penetration of storms which deliver greater amounts of precipitation farther inland than in East Antarctica [*Lettau*, 1969]. Thus, the cross-sectional profile of the West Antarctic ice sheet differs significantly from the classic near-parabolic profiles of either the Greenland or the East Antarctic ice sheets.

Ice-stream flow is radically different from inland flow. The West Antarctic ice sheet lies in a deep, sediment-laden marine basin warmed from beneath by a relatively high geothermal heat flux [*Drewry*, 1983]. These factors combine to create a well-lubricated, water-saturated bed on which ice streams can maintain fast motion through the generation of sufficient quantities of water by basal friction and shear heating. The deformable aspects of the marine, subglacial till are important if not directly in the motion of the ice, then at least in their malleability or erosion into a smooth subglacial surface devoid of frequent protuberances which might hinder ice flow [*Alley*, 1990; *Cuffey and Alley*, 1996]. The rapid flow is also responsible for faster response times of West Antarctica as perturbations of flow are transmitted rapidly upstream into the inland ice and downstream onto the ice shelf along ice streams [*Bindschadler*, 1997].

Ice streams develop from the slower moving ice in the central regions of West Antarctica. Understanding what conditions must be present, or must develop, for the ice to make this major transition in its mode of flow is fundamental to understanding where and why ice streams form. Far less work has been completed on the study of streaming onsets than has been undertaken on the streaming process itself. Most of the work has been surface observations and inferences from them rather than direct measurements of subglacial processes in the vicinity of an onset. Nevertheless, it is worthwhile consolidating the work to date to assess what is known, what is speculated and whether earlier ideas about onset dynamics remain viable in the light of more recent work. This paper addresses these topics, along with clarifying the terminology associated with the onset region.

DEFINITION OF ONSET

The "ice-stream onset" or "streaming onset" (shortened hereafter as "onset") is defined here as the location of the transition between inland flow and streaming flow (Figure 1). This definition itself requires definitions of "inland flow" and "streaming flow". Inland flow is usually taken to describe flow resulting from internal deformation within the ice mass but, in West Antarctica, basal temperatures at the pressure melting point are likely so widespread that we expand the definition of inland flow here to allow some amount of basal sliding. The key characteristics of inland flow are that the gravitational driving stress (defined below) is balanced largely by basal shear stress and that the flow speed increases with increases in basal shear stress. The convex-up shape of ice sheets is a result of this relationship. A more detailed discussion of how much and what type of basal sliding occurs in inland flow is reserved for a later section.

Streaming flow, on the other hand, appears to be a distinctly different basal sliding process. Ice streams are the defining example, although any reference to their sliding mechanism was not included in the original definition of ice streams [*Swithinbank*, 1954]. Streaming flow appears to require both subglacial water pressurized to near the ice overburden pressure and subglacial till with a small shear strength. What fraction of the motion occurs by deformation within the till and how much is sliding of the ice over the till is still debated and is probably spatially variable [*Kamb*, 1991; *Kamb and Englehardt*, 1998; *Alley and others*, 1989]. What typifies this type of flow is that speed increases with decreases in

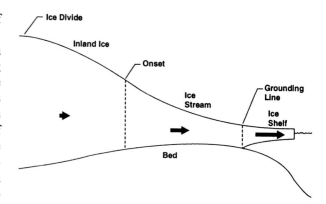

Fig. 1. Schematic cross-section of West Antarctica showing slow, inland ice, fast ice stream, and flat ice shelf. The transitions between these areas occur at the onset and at the grounding line. Length of arrows indicates flow speed. Surface and bed elevations illustrate progressively thinner ice from ice divide to shelf edge.

driving stress and that lateral shear, in addition to basal shear, provides an important resistance to flow [*Hughes*, 1977; *McIntyre*, 1985; *Whillans*, 1987; *Echelmeyer and others*, 1994; *Jackson and Kamb*, 1997; *Whillans and van der Veen*, 1997; *Harrison and others*, 1998]. The surface shape of an ice stream that results is concave-up as the slope decreases downstream from a steeper slope imposed by the inland flow regime, to a much shallower slope as the ice stream flows into the much flatter ice shelf (Figure 1).

These definitions of inland flow and streaming flow are shown in the next section to lead directly to means to locate the onset by examining either surface shape or the relationship between driving stress and flow speed.

LOCATION OF THE ONSET

Identifying fully developed ice streams is not difficult. They are many tens of kilometers wide, flow at speeds of many hundreds of meters per year, have heavily crevassed shear margins and have a more undulated surface topography than the adjacent, slower interstream ridges [*Bindschadler and Vornberger*, 1990]. Ice streams are fed through coalescence of an arborescent pattern of tributaries corresponding to a pattern of basal valleys [*Shabtaie and Bentley*, 1988; *Joughin and others*, 1999]. Each tributary has its own onset where it enters an ice stream. Multiple tributaries often feed a single ice stream forming larger ice streams [*Whillans and van der Veen*, 1993; *Scambos and Bindschadler*, 1991]. Downstream, the ice streams accelerate, they become thinner, and their beds become smoother on the multi-kilometer scale [*Retzlaff and Bentley*, 1993]. Most of these characteristics develop gradually making it difficult to trace an ice stream upstream and pinpoint the onset of each tributary.

Increasing the difficulty of onset location is the fact that, by our definition, the onset manifests a transition in flow mechanism at the base of the ice sheet, but measurements of basal conditions are sparse and often absent in potential onset areas. Yet, identification of the incipient ice stream is necessary to directing field studies of the physical processes critical to this flow transition. Thus, proxy indicators of the contrast in basal conditions have been sought to locate the onset. There may be no single best means to identify an onset region. We review various methods which have been employed, beginning with methods that rely on qualitative examination of the surface topography. There are two types of distinctive features that are numerous on ice streams: crevasses and flowstripes.

Crevasses

Crevasses are formed when surface horizontal stresses exceed the tensile strength of ice. They are common on ice streams, especially at the margins, due to large velocity gradients. Some ice streams, particularly tributaries of B and E, have extensive sets of very long crevasses as their most upstream surface expression of fast flow (Figure 2). Individual crevasses are tens of kilometers in length, suggesting a regional and rapid increase in velocity [*Vornberger and Whillans*, 1986; *Stephenson and Bindschadler*, 1990]. This pattern is also seen on many Antarctic outlet glaciers, such as Byrd and Pine Island Glaciers which are not ice streams [*Bentley*, 1987; *McIntyre*, 1985], so this characteristic does not always represent an ice-stream onset. Nevertheless, when long transverse crevasses exist upflow of an ice stream, the inference that the crevasses locate the onset is reasonable. *McIntyre* [1985] noted that outside West Antarctica, this type of onset is correlated with a subglacial escarpment. *Bell and others* [1998] have made a similar correlation for one location in West Antarctica, but generally the onsets of the Siple Coast ice streams do not correlate with basal escarpments.

Flowstripes

Another distinctive type of ice-stream surface feature is "flowstriping"—development of narrow ridges or troughs a few hundred meters across, tens to hundreds of

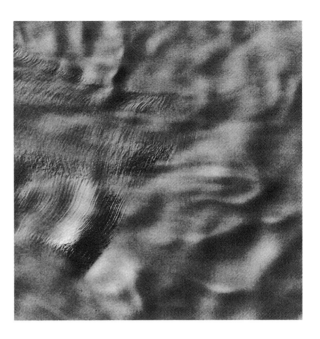

Fig. 2. Landsat image of an upstream area of ice stream E. Flow direction is toward bottom of image. Extensive transverse crevasses and flowstripe initiation both have been suggested as onset locators, however in this region they occur in different places. Image is 34.2 km on each side. Plate 1 shows approximate location.

kilometers long and with amplitudes of a few meters. Flowstripes also have been referred to as "flowbands", "flowlines" and "stripes" by various authors. Different theories have been proposed to explain their formation [*Gudmundsson and others*, 1998; *Casassa and Brecher*, 1993] but all agree that their considerable length is the result of rapid ice motion which displaces them great distances over a time period much less than their relaxation time.

The formation of flowstripes has been modeled by ice sliding over perturbations in either basal topography or basal friction [*Gudmundsson and others*, 1998]. Flowstripes also have been observed emanating from surface bumps and margins of ice streams [*Merry and Whillans*, 1993]. Where surface velocities are available, flowstripes correspond roughly with regions moving at speeds in excess of 100 m/a [*Scambos and Bindschadler*, 1993; *Chen and others*, 1998].

Implicit in the association of flowstripes with ice streams is the assumption that the boundary between flowstripes and no flowstripes represents the onset of the ice stream. In other words, the subglacial conditions necessary for flowstripe generation are the same conditions required for streaming flow. *Dowdeswell and McIntyre*

[1987] showed that flowstripes affected the power spectrum of the surface elevation field enough that a broad discrimination was possible between the ice-stream region which contained flowstripes (their "type 3" surface) and the inland region ("type 2") which did not contain flowstripes. They quantified type 2 topography as having horizontal wavelengths generally greater than 20 km and vertical amplitudes of up to 16 meters with a root-mean-square of 4-5 meters. By contrast, type 3 topography had a much broader power spectrum often dominated by horizontal wavelengths less than 10 kilometers. The boundary of these two topography types afforded an estimate of the onset position, but the construction of the power spectrum required a large spatial sample, so the boundary, and therefore the onset location, is not particularly sharp.

Flowstripes are resilient features that can persist for centuries [*Gudmundsson and others*, 1998]. However, the flow field can change on shorter time scales [*Stephenson and Bindschadler*, 1988; *Bindschadler and Vornberger*, 1998]. There are many examples of flowstripes misaligned with a measured flow field, indicating that conditions are out of equilibrium. This condition can lead to misinterpretation of present onset conditions and flow direction if the most upstream flowstripes are equated to a present onset of streaming flow. An excellent example of this has been documented in the upstream reaches of ice stream D, where a band of flowstripes was interpreted as an onset region [*Hodge and Doppelhammer*, 1996], yet later measurements of velocity indicated flow is not presently parallel to the flowstripes [*Bindschadler and others*, in press]. The analysis of the velocity field (presented later) suggests this region is not presently an onset.

The formation process of flowstripes may not be unique, and therefore their use for locating onsets may be questionable. Figure 2 shows flowstripes of the type in which the upstream end of each flowstripe is associated with a surface topographic bump. This situation is common at the margins of narrow ice streams in the catchment areas of ice streams D and E, where complete coverage by Landsat imagery permits detailed examination [*Scambos and Bindschadler*, 1991]. Ice streams often become wider by the incorporation of adjacent ice, but the flowstripes that accompany this process do not locate an onset because the ice stream already exists upstream. The example of Figure 2 also casts crevasse formation in doubt as a reliable means of onset determination because the flowlines just described occur upstream of the crevasses.

A second situation in which flowstripes often occur is shown in Figure 3. Here, the upstream ends of a set of parallel flowstripes are not associated with any obvious

Fig. 3. Possible failed onset. Landsat image show a region of flowstripes (center of image) which fails to continue downstream (toward bottom of image). Image is 68.4 km x 57 km. Plate 1 shows approximate location.

surface feature. Flowstripes are extremely subtle, indicating subdued surface relief (or, in some cases, a sun-parallel orientation), making their initiation difficult to locate. In most cases, flowstripes continue downstream and become more clearly defined as the ice stream accelerates [*Stephenson and Bindschadler*, 1990]. The flowstripes of Figure 3 are unique because the flowstripes shown do not connect with an ice stream. This suggests the occurrence of failed onsets, an important possibility of onset dynamics.

Surface Elevations

Surface elevations furnish an additional method of locating onsets. The general profile along an ice-stream flowline (Figure 1) illustrates that the boundary between inland ice and the ice stream corresponds to the maximum surface slope and an inflection point in the along-flow elevation profile. This is a result of the convex-up elevation profile of inland ice and the concave-up profile of ice streams mentioned earlier. In practice, topographic variations on smaller scales make it difficult to apply this method without smoothing the surface profile, which partially defeats the purpose of finding the precise location of maximum surface slope. Because this method is closely tied to the use of maximum driving stress for onset location, it is developed more fully in the "Driving Stresses" section.

Bed Elevations

All the above methods rely on surface topography alone. Introduction of information on ice thickness can usually improve the identification. *McIntyre* [1985] has noted that the West Antarctic ice streams do not follow the general correlation observed elsewhere in Antarctica that fast flow initiates at a step in the bed elevation. *Retzlaff and Bentley* [1993] also commented that their extensive bed-elevation mapping of ice streams A, B, and C failed to show any large gradients in basal topography in the regions where they believed onsets must be located. However, *Bell and others* [1998] found one deep bed channel whose distinct headwall coincided with the initiation of flowstripes. They inferred this location was also the onset of streaming flow.

Driving Stress

Ice moves in response to the gravitational force exerted on it. This driving stress, τ_d, is defined as

$$\tau_d = \rho \, g \, H \sin \alpha \qquad (1)$$

where ρ is the average column density, g is the gravitational acceleration, α is surface slope and H is ice thickness. Slope increases downstream within the inland-ice regime because driving stress must increase to transport an increasing flux of ice [*Paterson*, 1994]. The specific relationship for a non-linear flow law produces the near-parabolic elevation profile [*Vialov*, 1970; *Paterson*, 1994]. By contrast, ice streams exhibit a decreasing slope downstream while their velocity is increasing. The diminishing slope produces a smooth transition to the nearly flat ice shelves. Decreasing ice stream thickness downstream is not only a result of downstream-dipping surface slope, but is also accentuated by the generally lower basal elevations in the interior caused by increased crustal depression underneath the thicker inland ice. This slope pattern creates an inflection point in the elevation profile at the onset.

Generally, surface slope varies spatially far more along an ice-sheet flowline than ice thickness (and certainly more than ice density) [*Whillans*, 1987], so the pattern of driving stress mimics the pattern of surface slope. Thus, just as there is a maximum in surface slope at the onset, the driving stress maximum is also nearly coincident with the onset.

Because driving stress depends on ice thickness, it cannot be determined from surface measurements alone. Data from regional surveys of ice thickness and surface elevation were used to calculate the pattern of driving stress [*Cooper and others*, 1982]. These data were collected from airborne surveys in a roughly orthogonal grid pattern with a spacing of approximately 50 km. As expected, where ice flowed into ice shelves, a zone of maximum driving stress occurred between the flat ice divides and the flat ice shelves. This was particularly apparent for the heads of ice streams A, B and C.

The largest source of error in those driving stress calculations was identified as the surface-elevation measurement. For this reason, we present a recalculation of West Antarctic driving stress (Plate 1) using a more accurate elevation data set based on satellite altimetry and a regridding of the same ice thickness measurements used by Cooper and others [*Bamber and Huybrechts*, 1996]. The spatial averaging for surface slope used in Plate 1 is 20 times the local ice thickness to average out the effect of longitudinal stress gradients on the effective shear stress at the bed [*Budd*, 1969].

The ice divides are evident by their very low driving stresses (red in Plate 1), caused by their very low surface slopes. The zone of maximum driving stress upstream of ice streams A-E is evident, ranging from 80 kPa (blue) to 150 kPa (brown) with the highest maxima and greatest widths on the north and south ends of the zone. This pattern expresses the overall bowl-shaped configuration of the Ross embayment section of the West Antarctic— the north end (Executive Committee Range) and south end (Whitmore and Transantarctic Mountains) of the zone maintain higher elevations, while the central elevations of West Antarctica are somewhat lower. These features contrast with outlet glaciers in East Antarctica, and with Pine Island and Thwaites Glaciers in West Antarctica as well, where driving stresses achieve maxima exceeding 150 kPa and where these maxima occur much closer to the coast [*Bentley*, 1987; *Cooper and others*, 1982]. The location of a recently discovered subglacial volcano [*Blankenship and others*, 1993] lies well away from the zone of maximum driving stress, suggesting it plays no role in ice-stream initiation.

Figure 4 shows 500-km-long longitudinal profiles of driving stress along center flowlines of ice streams B-E. These all show the characteristic feature of low driving stress at the divide, rising to a maximum, and decreasing along the ice stream to a value of about 20 kPa at the entrance to the ice shelf. The magnitude of the maximum ranges from 80 kPa for D to 150 kPa for B. The maximum is distinct and single for these two cases. The profile for ice stream C exhibits a modest secondary maximum upstream of the largest peak. Thickening and a topographic bulge have been identified at the head of the

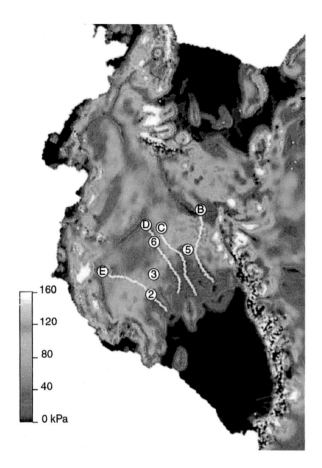

Plate 1. Distribution of driving stress in West Antarctica calculated from Equation 1 as described in the text. Large black areas correspond to Ross and Ronne/Filchner Ice Shelves. Solid white lines indicate positions of 500-km-long centerline profiles of Siple Coast ice streams B–E in Figure 4. Numbers indicate locations of Figures 2, 3, 6 and 7. Ice divides appear as narrow red regions in central West Antarctica. Band of maximum driving stress across Siple Coast ice streams appears as brown areas on B profile and north of E profile and connecting blue band across C and D.

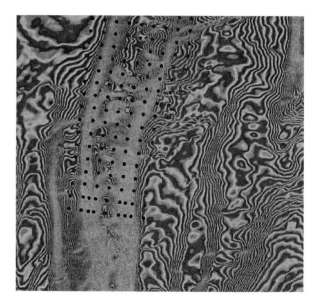

Plate 2. Synthetic aperture radar (SAR) interferogram from Radarsat data covering downstream one-third of ice stream D onset grid. Fringes correspond predominantly to relative velocity in direction toward satellite (toward top of figure). Points on figure correspond to downstream portion of gridpoints in Figure 6.

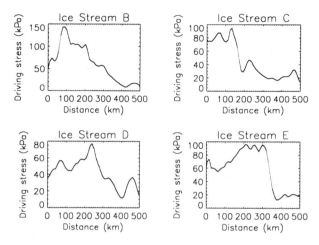

Fig. 4. Profiles of driving stress along ice streams B, C, D, and E. Profile locations are shown in Plate 1.

stagnant portion of this ice stream [*Joughin and others*, 1999]. The observed maxima in Figure 4 express this evolving condition and are probably not a reliable locator of the onset of this inactive ice stream. Ice stream E's profile displays a triple maximum suggesting a case where streaming flow begins, but is not sustained (c.f., Figure 3) until the position of the third maximum. This interpretation is supported by recent measurements of the flow pattern feeding the Siple Coast ice streams [*Joughin and others*, 1999].

DETAILED STUDIES OF ONSET AREAS

Recent fieldwork has focused on particular regions believed to be onset areas. One such study was based entirely on data collected remotely in the catchment area of ice stream C [*Bell and others*, 1998]. From the combined evidence of surface features, ice thickness, as well as magnetic and gravimetric signatures, it was concluded that initial streaming flow coincided with a 1000-m deep, 20-km wide bed channel in the basal topography. This correlation is reminiscent of the earlier correlations suggested by *McIntyre* [1985]. The onset's location was inferred from the appearance of flowstripes in 1-km resolution satellite imagery, but a new measurement of the velocity field shows that while ice does speed up at this location, the ice eventually slows as it meets the stagnant ice stream C [*Joughin and others*, 1999]. It is uncertain whether the ice does, or ever did, attain the streaming condition at this location. Nevertheless, the modeling of the gravity and magnetic signatures of this region demonstrated that the channel where speed increases contains a low density, non-magnetic sedimentary deposit at least one-kilometer thick.

The correlation of sedimentary units with fast flow is further advanced by surface field studies farther downstream along this channel. Reflection and refraction seismic surveys, in combination with measurements of surface ice velocity, demonstrate that a sharp contrast in ice speed coincides with the edge of a modeled subglacial sedimentary layer many hundreds of meters thick [*Anandakrishnan and others*, 1998]. Figure 5 shows the spatial correlation deduced by these authors, who also suggest that the thinner, narrower sedimentary unit (at km 40 of Figure 5) does not produce streaming flow because the width of this unit is narrow enough that side shear of adjacent slower ice effectively prevents streaming. This point is discussed more fully later.

A spatially more extensive field program designed specifically to locate and measure the surface deformation field associated with an onset was undertaken near Byrd Station on ice feeding into ice stream D [*Chen and others*, 1998; *Bindschadler and others*, in press]. Repeated surveys of surface markers placed in a regular grid (5-km spacing) quantified the surface velocity field and, from it, the surface-horizontal strain field [*Chen and others*, 1998]. A convergent, accelerating flow into the ice stream was apparent (Figure 6). As with the previous onset study, the developing stream coincided with a bed channel, although measurements to confirm the presence of a sedimentary layer were not made as part of the survey [*Bamber and Bindschadler*, 1997].

Analysis of the measured velocity field and associated ice-sheet geometry was conducted to locate the onset. The approach deemed most successful drew on the previously discussed difference of the driving stress versus velocity relationship for inland flow and for streaming flow. We now develop this subject more fully.

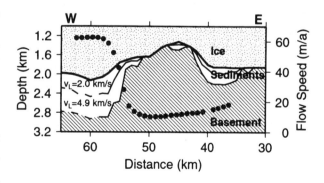

Fig. 5. Model of across-flow section near an onset calculated from inversion of seismic data and the GPS-measured surface velocity profile (from *Anandakrishnan and others*, 1998, Figure 4). Plate 1 shows approximate location.

130 THE WEST ANTARCTIC ICE SHEET: BEHAVIOR AND ENVIRONMENT

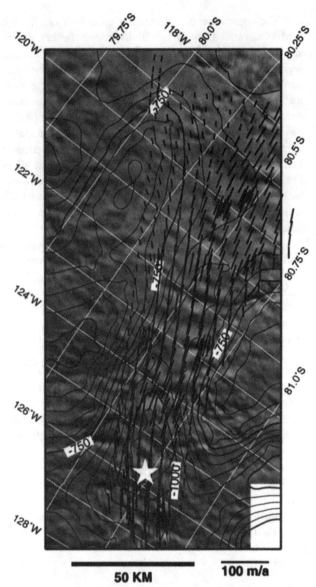

Fig. 6. GPS-measured surface velocities, Landsat imagery and bed elevation contours in the vicinity of the onset of ice stream D (from *Bindschadler and others*, in press). Data sources are discussed in text. Velocities are spaced 5 km apart. Image size is 152 km x 195.5 km. Plate 1 shows approximate location. Star is location of maximum driving stress.

The rheological properties of ice often are treated as a non-linear flow law of the form,

$$\varepsilon = A\tau^n \quad (2)$$

where ε is the octahedral strain rate, τ is the octahedral stress, A is the temperature-dependent flow-law coefficient, and n is the flow-law exponent [*Glen*, 1955]. In the simplified case of parallel-sided, laminar flow, where bed-parallel shear is the only non-zero component of the stress tensor, the surface velocity, U_s, can be expressed as

$$U_s = U_b + 2 A (\rho g \sin\alpha H)^n H / (n+1) \quad (3)$$

where U_b is the basal sliding velocity [*Paterson*, 1994]. Substituting driving stress (Equation 1), this can be re-written as

$$(U_s - U_b)/H = 2 A \tau_d^n / (n+1) \quad (4)$$

The ratio $(U_s - U_b)/H$ expresses the mean shear strain rate through the column [*Budd and Smith*, 1981]. Because the sliding component in the surveyed data is unknown, Figure 7 plots U_s vs. τ_d for the surveyed grid. Surface slope at each gridpoint was determined from a bilinear fit to surface elevations over a 40 km x 40 km area [*Bindschadler and others*, in press]. Open circles in Figure 7 follow the kinematic centerline, where the laminar flow approximation is most likely valid. These points show the expected pattern of increasing velocity versus increasing driving stress.

From the definition of streaming flow, the opposite relationship between driving stress and velocity is expected on the ice stream. Figure 8 includes points taken from data farther downstream on ice stream D and, as expected, an inverse relationship is evident [*Bindschadler*

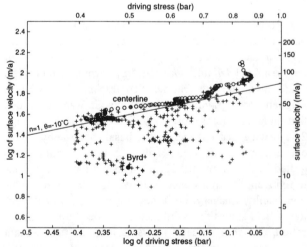

Fig. 7. Surface velocity versus driving stress. Data points are from ice stream D onset grid where bold circles are gridpoints along the kinematic centerline [*Bindschadler and others*, in press]. Solid line corresponds to n=1 at temperature of –10°C in Equation 4.

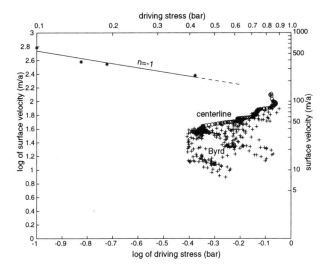

Fig. 8. Surface velocity versus driving stress. Crosses and circles are same data as in Figure 7. Additional data (4 asterisks) are taken from centerline positions distributed along ice stream D where velocities are given in *Bindschadler and others* [1996]. Solid line is n=-1 linear fit to ice stream data.

and others, in press]. Thus, along the centerline of the developing ice stream, the driving stress rises almost steadily, and reaches a maximum value of 87 kPa before reversing to decreasing stresses as the velocity continues to increase. *Bindschadler and others* [in press] used this result to characterize the position where the slope of the relationship between τ_d and U_s/H changes sign as the onset of streaming flow.

A more detailed measurement of surface motion in the vicinity of ice stream D's onset was produced from synthetic aperture radar data from Canada's Radarsat satellite [*Gray and others*, 1998]. Two images were combined into an interferogram giving a very sensitive measurement of differences in path length from the satellite to the surface. Plate 2 represents the path-length difference as a series of color fringes. Each full fringe represents approximately 6 cm of motion along the path from the satellite to the surface (up from the bottom of the figure). Displacements are caused by both surface topography (due to parallax) and ice motion, but because the parallax component is rather small, most of the densely spaced fringing represents velocity gradients in the surface motion field. The fringe pattern confirms the general pattern of the grid survey measurements and includes more detail within the survey area. Most importantly, this study shows the complexity of the motion field within, and beyond, the survey grid. In particular, there is a region to the north (left in the figure) of the main tributary that illustrates a secondary flow parallel to the main

tributary. Future interferometric results undoubtedly will prove extremely valuable in assessing the location of and patterns in the onset of streaming flow.

PHYSICAL PROCESSES AT THE ONSET

There have been no direct observations of the basal processes within an onset area. Nevertheless, the surface measurements discussed above, combined with known processes extant under active ice streams and at Byrd Station, in the inland ice regime, provide some basis for speculation of what may occur within an onset area.

Basal Temperatures

A necessary condition for streaming to begin is that the basal ice be at the pressure-melting temperature. The high probability of temperate basal conditions throughout much of West Antarctica has been mentioned earlier. A melting condition was observed at the base of the Byrd Station borehole more than 100 km upstream of ice stream D's onset, but the physical setting of Byrd—positioned on the upstream side of a large bed high—can be viewed as atypical [*Whillans*, 1979].

High accumulation rates will tend to lower the average temperature of the ice, but large geothermal heat flow will help offset this cooling [*Paterson*, 1994]. Thermal models of the West Antarctic ice sheet based on the steady state solution of *Robin* [1955] require improbably low values of geothermal heat flow to avoid a temperate bed [*Rose*, 1979; *Anandakrishnan and others*, 1998]. Steady state thermal conditions may not be reached on fast moving ice, but the journey of inland ice from the ice divide toward the onset area is measured in many thousands of years. Thus, through the combined processes of geothermal heating and strain heating in deforming basal layers, there is ample time for basal ice to be warmed to the pressure melting point.

Vertical stretching of ice will enhance the warming of basal layers contributing to the observed preference of streaming initiation in bed channels. This effect has been observed on Jakobhavns Isbræ, Greenland, where a thick temperate-ice layer created by vertical stretching as ice enters a deep bed channel is believed responsible for rapid shear deformation rates [*Iken and others*, 1993; *Funk and others*, 1994]. Vertical stretching experienced by ice flowing into bed channels also acts to decrease the vertical temperature gradients in the ice which, in turn, decreases the vertical conduction of heat upward through the column. The result would be a tendency to retain geothermal heat in the basal layers and deformation rates.

This effect is transient and would only last as long as it takes the temperature profile to re-equilibrate, but increased shear heating would delay equilibration time and maintain a thicker layer of temperate ice. If the basal ice is already temperate before entering the bed channel, the trapped heat would then be used to melt ice, producing water that could accentuate basal sliding.

Basal Sliding

The basal sliding process has been examined most extensively on temperate glaciers. In general, the sliding velocity, U_b, is related to the basal shear stress, τ_b, through a relationship of the form,

$$U_b = k\, \tau_b^m \qquad (5)$$

The coefficient, k, is not constant in many proposed sliding laws and is specified to depend on factors such as water pressure and, in some cases, the sliding velocity itself [*Paterson*, 1994]. Early theoretical treatments of basal sliding suggested m=2, but subsequent approaches produced values of m=1 [*Paterson*, 1994]. The relevant point here is that as long as the exponent, m, is positive, the form of this sliding relationship fits with the earlier definition of inland flow—i.e., increasing velocity with increasing driving stress. It is only when the exponent, either of deformation (Equation 4) or sliding (Equation 5) is negative, that ice velocity increases with decreasing driving stress—the defining characteristic of streaming ice.

Evidence already presented argues strongly for a temperate bed and possibly a thick layer of temperate ice as the streaming onset is approached. In Figure 7, ice upstream of the onset is best fit by a flow-law exponent of unity (ice thickness in this region is nearly constant at 2300 ± 100 m; Bindschadler and others, in press]. Without more observational data of temperature and basal conditions, the issues of how the measured surface velocity is distributed between internal deformation and basal sliding and what conditions are responsible for this distribution cannot be pursued productively.

Subglacial Water

Subglacial water is known to play a critical role in streaming flow. Thus, the hydrologic evolution of the subglacial environment is likely key to ice-stream initiation. Although the driving stress is a maximum at the onset, the rate of water production depends on many factors.

Geothermal heat flux is spatially variable, yet even a subglacial volcano, believed by some to be active, does not initiate an ice stream [*Blankenship and others*, 1993]. Strain heating also produces water in the lower layers of ice, but at rates much less than produced by basal friction. Frictional heating depends on the stress at the interface. Deforming basal tills could lower the efficiency of water production.

Surface slopes converge laterally toward ice streams and tributaries leading toward onsets occupy basal channels. These two conditions will tend to concentrate whatever water is produced toward the onset region. By contrast, the bulge now forming at the head of the stagnant ice stream C is diverting water from this site [*Anandakrishnan and Alley*, 1997; *Price*, personal communication]. How, and if, this ice stream reinitiates remains to be seen.

Lateral Margins

Theoretical analysis of shear-margin dynamics supports the idea that ice-stream initiation is controlled by the development of basal slip rather than by the development of lateral shearing. *Raymond* [1996] analyzed the delocalization of shear stress within an ice-stream margin, and showed that flow resistance extended a variable distance into the ice stream depending primarily on the sliding speed (expressed as a ratio of the deformation speed, and the flow law exponent). Three regimes were identified: low relative sliding speed, where the effect of the margin was local to the margin and the speed in the center of the stream was controlled by basal conditions; very high sliding speeds where the margin resistance controlled the speed across the entire ice stream; and an intermediate case where the ice stream speed was controlled by a combination of margin and basal resistance. The sliding speeds on active ice streams were found to place them in the category where both margin and basal processes are important. Applying his results to onset areas, lower relative sliding speeds are certain, placing the dynamics of the onset region in the category where margin resistance remains local.

In order for the dynamics of an onset area to be controlled by lateral drag from the margins, the onset would have to be no wider than a few ice thicknesses. The general observation from Landsat imagery is that tributaries form with characteristic widths of 20 km. Narrower cases do occur, but the widths increase shortly downstream by incorporation of ice across a margin. Figure 3 illustrates an example of a narrow flowstripe area believed to represent the early stages of streaming, but it does not widen and, by the disappearance of flowstripes, flow speed is inferred to decrease. Narrow tributaries feeding ice stream E from the south also illustrate a repeated pattern of acceleration and deceleration prior to becoming a fully developed ice stream [*Joughin and others*, 1999]. Thus, it is improbable that ice-stream initiation is controlled by the formation of "soft spots" of strain-softened ice creating a weak margin as has been hypothesized [*Merry and Whillans*, 1993; *Whillans and Bolzan*, 1987].

MIGRATION OF THE ONSET

The subject of onsets should not be left without discussing their possible migration, because this process has major implications for the evolution of the ice sheet. It is known that the West Antarctic ice sheet was much larger during the Last Glacial Maximum (LGM)—*Hughes* [1998] has modeled an ice sheet fully three times its present volume during this earlier epoch, with ancient ice streams extending to the edge of the larger ice sheet. A stratified ridge-trough system in the Ross Sea and under the Ross Ice Shelf also suggest that a number of ice streams extended to the ice-sheet's margin [*Anderson and others*, 1992].

What cannot be determined is whether those LGM ice streams originated at the same locations as the present ice streams. If so, they would have been many times longer than present and their thicknesses across the Ross Ice Shelf would have had to have been larger to be grounded. Higher inland elevations of a few hundered meters at most during the LGM [*Whillans*, 1979; *Borns*, 1995; *Hughes*, 1998] require that the average driving stresses for these ancient ice streams would have been considerably smaller. Yet, a characteristic of ice streams is that they move rapidly at low driving stresses. It is not known whether the fact that all the present ice streams feeding the Ross Ice Shelf are roughly the same length also is a significant characteristic.

A simple argument has been proposed to support the idea that the heads of ice streams are out of equilibrium [*Bindschadler*, 1997]. The lateral convergence of inland ice into ice streams feeding the Ross Ice Shelf is roughly a factor of two. In addition, ice thickness decreases by about a factor of two as ice streams form. Thus, based on mass continuity, ice speed should accelerate by a combined factor of four. In contrast, ice typically accelerates by a larger factor—ten in the case of the ice stream D grid. This non-equilibrium situation can result in a number of responses: the region of acceleration can thin; the upstream ice can accelerate; the downstream ice can decelerate; or the downstream width can narrow. (Note that the upstream reservoir cannot widen). Neither of the last two possibilities is observed. Either of the remaining first two possibilities lead to the upstream migration of the onset.

If ancient ice streams migrated, they did so at roughly the same rate as the ice sheet retreated. *Bindschadler* [1998] has calculated that for ice stream B, the retreat rate averaged 100 m/a for approximately the last 12,000 years. This rate compares favorably with two other attempts to calculate the inland migration rate of ice stream B. From the thinning rate at the onset [*Shabtaie and others*, 1988] and the regional surface slope, *Bindschadler* [1997] calculated a migration rate of 488 m/a. In a more local study, *Price* [1998] modeled the evolution of a peculiar crevasse pattern near the onset of one branch of the ice stream and calculated a mean upstream migration rate 230 m/a.

It is conceivable, even likely, that conflicts might arise between the requirement to migrate, from mass continuity, and the basal conditions necessary to support an onset. If only a limited number of sites capable of supporting an onset existed, as the onset region evolved, the suitability of the current onset might diminish as the suitability of nearby sites increased. Eventually, the onset might jump to a new position or, if no suitable sites were near, streaming flow might cease. Episodic migration was first suggested by *Whillans and Bolzan* [1987] and later supported by the analysis of *Price* [1998]. This effect would depend on the spatial heterogeneity of the basal conditions capable of supporting an onset—a parameter that cannot be quantified with our present understanding of onsets and basal conditions. Such a phenomenon might generate the large "rafts" of relatively undisturbed ice observed on ice stream B [*Whillans and Bolzan*, 1987; *Bindschadler and others*, 1988].

SUMMARY

Necessary conditions for streaming flow are subglacial water, a thawed ice/bed interface, and probably sediment with a low shear strength. Observations of onset locations are limited, but are consistent with the presence of sediment, either inferred from seismic or gravity and magnetic measurements and/or by the correlation of the incipient ice stream with a bed channel. They also are consistent with the likely presence of subglacial water because the shape of the ice sheet surface and the subglacial bed combine to concentrate subglacial water toward the onset region. The predominance of these factors and their correlation with known or strongly suspected onsets is further supported by modeling of the effects of lateral shear. This leads to the conclusion that the onset of an ice stream represents a radical transition in basal motion that is not controlled by internal weaknesses in the ice such as strain softening in lateral shear margins.

Proxy indicators based on surface features alone are the least reliable, yet the most convenient, means to identify the onset location of an ice stream. The absence of either flowstripes or crevasses is not reliable confirmation of inland flow. Thus, these are better used to confirm the existence of streaming flow than to pinpoint its initiation. Also, the longevity of flowstripes can lead to false interpretations if flow patterns have changed. Maximum driving stress is arguably the most reliable means to locate the onset of an ice stream in lieu of direct subglacial measurements, but is further enhanced if velocity and thickness measurements are available to examine more carefully the spatial relationship between driving stress, mean shear strain rate and inferred basal motion.

Acknowledgements. Review articles, such as this paper, owe nearly all their content to the published results of others. The synthesis of these collected works has taken place over time through discussions with colleagues too numerous to mention. Keith Echelmeyer, Shawn Marshall, Christina Hulbe and Richard Alley each gave this manuscript a careful, constructive review. Their comments significantly improved the final text. Patricia Vornberger helped with the preparation of many of the figures. This work was funded by NSF grant OPP-9616394.

REFERENCES

Alley, R.B., Mulitple steady states in ice-water-till systems, *Ann. Glaciol., 14*, 1-5, 1990.

Alley, R. B., D.D. Blankenship, S.T. Rooney, and C.R. Bentley, Water-pressure coupling of sliding and bed deformation: III. Application to Ice Stream B, Antarctica, *J. Glaciol., 35*, 130-139, 1989.

Anandakrishnan, S. and R.B. Alley, 1997. Stagnation of ice stream C, West Antarctica by water piracy, *Geophys. Res. Ltr., 24*, 265-268.

Anandakrishnan, S., D.D. Blankenship, R.B. Alley, and P.L. Stoffa, Influence of subglacial geology on the position of a West Antarcitc ice stream from seismic observations, *Nature, 394*, 62-65, 1998.

Anderson, J.B., S.S. Shipp, L. Bartek, and D.E. Reid, Evidence for a grounded ice sheet on the Ross Sea continental shelf during the late Pleistocene and preliminary paleodrainage reconstruction, *Ant. Res. Ser., 57*, 39-62, 1992.

Bamber, J.L. and P. Huybrechts, Geometric boundary conditions for modelling the velocity field of the antarctic ice sheet, International Symposium on Ice Sheet Modelling, Chamonix, France, Sept. 18-22, 1995 (ed. K. Hutter), 364-373, 1996.

Bamber, J.L. and R. A. Bindschadler, An improved elevation data set for climate and ice-sheet modelling: validation with satellite imagery, *Ann. Glaciol., 25*, 439-444, 1997.

Bell, R. E., D.D. Blankenship, C.A. Finn, D.L. Morse, T.A. Scambos, J.M. Brozena and S.M. Hodge, Influence of subglacial geology on the onset of a West Antarctic ice stream from aerogeophysical observations, *Nature, 394*, 58-62, 1998.

Bentley, C.R., Antarctic ice streams: a review, *J. Geophys. Res., 92* (B9), 8843-8858, 1987.

Bindschadler, R.A., Actively surging West Antarctic ice streams and their response characteristics, *Ann. Glaciol., 24*, 409-414, 1997.

Bindschadler, R.A., Future of the West Antarctic Ice Sheet, *Science, 282*, 428-429, 1998.

Bindschadler, R.A., X. Chen and P.L. Vornberger, The Onset of Ice Stream D, West Antarctica, *J. Glaciol.,* in press.

Bindschadler, R.A., P.L. Vornberger, S.N. Stephenson, E.P. Roberts, S. Shabtaie, and D.R. MacAyeal, Ice shelf flow at the boundary of Crary Ice Rise, Antarctica, *Ann. Glaciol., 11*, 8-13, 1988.

Bindschadler, R.A. and P.L. Vornberger, AVHRR imagery reveals Antarctic ice dynamics, *Eos, 17*, (23), 741-742, 1990.

Bindschadler, R.A., P.L. Vornberger, D.D. Blankenship, T. A. Scambos, R. Jacobel, Surface Velocity and Mass Balance of Ice Streams D and E, West Antarctica, *J. Glaciol., 42*, 461-475, 1996.

Bindschadler, R.A. and P. L. Vornberger, Changes in West Antarctic ice sheet since 1963 from Declassified Satellite Photography, *Science, 279*, 689-692, 1998.

Blankenship, D.D., R.E. Bell, S.M. Hodge, J.M. Brozena, J.C. Behrendt and C.A. Finn, Active volcanism beneath the West Antarctic ice sheet and implications for ice-sheet stability, *Nature, 361*, 526-529, 1993.

Borns, H.W., Jr., Evidence of thicker ice in interior west Antarctica, *Antarct. J. U. S. XXX*, (5), 100-101, 1995.

Budd, W.F., The Dynamics of Ice Masses, ANARE Scientific Reports, Series A (IV) Glaciology, *108*, 1969.

Budd, W.F. and I.N. Smith, The growth and retreat of ice sheets in response to orbital radiation changes, Sea level, ice and climatic change (Proceedings of 17th IUGG general assembly, September, 1979, Canberra), *IAHS, 131*, 369-409, 1981.

Budd, W.F. and D. Jensen, Numerical modelling of glacier systems, Snow and Ice (Proceedings of Moscow Symposium, August, 1971), *IAHS, 104*, 257-291, 1975.

Casassa, G. and H.H. Brecher, Relief and Decay of flow stripes on Byrd Glacier, Antarctica, *Ann. Glaciol., 17*, 255-261, 1993.

Chen, X., R.A. Bindschadler, and P.L. Vornberger, Determination of Velocity Field and Strain-rate Field in West Antarctica Using High Precision GPS Measurements,

Cooper, A.P.R., N.F. McIntyre and G. de Q. Robin, Driving stresses in the Antarctic Ice Sheet, *Ann. Glaciol., 3*, 59-64, 1982.

Cuffey, K. and R.B. Alley, Is erosion by deforming subglacial sediments significant? (Toward till continuity), *Ann. Glaciol., 22*, 17-24, 1996.

Dowdeswell, J.A. and N.F. McIntyre, The surface topography of large ice masses from Landsat imagery, *J. Glaciol., 33*, 16-23, 1987.

Drewry, D., Antarctica: Glaciological and Geophysical Folio, Scott Polar Research Institute, University of Cambridge, 9 sheets, 1983.

Echelmeyer, K.A, W.D. Harrison, C. Larsen and J.E. Mitchell, The role of the margins in the dynamics of an active ice stream, *J. Glaciol., 40*, 527-538, 1994.

Engelhardt, H. and B. Kamb. Basal Sliding of Ice Stream B, West Antarctica, *J. Glaciol., 44*, 223-230, 1998.

Funk, M., K. Echelmeyer, A. Iken, Mechanisms of fast flow in Jakobshavns Isbræ, West Greenland: Part II. Modeling of englacial temperatures, *J. Glaciol., 40*, 569-585, 1994.

Glen, J., Creep of polycrystalline ice, *Proceedings of the Royal Society* (London), *228A*, 519-538, 1955.

Gray, A.L., K.E. Mattar, P.W. Vachon, R. Bindschadler, K.C. Jezek, R. Forster and J.P. Crawford, InSAR Results from the RADARSAT Antarctic Mapping Mission Data: Estimation of Glacier Motion using a Simple Registration Procedure, *IGARSS'98*, Seattle, Washington, July 1998.

Gudmundsson, G.H., C.F. Raymond and R.A. Bindschadler, The origin and longevity of flow-stripes on Antarctic ice streams, *Ann. Glaciol., 27*, 145-152, 1998.

Harrison, W.D., K.A. Echelmeyer and C. Larsen, Measurement of temperature in a margin of Ice Stream B: implications for margin migration and lateral drag, *J. Glaciol.*, in press.

Hodge, S.M. and S.K. Doppelhammer, Satellite imagery of the onset of streaming flow of ice streams C and D, West Antarctica, *J. Geophys. Res., 101*, 6669-6677, 1996.

Hughes, T.J., West Antarctic Ice Streams, *Rev. Geophys. Space Phys., 15*, 1-46, 1977.

Hughes, T.J., *Ice Sheets*, Oxford University Press, New York, NY, 360 p, 1998.

Iken, A., K. Echelmeyer, W. Harrison and M. Funk, Mechanisms of fast flow in Jakobshavns Isbræ, West Greenland: Part I, Measurements of temperature and water level in deep boreholes, *J. Glaciol., 39*, 15-25, 1993.

Jackson, M. and B. Kamb, The marginal shear stress of Ice Stream B, West Antarctica, *J. Glaciol., 43*, 415-426, 1998.

Joughin, I., L. Gray, R.A. Bindschadler, S. Price, C. Hulbe, D. Morse, K. Mattar and C. Werner, RADARSAT interferometry reveals tributaries of West Antarctic ice streams, *Science*, in press.

Kamb, B., Rheological nonlinearlity and flow instability in the deforming bed mechanism of ice stream motion, *J. Geophys. Res., 96*, 16,585-16,595, 1991.

Lettau, B., The transport of moisture into the Antarctic interior, *Tellus, 21*, 331-340, 1969.

McIntyre, N.F., The dynamics of ice-sheet outlets, *J. Glaciol., 31*, 99-107, 1985.

Merry, C.J. and I.M. Whillans, Ice-Flow Features on Ice Stream B, Antarctica, Revealed by SPOT HRV Imagery, *J. Glaciol., 39*, 515-527, 1993.

Paterson, W.S.B., *The Physics of Glaciers* (Third Edition), Pergamon Press, Oxford, UK, 480 p, 1994.

Price, S., Studies in the catchment and onset region of Ice Stream B, West Antarctica, M.S. Thesis, Ohio State University, 1998.

Raymond, C.F., Shear margins in glaciers and ice sheets, *J. Glaciol., 42*, 90-102, 1996.

Retzlaff, R., N. Lord and C.R. Bentley, Airborne radar studies:ice streams A, B and C, West Antarctica, *J. Glaciol., 39*, 495-506, 1993.

Robin, G. de Q., Ice movement and temperature distribution in glaciers and ice sheets, *J. Glaciol., 2*, 523-532, 1955.

Rose, K.E., Characteristics of Ice Flow in Marie Byrd Land, Antarctica, *J. Glaciol., 24*, 63-74, 1979.

Scambos, T.A. and R. A. Bindschadler, Feature maps of ice streams C, D, and E, West Antarctica. *Antarct. J. U. S.*, XXVI, 312-314, 1991.

Scambos, T.A. and R. A. Bindschadler, Complex ice stream flow revealed by sequential satellite imagery, *Ann. Glaciol., 17*, 177-182, 1993.

Shabtaie,S. and C.R. Bentley, Ice-thickness map of the West Antarctic ice streams by radar sounding, *Ann. Glaciol.*, 11, 126-136, 1988.

Shabtaie, S., C.R. Bentley, R. A. Bindschadler, and D.R. MacAyeal, Mass-Balance Studies of Ice Streams A, B, and C, West Antarctica, and Possible Surging Behavior of Ice Stream B, *Ann. Glaciol., 11*, 137-149, 1988.

Swithinbank, C.W.M., Ice streams, *Polar Record, 7* (48), 185-186, 1954.

Stephenson, S.N. and R.A. Bindschadler, Observed velocity fluctuations on a major Antarctic ice stream, *Nature, 334*, 695-697, 1988.

Stephenson, S.N. and R.A. Bindschadler, Is ice-stream evolution revealed by satellite imagery?, *Ann. Glaciol., 14*, 273-277, 1990.

Vialov, S.S., Visco-plastic flow of glacial covers and the laws of ice deformation, *Cold Region Research and Engineering Laboratory (CRREL) Report*, 30 p, 1970.

Vornberger, P.L. and I.M. Whillans, Surface features of Ice Stream B, Marie Byrd Land, West Antarctica, *Ann. Glaciol., 8*, 168-170, 1986.

Whillans, I.M., Ice flow along the Byrd Station Strain Network, Antarctica, *J. Glaciol., 24*, 15-28, 1979.

Whillans, I.M., Force budget of ice sheets, in *Dynamics of the West Antarctic Ice Sheet*, C.J. van der Veen and J. Oerlemans, eds., D. Reidel Pub., 17-36, 1987.

Whillans, I.M. and J.Bolzan, Velocity of Ice Streams B and C, Antarctica, *J. Geophys. Res., 92* (B9), 8895-8902, 1987.

Whillans, I.M. and C.J. van der Veen, Patterns of calculated basal drag on Ice Streams B and C, Antarctica, *J. Glaciol.*, *39*, 437-446, 1993.

Whillans, I.M. and C.J. van der Veen, The role of lateral drag in the dynamics of Ice Stream B, Antarctica, *J. Glaciol.*, *43*, 231-237, 1997.

S. Anandakrishnan, Department of Geology, University of Alabama, Box 870338, Tuscaloosa, AL 35487-0338.

J. Bamber, Bristol Glaciology Centre, Department of Geography, University of Bristol, Bristol, BS8 1SS, United Kingdom

R. A. Bindschadler, Oceans and Ice Branch (Code 971), Laboratory for Hydrospheric Processes, NASA Goddard Space Flight Center, Greenbelt, MD 20771.

ICE STREAM SHEAR MARGINS

C. F. Raymond,[1] K. A. Echelmeyer,[2] I. M. Whillans,[3] C. S. M. Doake[4]

Shear zones in the West Antarctic Ice Sheet (WAIS) mark the boundary between fast-moving ice streams and slow-moving ice lying in the interstream areas. The shear margins support high shear stress (exceeding 100 kPa) with consequent large strain rates (order 0.1 /yr) and pervasive crevasses. The shear stress in the margins can greatly exceed the basal shear stress (order 10 kPa). One consequence is that a significant fraction of the down-slope weight of an ice stream can be supported from the sides (typically 50%, in some cases approaching 100%). This redistribution of force causes the speed of ice stream flow to be sensitive to the width of the flow. It also causes the heating associated with dissipation of potential energy loss from the down-slope motion of the ice stream to be focused in the margins. In some circumstances the high stress and energy dissipation in the margins may tend to drive the margins of an ice stream outward both allowing increase of the discharge velocity and widening the ice stream discharge gate. In other circumstances, reduction of stress and energy dissipation on the ice base inside of the margins may cause the opposite tendencies, thus lowering ice stream discharge. Understanding the processes that limit the forces acting at ice stream margins and control their positions is crucial to assessing whether WAIS is susceptible to rapid disintegration by acceleration of the ice streams.

1. INTRODUCTION

An ice stream is characterized by a cross-flow transition from rapid motion in the ice stream to the slowly moving ice on either side. The related high rates of marginal shear strain rate cause zones of spectacular fracturing that mark the lateral margin of the ice stream (Fig. 1). These zones are the most obvious identifiers of ice stream activity on the ground, from the air or in satellite images, and they are amongst the most prominent morphological features on the surface of the West Antarctic Ice Sheet (WAIS) [*Bindschadler and Vornberger*, 1990]. The dynamic importance of margins for the flow of ice streams was anticipated long ago by *Hughes* [1975].

[1] University of Washington, Geophysics Program, Seattle.

[2] Geophysical Institute, University of Alaska, Fairbanks.

[3] Byrd Polar Research Center and Department of Geological Sciences, The Ohio State University, Columbus.

[4] British Antarctic Survey, Natural Environment Research Council, Cambridge, England

Ice streams are important because they are the main pathways for discharge of ice from the interior of WAIS to the ocean. The discharge of ice volume through an ice stream depends on mean speed and the cross sectional area. The forces generated by the shearing in the margin appear to be large enough to support a substantial fraction of the down-slope weight of an ice stream and thereby have a strong influence on the velocity. The margins define the width of ice stream flow and the discharge gate. Thus, there are two important questions concerning the role of ice-stream margins in the discharge of ice: (1) How do the margins affect speeds in the ice streams? (2) What determines the locations of the ice stream margins and the width of ice stream flow? These questions are strongly motivated by recent observational evidence for changes in both speed and margin position [*Clarke and Bentley*, 1995; *Bindschadler and Vornberger*, 1998; *Harrison et al.*, 1998; *Echelmeyer and Harrison*, 1999; *Clarke and Bentley*, 2000].

This paper gives an overview of morphological and dynamical characteristics observed in the margins of

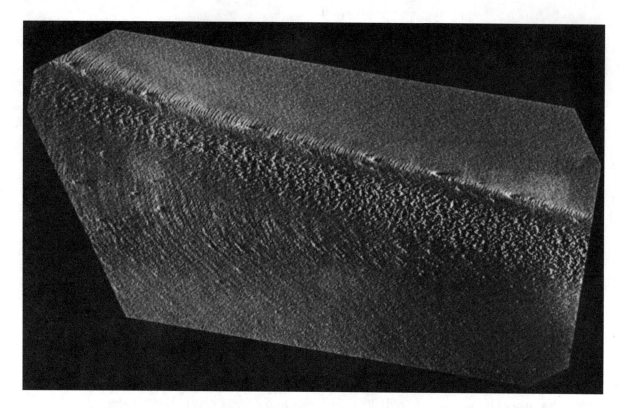

Fig. 1. Satellite image showing the crevasse band defining the south margin of Ice Stream B (branch B2). This margin is called the Dragon. Ice Stream B2 is in the lower part of the image. The inter-ice-stream area called the Unicorn is in the upper part of the image. Flow direction is from left to right.

WAIS ice streams and the methods used to analyze them. On the basis of this overview, we examine the status of answers to the above questions concerning the effect of ice stream margins on the discharge of the ice streams.

2. CHARACTERISTICS OF ICE STREAM MARGINS

2.1. Types of Shear Margins

A shear margin is an along-flow zone separating slow flow on one side and fast flow on the other. The cross-flow velocity gradient corresponds to strong shear across vertical surfaces parallel to the flow. This zone is also the boundary of the fast flow, and in this sense it is a margin.

Shear margins may be caused by jumps at the bed either in the elevation or in the slip-condition (Fig. 2). Shearing near the edges of valley glaciers or at the margins of deep sub-glacial troughs such as that found beneath Jakobshavns Isbrae in Greenland [*Clarke and Echelmeyer*, 1996] illustrates control by bed topography. Examples in Antarctica are the rock-bounded outlet glaciers passing from the East Antarctic Ice Sheet through the Trans Antarctic Mountains to WAIS and the Ross Ice Shelf. The well-developed ice streams of the Siple Coast approach the other extreme. The margins of these ice streams are located above a sharp boundary between slowly- and rapidly-slipping portions of the bed associated with a jump in the lubrication controlling the slip [*Bentley*, 1987].

Most real cases involve some combination of topography and lubrication. Rutford Ice Stream, bounded on the south by the escarpment of the Ellsworth Mountains and on the north by the slowly moving ice of Carlson Inlet and Fletcher Promontory, is an example where one margin is distinctly controlled by topography and the other is not [*Doake et al.*, this volume]. The onset zones of the Siple Coast Ice Streams may be associated with deep sediment filled basins where both bed topography and rock type appear to be crucial [*Anandakrishnan et al.*, 1998; *Bell et al.*, 1998; *Bindschadler et al.*, this volume]. Further down flow, these ice streams roughly follow sub-glacial troughs that are a few hundred meters deep

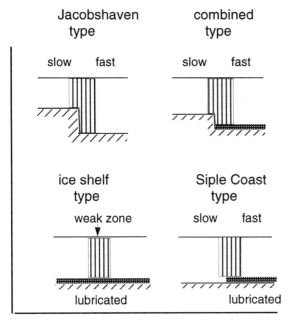

Fig. 2. Schematic of shear margins caused by an across flow jump in bed elevation (Jacobshaven type), by an across flow jump in lubrication (Siple Coast type) or by a mechanically weak zone in a floating ice plate responding to a remotely applied shear stress(ice shelf type).

and that probably determine their general locations [Shabtaie and Bentley, 1987].

Shear may also be localized in mechanically soft zones in the ice without a jump in basal boundary condition. For this to happen there must be shear stress driving deformation and sufficient lubrication of the base to allow differential motion. Such shear zones may be found locally in floating ice shelves [Bindschadler et al., 1996], within ice streams [Hulbe and Whillans, 1997] or bridging between rock promontories on the irregular margins of glaciers. In such cases, the properties of the ice may play the dominant role, rather than local changes in basal boundary conditions.

This paper focuses primarily on ice stream margins where a jump in slip condition is the main factor in generating the margin. Bed topography evolves only slowly by erosion and deposition, and it is essentially fixed over the time intervals of primary interest with regard to the future of the WAIS over the next century to millenium. Vertical movement of the bed due to changing ice load and tectonic activity is likely to be regional and not focused in ice stream margins and so affect the local topography. In contrast, slip conditions at the base are potentially rapidly varying as thermal and hydrological conditions change. It is this kind of margin that presents the greatest difficulty in predicting the near-term evolution of WAIS.

2.2. Surface and Bed Topography

The surfaces of the Siple Coast ice streams have low slopes (order 10^{-3}) along their lengths [Alley and Whillans, 1991]. The lower reaches of Siple Coast ice streams are separated by distinct inter-stream ridges with crests up to several hundred meters above the adjacent ice streams and relatively large slopes (order 10^{-2}) toward the ice stream margins. These slope changes are generally evident in satellite images [Fahnestock and Bamber, this volume] and have been mapped locally using photogrametry [Whillans et al., 1993], radar altimetry from aircraft [Shabtaie and Bentley, 1987; Retzlaff et al., 1993] and satellite [Bamber and Bindschadler, 1997] interpolated with photoclinometry [Bindschadler and Vornberger, 1994; Scambos et al., 1998].

While the margins of ice streams are often clear on the surface, no obvious consistent relations amongst the position of shear margins, the edges of the main sub-glacial troughs or local bed topography have been identified [Shabtaie and Bentley, 1987; Retzlaff et al., 1993]. Nevertheless, the published radar traverses across margins [Shabtaie et al., 1987; Bindschadler et al., 1996; Echelmeyer and Harrison, 1999] show that the outer edge of margins lie preferentially over parts of the bed that slope inward toward the ice stream center. (Of total 26 margin crossings for Ice Streams A, B, C, D and E, the underlying bed is locally inward sloping for 15 crossings, flat for 10 crossings and outward sloping for only 1 crossing.) Even though there are non-zero slopes of the bed, in all cases the bed slopes are small. The corresponding differences in thickness of 20% or less and associated stress-driven deformational velocity do not play a major, direct role in generating the more than 1 order of magnitude velocity difference characteristic of these margins.

2.3. Surface Morphology

The bands of intense crevasses generated in shear margins are the most obvious surface features on WAIS. They can be detected using aerial photographs [Whillans et al., 1993], airborne radio echo sounding [Rose, 1979; Shabtaie and Bentley, 1987] and satellite images including visible [Merry and Whillans, 1993], near infra red AVHRR [Bindschadler and Vornberger,

1990; *Scambos and Bindschadler*, 1991] and SAR [*Frolich and Doake*, 1998]. Mapping of shear margins using these methods has shown that the margins can begin rather abruptly [*Frolich and Doake*, 1998] deep in the interior of the ice sheet [*Hodge and Doppelhammer*, 1996], run to the grounding line and leave tracks in the ice shelf beyond [*Fahnestock and Bamber*, this volume]. The map trace is usually smooth.

Where the crevasse bands first appear in the ice-sheet interior, individual crevasses are distinct and in complex arrangements sometimes associated with a series of folds and buckles in the intervening ice surface [*Merry and Whillans*, 1993]. (See also *Whillans et al.*, [this volume].) These sites mark one definition for the onset of ice-stream flow, where the lateral shear stress between the fast-moving ice stream and the slower inter-stream ridge ice is large enough to develop crevasses.

Downstream along much of their lengths, the bands of crevasses display a characteristic zonation with sub-bands of distinctive structure (Fig. 1). The outer edge of the band is composed of relatively uniformly-spaced, arcuate crevasses spanning a width of several hundred meters. Inward toward the ice-stream center, there is a chaotic zone several kilometers wide where the ice surface takes on a granular appearance formed from multiply intersecting crevasses and associated snowdrift mounds. Back scatter from these features causes the margins to appear bright in SAR images, especially L-band. The back scatter along the margin of Rutford Ice Stream has been found to be roughly proportional to the greatest principal strain rate [*Vaughan et al.*, 1994]. Development of the chaotic fracturing appears to be restricted to locations of very high shear-strain rate [*Echelmeyer et al.*, 1994]. Yet further toward the center, there is an interior zone, where lateral strain rate is smaller and crevasses are again distinct with an orientation related to lateral shear or longitudinal stretching.

In addition to crevasses and local disturbance of the surface topography at the scale of crevasse spacing, there is commonly a margin-parallel furrow close to the inward edge of the arcuate crevasses or in the outer part of the chaotic zone [e.g. *Echelmeyer et al.*, 1994]. Other ridges and furrows running diagonally across a margin [*Merry and Whillans*, 1993] may represent paths of ice entering the ice stream from the adjacent inter-stream ridge.

2.4. Velocity and Strain Rate

Measurement of velocity and strain rate in shear margins is confronted with special difficulties. The high strain rate and associated crevasses make access on the ground and placement of survey markers difficult. These traditional methods still give the highest point accuracy [*Echelmeyer et al.*, 1994; *Echelmeyer and Harrison*, 1999; *Harrison et al.*, 1998], but spatial sampling is very difficult to obtain. The numerous crevasses provide abundant natural features that can be tracked in sequential satellite images [*Bindschadler and Scambos*, 1991; *Bindschadler et al.*, 1996], aerial photographs [*Whillans et al.*, 1993] or using INSAR [*Frolich and Doake*, 1998] to get broad area coverage. However, distortion of the patterns over time by the straining limits the time over which these features can be accurately positioned or even identified in strong shear margins.

The main component of velocity (u) is directed down the ice steam along its length (x). The strong gradient in u in the margin-perpendicular direction (y) is the defining characteristic of a shear margin. The speeds of the active ice streams increase down stream and the shearing is correspondingly more intense (e.g. *Scambos et al.* [1994]; *Bindschadler et al.* [1996]). Commonly, the full cross-flow profile $u(y)$ is like plug flow, for example most notably in the measurements from Ice Stream B (Fig. 3). The symmetric, plug-flow character suggests rather uniform conditions on the base with the primary control on $u(y)$ coming from the margins. Many profiles, however, show irregular variations of velocity inward from the outer margin indicative of local obstructions to the flow within the ice stream that affect $u(y)$ as is common in Ice Streams D and E (Fig. 3). (Also see images in *Fahnestock and Bamber*, [this volume].)

Characteristics of the velocity pattern and associated shear strain rate are illustrated in more detail by the specific example in Figure 4 (bottom panel), which is based on measurements from the south margin of Ice Stream B2 (the Dragon) [*Echelmeyer et al.*, 1994]. About 90 to 95% of the 430 m/yr velocity jump from the nearly motionless ridge ice to the fast ice stream occurs over a 3 to 5 km wide zone. The shear strain rate $\partial u/\partial y$ reaches a maximum $\gamma_m = 0.15$ /yr. Even higher shear strain rates have been found in other margins (Table 1, column 6). The maximum $\partial u/\partial y$ is reached in the outer part of the margin with a very abrupt decay to 0 in the ridge over a relatively short distance of about 2 km and a more gradual decay to low values inward into the ice stream over several kilometers.

The top panel of Figure 4 shows how the sub-zones of the crevasse pattern (Sec. 2.3) and the strain rate are related. The arcuate crevasses occur on the outer

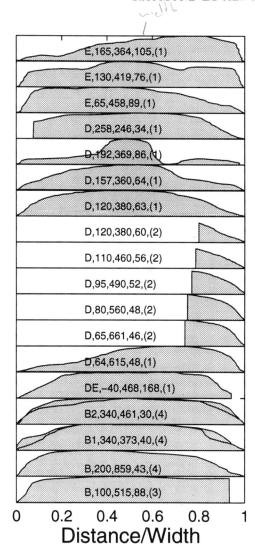

Fig. 3. Compilation of velocity profiles measured across ice streams. The labels give [ice stream name, nominal distance from the recent grounding zone in km, centerline speed in m/yr used to normalize u, full width in km used to normalize y, (published source for data)]. The doubled curves for locations B1 and B2 represent the limits in a composite profile formed from several cross profiles. Sources are (1) [*Bindschadler et al.*,1987], (2) [*Scambos et al.*,1994], (3) [*Bindschadler et al.*,1996], (4) [*Whillans and van der Veen*, 1997].

edge of the margin, where $\partial u/\partial y$ is in the range 0.02 /yr to 0.10 /yr. Chaotic crevassing occurs where $\partial u/\partial y$ exceeds about 0.12 /yr.

It is generally assumed that ice flows from the inter-stream areas into the ice streams. Thus, there is a margin-perpendicular component of velocity (v) along direction y. Near the Siple Coast, slopes of the ridges toward the ice streams and mass balance constraints suggest v in the range 1 to 10 m/yr at the

margins. This is verified at several locations by direct survey [*Whillans and van der Veen*, 1993]. Little is known about how v varies within a shear margin because of the difficulties for measurement and for accurate definition of the cross-flow direction.

Information about the time variation of velocity in shear margins is very limited. Some initial indications are provided by observations in the north margin of the Rutford Ice Stream using sequential INSAR [*Frolich and Doake*, 1998]. Changes in speed were detected within this margin that were not correlated with any changes in the central part of the ice stream. (See discussion by *Doake et al.*, [this volume].)

3. DYNAMIC ANALYSIS

An ice stream margin is a transition zone between two distinctly different dynamic regimes. Within an ice stream, the shearing over its thickness makes a negligible contribution to the motion. Essentially all of the motion is produced at the base. Thus, velocity and associated strain rate components vary hor-

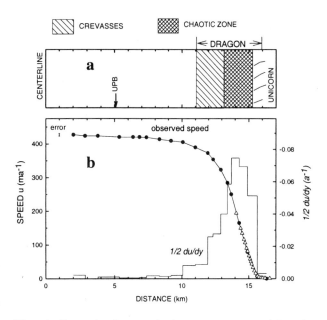

Fig. 4. Zones in the marginal crevasse pattern (a) and the profile of longitudinal component of velocity $u(y)$ (b) across the south margin of the B2 branch of Ice Stream B (the Dragon) from *Echelmeyer et al.* [1994] and *Echelmeyer and Harrison* [1999]. The profile extends from the inter-stream ice (the Unicorn) to the center of the ice stream flow. Filled circles and open triangles in (b) are measurements from two different one-year intervals spaced 5 years apart.

TABLE 1. Force Balance Terms

Transect (Source)	H (km)	W (km)	U (m/yr)	$\bar{\tau}_d$ (kPa)	γ_m (/yr)	E	τ_m (kPa)	$\bar{\tau}_b$ (kPa)	side fraction
DwnB (1, 6)	0.8	36	510	3	0.07	10	80	1.4	0.5
B 40block (1)	0.925	24	820 to 880	9	0.26	1.2	185 to 250^f	4 to 2^f	0.5 to 0.8^f
B1 comp (1)	1.50	21	400	21	0.10	1.2	150 to 200^f	+5 to 0^f	0.8 to 1.0^f
B2 comp (1)	1.00	16	400 to 450	12	0.14	1.2	165 to 225^f	+2 to -2^f	1
Dragon (2a)	1.10	15	430	13	0.15	12	180	6	0.5
Dragon (2b)	1.10	15	430	13		12	110	5	0.6
Dragon (3)	1.10	15	430	13		1 to 2	220		> 0.6
Dragon (4)	1.10	15	430	13		na	200		> 0.6
D1 (5)	0.8^e	14^b	660	9	0.16	10	< 150	4.5	0.5
D5 (5)	0.8^e	30	420	9	0.03	1	< 145		0.3
E (7 to 9)	0.9 to 1.4	35 to 50	250 to 550	49	> 0.01	1^a	c	14	< 0.7^d
Rutford (10)	1.8 to 2.3	10 to 15	360 to 400	40	0.08	1^a	< 160	> 20	< 50

Sources: 1. [*Whillans and van der Veen*, 1997], 2a. [*Echelmeyer et al.*, 1994, analytical], 2b. [*Echelmeyer et al.*, 1994, numerical],3. [*Jackson and Kamb*, 1997], 4. [*Harrison et al.*, 1998] 5. [*Scambos et al.*, 1994], 6. [*Bindschadler et al.*, 1987], 7. [*MacAyeal*, 1992], 8. [*MacAyeal et al.*, 1995, standard inversion], 9. [*Bindschadler et al.*, 1996], 10. [*Frolich and Doake*, 1988]
^{a}Assumed value.
^{b}Physical half width is 23 km, but the maximum in velocity is at 14 km.
^{c}Source does not give this quantity.
^{d}Maximum value assumes no mean longitudinal forces.
^{e}Extrapolated from measurements at a different location in IS D.
^{f}Range results from plausible range of flow law parameter A (Eq. 5).

izontally but not vertically. These conditions allow rigorous analysis of the interior ice stream flow using vertically integrated conservation equations leading to a practical 2-D map plane model [*MacAyeal*, 1989]. In an inter-stream ridge the flow is dominated by vertical gradients and can be analyzed based on "laminar" flow assumptions [*Paterson*, 1994]. Neither of these two approximations holds in a shear margin, where both lateral and vertical gradients can be large. However, longitudinal gradients are typically much smaller and secondary to the main processes associated with a shear zone. Thus, a two-dimensional analysis in a transverse section perpendicular to the ice stream flow direction can be used to capture the primary dynamic characteristics of the transition.

The following gives a two-dimensional formulation that includes the most important physical processes identified by present understanding and forms the foundations for analyzing measurements in margins to date.

3.1. Mathematical Model of a Cross Section

The purpose is to calculate the distributions of down-stream velocity (u), shear-stress components (τ_{xy}, τ_{xz}) and temperature (T) in a transverse cross section (y, z plane) of an ice stream and closely adjacent ridge ice. The motion is driven by the component of gravity g aligned along the ice stream surface slope α. Field equations are conservation of momentum in the longitudinal direction x

$$\frac{\partial \tau_{xy}}{\partial y} + \frac{\partial \tau_{xz}}{\partial z} + \rho g \alpha = 0 \quad (1)$$

and conservation of energy

$$\frac{\partial}{\partial y} k \frac{\partial T}{\partial y} + \frac{\partial}{\partial z} k \frac{\partial T}{\partial z} - \rho c \left(u \frac{\partial T}{\partial x} + v \frac{\partial T}{\partial y} + w \frac{\partial T}{\partial z} \right) + \tau_{xy} \frac{\partial u}{\partial y} + \tau_{xz} \frac{\partial u}{\partial z} = \rho c \frac{\partial T}{\partial t} \quad (2)$$

These are completed by a flow law, which we describe by

$$\tau_{xy} = \eta \frac{\partial u}{\partial y} \quad (3)$$

$$\tau_{xz} = \eta \frac{\partial u}{\partial z} \quad (4)$$

These 4 equations can be solved for u, τ_{xy}, τ_{xz}, T as a function of y and z subject to specification of constituitive variables (effective viscosity η, thermal conductivity k, heat capacity c and density ρ), boundary conditions around the perimeter of the section and surface slope (α). It is assumed that $(v(y,z), w(y,z), \partial T/\partial x)$ are specified independently from the three dimensional context using measurements and/or simple analysis.

Based on a power creep law for ice $\epsilon = A\tau^n$, effective viscosity for simple shear is

$$\eta(\tau_{xy}, \tau_{xz}, T) = \frac{(\tau_{xy}^2 + \tau_{xz}^2)^{0.5(1-n)}}{2EA(T)} \quad (5)$$

Here $A(T)$ and n are the temperature dependent softness and power [Paterson, 1994]; E is an enhancement factor commonly employed to represent effects from c-axis fabric, texture and chemistry for simple shear deformation [Dahl-Jensen and Gundestrup, 1987]. We discuss the limitations of this representation of the flow law below. The remaining constitutive variables (k, c and ρ) are known [Paterson, 1994].

Mechanical and thermal boundary conditions must be applied on all surfaces. The top surface is shear stress free and fixed at the mean-annual temperature T_s just beneath the surface, which can be influenced by the presence of crevasses (Sec. 3.4). The lateral boundaries can be located well into the adjacent ridges, where a normal flow and thermal regime can be imposed. Alternatively, reflection symmetry ($\partial/\partial y = 0$) can be imposed at the center of the ice stream or ridge. Heat generation in the margin may be substantial, and the possibility of a temperate zone can not be ruled out, which would introduce another boundary with associated conditions between cold and temperate ice [Hutter et al., 1988].

The mechanical boundary condition at the bed is especially crucial, since it is the cause of the streaming flow. Most formulations involve the basal shear stress τ_b and/or the basal velocity u_b. They can be cast in the form

$$u_b = \left(\frac{\tau_b}{\tau_0}\right)^m \quad (6)$$

where τ_0 is inversely related to the amount of lubrication of the bed depending on the structure of the bed (type of rock, roughness, till thickness and granulometry, ...) and hydrological state (effective normal pressure and/or the water content on the ice bed interface and/or in basal till). When power $m = 1$, the slip law is linear as would arise from a linearly viscous till shearing over a finite till thickness [Boulton and Hindmarsh, 1987]. When $m \to \infty$ the slip becomes plastic-like as would arise from a plastic till with yield stress τ_0 [Kamb, 1991].

The thermal boundary condition at the ice base can be described by

$$\dot{M} = B + \tau_b u_b \quad (7)$$

$\dot{M}$ is the rate of melting (freezing if negative). The product $\tau_b u_b$ is the heating rate per unit area from basal motion by slip at the ice base and deformation of till beneath. $B = [k\partial T/\partial z]_b$ represents the jump in upward heat flux density arriving at the base of the deforming zone from the rock and departing from the ice base through the ice. We will refer to B as the conductive heat balance. The upward heat flux from the rock will be controlled by the geothermal heat flux G. To account for changes in heat storage and horizontal conduction in the rock near the ice stream margin where bed temperature can vary spatially, G needs to be imposed at the base of a rock layer of thickness similar to the width of the marginal zone or more. If the local basal temperature T_b is below the melting point, it is presumed $\dot{M} = \tau_b u_b = 0$, thus implying continuity of the conductive heat flux and zero conductive heat balance ($B = 0$).

This formulation (Eqs. 1 to 7) neglects the following factors: conservation of mass, longitudinal stress gradients (in Eq. 1), diffusion of heat in the x direction (in Eq. 2), conservation of $y-$ and $z-$ momentum, and ice anisotropy (Eqs. 3 and 4). Of these various omissions, the neglect of ice anisotropy by fabric development is probably the most serious (Sec. 3.3). Because of flow into ice streams at their heads and across their margins, the ice in different parts of the cross section will have strain histories of varying complexity. This is especially so in the margin, where ice flowing into it has been sheared across planes parallel to the upper surface and is then subjected to shearing on vertical surfaces on entry into the margin. Thus, a margin is a zone of fabric evolution and Equations 3 and 4 may not be an adequate description of ice deformation. A formulation accounting for spatially varying fabric and its evolution other than can be described by E in Equation 5 is not attempted here.

3.2. Theoretical Patterns of Deformation

The cause of a shear margin considered here (Sec. 2.1) is a variation of τ_0 in the cross-flow direction (y). The primary dynamic consequence is that the second term in Equation 1, the lateral stress gradient $\partial \tau_{xy}/\partial y$, is non negligible or even dominant.

It is useful to consider the situation in the absence of lateral stress ($\tau_{xy} = 0$) to establish a reference. Then, integration of Equation 1 with respect to z downward from the free surface z_s to the bed z_b gives $\tau_{xz} = \tau_d(z_s - z)/H$, where $H = (z_s - z_b)$ is the ice thickness and

$$\tau_d = \rho g H \alpha \qquad (8)$$

is the gravitational driving stress. Integration of Equation 4 using Equation 5 (with $\tau_{xy} = 0$ and specified distributions of T and E) gives velocity from shearing through the ice thickness that would be caused by τ_d acting at the base, here denoted by U_d. Similarly, τ_d would cause a basal motion $U_b = (\tau_d/\tau_0)^m$ as given by Equation 6. The quantities τ_d, U_d and U_b will be used to scale stress and velocity in the following. The margin produces large departures from these reference values, in some circumstances extending across the full ice stream width.

Figure 5 shows a theoretical steady-state pattern of velocity, shear strain rate, shear stress, heat generation rate and temperature that arises from the coupled solution of Equations 1 to 6 based on the work of *Jacobson and Raymond* [1998]. This calculation assumes a standard T-dependent flow law [*Paterson*, 1994] with no enhancement ($E = 1$) and a large jump in τ_0 to generate the margin. The driving stress ($\tau_d = 11.4$ kPa) and assumptions about the flow and slip laws correspond to $U_d = 0.032$ m/yr and $U_b = 816$ m/yr. The assumed surface temperature (-26 °C), accumulation-controlled downward vertical velocity at the surface ($w = 0.1$ m/yr) and cross flow into the ice stream ($v = 5$ m/yr at the surface) are consistent with known environmental conditions on Siple Coast. The assumed geothermal heat ($G = 73$ mw/m^2 applied 1 km below the ice base) leads to a steady-state bed temperature of about -3 °C in the absence of heat generated by the motion. This temperature is similar to basal temperatures found in inter-stream areas (Engelhardt and Kamb, oral presentation at the Chapman Conference on the West Antarctic Ice Sheet, 1998).

The predicted τ_b in the center (8 kPa) is less than τ_d (11.4 kPa). Similarly, the predicted velocity at the surface in the center (535 m/yr) is less than $U_b + U_d$ (816 m/yr). Both differences arise because of interaction with the margins that transfers force from the ice stream base to the adjacent ridge. In the ice stream, heating generated by motion over the bed raises it to the melting temperature and warms the overlying ice. Shear induced heating in the margin locally elevates the isotherms there. The cross flow into the ice stream carries the ice warmed in the margin into the ice stream and tends to suppress development of localized high temperatures in the margin and associated thermal softening of the marginal ice. In particular, note the vertical structure of the strain rate, stress and heat generation that reflects the transition from dominantly horizontal to vertical gradients associated with the margin.

The temperature field depends sensitively on the specific assumptions about geothermal heat G, basal lubrication τ_0, accumulation on the surface affecting w and cross-margin speed v. Because of the general limitations of the two dimensional formulation (Eqs. 1 and 2) and other simplifications described above, these calculations can not be considered as quantitative predictions, but rather as illustrating the basic characteristics expected for the distributions of u and T in a cross section.

The basic features of the momentum balance, as illustrated in the specific example of Figure 5, were examined more generally by *Raymond* [1996] with the simplification of isothermal ice (thus allowing neglect of Eq. 2). This analysis showed that the interaction between an ice stream and the adjacent slow ice may be restricted to a boundary layer with a limited cross-flow width depending on the assumptions about the slip law (Eq. 6). Figure 6 illustrates the primary features. In this particular case, the lubrication level ($1/\tau_0$) in the model ice stream (top panel) is set so that $U_b/U_d = 17$. The main lateral shearing and associated cross flow gradients in speed and base stress (center two panels) are focused near the jump in lubrication (top panel) extending a few times H into the slow ice and about 10 H into the ice stream. Near the center, $\tau_b = \tau_d$, $u_b = U_b$ and $u_s = U_d + U_b$ with negligible influence from the margin.

The boundary layer has two parts. Under the edge of the ice stream, there is a zone of stress relaxation ($\tau_b < \tau_d$). Under the edge of the slow

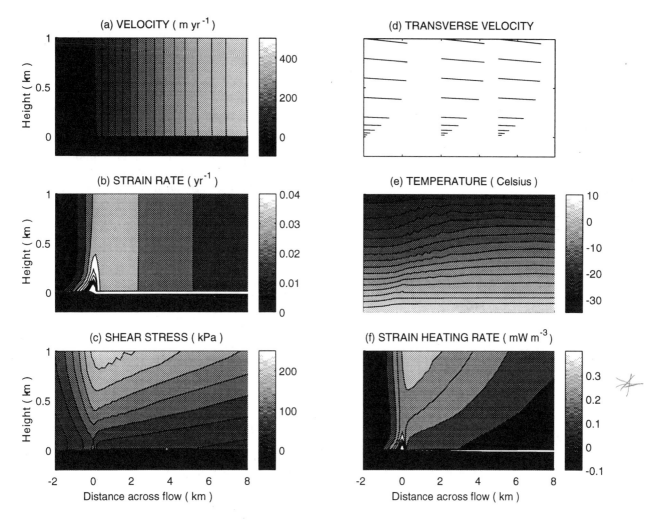

Fig. 5. Theoretical distributions of down-stream velocity u (panel a), strain rate $\partial u/\partial y$ (panel b), shear stress (panel c), temperature (panel e) and heat generation rate (panel f) in a schematic ice stream-cross section after *Jacobson and Raymond* [1998, case 4c]. Calculation assumes ice thickness $H = 1\ km$, half-width $W = 30\ km$, surface temperature -26 °C, geothermal heat flux $G = 73\ mw\ m^{-2}$, the recommended temperature dependent flow law of *Paterson* [1994] and $m = 3$ in (Eq. 6). τ_0 was chosen to give a basal velocity u_b of about 500 $m\ yr^{-1}$ in the center. The assumed transverse velocity (components v and w) is shown in Panel d based on $v = 5\ m\ yr^{-1}$ and $w = 0.1\ m\ yr^{-1}$ at the surface and $v = w = 0$ at the bed. Heat flow calculation includes 1 km thick rock layer beneath the ice. The panels show only the near margin part of the solution, which was carried out using finite elements from -20 km to 30 km.

ice, there is a narrower zone of stress concentration ($\tau_b > \tau_d$). Overall force balance (described in Sec. 3.3 below) requires that the total basal force reduction F_R in the relaxation zone is transferred from the bed through the margin F_M and reacted as an increase in basal force in the concentration zone such that $F_R = F_M = F_C$ (Fig. 6, bottom panel). The maxima in τ_{xy} and associated $\partial u/\partial y$ on the upper surface are located above the jump in lubrication on the bed and identify the boundary between the stress relaxation and concentration zones.

Raymond [1996] showed that the theoretical width of the relaxation zone increases with lubrication (approximately in proportion to $H(1 + U_b/U_d)^{1/(n+1)}$ when $m = 1$ in the slip law (Eq. 6). In the example of Figure 6, the assumed level of lubrication is modest ($U_b/U_d \sim 17$) and the width of the relaxation zone is limited. The fast moving parts of Siple Coast

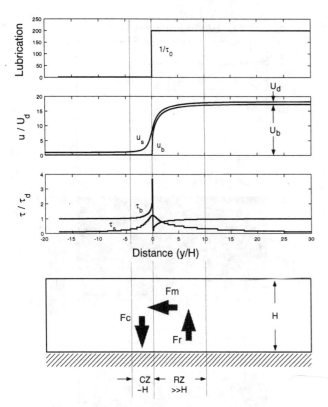

Fig. 6. Predicted variation of velocity on the surface (u_s) and bed (u_b) and associated basal shear stress ($\tau_b = \tau_{xz}(y,H)$) and lateral shear stress at the surface ($\tau_s = \tau_{xy}(y,0)$) in response to a jump in lubrication ($1/\tau_0$). Distance y across flow is normalized by thickness H. Stress is normalized by driving stress $\tau_d = \rho g H \alpha$. Velocity is normalized to deformational velocity $U_d = 2A\tau_d^{n+1}H/(n+1)$. Calculation assumes isothermal ice, $m = 1$ in Eq. 6 and a jump in τ_0 that gives $U_b/U_d = 17$. Bottom panel shows a schematic representation of force redistribution from a zone of stress relaxation under the edge of streaming flow (RZ) across the margin to the zone of stress concentration (CZ) under the edge of the inter-stream ridge.

ice streams are highly lubricated ($U_b/U_d \geq 10^3$) [*Bentley*, 1987; *Echelmeyer et al.*, 1994; *Whillans and van der Veen*, 1997], and in these circumstances (as in the example of Figure 5) the theoretical width of the relaxation zone becomes comparable to the full ice-stream width and the margin affects the ice stream center.

These qualitative features of shear zone width and associated force redistribution are general, but the quantitative predictions would be modified by different assumptions. The boundary layer is broader with increasingly non-linear slip (higher m in Eq. 6) becoming effectively infinite for a plastic bed ($m = \infty$,

$\tau_0 < \tau_d$). A broadening would also be expected, if the transition from lubricated to sticky bed were spread out rather than the assumed discontinuity. On the other hand, local softening in the margin associated with heating or fabric development would more strongly concentrate the shearing.

3.3. Force Balance and Side Drag

A fundamental question is what supports the down-slope gravitational weight of an ice stream? The weight of an ice stream must be balanced by forces transmitted from the bed, from its sides and along its length. On average, except possibly in the onset and grounding zones, longitudinal forces play a minor role [*Whillans and van der Veen*, 1993]. The bed and sides supply the main resistive forces. These are commonly termed bed- and side- drag respectively. Global force balance requires that as the base of an ice stream becomes more lubricated, basal drag drops and support of the down-slope weight of the ice stream must be shifted from its bed through the margins to the less lubricated bed under the adjacent ridge (Fig. 6). Thus, there are two end member regimes: bed drag and side drag. The partitioning between the bed and side can be investigated using measurements of the variation of longitudinal velocity u with distance y from the margin (Fig. 3), which gives the lateral shear strain rate $\partial u/\partial y$ and the associated lateral shear stress τ_{xy}.

A practical way to describe the relative importance of side drag is to examine Equation 1 integrated vertically to give the force balance of a column through the ice thickness or integrated over the area of the cross section of the ice stream to give a global force balance [*Echelmeyer et al.*, 1994; *Jackson and Kamb*, 1997; *Whillans and van der Veen*, 1997; *Harrison et al.*, 1998]. We consider only the global force balance by integration of Equation 1 with respect to z over $H(y)$ and then with respect to y over the width between the ice-stream centerline ($y = 0$, where $\tau_{xy} = 0$) and the margin ($y = W$) to find

$$\tau_m H_m + \bar{\tau}_b W = \bar{\tau}_d W \quad (9)$$

where

$$\tau_m \equiv \; <\tau_{xy}(y=W)>_H \quad (10)$$

$$H_m \equiv \; H(y=W) \quad (11)$$

$$\bar{\tau}_b \equiv \; <\tau_b>_W \quad (12)$$

$$\bar{\tau}_d \equiv \; \rho g <\alpha H>_W \quad (13)$$

and where $<>_{(H,W)}$ signify average over H and W respectively. This gives the force balance for the half-width W of the ice stream. $\bar{\tau}_d W$ is the total gravitational force driving down stream motion. $\tau_m H_m$ is the side drag (F_M in Fig. 6). $\bar{\tau}_b W$ is the bed drag. The ratios $\tau_m H_m/\bar{\tau}_d W$ and $\bar{\tau}_b W/\bar{\tau}_d W$ give the fraction of support from side and basal drag respectively. These terms can be estimated using inputs from the ice stream cross section (H and W), surface slope (α), marginal strain rate $\partial u/\partial y \equiv \gamma_m$ and a flow law (Eqs. 3, 4 and 5), which allow calculation of τ_d (Eq. 13), τ_m (Eq. 10) and then τ_b (Eq. 9). However, there is in most cases substantial uncertainty in α (affecting estimate of τ_d from Eq. 13) and E (affecting estimate of τ_m using Eqs. 3, 5 and 10). The considerable difficulty in estimating E is discussed in more detail below. Furthermore, γ_m (and the related estimate for τ_m) is accessible for measurement on the upper surface, but the relevant average over depth must be inferred.

Table 1 summarizes estimates of the terms defined above. Of particular interest is the fraction of support from the side ($\tau_m H_m/\bar{\tau}_d W$, column 10). The sources of these deductions about force balance are reviewed below. The discussion focuses on the several sets of measurements made at nearly the same location along the south margin of Ice Stream B2 (the Dragon), which illustrate the range of methods and uncertainties for estimating side drag.

Echelmeyer et al. [1994] analyzed the detailed shape of the velocity profile that they measured in the Dragon (Fig. 4) using Equations 1, 3, 4 and 5 or equivalents. They attempted to adjust the distribution of τ_b (equivalent to using Eq. 6 with $m = \infty$ and adjusting τ_0). No distribution of τ_b could predict the narrow width of the observed profile, when laterally homogeneous parameters were assumed for the ice viscosity (Eq. 5). They were able to fit the observations by introducing a local softening ($E > 1$ in Eq. 5) for the ice in the zone of largest $|\partial u/\partial y|$ that concentrated the shear strain into a narrower zone. They concluded that a combined temperature/fabric enhancement of about 10 is required, which results in a marginal stress of 110 to 180 KPa. The crucial information forcing this conclusion is the narrowness of the shear zone revealed by the detailed shape of the lateral profile of longitudinal velocity within the margin. This reasoning is supported by the general analysis of *Raymond* [1996].

A large number of profiles were examined by *Whillans and van der Veen* [1997] from several locations on Ice Stream B (Fig. 3). Collectively, these profiles also show a concentration of shear in the margin that Whillans and van der Veen interpret to mean that there is softening of the marginal ice. Their analysis indicates an enhancement E of only about 1.2 would be adequate to explain the observations. They estimate a marginal shear stress of 150 to 200 Kpa, where the range arises from uncertainty in the flow law.

Jackson and Kamb [1997] sampled the ice in the Dragon from a depth of about 300 m within the area of measurements by *Echelmeyer et al.* [1994] in order to directly measure the strength of the ice and role of the fabric. They did not find a strongly developed single maximum fabric that would be expected in such a shear zone [*Jacka and Budd*, 1989]. In situ strain rate and temperature were imposed in the samples using a deformation apparatus, and the required stress was measured to be slightly larger than found by *Whillans and van der Veen* [1997]. Both the deformation experiments and examination of the fabric indicated a rather small fabric-induced enhancement factor of 1 to 2. Alteration of the structure of samples during collection and subsequent storage was considered unlikely, but possible.

Harrison et al. [1998] made another estimate of the marginal shear stress τ_{xy} that is essentially independent of the flow law of ice. They determined the volumetric heating rate of the ice (approximately $\tau_{xy} \partial u/\partial y$) estimated from temperature measurements at depth in the margin, as discussed below (Sec. 3.4). They concluded that τ_{xy} was about 200 KPa, as averaged over the top third of the ice column. This value is close to that estimated by Whillans and van der Veen [1997] and about 10% lower than that estimated by *Jackson and Kamb* [1997] at the same location.

The range of estimates for the marginal shear stress τ_m from 110 to 225 Kpa at the same location (Table 1, column 8, Dragon and B2) illustrates the problems for estimating the side drag. Uncertainty in the flow law including the effects from temperature and fabric is clearly a factor that the observations have not unraveled. Although localized softening in the margins is evident, the relative contribution from elevated temperature or c-axis fabric is unclear. If fabric does play a strong role, it is likely that its effects will vary from place to place, possibly with some trend along the length because of differing strain history of the ice [*Scambos et al.*, 1994]. Since measurements so far have been based on the surface or in the upper third of the thickness, the vertical averaging of stress over thickness in the margin is a consideration that is not necessarily straight forward (Fig. 5). The consideration of side drag is somewhat muddied in those instances where there are major disturbances to the flow in the central portions of ice streams, for example E [MacAyeal et al., 1995] and the lower part of B [Bindschadler et al., 1987] that disrupt the cross flow gradient. These disturbances arise from obstruction of the flow within the ice stream by sticky spots or basal topography [MacAyeal et al., 1995]. In this regard, it is important to keep in mind that, on average, longitudinal stress gradients are not important. However, this does not guarantee that they can be neglected for interpretation of measurements at a given location. In spite of these various uncertainties, the observations clearly show that side drag can be a very significant or dominant resistance to ice stream motion.

3.4. Energy Balance

The energy balance of an ice stream is crucial for maintaining the water-generated lubrication responsible for the fast motion. Furthermore, as discussed in Section 3.3, heating of the margin and associated reduction in effective viscosity of the marginal ice is coupled to the force balance.

The redistribution of force associated with the transition from bed-drag to side-drag regime causes a corresponding shift in the location of heat generation. In a progression of increasingly more lubricated bed conditions, the heat generation per unit area on the bed ($u_b \tau_b$) in the central part of an ice stream reaches a maximum at an intermediate level of lubrication corresponding roughly to the transition between bed-drag and side-drag [Raymond, 1995]. This behavior arises because with low lubrication u_b is small and with high lubrication τ_b is small, either of which causes the heat generation on the bed ($\tau_b u_b$) to be small. Although the heat generation at the bed in the center falls off as the level of lubrication rises to high values, the total heat generated by the moving ice stream continues to increase. Heat not dissipated on the bed appears in the ice volume, primarily in the marginal shear zone and the stress concentration zone just outside (Fig. 5). In the side drag regime, a large fraction of the total energy dissipated by the ice stream motion is focused in the margins.

The analysis of temperature measurements in the Dragon by Harrison et al. [1998] illustrates the main principles of energy conservation in the margin of an ice stream. One of their goals was to measure the heating rate $\tau_{xy}\partial u/\partial y + \tau_{xz}\partial u/\partial z \approx \tau_{xy}\partial u/\partial y$ in the margin. The heating rate was estimated from the lateral variation of temperature $\partial T/\partial y$ that arises from advection of ice at a known speed v through the zone of heating. Harrison et al. [1998] assumed negligible transverse conduction, linear variation of vertical advection velocity w from the estimated accumulation rate at the surface to zero at the bed and a lateral velocity v that was independent of depth. These assumptions lead to a simplified form of Equation 2. Based on the trajectory of ice in the shear margin, they estimated that the ice near the center of the chaotic zone had been within the zone of heating for 50 to 70 years, subject to a heating rate of about 1 to 2 K per century. As mentioned above (Sec. 3.3 and Table 1), the measured value of $\partial u/\partial y$ then implies a stress τ_{xy} of about 200 KPa, providing further evidence for the importance of side drag in the force balance of Ice Stream B.

The measurements and analysis of Harrison et al. [1998] illuminate another important feature of the thermal regime of ice stream margins caused by the many crevasses (Sec. 2.3). Crevasses allow cold winter air to pool beneath the surface without a compensating circulation of buoyant warm air during summer. This convective heat pump has an immediate strong cooling effect on the near surface ice as it enters a margin and becomes fractured. The resulting heat deficit is conducted and advected downward with time into the coherent ice below. Temperatures at a depth of 30 m were found to be 12 K below the mean surface temperature. By mapping the lateral and depth variation of temperature beneath the crevasses and analysis of the vertical penetration of the cooling wave, they were able to estimate the rate of cross-flow motion relative to the crevasses. This rate is related to the cross flow speed v and the migration rate of the margin assuming that the crevasses have a fixed location in the margin (Sec. 4.2).

These notions illustrate the great importance of strain heating, cross-margin flow and the margin surface morphology on the temperature of the ice. However, the elevation of the temperature within the margin observed by Harrison et al. [1998] was not large (≤ 2 K). This small increase in temperature is unlikely to cause substantial thermal softening of the ice within the margin (Sec. 3.3) and, in fact, the strong cooling of the upper 150 m of the ice column is likely to cause a stiff "lid" to the marginal ice.

4. MIGRATION OF SHEAR MARGINS

Shifting ice stream margins would alter the velocity and width of streaming flow and thus the discharge. The control on the positions of ice stream margins is therefore a question of central importance with regard to upper limits on the discharges of ice streams and stability of the WAIS.

4.1. Identification of Former Margin Positions

When an ice stream narrows suddenly, its margin is abandoned. For sometime thereafter, the relic margin may be discerned as a topographic "scar" that is visible in satellite images [*Bindschadler and Vornberger*, 1990; *Fahnestock and Bamber*, this volume]. Subsurface indicators of shearing in a former margin are disruption of internal layers [*Jacobel et al.*, 1996] and buried crevasses [*Rose*, 1979; *Shabtaie and Bentley*, 1987; *Retzlaff and Bentley*, 1993]. These subsurface features are visible in radar traverses across relic margins (Fig. 7). Rates of burial by accumulation and associated ages of internal layers allow estimates of timing of shut down of the earlier streaming flow. The depth of burial of crevasses in the inactive margins of Ice Stream C as detected by high frequency radar was the key to establishing that this ice stream stopped nearly simultaneously over its lower reaches about 100 to 200 years ago [*Retzlaff and Bentley*, 1993]. *Clarke and Bentley* [1995, 2000] discovered buried crevasses outside the present south margin of Ice Stream B2 (the Dragon) extending southward beyond a scar (the Fishhook), showing that B2 narrowed up to 13 km starting about 200 years ago. In these cases of abandoned relic margins, the lack of nested sets of scars indicates that the inward rate of migration was continuous and/or rapid. Multiple scars on the north margin of Ice Stream C indicate that there were apparently some temporarily stable positions as if the margin moved inward in jumps before final termination of activity of Ice Stream C over its central parts [*Jacobel et al.*, 2000].

The mapping of scars from satellite images [*Scambos and Bindschadler*, 1991] has the potential for identifying the limits of recent ice stream activity. However, interpretation of scars must be done cautiously with support from subsurface measurements, because all scar-like features are not former shear margins. For example, the prominent scars on Roosevelt Island are generated by flow off the island over steps in the bed (Conway and Gades, oral presentation at the Chapman Conference on the West Antarctic Ice Sheet, 1998). It should also be kept in mind that a scar can be displaced both horizontally and vertically subsequent to its abandonment by the active ice stream depending on the flow from the adjacent inter-stream ridge [*Nereson*, 2000].

4.2. Rates of Migration

The margins, as we have described them thus far, are rather broad zones several kilometers wide, either in terms of their kinematic (Sec. 2.4) or morphological (Sec. 2.3) features, which complicates observation of migration. If the displacement of a margin is large enough, it can be obvious. By comparing satellite images taken 29 years apart, *Bindschadler and Vornberger* [1998] discovered a 4 km outward displacement of the north margin of Ice Stream B (the Snake) near the grounding zone, which revealed a mean outward migration rate of 137 m/yr. *Clarke and Bentley* [2000] estimated an inward migration rate of about 100 m/yr for the south margin of B2 as it was displaced northward from the region of the Fishhook. This rate was deduced from the variation in burial depth of crevasses.

Measurement of migration over shorter time scales requires a precise definition of the location of the margin by identification of some sharp feature tied to the margin that can be used as a reference for the margin location. In practice it may be easier to measure the motion of margin features relative to the ice than directly relative to earth-fixed coordinates. From a kinematic perspective, the margin migration rate $\partial W/\partial t$ (assuming only one margin is moving) may be described by

$$\frac{\partial W}{\partial t} = v_e - v \qquad (14)$$

where v_e is the rate at which ridge ice is entrained by the ice stream [*Alley and Whillans*, 1991; *van der Veen and Whillans*, 1996]. When cross-margin speed v and v_e are equal, they balance each other, and there is no displacement of the margin in space. The instantaneous rate of margin migration can be determined by measuring both v and v_e.

So far, these measurements have been done only for the south margin of Ice Stream B2 (the Dragon). There $v \approx 1$ m/yr as determined from GPS surveying of markers [*Whillans and van der Veen*, 1993; *Echelmeyer and Harrison*, 1999]. The motion of the ice relative to various features of the margin has been used to determine v_e.

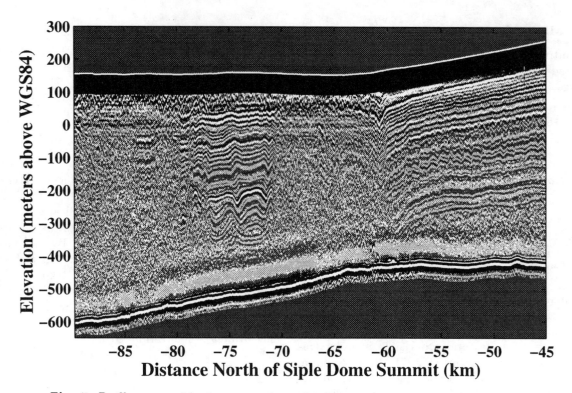

Fig. 7. Profile measured by ice penetrating radar (frequency about 4 MHz) across the north margin of Ice Stream C in its lower portion where it is presently inactive [*Gades et al.*, 2000; *Jacobel et al.*, 2000]. Siple Dome lies to the north (right side of figure). Line of sight is down stream to the west. Disturbance of internal layering at -60 km marks the outer limit of the streaming motion. The most recent margin apparently was at -83 km. The pattern of disruption of layers in the intervening zone, together with the map pattern of scars on the surface suggests a stepped inward migration of the margin [*Jacobel et al.*, 2000] before the ultimate termination of activity between 100 and 200 years ago [*Retzlaff and Bentley*, 1993]. Note the diffractors near the surface and locally on the bed and large changes in the power return from the bed indicative of substantial changes in bed reflection coefficient probably associated with variations in the amount of water at the bed [*Gades et al.*, 2000].

Because shear margins are defined in terms of kinematics, the most direct identifier of margin position is the steep velocity gradient $\partial u/\partial y$. This was exploited by *Echelmeyer and Harrison* [1999], who analyzed measurements of the cross-margin profile of velocity in the Dragon determined for two different time intervals. Markers in the high shear-strain-rate zone sped up from the first to the second epoch. With the assumption that the velocity profile has a fixed shape and position relative to a moving margin, the speed up of these markers implied $v_e \approx 11$ m/yr ($\partial W/\partial t = 10$ m/yr).

Harrison et al. [1998] estimated $v_e \approx 8$ m/yr ($\partial W/\partial t = 7$ m/yr) using the variation of sub-surface ice temperature with distance through the margin that arises from localized cooling from the crevasses (3.4). Their analysis was based on the theoretical time scale for downward propagation of the cooling wave from the base of the crevasse zone and the observed y-dependent depth of penetration. Their estimate represents an average rate over the last 50 years.

Hamilton et al. [1998] estimated v_e to be in the range 5 to 28 m/yr based on the curvature of the arcuate crevasses (Sec. 2.3) in the outer part of the Dragon. The curvature is a result of differential rotation in the shear field $\partial u/\partial y$ arising from gradients in $\partial u/\partial y$ and time of residence [*Vornberger and Whillans*, 1990].

All of these methods indicate that the Dragon is presently migrating into the Unicorn at a rate of order 10^1 m/yr. It is interesting to note that the Unicorn is presently thinning [*Hamilton et al.*, 1998] as the Dragon slowly advances into it. The causal relationship is unknown.

Evidently the south margin of Ice Stream B2 has a complex history over the last few centuries with the margin migrating inward at speed of order 100 m/yr starting about 200 years ago with subsequent outward migration at a slower decameter per year rate. We also know that more rapid outward migration of order 100 m/yr is possible from the outward shift of the north margin of Ice Stream B. The configuration of scars indicates that similar shifting of margins probably occurs on the other ice streams.

4.3. Physical Controls

The concentration of stress and focused heating just outside an active margin (Figs. 5 and 6) would tend to promote rapid basal motion and allow the margin to migrate outward by direct mechanical action, thawing or a combination. On the other hand, reduction in basal velocity, stress and associated heat production on the bed locally inside the margin (Sec. 3.4) could promote sticking of the base and incorporation of the ice stream ice into the ridge. With these general concepts in mind, shifting of margins is not unexpected.

For this discussion, we regard the location of the margin to be the location of the jump in lubrication ($1/\tau_0$), which defines the mechanical boundary on the ice base and sets the location of shear separating the fast and slow motion. If the transition is gradational, then the location of maximum gradient in $1/\tau_0$ can serve to define the margin position. To discuss the physical controls in more detail, we identify other types of boundaries in the basal environment that could be important in relation to the location of this lubrication boundary (Fig. 8). These could be geologically imposed bed morphology (e.g. the presence or absence of basal till or smooth versus rough rock surface), temperature (melting point or sub freezing), heat balance (actively melting or freezing) and hydrological (high or low pore pressure). We will assume that the mechanical (weak bed), lubrication and hydrological (high pore pressure) boundaries coincide and refer to all as lubrication with the proviso that they do not lie outside of a morphological boundary, where it is presumed that high pore pressure can not produce effective lubrication. These boundaries may be overlaid on topography of the bed, which may influence all of them.

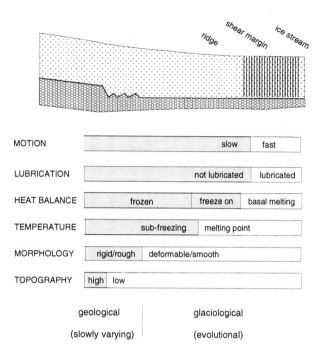

Fig. 8. Schematic of boundaries separating locations on the bed between low and high elevation (topography), between geologically controlled bed morphology that is or is not susceptible to weakening (morphology), between sub-freezing and melting temperature (temperature boundary), between freeze-on and melting (heat-balance boundary), between low and high water pressure (lubrication) and between slow and fast (motion).

The actual arrangement of these boundaries is not known from direct observations, but some characteristics can be inferred.

It is clear that enabling morphological (e.g. till or smooth bed) and thermal (bed at melting point with water present) conditions are necessary to allow streaming motion. However, these conditions are not sufficient, which is shown by the thawed condition beneath some locations on quiescent ice streams even 100 years or more after fast motion stopped [*Engelhardt and Kamb*, oral presentation at the Chapman Conference on the West Antarctic Ice Sheet, 1998; *Bentley et al.*, 1998; *Gades et al.*, 2000]. Thus, the lubrication boundary must be at or inside the morphological and temperature boundaries (Fig. 8).

The role of heat balance is less clear. Where melting is occurring, one may argue that the bed would eventually become lubricated, since water can not be drained away without reaching or exceeding the pore pressure in the actively lubricated part of the ice stream. On the other hand where freezing is occurring, eventual locking of the bed would be expected either by elimination of free water or lowering of pore pressure required to draw replacement water from the surroundings. Very high sensitivity of lubrication to freezing rate is plausible, because of the extreme sensitivity of till strength to pore pressure or the related water content [*Kamb*, 1991; *Tulaczyk*, 1997; *Jacobson and Raymond*, 1998; *Tulaczyk et al.*, 2000a, b]. If these arguments are correct, then the heat balance and lubrication boundaries coincide.

These relationships can be explored theoretically using a coupled mass/heat flow model of a cross section (Sec. 3.1). If water flow along the bed is neglected, the heat flow and temperature boundaries eventually coincide in a single thermal boundary as temperature adjusts to a steady state. For certain combinations of geothermal heat G, cross margin flow v, accumulation rate w, lubrication $(1/\tau_0)$ and corresponding ice stream speed, this thermal boundary lines up with the imposed lubrication and associated mechanical boundary (e.g. Fig. 5), thus appearing to explain the location of the margin as the boundary between frozen and thawed bed. However, more generally the thermal and lubrication boundaries do not align in the predicted steady state. With higher speed (lubrication), higher G, or lower v into the ice stream, the melting zone on the base extends outside beyond the imposed lubrication boundary. With lower lubrication, lower G and/or larger v, an opposite effect occurs, in which the bed is predicted to be below freezing inside the imposed lubrication boundary. These theoretical steady states are not consistent with the physical connection between lubrication and thermal conditions that is expected, which would indicate outward or inward displacement of the lubrication boundary and corresponding changes in width.

On this basis, *Jacobson and Raymond* [1998] concluded that a lubrication boundary controlled only by thermal effects would be unstable. Typical speeds of active ice streams (e.g. $u = 0.5\ km/yr$), cross-margin flow (e.g. $v = 2\ m/yr$) and geothermal heat flux $(G > 60\ mw\ m^{-2})$ predict melting outside the margin. Thus, the active ice streams would be widening unless impeded by a bed morphology resistant to fast motion. On the other hand, as long as $G < 80\ mw\ m^{-2}$, there could be freezing just inside the margin of an ice stream moving too slowly, which would cause inward migration and eventually stop the streaming flow. These notions provide a potential explanation for the shifts in margins that have been identified. Furthermore, they suggest a fundamental role for bed morphology in establishing a stable width and outer limits to expansion.

With the possible direct effect of the stress concentration on the bed outside the margin in mind, *van der Veen and Whillans* [1996] modeled margin migration (Eq. 14) assuming that $v_e \propto \tau_m$. Unstable ice-stream width is also predicted with this assumption in a simple model of the ice stream force balance. The basic cause is clear. As the model ice stream widens, τ_m and the associated v_e increase, thus increasing the rate of widening. The van der Veen and Whillans model examines the mass balance of the ice stream to determine whether a widening ice stream and associated increase in discharge decreases τ_d and thus τ_m sufficiently to counter runaway widening. According to their analysis, it does not.

While thermo-mechanical considerations provide some basis for understanding the direction of margin migration, the prediction of rates remains unresolved. Some combination of basal heat and water balance along the lines considered above is likely to play a role. Whether certain areas can not be accelerated to ice stream speeds because of constraints from morphology and topography of the bed also remains unanswered. The existence of undisturbed stratigraphy in the present Siple Coast inter-stream ridges (e.g. Siple Dome) suggests that the adjacent ice streams (e.g. C and D) have not been able to migrate into them for many millennia [*Nereson et al.*, 1996, 1998]. The base of Siple Dome is presently high and till, if present, is probably very thin [*Gades et al.*, 2000], thus possibly excluding streaming flow for topographic and/or geological reasons. If those conditions now preclude ice stream motion, it remains unclear why these conditions hold in view of the long-term evolution of the ice stream system and its imprint on the sub-glacial terrain.

5. SYNTHESIS

To characterize the effects of margins on discharge of ice streams, it is instructive to examine the motion of an ice stream under the assumption that the bed is plastic (Eq. 6 with $m = \infty$ and $\tau_0 =$ constant). The evidence (Sec. 3.3) suggests $\tau_0 < \tau_d$. Furthermore, the velocity in the ice stream is nearly independent of depth z, and $\partial \tau_{xz}/\partial z$ in Equation 1 can be approximated by τ_0/H. Integration of Equation 1 with respect to y from the ice stream center (taken to be $y = 0$) toward the margin $(y = W)$ gives

$\tau_{xy} = -(\tau_d - \tau_0)y/H$. Then integration of Equation 3 using Equation 5 yields

$$u(y) = \frac{2\tilde{E}A(\tilde{T})}{n+1} \frac{(\tau_d - \tau_0)^n}{H^n}[W^{n+1} - y^{n+1}] \quad (15)$$

(For example, see *Raymond* [1996].) Accounting for thickness H and integration over half-width W gives discharge

$$Q = \frac{2\bar{E}A(\bar{T})}{n+2} \frac{(\tau_d - \tau_0)^n}{H^n} HW^{n+2} \quad (16)$$

Differentiation of Eq. 16 gives the sensitivity of Q to changes in τ_0, W, E and T, which expressed as fractional change is

$$\frac{\Delta Q}{Q} = -n\frac{\Delta \tau_0}{\tau_d - \tau_0} + (n+2)\frac{\Delta W}{W} + \frac{\Delta \bar{E}}{\bar{E}} + \frac{\partial A}{\partial T}\frac{\Delta \bar{T}}{\bar{T}} \quad (17)$$

Here $\tilde{E}$, $\bar{E}$, $\tilde{T}$ and $\bar{T}$ represent appropriate spatial averages of E and T that will be weighted more heavily toward the margins. We omit explicit changes in α and H, which will be driven at larger scale than the margins and the local cross section [e.g. *Hulbe and Payne*, this volume].

The first term on the right illustrates the effect on Q of changing lubrication (as represented by $1/\tau_0$) which alters the velocity (Eq. 15). As τ_0 becomes very small, both the potential amount for further enhancement of lubrication (decrease in τ_0) and the sensitivity are decreased, which illustrates how the margins can limit the speed. If fabric is developed with strains as small as 0.1 (e.g. *Jacka and Budd* [1989]), then the present margins are probably softened by this mechanism ($E > 1$) as much as can be already, which limits the potential ΔE. The cross-margin advection of heat limits ΔT and the associated thermal softening (Sec. 3.4).

Consequently, the potential for further velocity and discharge increase lies in widening (second term on the right hand side of Eq. 17) for those ice streams of Siple Coast that are already highly lubricated and largely supported by side drag. If in addition the ice stream margins are presently at or close to the outer limits allowed by geological constraints, then short term potential for very large increases in discharge of the active ice streams of the Siple Coast is limited. Major increases would have to be driven by evolution of the large-scale geometry affecting thickness and slope. These circumstances would suggest considerable stability to the ice stream system.

On the other hand, there are substantial uncertainties about the size of side drag in most cases. If side and bed drag are roughly equal ($\tau_0 = 0.5\tau_d$), as is indicated by many of the measurements, then Equation 16 shows that there is still potential for nearly order of magnitude increase in discharge by additional lubrication of the bed ($\tau_0 \to 0$). The potential for speed up by enhanced lubrication is even larger for those sections of ice streams with less side drag. Present knowledge does not allow the conclusion that the sides of ice streams place stringent constraints on the amount of speed up relative to present conditions.

While we do know already that the side drag can be quite large (Sec. 3.3), more data and analyses of the velocity fields in ice streams are needed to narrow the uncertainty about the stabilizing role of margins in the dynamics. The question concerning the geological limits of streaming flow (Sec. 4.3) is clearly important and less explored. Remote sensing methods to identify various boundaries in physical properties on the bed beneath the margins will be very important. Models of the evolution of WAIS and its ice streams need to include the redistribution of forces and mechanical energy dissipation from the ice streams across their margins to the adjacent slow ice (Sec. 3) as first order processes in the dynamics.

Acknowledgments. Preparation of this paper was supported by National Science Foundation Grant number OPP9726704.

REFERENCES

Alley, R. B., and I. M. Whillans, Changes in the West Antarctic Ice Sheet, *Science*, *254*(5034), 959–963, 1991.

Anandakrishnan, S., D. D. Blankenship, R. B. Alley-R.B., and P. L. Stoffa, Influence of subglacial geology on the position of a West Antarctic ice stream from seismic observations., *Nature*, *394*(6688), 62–65, 1998.

Bamber, J. L., and R. A. Bindschadler, An improved elevation data set for climate and ice sheet modelling: validation with satellite imagery, *Annals of Glaciology*, *25*, 439–444, 1997.

Bell, R. E., D. D. Blankenship, C. A. Finn, D. L. Morse, T. A. Scambos, J. M. Brozena, and S. M. Hodge, Influence of subglacial geology on the onset of a West Antarctic ice stream from aerogeophysical observations, *Nature*, *394*, 58–62, 1998.

Bentley, C. R., Antarctic Ice Streams: A Review, *Journal of Geophysical Research*, *92*(B9), 8843–8868, 1987.

Bentley, C. R., N. Lord, and C. Liu, Radar reflections reveal a wet bed beneath stagnant Ice Stream C and a frozen bed beneath ridge BC, West Antarctic, *Journal of Glaciology*, *44*(146), 149–156, 1998.

Bindschadler, R., and P. Vornberger, Changes in the West Antarctic Ice Sheet since 1963 from declassified satellite photography, *Science*, *279*(5351), 689–692, 1998.

Bindschadler, R., P. Vornberger, D. Blankenship, T. Scambos, and R. J. Jacobel, Surface velocity and mass balance of Ice Streams D and E, West Antarctica, *Journal of Glaciology*, *42*(142), 461–475, 1996.

Bindschadler, R., J. Bamber, and S. Anandakrishnan, Onset of Streaming Flow in the Siple Coast region, West Antarctica, this volume.

Bindschadler, R. A., and T. A. Scambos, Satellite-image-derived velocity field of an Antarctic Ice Stream, *Science*, *252*(5003), 242–246, 1991.

Bindschadler, R. A., and P. L. Vornberger, AVHRR imagery reveals Antarctic ice dynamics, *EOS*, *71*(23), 741–742, 1990.

Bindschadler, R. A., and P. L. Vornberger, Detailed elevation map of Ice Stream C, Antarctica, using satellite imagery and airborne radar, *Annals of Glaciology*, *20*, 327–335, 1994.

Bindschadler, R. A., S. N. Stephenson, D. R. MacAyeal, and S. Shabtaie, Ice dynamics at the mouth of Ice Stream B, Antarctica, *Journal of Geophysical Research*, *92*(B9), 8885–8894, 1987.

Boulton, G. S., and R. C. A. Hindmarsh, Sediment deformation beneath glaciers: Rheology and geological consequences, *Journal of Geophysical Research*, *92*(B9), 9059–9082, 1987.

Clarke, T. S., and C. R. Bentley, Evidence for a recently abandoned ice stream shear margin, *EOS*, *76*(46), F194, 1995.

Clarke, T. S., and C. R. Bentley, Evidence for a recently abandoned ice stream shear margin adjacent to Ice Stream B2, Antarctica, from ice-penetrating radar measurements, *Journal of Geophysical Research*, in press, 2000.

Clarke, T. S., and K. Echelmeyer, Seismic-reflection evidence for a deep subglacial trough beneath Jakobshaven Isbrae, West Greenland, *Journal of Glaciology*, *43*(141), 219–232, 1996.

Dahl-Jensen, D., and N. Gundestrup, Constituive properties of ice at Dye 3, Greenland, in *The physical Basis of Ice Sheet Modelling*, vol. 170, pp. 31–43, International Association of Hydrological Sciences, 1987.

Doake, C. M. S., H. F. J. Corr, A. Jenkins, K. Makinson, K. W. Nicholls, C. Nath, A. M. Smith, and D. G. Vaughan, Rutford Ice Stream, Antarctica, this volume.

Echelmeyer, K. A., and W. D. Harrison, Ongoing margin migration of Ice Stream B, Antarctica, *Journal of Glaciology*, *45*(150), 361–369, 1999.

Echelmeyer, K. A., W. D. Harrison, C. Larsen, and J. E. Mitchell, The role of the margins in the dynamics of an active ice stream, *Journal of Glaciology*, *40*(136), 527–538, 1994.

Fahnestock, M. A., and J. Bamber, Morphology and surface characteristics of the West Antarctic Ice Sheet, this volume.

Frolich, R. M., and C. M. S. Doake, Relative importance of lateral and vertical shear on Rutford Ice Stream, Antarctica, *Annals of Glaciology*, *11*, 19–22, 1988.

Frolich, R. M., and C. M. S. Doake, Synthetic aperature radar interferometry over Rutford Ice Stream and Carlson Inlet, Antarctica, *Journal of Glaciology*, *44*(146), 77–92, 1998.

Gades, A. M., C. F. Raymond, H. B. Conway, and R. Jacobel, Bed properties of Siple Dome and adjacent ice streams, West Antarctica inferred from radio echo-sounding measurements, *Journal of Glaciology*, in press, 2000.

Hamilton, G. S., I. Whillans, and P. Morgan, First point measurements of ice-sheet thickness change in Antarctica, *Annals of Glaciology*, *27*, 125–129, 1998.

Harrison, W. D., K. A. Echelmeyer, and C. F. Larsen, Measurement of temperature within a margin of Ice Stream B, Antarctica: implications for margin migration and lateral drag, *Journal of Glaciology*, *44*(148), 615–624, 1998.

Hodge, S. M., and S. K. Doppelhammer, Satellite imagery of the onset of streaming flow of ice streams C and D, West Antarctica, *Journal of Geophysical Research*, *101*(C3), 6669–6677, 1996.

Hughes, T., The West Antarctic Ice Sheet: Instability, disintegration and initiation of ice ages, *Reviews of Geophysics and Space Physics*, *13*(4), 502–526, 1975.

Hulbe, C. L., and A. J. Payne, The contribution of numerical modelling to our understanding of the West Antarctic Ice Sheet, this volume.

Hulbe, C. L., and I. M. Whillans, Weak bands within Ice Stream B, West Antarctica, *Journal of Glaciology*, *43*(145), 377–386, 1997.

Hutter, K., H. Blatter, and M. Funk, A model computation of moisture content in polythermal glaciers, *Journal of Geophysical Research*, *93*(B10), 12,205–12,214, 1988.

Jacka, T. H., and W. F. Budd, Isotropic and anisotropic relations for ice dynamics, *Annals of Glaciology*, *12*, 81–84, 1989.

Jackson, M., and B. Kamb, The marginal shear stress of Ice Stream B, West Antarctica, *Journal of Glaciology*, *43*(145), 415–426, 1997.

Jacobel, R. W., T. A. Scambos, C. F. Raymond, and A. M. Gades, Changes in the configuration of ice stream flow from the West Antarctic Ice Sheet, *Journal of Geophysical Research*, *101*(B3), 5499–5504, 1996.

Jacobel, R. W., T. A. Scambos, N. A. Nereson, and C. F. Raymond, Changes in the margin of Ice Stream C, Antarctica, *Journal of Glaciology*, in press, 2000.

Jacobson, H. P., and C. F. Raymond, Thermal effects on the location of ice stream margins, *Journal of Geophysical Research*, *103*(B6), 12,111–12,122, 1998.

Kamb, B., Rheological nonlinearity and flow instability in the deforming bed mechanism of ice stream motion, *Journal of Geophysical Research*, *96*(B10), 16585–16595, 1991.

MacAyeal, D. R., Large-scale ice flow over a viscous basal sediment: Theory and application to Ice Stream B, Antarctica, *Journal of Geophysical Research*, *94*(B4), 4071–4078, 1989.

MacAyeal, D. R., The basal stress distribution of Ice Stream E, Antarctica, inferred by control methods, *Journal of Geophysical Research*, *97*(B1), 595–603, 1992.

MacAyeal, D. R., R. A. Bindschadler, and T. A. Scambos, Basal friction of Ice Stream E, West Antarctica, *Journal of Glaciology*, 41(138), 247–262, 1995.

Merry, C. J., and I. M. Whillans, Ice-flow features on Ice Stream B, Antarctica, revealed by SPOT HRV imagery, *Journal of Glaciology*, 39(133), 515–552, 1993.

Nereson, N. A., The evolution of ice domes and relict ice streams, *Journal of Glaciology*, in press, 2000.

Nereson, N. A., E. D. Waddington, C. F. Raymond, and H. P. Jacobson, Predicted age-depth scales for Siple Dome and inland WAIS ice cores in West Antarctica, *Geophysical Research Letters*, 23(22), 3163–3166, 1996.

Nereson, N. A., C. F. Raymond, E. D. Waddington, and R. W. Jacobel, Recent migration of Siple Dome ice divide, West Antarctica, *Journal of Glaciology*, 44(148), 643–652, 1998.

Paterson, W. S. B., *The Physics of Glaciers*, Pergamon, 3rd edn., 1994.

Raymond, C. F., Constraints on the velocity of ice streams imposed by their widths, *EOS Trans. AGU*, 67(46), Fall Meet. Supl., F208 – F209, 1995.

Raymond, C. F., Shear margins in glaciers and ice sheets, *Journal of Glaciology*, 42(140), 90–102, 1996.

Retzlaff, R., and C. R. Bentley, Timing of stagnation of Ice Stream C, West Antarctica, from short-pulse radar studies of buried surface crevasses, *Journal of Glaciology*, 39(133), 553–561, 1993.

Retzlaff, R., N. Lord, and C. R. Bentley, Airborne-radar studies: Ice Streams A, B and C, West Antarctica, *Journal of Glaciology*, 39(133), 495–506, 1993.

Rose, K. E., Characteristics of ice flow in Marie Byrd Land, Antarctica, *Journal of Glaciology*, 24(90), 63–75, 1979.

Scambos, T. A., and R. A. Bindschadler, Feature maps of ice streams C, D, and E West Antarctica, *Antarcitic Journal of the United States*, XXVI(5), 312–314, 1991.

Scambos, T. A., K. A. Echelmeyer, M. A. Fahnestock, and R. A. Bindschadler, Development of enhanced flow at the southern margin of Ice Stream D, Antarctica, *Annals of Glaciology*, 20, 313–318, 1994.

Scambos, T. A., N. Nereson, and M. A. Fahnestock, Detailed topography of Siple Dome and Roosevelt Island, West Antarctica, *Annals of Glaciology*, 27, 61–67, 1998.

Shabtaie, S., and C. R. Bentley, West Antarctic ice streams draining into the Ross Ice Shelf: Configuration and mass balance, *Journal of Geophysical Research*, 92(B2), 1311–1336, 1987.

Shabtaie, S., I. M. Whillans, and C. R. Bentley, The morphology of ice streams A, B, C, West Antarctica, and their environs., *Journal of Geophysical Research*, 92(B9), 8865–8883, 1987.

Tulaczyk, S., Freeze-on-driven consolidation of till; a general mechanism for stopping an ice stream, *EOS Trans. AGU*, 78(46), Fall Meet. Supl., F252, 1997.

Tulaczyk, S., B. Kamb, and H. Engelhardt, Basal mechanics of Ice Stream B, West Antarctica. I. till mechanics, *Journal of Geophysical Research*, in press, 2000a.

Tulaczyk, S., W. B. Kamb, and H. F. Engelhardt, Basal mechanics of Ice Stream B, West Antarctica. II. till undrained-plastic-bed model, *Journal of Geophysical Research*, in press, 2000b.

van der Veen, C. J., and I. M. Whillans, Model experiments on the evolution and stability of ice streams, *Annals of Glaciology*, 23, 129–137, 1996.

Vaughan, D. G., R. F. Frolich, and C. S. M. Doake, Ers-1 SAR: stress indicator on Antarctic ice streams, in *Space at the service of our environment. Proceedings of 2nd ERS-1 Symposium*, no. 361 in ESA SP, pp. 183–186, European Space Agency, 1994.

Vornberger, P. L., and I. Whillans, Crevasse deformation and examples from Ice Stream B, Antarctica, *Journal of Glaciology*, 36(122), 3–9, 1990.

Whillans, I. M., and C. J. van der Veen, Patterns of calculated basal drag on Ice Streams B and C, Antarctica, *Journal of Glaciology*, 39(133), 437–446, 1993.

Whillans, I. M., and C. J. van der Veen, The role of lateral drag in the dynamics of Ice Stream B, Antarctica, *Journal of Glaciology*, 43(144), 231–237, 1997.

Whillans, I. M., M. Jackson, and Y.-H. Tseng, Velocity patterns in a transect across Ice Stream B, Antarctica, *Journal of Glaciology*, 39(133), 562–572, 1993.

Whillans, I. M., C. R. Bentley, and C. van der Veen, Ice Streams B and C, this volume.

C. S. M. Doake, British Antarctic Survey, Natural Environment Research Council, High Cross, Madingley Road, Cambridge CB3 0ET, England

K. A. Echelmeyer, Geophysical Institute, University of Alaska - Fairbanks, Fairbanks, Alaska 99775-0800

C. F. Raymond, Geophysics Program, Box 351650, University of Washington, Seattle, WA 98195-1650

I. M. Whillans, Byrd Polar Research Center and Department of Geological Sciences, The Ohio State University, Columbus, Ohio 43210

BASAL ZONE OF THE WEST ANTARCTIC ICE STREAMS AND ITS ROLE IN LUBRICATION OF THEIR RAPID MOTION

Barclay Kamb

Division of Geological and Planetary Sciences, California Institute of Technology, Pasadena, California

Basal processes and conditions at the bottom of ice streams B, C, and D and under adjacent parts of interstream ridges B1-B2 and C-D have been studied via boreholes drilled to the base of the ice by the hot-water jet drilling method. The objective is to reveal on an observational basis the mechanism of rapid ice-streaming motion as a guide to theoretical models of the ice-stream phenomenon in the West Antarctic Ice Sheet. Whereas the ice sheet outside the ice streams is frozen to its bed, the base of the ice streams is at the melting point and water is available there at a pressure close to the ice overburden pressure, so that the effective pressure is near zero. These conditions are favorable for both basal sliding and deformation of soft basal sediments as mechanisms for lubrication of rapid ice-stream motion. Sediments consisting of unfrozen glacial till are present at the base of the ice streams in a layer ≤10 m thick. Frozen basal till is present under interstream ridges with a prior history of streaming motion. The till's lithological characteristics reflect derivation from Tertiary glacimarine sediments of the Ross Sea sequence, which are thought to underlie the till layer. The water content of the unfrozen till is high, corresponding to water-saturated bulk porosities ranging from 26% to 58%. The high porosity is compatible with till deformation under low effective pressure. Except for extensive breakage of diatom tests and mixing of diatom ages, and except possibly for the occurrence of prominent "skelsepic plasmic" fabrics, the till does not show unambiguous internal evidence of deformation. This is probably because a "cushioning" action of the clay-rich matrix under low effective pressure shields the rock particles from abrasion and comminution. Till deformation is however indicated by measurements of basal sliding made with the tethered stake instrument. They show that in Ice Stream D about 80% of the motion was by till deformation whereas in Ice Stream B only about 25% was. The large, as-yet unexplained difference between the two ice streams in the relative contributions of till deformation and basal sliding to their streaming motions constitutes a challenge to understanding the ice stream mechanism, as does the time variation in these contributions. Mechanical tests of the till in the field and in the laboratory (direct shear, ring-shear, and triaxial tests on till cores, plus in-situ tests with a torvane instrument) were used to investigate the till rheology, which probably plays an important role in the ice stream mechanism. The need is to distinguish among quasiviscous rheology, treiboplastic (Coulomb-plastic) rheology, and possibly a hybrid of the two. The till has a well-defined yield stress that depends linearly on the effective pressure and only slightly on the shear strain rate. These are characteristics of treiboplastic rheology. The internal friction is 0.45±0.02 and the apparent cohesion is small (~ 1 kPa). The strength depends inversely on till porosity and tends to increase at deeper levels in the till. The slight strain-rate dependence amounts on average to a 5% increase in strength per decade increase in strain rate. This type of dependence can be represented empirically as a power-type flow law with a very large value of the exponent, $n \sim 40$, very much larger than the $1 \leq n \leq 5$ typical of viscous or quasiviscous rheologies. The strain-rate dependence can be explained as the result of measurable pore-pressure changes induced by changes in strain rate. Observations in Ice Stream D establish the existence of a basal water conduit system that in its natural state, unmodified by effects of borehole drilling, is capable of delivering a substantial quantity of water at a pressure approximately equal to the ice overburden pressure. Damped oscillations of the borehole water level are compatible with a "gap conduit" model in which the conduits form a 2.5-mm water-filled gap between the base of the ice and the top of the till. A diagnostic parameter for the gap width is the time that it takes for the borehole water level to drop from an initial shallow depth to a final depth near 105 m when a borehole being drilled breaks through into the basal water system. For Ice Streams B and D this drop time is mostly in the range 1-3 minutes, whereas the boreholes in Ice Stream C tend to have drop times distinctly longer, in the range 5-30 minutes, suggesting a tendency for thinner gap-conduits in its basal water system. An extreme case is a borehole drilled in the fossil marginal shear zone, in which the drop time was 5 hours. Problems in interpretation of the basal water system and ice streaming mechanism in terms of the observed basal water pressures and calculated ice overburden pressures are discussed. A large difference in some well defined flow-controlling variable is expected on comparing Ice Streams B and D with Ice Stream C, whose rapid streaming motion came nearly

to a stop about 150 years ago. But despite the great difference in flow velocity (B and D ca. 1 m d^{-1}, C ca. 0.04 m d^{-1}), these three ice streams appear quite similar in terms of the physical conditions and structures (till layer, water conduit system) found by borehole observation of their basal zones. The a-priori most likely controlling variable, the basal effective pressure, has average values (0.2×10^5 Pa for C, 0.6×10^5 Pa for B, 0.4×10^5 Pa for D) that differ in the wrong way to explain the slowing of Ice Stream C. However, there are a few subtle differences that may point indirectly to what has caused the slowdown: (1) indications of freeze-on at the base of Ice Stream C; (2) thinner gap-conduits in the basal water system of Ice Stream C; (3) in-situ shear strength of basal till under Ice Stream C (5 kPa) is higher than that under Ice Streams B and D (2 kPa, 1 kPa). According to a new theory of ice-stream soft-bed mechanics by S. Tulaczyk, the increased till strength under Ice Stream C is about great enough to switch the ice stream from a stable state of high velocity to an unstable state of lower velocity that decays with time and leads to ice-stream shut down. It appears that the streaming flow of Ice Stream C has not completely shut down in the area studied, which to some extent explains the similarity in physical conditions with those of Ice Streams B and D.

1. INTRODUCTION

Ever since the discovery of the great ice streams in the West Antarctic Ice Sheet (WAIS), with flow speeds ~10-100 times faster than the non-streaming adjacent parts of the ice sheet [*Bentley*, 1987; *Bindschadler and Scambos*, 1991; *Whillans and van der Veen*, 1993; *Bindschadler et al.*, 1996; *Joughin et al.*, 1999; *Doake et al.*, 1987; *Smith*, 1997], there has been a pressing need to understand the ice stream flow mechanism so that the role of the ice streams in the configuration and stability of WAIS could be rationally assessed. Such an assessment is needed as part of the overall effort to predict the future behavior of WAIS in relation to possible climate change and possible change of world-wide sea level [*Mercer*, 1978; *Alley and Whillans*, 1991; *Alley and MacAyeal*, 1993; *Oppenheimer*, 1998; *Bindschadler et al.*, 1998; *Bentley*, 1998]. The flow mechanism is recognized also as a fundamental problem in glaciology, possibly involving phenomena quite different from those in normal glacier flow. Three candidate mechanisms have been proposed for the rapid ice streaming flow: (1) enhanced basal- and marginal-ice shear deformation due to stress-induced recrystallization [*Hughes*, 1977]; (2) basal sliding due to basal melting [*Rose et al.*, 1979]; and (3) lubrication of the bed by soft basal sediment [*Alley et al.*, 1986]. (1) and (2) are conventional mechanisms, already much studied in glaciology, whereas (3) was put forward as a new paradigm of glacier mechanics by *Boulton* [1986] and is only now getting full attention as a glacier flow mechanism. It was proposed as the mechanism for ice stream flow by *Blankenship et al.* [1986] on the basis of reflection seismic data from Ice Stream B, and it has been extensively discussed in a series of papers by these authors [*Blankenship et al.*, 1987; *Rooney et al.*, 1987; *Alley et al.*, 1987a, 1987b, 1989, 1994; *Alley*, 1989a, 1989b, 1993, 1994; *Cuffey and Alley*, 1996]. These papers paint an attractive, largely theoretical picture of mechanism (3), which has led to a rather general acceptance of (3) over (1) and (2) by the WAIS research community.

The present paper summarizes the results of a complementary, largely observational approach that endeavors to make direct in-situ observations of the ice stream mechanism and the physical conditions and processes that make it function and control it. The objective is to distinguish among (1), (2), and (3) on a comprehensive observational basis. Since the proposed mechanisms all involve processes at or near the base of the ice, to make direct in-situ observations of them requires drilling boreholes to the bed and deploying instruments there. Although this may appear to be a straightforward approach, in practice various complications and apparent contradictions in data interpretation have arisen, such that, as will be evident below, an unambiguous observational concept of the ice stream mechanism is yet to be achieved. For this reason the paper should be viewed as a progress report on this effort rather than a final statement on the mechanics of ice stream motion. Some of the results have been reported by *Engelhardt et al.* [1990], *Kamb* [1991], *Scherer* [1991, 1994], *Jackson and Kamb* [1997], *Engelhardt and Kamb* [1993, 1997, 1998], *Jackson* [1998] *Scherer et al.* [1998], *Tulaczyk* [1999], *Jackson* [1999], *Tulaczyk et al.* [1998, 2000a, 2000b, 2000c], and Engelhardt et al [2000]. The reference *Engelhardt and Kamb* [1997] is used so frequently that it is here abbreviated "E & K".

2. OBSERVATIONAL METHODS AND STUDY SITES

In our program, boreholes to the base of the ice, typically 1000-1200 m deep, are drilled by the hot-water jet drilling method with thermal power of approximately 480 kW, producing water-filled holes of nominal 10 cm diameter in approximately 15 hours of drilling. Completed holes are normally reamed for safety's sake to a minimum diameter of 10 cm by a hot-water reaming tool, requiring about 12 hours of reaming time. Instruments of diameter 5 cm are deployed in such holes for periods of up to about 4 hours without getting caught by freeze-in of the borehole wall under the ambient ice temperature of about –20 C. For more extended access the instruments are either allowed to freeze in or are removed and the hole rereamed to diameter 10 cm. The alternative of stabilizing the hole against refreezing by introducing antifreeze (ethanol) has been tried but was not generally successful. The individual borehole instruments are indicated in the following survey of the types of observations made and the results obtained.

Borehole observations have been made at five nearby "study sites" on Ice Stream B2 (hereafter called simply B), at six nearby sites on Ice Stream C, and at one site on Ice Stream D; also at three nearby sites on interstream Ridge B1-B2 (the Unicorn) and at two nearby sites on Ridge C-D (Siple Dome). The location of these sites is shown in Figure 1, in which the nearby sites are plotted together as single locations. "Nearby" in this context means within about 10 km of one another. On Ice Stream C, one of the sites is about 25 km distant from the other five and has other special characteristics discussed later. Four of the sites on Ice Stream B were close to the surface location of Camp Up B in different years, but because of the 0.44 km a^{-1} ice stream motion these sites sample basal conditions at locations 0.4-0.9 km apart; the fifth site was located about 10 km upstream from Up B, close to Camp New B. At each of these sites, usually several (1 to 13) individual boreholes were drilled, the number depending on the number and types of borehole observations wanted and on the successfulness of the holes in attaining specified objectives. Figure 2 shows in map view the borehole pattern at the study sites on Ice Stream C, drilled in the 1996-97 season; the pattern of boreholes in Ice Stream B (at Up B and New B) is given in E & K, Figure 2. The individual boreholes are identified by a number of the form "96-1", where "96" is the field season when the hole was drilled (1996-97) and "1" is the number of the borehole counted sequentially in time in that field season. The boreholes are listed in Table 1. The appearance of a typical drill site during drilling is shown in Figure 3.

The rationale for choosing the foregoing study sites was as follows. Initially the choice was Ice Stream B near Up B because this is where the seismic work that indicated a soft basal sediment layer had been done. The 10-km move to Up B '95 was made in order to sample basal conditions in an area where, according to the seismic work, the basal sediment layer was thin or absent and therefore a "sticky spot" with slow ice-stream motion might be expected. Ice Stream D was chosen to provide observations in a separate rapidly moving ice stream, for comparison with Ice Stream B. Ice Stream C was chosen to provide comparison with an ice stream whose rapid motion slowed down greatly about 150 years ago [*Shabtaie and Bentley*, 1987]. The sites on interstream ridges (Unicorn and Siple Dome) provide information on basal conditions in the non-streaming ice sheet away from the ice streams, although the location of these particular sites was chosen primarily for other reasons.

3. PHYSICAL CONDITIONS IN THE BASAL ZONE

3.1 Temperature

Basal and englacial temperatures were measured in selected boreholes, normally one per study site, with thermistors enclosed in pressure-tight cases and calibrated to an accuracy of ±0.02 C. The thermistors were introduced into the measurement hole at predetermined depths via cables, the hole was allowed to freeze in, and the temperature of each thermistor was followed with time as it asymptotically approached a final temperature equal to the ambient temperature at the thermistor's depth. Details of the technique are given by *Engelhardt and Kamb* [1993]. Typical results are shown in Figure 4, which compares the vertical profile of temperature in an ice stream (at Up B) and in an interstream ridge (B1-B2, known as the Unicorn). The main difference is that the measured basal temperature in the ice stream (–0.82 C) is close to the pressure melting point (calculated as –0.71 C for ice 1057 m thick), whereas outside the ice stream the basal temperature is below the pressure melting point by amounts ranging from 0.6 to 2.7 degrees C at different study sites.

An independent indication of the distinction in basal temperature between the ice streams and the non-streaming ice sheet is that free water in a conduit system is almost always encountered at the base of the ice streams but is not encountered under the inter-stream ridges. This is further discussed in Section 7.

The base of ice Stream C presents a special situation, the basal temperature (measured in boreholes 96-2 and 96-12) lying 0.35 C below the calculated melting point. While the 0.1-degree discrepancy between measured basal temperature and calculated basal melting point for Ice Stream B can possibly be ascribed to measurement and/or calculation error, this does not seem possible for the 0.35 degree discrepancy for Ice Stream C in relation to the thermistor calibration accuracy of ± 0.02 C. This situation is discussed further in Sections 4.6 and 8.3.

The simple, straightforward conclusion is that ice streams form only where basal ice reaches the melting point. This conclusion was drawn by *Bentley et al.* [1998] from radar reflection data and was originally advocated on a theoretical basis by *Rose* [1979]. In Section 4.6 it will be shown that the basal temperature situation is somewhat more complicated but that the simple conclusion still holds, although with possible question as to its applicability to Ice Stream C.

3.2 Pressure

Three basal pressure quantities are important in ice stream mechanics: the ice overburden pressure (ice pressure) at the bed, the pressure of water having access to the bed, and the pore water pressure in basal sediments.

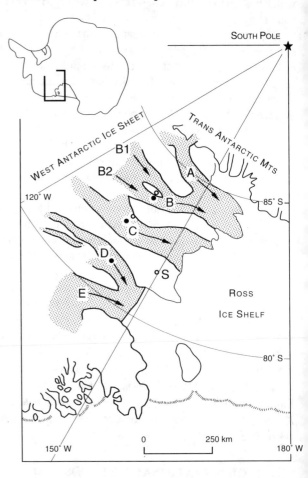

Figure 1. Sketch map showing West Antarctic ice streams (stippled) and location of study sites (solid dots on ice streams, open dots on interstream ridges and marginal shear zones). Arrows indicate the general ice flow. Ice streams are labelled A, B, C, D, E. These letters are placed so as to identify the study sites, from south to north as follows:

B, solid dot: Up B and New B
B, open dot: Unicorn (Ridge B1-B2)
C, solid dot: Up C
C, open dot: borehole 96-12, in shear margin of C.
D, solid dot: Up D
S, open dot: Siple Dome (Ridge C-D)

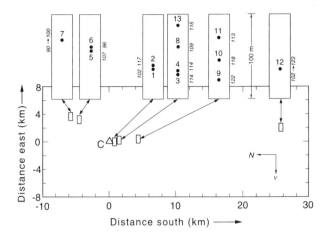

Figure 2. Map of the study site area on Ice Stream C, showing location of individual boreholes (black dots, identified by borehole number, with the prefix "96" omitted). On main map, in lower part of figure, locations of clusters of boreholes are shown schematically with small rectangles. Boreholes in each cluster are shown in strip maps (large rectangles) inserted along top of main map. Each strip map corresponds to a small rectangle on the main map via a two-headed arrow. Actual length of area shown in each strip map is 100 m as indicated. Italicized numbers along the sides of the strip maps give the water-level depth (in meters) in each borehole upon completion of breakthrough (Table 1). Major changes in water level are indicated with small arrows. Open triangle at origin of distance coordinate system is camp Up C, located at latitude 82° 26.5′ S, longitude 135° 58.0′ W. The arrow labeled v is the general westerly ice-flow direction in Ice Stream C.

The basal water pressure is measured in boreholes that have made hydraulic connection with a basal water-conduit system (see Section 7.1). During borehole drilling and for several hours after reaching the bed, the pressure measurements are made by measuring the depth of the borehole water level with a sounding float or by means of a pressure transducer placed at a depth of about 120 m in the borehole. Thereafter in some of the holes a pressure transducer is placed at the bottom, where it measures the basal water pressure directly. After the borehole freezes up and pressure transients have died away, and after sufficient time has elapsed for equilibrium between the sediment pore-water pressure and the pressure in the local basal water system, the pressure reported by the transducer represents the pore pressure. The pore pressure thus may manifest itself in long-term records of basal water pressure.

Ice overburden pressure is here estimated by calculation from an assumed ice density as a function of depth (E&K, Section 4). As a check, it can be measured by enclosing the pressure entry port of the transducer in an antifreeze-filled rubber gland (to prevent formation of ice inside the transducer body) and placing the transducer ca. 1 m above the bottom, where it comes to bear the ice pressure after the borehole has frozen up and the resulting pressure transients have relaxed. This check was made in borehole 93-9. The measured and calculated ice pressures agree to within about 0.4×10^5 Pa, the calculated pressure being the higher of the two (E & K, p. 213). For the other boreholes the ice pressure was obtained by the same type of calculation, with the ice density profile held fixed relative to the ice surface. A possible systematic error of 0.1 to 0.6 $\times 10^5$ Pa in these calculated pressures is indicated by the comparison for hole 93-9.

Basal water pressures are reported and discussed here mostly in terms of the equivalent borehole water-level depth below the ice surface; a 10 m increment in water level corresponds to a 1.0 bar (10^5 Pa) increment in water pressure. In general the "post-breakthrough water-level depth" given in Table 1 for each borehole is the deepest level recorded within several hours after the borehole makes connection. The water-level depths have a conservatively estimated accuracy of ± 1 m (E & K, p. 209). Ice overburden pressure is expressed in terms of flotation level, which is the water level that produces a basal water pressure equal to the ice overburden pressure.

Results of the pressure measurements (Table 1) are as follows. Outside the ice streams, boreholes do not make hydraulic connection with a basal water system, and no meaningful basal water pressure can be measured. In the ice streams, boreholes almost invariably make hydraulic connection with a basal water system (Section 7). The pressure in the system corresponds to borehole water level depths mostly in the range 100-115 m, with extreme values 96 and 123 m. Ice overburden pressures, calculated as explained above, correspond to flotation level depths in the range 98-101 m in Ice Stream B, 103-113 m in Ice Stream C, and 103 m in Ice Stream D. These figures show that the basal water pressures are in general quite high, close to the ice pressure—a situation that can result in significant enhancement of basal sliding and/or basal sediment deformation. This is best considered in terms of effective pressure—the difference between ice overburden pressure and basal water pressure, whose value in bars (10^5 Pa) is readily obtained from the difference between water level depth and flotation level depth (in meters), times 0.1. Basal effective pressures in Ice Stream B are mostly in the range –0.2 to +1.2 bar, with extreme values –0.3 and +1.6 bar; In Ice Stream C they are mostly in the range –0.9 to +1.1 bar with extremes –1.0 and +1.7 bar; and at the one site (with four boreholes) in Ice Stream D the effective pressure is uniform at + 0.4 bar. The mean effective-pressure values and standard deviations of the individual values about the means are +0.6 ± 0.6 bar for Ice Stream B, + 0.2 ± 0.7 bar for Ice Stream C, and +0.4 ± 0.0 bar for Ice Stream D. These mean values of effective water pressure

Table 1. Boreholes to Bottom in Ice Streams B, C, and D, and on Ridges B1-B2 and C-D; Water-Level Depth Before and After Breakthrough; Flotation Level and Basal Effective Pressure P'; Water-Level Drop Time (WLDT) and Time Constant T

Study site	Borehole Year-No.	Hole depth m	Water-level depth pre-breakthrough m	Water-level depth post-breakthrough m	Flotation level depth m	P' 10^5 Pa	WLDT min	T min
Up B	88-1	1035	-[a]	102	99	+0.3		
	2	(1035)[b]	-	111	(99)[b]	+1.2	540[q]	
	3	(1035)	-	105	(99)	+0.6		
	5	(1035)	-	109	(99)	+1.0		
	6	(1035)	-	115	(99)	+1.6		
	89-1	1058	-	113	101	+1.2		
	2	(1058)	-	115	(101)	+1.4		
	3	(1058)	-	112	(101)	+1.1		
	4	1057	-	99	(101)	–0.2		
	5	1057	28	98	(101)	–0.3	2	1.6
	6	1057	28	98	(101)	–0.3	3	2.7
	91-1	1055	-	112	101	+0.7		
	2	(1055)	-	108	(101)	+0.7		
	3	(1055)	16	109	(101)	+0.8	3	2.9
	92-1	1052	85	117	101	+1.6		
	2	(1052)	-	115	(101)	+1.4		
	4	(1052)	-	112	(101)	+1.1		
Unicorn	93-9[r]	910	30	30[g]				
	10[r]	914	24	29[g]				
	12[s]	993	38	39[g]				
	14[t]	1,092	30	30[g]				
New B	95-1	1,026	29	99	98	+0.1	2	1
	2	≈1026	82	96	-98	–0.2	5	6.4
	3	1,025	107[c]	107[d]	98	+0.9[l]	-	
	4	≈1029	21	98	-98	0	7[f]	
	5	1,028.5	22	100[m]	-98	+0.2	2.5	1.1
	6	1,029.5	62	97	98	–0.1	2	1.7
	7	1,026	33	105	98	+0.7	3	2.5
	8	1,027	51	104	98	0.6		
Up C	96-1	1,189	33	102	111	−0.9		
	2	-1,189	71	117	111	+0.8	2	2.6

Table 1, continued

Study site	Borehole Year-No.	Hole depth m	Water-level depth pre-breakthrough m	Water-level depth post-breakthrough m	Flotation level depth m	P' 10^5 Pa	WLDT min	T min
Up C	96-3	1,184	111[c]	114	111	0.3		
	4	-1,184	40	114	111	+0.3	2→30[h]	1.9[h]
	5	1,116	33	107	105.5	+0.15	15	10
	6	-1,116	37	96	105.5	-1.0	15	7
	7	1,088	27	90→100	103	-0.3	6[p]	1.4[p]
	7[i]	-1,088	65	107	103	-0.3	40[p]	6.5[p]
	8	(1184)	39	109	(111)	-0.2	5	2.3
	9	1194	58	122	111	+1.1	1.3	2
	10	1198	31	118	111	+0.7	2	1.3
	11	1201	55	113	111	+0.2	6	2
	12	1124	32	102→123	106	+1.7	~300[h]	75[h]
	13	1205	36	113	113	0.0	~2	3.3
Siple[u]	97-1	1004	-	-[g]				
	2	955	-	-[g]				
Up D	98-1	(1086)	120	107	(103)	+0.4	3[k], 2[n]	
	2	1086	87	107	103	+0.4	1	
	3	(1086)	54	107	(103)	+0.4	1	0.6
	4	(1086)			(103)			
	6	1,086	35	107→105	-103	+0.2	1.4	

a A dash in this column means that water level was not recorded but was generally in the range 20-30 m depth.
b Parentheses are used where an estimate of hole depth or flotation level is assumed to apply to nearby holes.
c Water level pumped down to this depth by the time the drill reached bottom.
d Water level remained at this depth after the drill reached bottom.
e This level was reached 36 min. after drill reached bottom; water remained at this level for 24 min., then began to rise.
f Estimated from measured drop rate from 20 to 47 m depth, extrapolated to 98 m.
g No breakthrough evident
h Complex drop curve; T value applies only to initial part.
i Redrilling of hole 96-7 after original hole lost connection to basal water system.
j Borehole at center of fossil marginal shear zone of Ice Stream C.
k Rise time (see Section 7.1).
l Valid if hole 95-3 did not bottom in frozen till (Sections 4.5 and 7.1). Not included in average effective pressure.
m After 1.7 days; initial level was 97 m.
n Drop time, after initial rise.
p Non-exponential drop curve causes large discrepancy between WLDT and T
q Water level dropped after a 9 hour delay.
r Station "Dragon Drill Pad" at edge of Unicorn (Ridge B1-B2)
s Station "Staging Area" between r and t
t Station "Fish Hook" near center of Unicorn
u Siple Dome, on Ridge C-D.
→ Indicates a change with time.

Figure 3. Photo of drill rig. From right to left the main components are: computer/personnel shelter, derrick, capstan, water-well pump winch (in front), four drilling-hose spools, eight water heaters, four high pressure pumps, two power plants, two snow-melting tubs (in front). Photos by H. Engelhardt.

are the lowest that have been reported for glaciers in which extensive pressure measurements have been made [*Kamb et al.,* 1994, p. 15,232]. Since the usually assumed basal sliding laws and till deformation laws [see e.g. *Bentley,* 1987; *Kamb,* 1993, pp. 28 and 71; *Kamb,* 1991, p. 16,585] predict basal shear motion increasing without limit as the effective pressure goes to zero under given basal shear stress, it stands to reason that rapid basal motion due to basal sliding and/or the shearing of basal sediments should be expected in the ice streams.

In Ice Stream B the pore water pressure in the basal sediment (Section 4), obtained from long-term water-pressure records and expressed as a borehole water-level depth, appears to lie mostly in the range 98-112 m (E & K, Figure 8), similar to the depth range 100-115 m quoted above for water levels observed shortly after borehole completion. Similarly, for Ice Stream C a two-year water pressure record (1997 and 1998) from borehole 96-8 (Figure 5) shows water-level depth varying over the range from 109 to 116 m. Thus the above expectation of water-

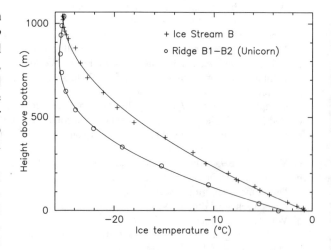

Figure 4. Vertical profiles of ice temperature in Ice Stream B (crosses) and in Ridge B1-B2 (circles). The Ice Stream B profile is near camp Up B, and the Ridge B1-B2 profile is at "Fish hook Station" (borehole 93-14). The smooth curves are fourth-order polynomial functions of depth, fitted to the temperature data.

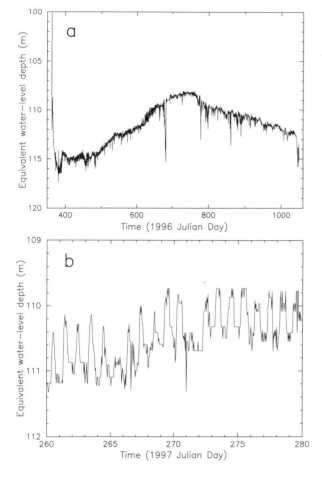

Figure 5. (a) Two-year record of basal water pressure under Ice Stream C (borehole 96-8). (b) Enlargement of a portion of (a), showing diurnal pressure fluctuations.

pressure enhancement of basal sliding and sediment shear seems to be generally valid whether pressure in a basal water system or in the pore water of the sediment is considered.

At times in the long-term water pressure records there occur periods of diurnal fluctuations in pressure, illustrated for Ice Stream C by Figure 5b and for Ice Stream B by E & K, Figures 12 and 13.

The cause of the pressure variations in Figure 5 is not known.

In summary, the basal melting of the ice streams makes basal sliding and bed deformation possible, and the high basal water pressure, producing unusually low effective pressure, is favorable for high rates of sliding and bed deformation. Thus the temperature and pressure observations support mechanisms (2) and/or (3) for ice-stream flow.

Further details concerning the pressure measurements are given in Section 7 and in E & K.

4. BASAL MATERIALS AND STRUCTURE

Samples of subglacial material were taken or attempted by piston coring at the bottom of one or more boreholes at each study site. The cores are listed with their individual lengths in Table 2. Each core is given a number of the form "96-6" or "96-7-3", in which "96-6" or "96-7" is the borehole number (Section 2 and Table 1). The additional cipher, such as the "3" is this example, is added when more than one core was taken in the indicated borehole, and designates (in this example) the third core taken, in order of coring. The early piston cores (to 1992) are 4.3 cm in diameter, while the later ones are 5.6 cm. The 4.3-cm core tube was 4 m long, the 5.6-cm tube 3 m or, in a few cases, 1 m long. The length of the core was never limited by the length of the core tube. A few small samples were obtained by their adhering to borehole instruments as illustrated in Figure 6. All cores were kept in sealed containers to minimize water loss, were hand carried from McMurdo to Pasadena, and were there stored in a refrigerator at + 1 C.

4.1 Basal Till

In the ice streams (including Ice Stream C) piston cores are commonly obtainable from boreholes that reach the bed. Overall, about 70% of the coring attempts were successful in obtaining a core 0.2 m or more long. With one exception (from borehole 95-3), the cores from the ice streams invariably consist of dark gray, wet, very sticky, clay-rich diamicton that shows no grading, bedding, or other structure as seen by core x-radiographs, by visual inspection of cores in the clear plastic core tubes, and by visual inspection of sediment adhering to instruments that have been lowered into the diamicton in situ (Figure 6). The diamicton is practically unsorted, the particle size distribution (Figure 7) extending from clay size ($\leq 1\ \mu m$) to almost the largest pebble size that can enter the core tube. The largest clast found so far measures 5.5×3.5×2 cm. Evidence for larger clasts in the diamicton or underlying bedrock is the considerable mechanical damage sometimes suffered by the core cutter. The clasts consist mainly of crystalline rocks—mainly granites and other plutonics, granitic gneisses, and schists, with rare volcanics (3%); there are ~10% sedimentary rocks, mainly well indurated graywacke. Exotics include small pieces of coal, the fine-grained fraction of which sometimes gives a coffee color to turbid water above the core in the core tube. Also, one spheroidal marcasite nodule in gray phyllite was recovered. The clasts are matrix supported, and the matrix contains about 35% mineralogical clay. The matrix also contains diatoms and other microfossils of a mixture of

Table 2. Subglacial Till Cores from Ice Streams B, C, and D, and from Interstream Ridge B1-B2 (Unicorn)

Study Site	Core No.	Core length m	Study site	Core No.	Core length m
UpB	89-1-2	0.6[d]	NewB	95-6	0.25
	89-1-3	0.3[d]		95-7-1	1.0
	89-1-4	1.95		95-7-2	0.25
	89-3-6[c]	1.45		95-8	0.4
	89-6-7[c]	3.1	UpC	96-3	0.2
	89-6-8[c]	1.3		96-6	0.35
	92-1	2.85		96-7-1	0.2
Uni[a]	93-10	0.3[e]		96-7-3	0.25
Uni[b]	93-14	0.3[e]		96-8	0.2
NewB	95-1	3.0		96-9-1	0.05
	95-3-1	0.9[e]		96-10	0.3
	95-3-2	1.9[e]		96-12	0.7
	95-5-1	0.75		96-13	0.3
	95-5-2	0.25	UpD	98-2	0.85
	95-5-3	0.5			

a Unicorn, Station "Dragon Drill Pad".
b Unicorn, Station "Fish Hook"
c For cores from UpB '89, the last cipher of the core number is counted up in sequence.
d Core taken with split-tube corer.
e Sediment is size-sorted and graded.

ages ranging from Eocene to Quaternary [*Scherer*, 1991]. On the basis of these and other petrological characteristics, and despite the absence of certain indications of deformation typical of tills (Section 4.7), *Tulaczyk et al.* [1998, p. 490] conclude that the diamicton is a glacial till—a sediment that has been transported and deposited by glacier ice without detectable involvement of running water [*Dreimanis*, 1988, p. 34]. We will thus refer to it henceforth as till. This is in harmony with the assumption of *Alley et al.* [1986] and of numerous later authors. The till was probably derived from Tertiary glacimarine sediments of the Ross Sea sequence, whose sedimentological and petrologic features are similar [*Tulaczyk et al.*, 1998, p. 492], and which are inferred from seismic data to underlie the till with erosional truncation in the region of the West Antarctic ice streams studied here [*Rooney et al.*, 1987, 1991]. None of our cores recovered bedded and/or indurated sediment that could represent the inferred underlying sedimentary bedrock. The foregoing information is summarized from much more extensive presentations by *Tulaczyk et al.* [1998] and *Tulaczyk* [1999, Chapter 2].

The near complete lack of volcanic clasts in the till raises doubt as to the widespread occurrence of volcanic rocks in the catchment regions of Ice Streams B, C, and D, contrary to a widely held view [e.g. *Blankenship et al.*, 1993; *Behrendt et al.*, 1994].

The till is unfrozen and ice-free as the cores are brought to the surface. This confirms the conclusion from temperature measurements (Section 3.1) that the melting isotherm lies at the base of the ice.

The basal till is soft and deformable, and, as detailed below, it is water saturated. It thus corresponds rather well with the deformable basal till visualized by Alley et al. (1986) in their interpretation of seismic data [*Blankenship et al.*, 1986]. More information on till deformability is given in Section 6.

4.2 Water Content of Till

Measurements of the water-saturated porosity of the till, which is related to the till's strength and loading history (Section 6.5), are plotted in Figure 8 and summarized in Table 3. Porosities are determined by weight-loss on drying and are of two types: bulk porosity, which is the porosity of each sample as taken from the core, and matrix porosity, which is the porosity indicated by the weight loss measurements when all clasts greater than 4 mm in size are removed. Matrix and bulk porosities differ by 1-3%, as shown by the data in Figure 8b and Table 3. Mean porosities are nearly the same in Ice Streams B and C, about 40%, but the scatter of measured values is greater in C (Table 3). From seismic data from Ice Stream B *Blankenship et al.* (1987) inferred a porosity of 40%. Porosities in Ice Stream D (Figure 8b) are on average about 8% higher than in B or C, a considerable difference. This may be due to the fact that the Ice-Stream-D samples were better protected from water loss after core recovery, but it may also be related to the smaller strength of the Ice-Stream-D till (Section 6.1). The highest porosities are about as high as what is typically encountered in the most porous soils, with porosities up to about 45-50% [*Lambe and Whitman*, 1969, Table 3.2]. Particularly noteworthy is the high matrix porosity of 60% for the specimen at the top of core 98-2 (Figure 8b; Table 3, note 3). It may indicate some incorporation of water into the top of the core in the piston-coring process.

Figure 8b shows that for Ice Stream D there is a general decrease in porosity with depth below the top of the till, interrupted by a gentle peak in porosity at a depth of about

Figure 6. Photos of till adhering to torvane instrument. Instrument body is to the left, with main vane to the right, connected to the body by a 0.25-inch shaft, visible in (b). In (a) the scale is shown by the hand. In (b) and (c) the counter-torque vane is to the left. In (c) the till surface is wet, while in (b) the surface is freeze-dried. Photos by H. Engelhardt.

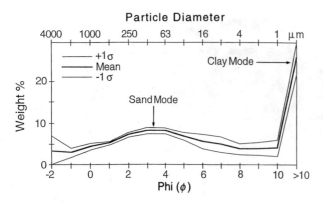

Figure 7. Particle size distribution in till, averaged from 35 till samples taken from 5 piston cores. Ordinate is weight percent of till solids per 1.0 phi interval. (Phi is minus the logarithm to the base 2 of the particle size in millimeters.) "Clay mode" is all particles smaller than 0.5 μm. σ is estimated standard deviation. From *Tulaczyk et al.* [1998, Figure 3A].

26 cm. Porosities of Ice-Stream-B till do not show a trend of this kind, except for the top two values from core 92-1 (Figure 8a). A decrease in porosity with depth is qualitatively what is expected due to the difference between the lithostatic and hydrostatic pressure gradients, which results in an increase in effective pressure from the top of the till downward. (The assumption of a hydrostatic gradient does not strictly apply if there is pore water flow in the vertical direction.)

The Atterberg liquid limit [*Lambe and Whitman*, 1969, p. 33], measured on a till matrix sample from Ice Stream B and expressed as a saturation porosity, is 42.5%. Although by this standard the measured porosities are quite high, there is no indication that the till approaches a slurry in its mechanical behavior.

The pore space of the till at depth, in-situ, is fully water saturated (see argument below). When the till is brought up from depth to the surface, some of the air that at depth was dissolved in the pore water exsolves, forming numerous small air-filled tubules ~1 cm long and ~1 mm in diameter within the till. Some of the tubules break through visibly to the core surface and emit bubbles of air. Many of them are aligned subparallel to the core axis and can be seen in x-radiographs of the core. Because of the air tubule content the cores tend to lengthen somewhat within the core tube and swell slightly on extraction from the core tube. As long as pore water from within a till specimen does not seep out of the till and get separated from the till in the air exsolution process, the in-situ water-saturated porosity measured by weight loss on drying will be unaffected by the exsolution.

In the triaxial tests (Section 6.3) a pore water pressure of $3.8\text{-}6.9 \times 10^5$ Pa (called the saturation pressure) is sufficient to eliminate the air bubbles in the pores of the till test specimens (Tulaczyk, 1999, Chapter 4, Appendix 4.A). The saturation pressures are so much smaller than the ambient hydrostatic pressure at depth (ca. 90×10^5 Pa) that complete water saturation of the till in situ is assured.

4.3 Thickness of Till

From seismic data *Blankenship et al.* [1987, p. 8907] interpreted the till at its discovery site near Up B as forming a basal layer 8.1 ± 0.3 m thick, and from seismic profiles *Rooney et al.* [1987, pp. 8915, 8918] judged the

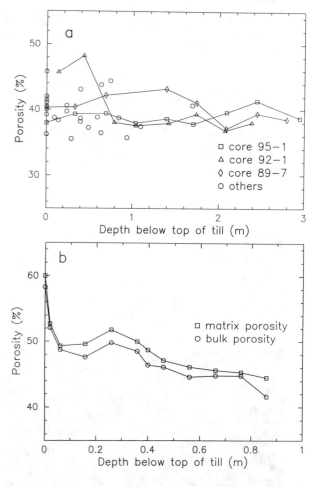

Figure 8. Porosity of till as a function of depth below the ice-till interface, (a) in Ice Stream B, and (b) in Ice Stream D. (b) shows both bulk and matrix porosities, while in (a) only matrix porosities are shown. In (a), the core designated 89-7 is the one labelled 89-6-7 in Table 2.

Table 3. Summary of Measured Till Porosities

Ice Stream	Type	Range %	Mean %	s.d. %	s.d. of mean %	No. of samples	Depth range (m)	Notes
B	matrix	33-44	40	3	0.5	48	0-3.0	1
C	matrix	28-60	42	7	1.5	19	0-0.35	2
Cm	matrix	45-52	48	4	2	5	0-0.7	3
D	matrix	44-53	48	3	1	11	0.02-0.8	4
C	bulk	26-58	40	8	2	19	0-0.35	2
Cm	bulk	44-52	47	4	2	5	0-0.7	3
D	bulk	41-50	46	3	1	11	0.02-0.8	4

1. Data of Figure 7a, from Up B and New B.
2. Unpublished data.
3. "Cm" means marginal shear zone of Ice Stream C, core 96-12.
4. Data of Figure 7b. Omits sample from depth 0, of bulk porosity 58%, matrix porosity 60%.

layer to vary in thickness from 13 m to zero (or to the minimum seismically resolvable thickness of 2 m). The lengths of the till cores (Table 2) represent minimum till thicknesses—minimum because the piston corer is subject to being stopped prematurely by running into large rock clasts in the till. The minimum till thickness for Ice Stream B near Up B is about 3 m, except in the "till-free swath" (Section 4.4), where the minimum is about 0.7 m. For Ice Stream C the minimum is about 0.3 m, except in the fossil marginal shear zone (borehole 96-12), where it is 0.7 m. The core lengths are probably more a measure (inversely) of the concentration of large clasts in the till than of the thickness of the till layer.

On several occasions, drilling tests in the till were carried out to see if the water-jet drill could penetrate detectably to the bottom of the till and reveal its thickness. This might be possible if the underlying sedimentary "bedrock" (Section 4.1) is sufficiently indurated to be detectably less penetrable than the till. A test on the Caltech campus, in which the water jet was used to drill down into alluvial sand and gravel, achieved a penetration of 5.5 m in 30 minutes of drilling time, showing that penetrations of this order are achievable. Tests in ice-stream boreholes seem to indicate drill penetrations of 5 to 15 meters, without an indication that the drill reached the bottom of the till in any of these holes (hence the apparent drill penetrations are minimum apparent till thicknesses.)

The apparent penetrations of 5-15 m are much larger than those achieved by other means—up to 3 m by piston coring (discussed above) and up to 0.7 m by slide-hammer penetrometer, tethered stake (Section 5), and torvane (Section 6), based on till coatings on them (Figure 6). Such coatings are generally not present on the hot-water drill stem, but in one rare case (borehole 95-4) the drill had a till coating up to 1.1 m above the tip. The coatings give only a minimum penetration because of their tendency to be washed off in the process of hauling the instruments to the surface.

The above results suggest a till thickness varying from ~0.5 to ~10 m, and are compatible with the 6.5-m average thickness inferred by *Rooney et al.* [1987, p. 8918] from seismic data.

4.4 Search for Till-free Bed

On Ice Stream B the L2 and L3 seismic profiles of *Rooney et al.* [1987, p. 8918] are inferred by these authors to cross a 300-m-wide swath of the bed over which the till layer is either absent or is thinner than the seismic resolution of 2 m. To test the ability of borehole observations to confirm the seismic interpretation, four boreholes were drilled on the original position of profile L2 and near the center of the swath (holes 95-3, -4, -5, -6); also two boreholes (95-7, -8) on its north flank where the till should be thin but not absent according to the seismic profiles. Borehole 95-3 behaved in a very peculiar way as discussed in Section 4.5, whereas boreholes 95-5 to -8 yielded till cores of normal lithology, showing that till is not absent. The till cores from the swath (95-5-1, -2, -3, and 95-6) are rather short, 0.2-0.7 m (Table 2), suggesting that the till is thin (~0.7 m) in the area sampled. Hole 95-4, which was about 12 m from holes 95-5 and 95-6, penetrated the till to the depth of at least 1.1 m as noted in Section 4.3, so the minimum till thickness in the swath center must be taken as 1.1 m despite the shorter cores. Drilling penetration tests of the kind discussed in Section 4.3 indicated an apparent minimum till thickness of 10 meters in borehole 95-6, at the swath center, and 16 meters in hole 95-7, on the north flank. The cores from the north flank (95-7-1, 95-7-2, 95-8) have a range of lengths 0.25-1.0 m (Table 2), similar to the cores from the swath center. In these till thickness data there is not a clear distinction between center and flank, contrary to what is portrayed in the seismic section.

4.5 Frozen Till

Boreholes outside the ice streams, in particular on Ridge B1-B2 (the Unicorn), bottom in frozen till. This is established as follows. The boreholes yield basal sediment cores, which are very different from the till cores from the base of the ice streams. The sediment is a diamicton of mineralogical/petrological/paleontological characteristics similar to the Ice Stream B till, except that it lacks most of the clay component that is so abundant in the till, and it is sorted and graded, fining upward from pebbles at the bottom to fine sand and silt at the top. These characteristics, together with the sub-freezing basal temperatures (Section 3.1), indicate that prior to drilling, the sediment occurs frozen in the ice, and is melted out of the ice by the hot-water drill. The water jet blasts the finer particles into suspension, from which they then settle out in the borehole, producing the sorting and grading. Clay is mostly winnowed out of the sediment column as sampled a few hours after drilling, because the settling velocity of the clay particles is small. A repeated cycle of attempted drilling followed by piston coring yields additional sediment of similar characteristics. It appears that the drill cuts slowly down into frozen, ice-saturated till, and also melts out sediment particles from the borehole wall. The advance of the drill is slowed and finally stopped by the accumulation of large rock clasts in the hole, too large to be blasted out of the way by the water jet or picked up in the core barrel. Another sign of drilling into frozen till is numerous fresh scratches along the length (4 m) of the brass drill stem, especially along the lowermost 1 m.

The available borehole observations do not provide a sufficient basis for distinguishing an abrupt from a gradational contact between frozen till and clean ice, or for recognizing the presence of a moderate amount of rock debris in the ice overlying the frozen till. (A heavy loading of debris would constitute essentially frozen till.) Observations of the slowdown in drilling speed as the bottom is approached, somewhat like the observations presented in Section 7.7 might allow these distinctions to be made.

At Siple Dome, on Ridge C-D, the bed appears to be solid bedrock without overlying till. About 0.1 kg of fine sediment was recovered from just above the bed in a sediment trap carried by the hot-water ice-coring drill. In the lowermost 30 cm of the deepest ice core, just above the bottom, a few small (~1 cm) rocks were found imbedded in the ice. Only a small amount (~ 1 cm^3) of sediment was recovered by the piston corer, which was heavily damaged by impact with solid rock(s).

The difference between the development of frozen till under Ridge B1-B2 and the lack of till development under Ridge C-D may be related to evidence of former streaming movement in Ridge B1-B2 and the lack of such evidence for Ridge C-D (H. Engelhardt, personal communication, 1999). The implication is that ice-stream motion is necessary to generate basal till, as is generally thought [*Cuffy and Alley*, 1996].

Surprisingly, the behavior of borehole 95-3 in Ice Stream B was quite similar to the Ridge B1-B2 boreholes both in terms of basal sediment cores as described above and in terms of the lack of breakthrough to a basal water system (Section 7). Hole 95-3 thus appeared to have bottomed in frozen till, contrary to the overwhelming expectation from drilling experience in the ice streams (E & K, p. 210). Boreholes only 3.5 and 7 m away (95-4, 95-5, and 95-6) bottomed in wet unfrozen till in what seemed the normal manner. However, details of the drilling records discussed in Section 7.7 imply that holes 95-4 and 95-5 penetrated through a thin frozen till layer and into normal unfrozen till below, indicated by piston cores. It seemed that the mass of frozen basal till that apparently stopped the drill in hole 95-3 became progressively thinner (or less debris-rich) as traced laterally past holes 95-4 and 95-5, where it slowed but did not stop the drill.

An alternative to the foregoing interpretation of borehole 95-3 is given in Section 7.1.

If the ice is frozen to the bed in the vicinity of hole 95-3, a sticky spot [*Alley*, 1993] would be expected there on the basis of Section 3.1. A similar expectation might be entertained if the till layer were absent in the "till-free swath" (Section 4.4). But no indication of a sticky spot has been found in this vicinity, either by flow-velocity measurement [*Whillans and van der Veen*, 1993; *Hulbe and Whillans*, 1994] or from basal microseismicity such as that observed in Ice Stream C [*Anandakrishnan and Bentley,* 1993]. A few microearthquakes were observed in Ice Stream B, but not near borehole 95-3.

In the expectation that the till might be absent or the bed might be frozen in the zone of microearthquake activity in Ice Stream C, boreholes 96-5, -6, and -7 were drilled at two sites where clusters of microearthquake epicenters had been located by *Anandakrishnan and Bentley* [1993]. However, these boreholes revealed normal ice-stream basal conditions—an unfrozen bed with till present. This might be explained by the seismic interpretation of *Anandakrishnan and Alley* [1994] that the probability that a randomly located borehole will hit a sticky spot is ~10^{-3}.

A recently measured flow-velocity profile across Ice Stream C [*Engelhardt*, unpublished data] reveals that the microearthquake zone lies at the southern edge of a large sticky area, ~10km x 20km in dimensions, which is shown also by a satellite image [*Engelhardt*, unpublished data] and a radar profile of basal topography and ice layering [*Conway et al.*, submitted, 2000]. Whether the microearthquakes occur at "stickier spots" within the sticky area remains to be seen.

4.6 Relation of Till and Bedrock to the Melting Isotherm

The presence of frozen till at the base of the ice sheet implies a thermally layered basal zone in which cold glacier ice overlies frozen till which in turn overlies either unfrozen till or bedrock, frozen or unfrozen, depending on the location of the melting isotherm relative to the till/bedrock contact. When a borehole terminates in frozen till, the drill falls short of reaching the melting isotherm. The basal temperature that is measured is the temperature at the depth reached by the drill, but the mechanically most significant level is the base of the frozen till, where the temperature is at the freezing point if unfrozen till underlies frozen till, or below freezing if frozen till directly overlies frozen bedrock. In the latter case there will be no appreciable basal sliding or till deformation, whereas in the former case the base of the frozen till will probably act as the glacier sole in any basal sliding that occurs and will be the upper limit of any shear zone that forms in the unfrozen till below. This seems to be the situation at boreholes 95-3, -4, -5, and -6, as discussed in Section 4.5. In the ice streams (except at these boreholes) thermal layering of the basal zone appears always to be simply cold ice over unfrozen till over bedrock, so that the basal temperature measurement always gives the pressure melting point or a value close to it. But in the ice sheet of Ridge B1-B2 there is a thickness of frozen till sufficient to stop the drill before it reaches the melting isotherm, so that sub-freezing basal temperatures are measured (Section 3.1). "Basal" here refers to the bottom of the boreholes rather than the base of the frozen till. From these basal temperatures we can calculate an estimated depth of the melting isotherm beneath the bottom of each borehole by assuming that the vertical temperature profile measured just above the base (as in Figure 4) continues downward with unchanged thermal gradient. For the four temperature profiles in the Unicorn the results are 10, 10, 24, and 41 m. If the total thickness of till (frozen, plus unfrozen if any) is 6.5 m, which is the average thickness estimated from seismic data (Section 4.3), and if the thickness of frozen till penetrated by the drill in each hole was ~1 m, as estimated from the total length of piston core recovered from each hole (Table 2), then these figures indicate that the ice sheet is in fact frozen to the bed there. At Siple Dome (basal temperature –2.35 C) the estimated depth of the melting isotherm beneath the bottom of the borehole is 56 m, by the same type of calculation. Thus even if, contrary to the interpretation in Section 4.5, there were a ~10 m thickness of till beneath the bottom of the hole, the till would be frozen and the ice sheet would be frozen to the bed. These results substantiate the conclusion in Section 3.1 that the ice sheet outside the ice streams is frozen to its bed.

As noted earlier (Section 3.1), in Ice Stream C the measured basal temperature, from boreholes 96-2 and 96-12, is 0.35 degrees C below freezing, which by the above reasoning indicates a melting isotherm 7 m below the depth level reached in the boreholes. This allows a frozen or unfrozen bed, depending on whether the bedrock lies less than or more than 7 m below the bottom of the boreholes.

If the indication of 0.3-m till thickness (Section 4.3) is correct, then Ice Stream C would appear to be frozen to its bed, which would account for the cessation of its streaming motion (Section 2). However, this conclusion is contradicted by the fact that only piston cores of unfrozen till were obtained from the boreholes in Ice Stream C (Table 2), and by the fact that all deep boreholes in Ice Stream C made connection with a basal water system (except possibly hole 96-3: see Section 7.1). This situation is considered further in Section 8.2. Geophysical evidence for an unfrozen bed under Ice Stream C is summarized by *Anandakrishnan et al.* [2000].

4.7 Lithologic Evidence re Till Deformation

Structural features indicative of soft-sediment deformation are found in many northern-hemisphere tills that are thought to have lubricated the motion of the

Laurentide and Scandinavian ice sheets [*van der Meer*, 1993], and such structures would be expected in till of the Antarctic ice streams if the till-lubrication model of the ice-stream mechanism is valid. A search has been made of the core-x-radiographs, individual till clasts, and till thin sections in an effort to identify lithologic and structural features produced by till deformation.

In the ice-stream till cores a vertical preferred orientation of the larger clasts is seen in radiographs, whereas a horizontal preferred orientation would be expected for extensive shear across horizontal planes. There is also in general a pronounced vertical orientation of the air-filled tubules (Section 4.2). These vertical preferred orientations are probably caused by a vertical extension of the till within the core tube as the air tubules form and expand. In core 89-4 the vertical orientation of the tubules and elongated clasts changes to horizontal within the upper 5 cm of the core, suggesting a horizontal planar anisotropic structure (planes of weakness) that would be expected for a zone of basal shear deformation in the till. Five more features of the same type, of thicknesses 2, 2.5, 6, 7, and 11 cm, have been found in cores 92-1, 95-5, and 95-7, but none of these are at the top of the core, which suggests that shear zones form at various depths in the till. However, the lack of microscopic evidence for such shear zones (see below) weakens their significance as evidence for till deformation.

Although evidence for subglacial comminution (i.e. clast crushing and abrasion) is not taken to be a criterion in the definition of till [*Dreimanis*, 1988, p. 34], such evidence is so common in tills that it is widely considered one of their most important distinguishing features [*Dreimanis*, 1990; *Harlan et al.*, 1966]. Striations are the most dependable indicator of clast abrasion [*Harlan et al.*, 1966] and are present on many clasts from North American Pleistocene tills [*Anderson*, 1955; *Drake*, 1972; *Holmes*, 1952]. But they are almost completely absent on clasts from the Ice-Stream-B till: only two clasts, 0.9%, showed (questionable) striations [*Tulaczyk et al.*, 1998, p. 489]. The clasts are mainly subangular to subrounded, and although rounding and rounded edges sometimes result from glacial abrasion [*Flint*, 1971, p. 165], it appears that in the Ice-Stream-B till they are associated with chemical-weathering features such as etch pits and therefore did not result from abrasion.

Extensive SEM examination of the surfaces of small (125-250 μm) clastic particles from the till has revealed that, contrary to what is typical in tills, there are very few microscopic features of particle fracture, crushing, grinding, abrasion, or comminution [*Tulaczyk et al.*, 1998, p. 491]. The few recognizable features of this kind are generally overprinted by features of chemical weathering, which might be taken to imply that a long period of weathering has intervened since the last recorded deformation of the till. A grain-scale indication of current or recent till deformation is thus lacking.

Thin sections of impregnated till that was not disturbed in core sampling show no discrete shears or other visible macroscopic or microscopic fabric suggestive of deformation. Clay particles are aggregated into thin, microscopically visible "plasma", which show no preferred orientation in the till matrix, contrary to what is expected as a manifestation of till deformation. Some of the plasma conspicuously coats the surfaces of the clastic grains (mainly quartz and feldspar). Such grain coatings are called "skelsepic plasmic fabrics" by *van der Meer* [1993, p. 555 and Figure 5] and are attributed by him (p. 559) to "rotational deformation", which he explains as follows: "A deforming till bed may be regarded as consisting of stacked, rotating wheels" of the till sediment [*van der Meer*, 1997, p. 828 and Figures 3 and 4]. If that concept is valid, and if "rotational deformation" can be related mechanistically to development of the plasmic coatings on grains and to the till lubrication process, one may be able to recognize the coatings as evidence for till deformation and therefore modify the first sentence of this paragraph.

Definite lithologic evidence for till deformation is present in the mixing of diatom ages in the till (Section 4.1), which requires some kind of stirring action that involves till deformation. The relative scarcity (by a factor $\sim 10^3$) of whole diatoms in the Up B till by comparison with glacimarine sediments of the Ross Sea may be an indication of mechanical disintegration of the fragile diatom tests by till deformation, but it could also result from chemical weathering [*Tulaczyk et al.*, 1998, p. 491].

Although lithologic evidence for till deformation is thus mostly lacking, evidence of other kinds (Sections 5 and 6) is fairly convincing that till deformation plays a role in the ice-stream mechanism. The lack of lithologic evidence is taken as an indication of the validity of the hypothesis that the subglacial deformation of clay-rich till produces no significant particle comminution because of a "cushioning" action of the clay-rich matrix, which is particularly likely under low effective pressure [*Tulaczyk et al.*, 1998, pp. 493-494].

5. BASAL SLIDING VS. TILL DEFORMATION

Since physical conditions at the base of the ice streams favor both basal sliding and till deformation as mechanisms for rapid ice-stream motion (Section 3), measurements are needed to evaluate observationally their actual contributions to the ice stream motion. This is important for any predictive models of ice stream behavior because the till flow law and the basal sliding law are probably quite different. We have made two measurements of basal sliding (and indirectly of till deformation), one on Ice Stream B (borehole 95-1) and one on Ice Stream D (borehole 98-3). The method of measurement (called the "tethered stake") and the results for Ice Stream B are given by *Engelhardt and Kamb* [1998].

These results indicate that at the New B site, sliding contributed 80% to 100% of the total motion over an observational period of 22.5 days, which was interrupted by a 3.5-day period of slow apparent sliding (8% of the full motion). From certain details in the sliding-vs.-time curves, *Engelhardt and Kamb* [1998, p. 228] concluded that the slow apparent sliding was an artifact caused by the tethered stake getting temporarily caught on rock clasts protruding from the ice sole. Even if this conclusion is incorrect, the average basal sliding motion is still large, 67% of the full motion. This sliding motion is basal sliding *sensu lato*: it is the sum of any sliding across the ice-till interface (sliding *sensu stricto*) plus any till shear from the top of the till down to the level of the tethered stake, which *Engelhardt and Kamb* [1998, p. 227] estimate to have been only about 3 cm below the sole. These results seem to indicate a dominant role for basal sliding in ice stream motion, or else they indicate that if till deformation dominates the motion, it is concentrated in a narrow shear zone at the top of the till and not distributed uniformly through the ca. 5-m inferred thickness of the till layer (Section 4.3), as has commonly been assumed.

In the last 2 days of tethered-stake observation at the New B site the measured apparent sliding reached 1.17 m d^{-1}, essentially 100% of the full ice-stream motion. This implies not only that the till-deformation contribution to the motion was zero, but also that there was a negligible contribution from ice deformation. This rules out enhanced ice deformation (Section 1, item (1)) as a mechanism of ice-stream motion, at least in this case. It is worthy of note since direct measurements of ice deformation that are needed to check mechanism (1) have not been made.

The basal-sliding results from Ice Stream D were obtained with a tethered-stake instrument improved over the original instrument in two ways: (1) it carries a greater length of tether line (300 vs. 21 m), which greatly increases the total amount of basal sliding that can be recorded; (2) the stake can be locked so that in a preliminary run the instrument can be lowered into the till without releasing the stake, and the depth of penetration into the till registered by the coating of till on the instrument. This was done in borehole 98-2. The depth of penetration of the bottom tip of the stake into the till was 63 cm, and the corresponding penetration of the tether attachment point at the top of the stake was 34 cm. The instrument with latch pins unlocked was then emplaced in the till at the bottom of hole 98-3, the stake was released, and its motion has been tracked continuously by the instrument and telemetered to Pasadena via the ARGOS data system since 22 January 1999. The record of apparent sliding distance with time (Figure 9) shows an initial apparent sliding speed of about 0.8 m d^{-1}, decreasing in a few days to about 0.2 md^{-1}, and decreasing gradually further to about 0.1 md^{-1} toward the end of March, with continued slow decrease thereafter. These speeds (except the initial 0.8 md^{-1}) are a small fraction of the surface velocity of 1.0 md^{-1} measured by GPS, and appear to indicate that most of the ice-stream motion is accommodated by till deformation below the level at which the tethered stake was emplaced (34 to 63 cm beneath the top of the till according to the preliminary run with the stake locked). A similar conclusion is reached by *Truffer* (1999, p. 38) for deformation of a 7-m-thick till layer at the base of Black Rapids Glacier, Alaska.

The relative amounts of till deformation inferred above for Ice Streams B and D do not correspond well with the till porosity profiles in Figure 8. The profile for D (Figure 8b), with a sharp, high peak in porosity at the top of the profile (depth = 0), giving a narrow zone of till weakness according to Section 6.5, should correspond to a narrow shear zone at the top of the till; but such a shear zone is a possibility inferred above from the tethered-stake data for Ice Stream B, not Ice Stream D. Except perhaps for profile

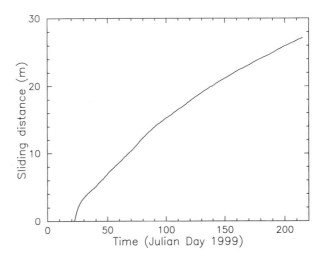

Figure 9. Displacement record for tethered stake in Ice Stream D, borehole 98-3. Pay-out of tethering line provides a measure of basal sliding (sensu lato). The slope of the curve is the sliding velocity.

92-1, the porosity profiles for Ice Stream B (Figure 8a) do not indicate a definite depth dependence of the till porosity and thus correspond better with the distributed till deformation inferred above from the tethered stake data for Ice Stream D, not Ice Stream B.

On the other hand, a concentration of till deformation close to the ice sole is compatible with a profile of till shear strengths from Ice-Stream-B cores (Section 6.2) and with an in-situ strength profile discussed in the last paragraph of Section 6.4.

The very contrasting results for Ice Streams B and D—basal sliding rapid and generally increasing with time in B, basal sliding slow and decreasing with time in D—illustrate the difficulty of making valid generalizations about the ice-stream mechanism with the amount of observational material so far in hand. The ratio of sliding to till deformation varies from one ice-stream site to another, and it appears also to vary with time. Until the problem of accounting for the amounts of basal sliding and till deformation is solved, we cannot consider the ice-stream mechanism to be understood.

6. TILL STRENGTH AND RHEOLOGY

Insofar as till deformation plays a significant role in ice-stream motion, the mechanical properties of the till are of much interest and importance. It has been generally assumed, following *Boulton and Hindmarsh* [1987, p. 9063], that the till behaves rheologically as a linear or slightly nonlinear viscous fluid [*Alley et al.*, 1987b; *Alley*, 1989b; *MacAyeal*, 1989, 1992; *Alley et al.*, 1989; *Hulbe*, 1998; *Hulbe and MacAyeal*, 1999]. However, since the till is a granular medium (Section 4.1) of composition and structure well within the range of granular media dealt with in soil mechanics, its rheological behavior should fall within the range exhibited by these materials. As such, it should show treiboplastic[*] (Coulomb-plastic) rheology: it should have a yield stress that is controlled by intergranular friction and that depends only slightly if at all on the strain rate [*Kamb*, 1991]. We have carried out four types of mechanical tests in an effort to resolve the question whether the till rheology is viscous or treiboplastic and to define its mechanical properties in relation to the contribution it may make to the ice stream mechanism.

Most of the tests were made on till samples obtained by piston coring. In order to provide information on the strength of the till in situ at the bottom of an ice stream

---

[*] I use the term "treiboplastic" in preference to "Coulomb-plastic" because it is more self-explanatory and more compact.

~1000 meters deep, the tests in Sections 6.1 and 6.2 depend on the "$\phi = 0$ concept" of soil mechanics [*Lambe and Whitman*, 1969, pp. 433, 440], according to which the shear strength of a water-saturated clay-rich soil of very low hydraulic permeability, in mechanical tests of modest duration, is not altered by changing the confining pressure from its in-situ value, so that the material behaves in this respect as though it had angle of internal friction (ϕ) equal to zero.

Applicability of the $\phi = 0$ concept depends on the time scale T^* for equilibration of the pore pressure with the external water pressure by flow of water out of or into a till specimen. This is described by consolidation theory [*Lambe and Whitman*, 1969, p. 406-412]. The equilibration time scale is $T^* = H^2/c_v$, where $2H$ is the specimen thickness (through which the pore water flows) and c_v is the till's hydraulic diffusivity (called "coefficient of consolidation" by *Lambe and Whitman* [1969, p. 407]). From oedometer tests on a till specimen 22 mm thick, by fitting the observed consolidation-vs.-time curve to the theoretical curve given by *Lambe and Whitman* [1969, Figure 27.3] Hermann Engelhardt determined $c_v = 7.4 \times 10^{-9}$ m^2 s^{-1}. For a 2-m-long core in its core tube, for which the pore water flow is along the length of the core, the time scale is thus $T^* = 1.0^2/7.4 \times 10^{-9} = 1.35 \times 10^8$ s = 4.3 years. Because the half hour needed to bring the core up from depth is only a small fraction of the equilibration time scale, only a small amount of pore-water flow will take place, and the $\phi = 0$ concept is applicable. This conclusion is linked to the till's very low hydraulic conductivity (2×10^{-9} m s^{-1} as measured by *Engelhardt et al.* [1990, p. 58]), since c_v is proportional to the conductivity.

When the recovered core reaches the surface and is removed from the core tube its pore pressure will usually be negative by a fraction of a bar because of the positive effective pressure inherited from its in-situ condition (Section 7.4). If it is then removed from the core tube and immersed in water, it will imbibe water on a time scale T^* of about a day, but if not immersed, capillary tension can maintain the negative pressure [see *Lambe and Whitman*, 1969, pp. 246, 315, 316]. This is probably why till core specimens retain a measureable shear strength in storage over times of many months (Sections 6.1, 6.2), contrary to what would be expected from the T^* limit on applicability of the $\phi = 0$ concept.

6.1 Shear Creep Tests on Freshly Sampled Till

At one location on Ice Stream B (borehole 89-1) and one on Ice Stream D (borehole 98-2) a sample of the till recovered by coring was subjected in the field to direct shear tests [*Lambe and Whitman*, 1969, p. 119] in the creep mode, in which the sample is subjected to a constant load in simple shear and the shear strain rate is observed as a function of time, for various values of the applied shear stress. The till from Ice Stream B (core 89-1-3), behaved as follows [*Kamb*, 1991, p. 16,587]. At stresses below about 2 kPa, the till showed transient creep, decreasing with time [*Singh and Mitchell*, 1968]. Above about 2 kPa it showed accelerating creep leading promptly to catastrophic failure, the promptness of which increased drastically with attempts to apply shear loadings greater than 2 kPa. This type of behavior is what is expected for a plastic material, with an indicated shear strength of $2 \pm .2$ kPa. The till from Ice Stream D (core 98-2) behaved similarly, with an indicated shear strength of $1 \pm .2$ kPa.

6.2 Direct Shear Tests in the Laboratory

Subsequent to their return to the U.S., till samples from Ice Stream B were extensively tested in direct shear under controlled shear rate, in apparatus kindly made available by Prof. Ronald Scott in the Caltech Engineering Division. The tests were carried out by Hermann Engelhardt. The diameter of the (circular) sample chamber (shear box) was 6.35 cm, and the shear gap between the upper and lower halves of the shear box was 4.6 mm. Clasts greater than ~10 mm in size, constituting less than 2% of the test sample volume, were removed from the till before testing, because they would tend to interfere with the tests. The samples tested were from core 89-1-4. Most of the tests were on till from near the top of the core. The tests were done about 8 months after original recovery of the core.

Examples of the test results are given in Figures 10, 11, and 12. These were nominally "drained" tests at atmospheric pressure, but because of the till's extremely low hydraulic conductivity (Section 6) its water content could have changed only slightly during each run (lasting at most 2 or 3 hours, whereas $T^* \approx 1.5$ days), so that the tests were effectively "undrained". To avoid evaporative water loss during testing, the surface of the test specimen exposed to the air (in the shear gap) was kept wet by administration of drops of water. The till in these tests was under a nominal normal stress of 9 kPa due to the weight of the test specimen and an overlying thin piston (steel plate). According to the $\phi = 0$ concept, this normal stress should not affect the measured strength. The confining pressure could not be increased above this level without producing extrusion of the sample through the shear gap of the testing machine.

Figure 10 shows tests at three different shear rates spanning a 58-fold range from 0.09 to 5.2 m d^{-1}. Essentially all of these tests reached a full "mobilization" of strength (i.e. reached essentially constant stress) within the 8 mm of shear displacement permitted by the testing machine. The average strengths are 1.62 ± 0.09 kPa at shear rate 0.09 m d^{-1}, 1.65 ± 0.08 kPa at 0.86 m d^{-1}, and 1.72 ± 0.08 kPa at 5.2 m d^{-1}. Thus the strength is nearly constant at about 1.7 kPa but increases slightly with strain rate. The strength 1.7 kPa is roughly the same as the value 2 kPa obtained in Section 6.1. Thus there was not a substantial loss (or gain) of pore water and a corresponding change in strength (Section 6.5) in the 8 months between the two sets of tests.

It is customary to report the dependence of the strength on the strain rate in terms of a quantity S, the percentage variation in strength per decade variation in strain rate (see

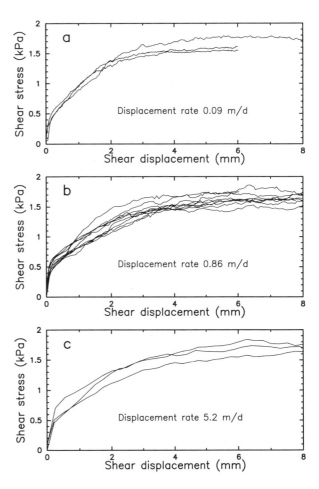

Figure 10. Direct shear tests of till from core 89-1-4, at three shear rates: (a) 0.09 m d^{-1}; (b) 0.86 m d^{-1}; (c) 5.2 m d^{-1}.

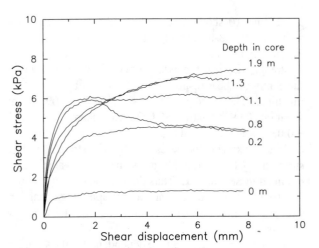

Figure 11. Direct shear tests of till samples from a succession of depths below the ice-till interface in core 89-1-4. Depths in meters as indicated. Shear displacement rate 1.25 m d.$^{-1}$

Section 6.6). In these terms, the dependence indicated by the above tests is $S = 3.4\%$ per decade. The S value is calculated from paired (shear strength, shear strain rate) data $(\tau_1, \dot{\gamma}_1)$ and $(\tau_2, \dot{\gamma}_2)$ by the relation

$$S = 100 \, \ln 10 \, \log(\tau_2/\tau_1)/\log(\dot{\gamma}_2/\dot{\gamma}_1)$$

which follows from equations (10) and (11), derived in Section 6.6.

In the 16 direct-shear tests plotted in Figure 10 there is little indication of any tendency for the strength to decrease at large strain to a "residual strength", as it does typically in tests on very clay-rich soils [*Skempton*, 1985]. A test of the till in a ring-shear device, which is capable of very large shear displacements, showed a 5% decrease in strength for a shear displacement of 200 m [*Tulaczyk et al.*, 1999a, Figure 3B]. Occasionally direct-shear tests showed a much larger decrease following a peak (e.g. one of the curves in Figure 11), but this appears to be an anomaly, caused perhaps by relatively large clasts getting caught between the edges of the shear gap and then working loose, or by overconsolidation of the till involved.

Figure 11 shows results of a series of direct-shear tests on till samples from a succession of locations down along the 89-1-4 core. The sample locations are identified in terms of their depth in meters below the top of the core. The sample at the top of the till (at depth 0 m) has a much lower strength than any of the others, which were from depths of 0.2 m or more below the top of the till. This suggests that if the till is undergoing shear deformation in situ, the shear is concentrated in or limited to a relatively thin (≤ 0.1 m) shear zone at the top of the till, which agrees

with the results of the basal sliding measurements on Ice Stream B (Section 5).

Figure 12 shows direct-shear test results for till specimens with a range of water-saturated porosities. They illustrate an important tenet of soil mechanics, that the strength of a water-saturated, clay-rich granular medium is a decreasing function of its porosity [*Lambe and Whitman*, 1969, p. 305]. In a general way this provides a relation between Figures 11 and 8—the increased strength of the deeper till samples being what is expected if the till porosity decreases with depth as in Figure 8b. This topic is considered in more detail in Section 6.5.

6.3 Triaxial tests

In relation to direct-shear tests, triaxial tests have the advantage that both the confining pressure and the pore water pressure can be controlled independently, in addition to the applied deviatoric stress (resulting in applied shear stress). Six sets of undrained triaxial tests, on six till samples from depth 1.5-2.5 m in core 92-1, were carried out by *Tulaczyk* [1999, Chapter 4] (see *Tulaczyk et al.* [2000a, p. 467]). Each test set began with a pressurization of the pore water to eliminate air bubbles, followed by preconsolidation at a chosen effective pressure. Deviatoric stress was then applied by axial compression, reaching specimen failure at axial strain ~2-4% and continuing to strains as large as 25%.

Results of three of the tests are plotted in the Mohr diagram in Figure 13. A Mohr circle is plotted for each test on the basis of the applied effective principal stresses

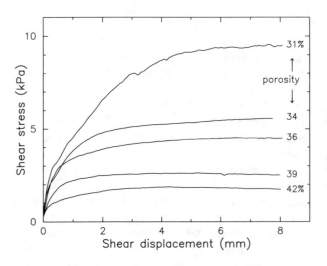

Figure 12. Direct shear tests of till with a succession of water-saturated porosities as indicated. The porosities were obtained by reconstituting dried till samples from core 89-1-4 with weighed amounts of water.

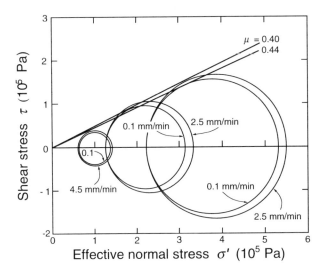

Figure 13. Mohr plot of results of three triaxial tests of till, each test at two strain rates. The strain rates are given in terms of shear displacement rates across the theoretical shear-failure plane for internal friction $\mu = 0.44$. Till is from near the top of core 92-1. Data from S. Tulaczyk (personal communication, 1999).

at failure (i.e. at full strength mobilization). The pattern of circles with common tangent lines is the pattern expected for a treiboplastic material. The indicated coefficient of internal friction is $\mu = 0.45$, the angle of internal friction $\phi = 24°$, and the cohesion ~1 kPa.

Figure 14 compiles all failure stress values (shear stress and corresponding effective normal stress at failure) for all of the six sets of tests. The shear stress and effective normal stress values are those acting across the theoretical shear plane for Coulomb failure, inclined at the angle $45°-\phi/2$ to the compression axis; they are calculated from the effective principal stresses at failure in each test. The results in Figure 14 are fit by the straight line with slope angle $\phi = 23.9°$, slope $\mu = \tan \phi = 0.443$, and intercept (cohesion) = 1.3 kPa. In relation to the range of shear stresses over which these parameters are experimentally determined, 20 kPa $\leq \tau_f \leq$ 150 kPa, the cohesion ~ 1 kPa is essentially nil.

The triaxial tests were carried out at many different strain rates over a wide range, so as to search for a strain-rate dependence of the yield strength of the till—that is, for viscous or quasiviscous behavior. An example is Figure 13, where the larger circle of each close pair is the result for a test at higher strain rate (higher by a factor of 25 to 75). From the strain-rate ratios (see figure caption) and the ratios of the circle diameters one finds $S = 4\%$, 5%, and 6% per decade for the three pairs of tests in Figure 13. Another example is in Figure 15, which shows the axial stress recorded for a specimen shortened alternatingly at four different axial strain rates. The variations in stress are $\approx 6\%$ for 5-fold variations in strain rate, which corresponds to $S \approx 4\%$ per decade. In Figure 16a all values of normalized shear strength, from runs generally like Figure 15, are plotted against the corresponding rates of shear strain across the theoretical shear plane, calculated from ϕ and the measured axial strain rate. The shear strength values are normalized with the preconsolidation effective pressure for each run, which suppresses the strong dependence of strength on effective pressure. The slope of the line in Figure 16a corresponds to a strain-rate dependence $S = 11\%$ per decade at $\dot{\gamma} = 2 \times 10^3$ a^{-1}.

6.4 In-situ Strength of Till

The primary interest is in the strength of the till in place, under the ice streams. As noted in Section 6, the $\phi = 0$ concept and the effects of negative pore pressure in principle permits the in-situ strength to be measured from till cores if the measurement is made soon enough after the core is taken. "Soon enough" is a somewhat vague requirement, although the reasoning in Section 6 and the data in Sections 6.1 and 6.2 suggest that the shear strength of till samples stored in sealed containers does not change much over time scales of a few days to 8 months. The triaxial tests do not directly indicate in-situ strength because of the preliminary steps of pore pressurization and preconsolidation (Section 6.3). They could provide an indirect indication from the Coulomb failure condition if the in-situ effective pressure were known, but this quantity is uncertain (see Section 7). To get around these

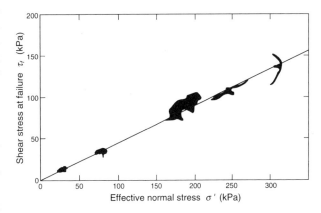

Figure 14. Triaxial test data from six sets of undrained tests on till from core 92-1 [*Tulaczyk et al.*, 2000a, Figure 5A]. Abscissa is effective normal stress across the calculated theoretical shear failure plane. The individual data points (numbering 3264) from the original work are not shown; instead, the distribution of data points is shown by dark shading.

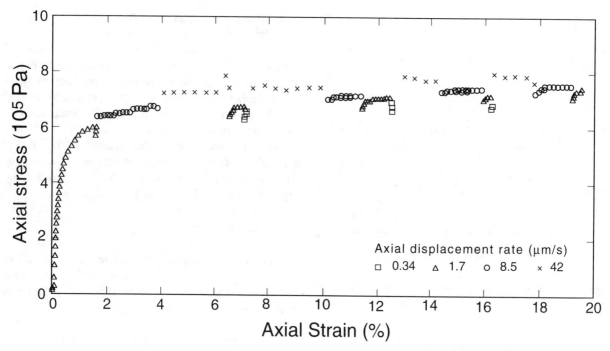

Figure 15. Triaxial test of till sample with alternation among four strain rates, as indicated at lower right. Axial strain rate is given in terms of the axial displacement rate in μms^{-1}. The test specimen is from core 92-1. Original specimen length was 6.5 cm. Data from S. Tulaczyk (personal communication, 1999).

difficulties we have developed and used a torque-vane instrument ("torvane") to measure the in-situ till strength directly. It is a version of an instrument used in soil mechanics [*Lambe and Whitman*, 1969, pp. 79, 451]. The torvane, which is pictured in Figure 5, measures the torque required to rotate a four-bladed (cross-shaped) vane that has been pushed down into the till at the bottom of a borehole. The rotation rate, and hence the strain rate of the till, can be varied over a 64-fold range.

Calibration of the instrument response (kPa of shear strength per volt of instrument output) was done by measuring a water-saturated test soil ("McMurdo silt") with the borehole torvane and with a commercially made and calibrated "hand vane tester" (EDECo "Pilcon" DR 3749). Calibration can also be done by measuring the torque/output-voltage response of the instrument and relating it to strength by a simple theoretical assumption as to the shear stress distribution around the rotating vane. This second calibration method gives a response factor (kPa V^{-1}) about twice as large as the first method, so the calibration has to be considered uncertain to this rather large extent. We report our results here in terms of the first method of calibration (which gives a response factor of 5 ± 1 kPa V^{-1}).

Results for borehole 91-1 at Up B are in Table 4. The indicated average shear strength of the basal till is about 2 kPa. Dependence of strength on shear strain rate is on average nil ($S = 0\%$ per decade) over 16- to 64-fold ranges in strain rate. These features confirm that the results of the direct-shear tests on till core material (Sections 6.1, 6.2) are applicable to in-situ conditions, and thus that reliance on the $\phi = 0$ concept is valid. The strength data in Table 4 and similar data from New B (borehole 95-1) and Up C (holes 96-4 and 96-6) are summarized in Table 5. They indicate that the in-situ till strength at New B is somewhat greater than at Up B, and is about 2.5 times greater at Up C. The strength of 1.0 ± 0.3 kPa for Up D in Table 5 is not from the torvane (which suffered an electrical malfunction) but from direct shear tests carried out on till from core 98-2 immediately upon core recovery (Section 6.1).

The till strength under Ice Stream C may be larger than the value measured with the torvane because of possible disturbance of the till by the hot-water jet once breakthrough has occurred (Section 7.1). The jet doubtless mixes some water into the upper part of the till, which weakens it (Section 6.5). The jet must also winnow the till and produce sorting and grading in a resedimentation process, much as it does in boreholes that bottom in frozen till (Section 4.5). These disturbances must affect the Ice-Stream-B till also, but the rapid basal sliding (Section 5) presumably displaces the disturbed area away from the bottom of the borehole by the time the torvane or piston

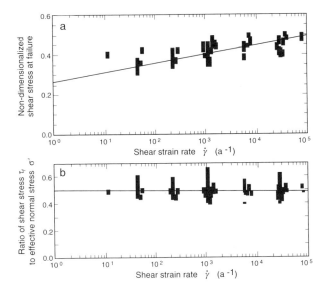

Figure 16. Triaxial test data showing dependence of till shear strength on strain rate. (a) Yield stresses from the data base for Figure 14 are here normalized by the preconsolidation pressure for each test series (such as Figure 15), which suppresses the strong dependence of strength on the general level of effective stress in the tests without bringing into consideration any variation of pore pressure or effective stress with strain rate. Shear strain-rates across the theoretical failure plane are as calculated from the measured axial strain rates. As in Figure 14, only the distribution of data points is shown, not the individual points. (b) Yield stresses from the data base for Figure 14 are here normalized by the effective normal stress in each test (each cluster of data points in Figure 15), which brings in the effect of the measured pore pressure on the effective stress in each test. Because the yield stresses normalized in this way show no dependence on strain rate, the dependence seen in (a) is demonstrated to be the result of a dependence of pore pressure on strain rate. Data from *Tulaczyk et al.* [2000a, Figures 5B and 5C].

corer reaches it. Such replacement of disturbed till by undisturbed till should not occur in Ice Stream C, because it presumably has little basal sliding, its streaming flow having nearly stopped. Most of the piston cores from Ice Stream C appear to be normal till, undisturbed by winnowing or resedimentation but core 96-6 gave an x-ray image that seems to show fining- upward grading in the uppermost 0.2 m of the core. Also, three of the Ice Stream C cores contained a sorted sand layer ~1 cm thick at the top of the core (S. Tulaczyk, personal communication, 1999). These facts suggest that some winnowing of till sediment by the hot-water drill is recorded in these cores.

There might be concern that the drilling of the borehole in which a torvane measurement is made has relieved the ice overburden stress and therefore the effective pressure, resulting in a falsely weak shear strength, so that the true in-situ strength is not measured. There are three reasons why this is not a concern: 1. When the ice is melted in drilling the borehole the ice overburden stress is replaced by the basal water pressure, which is nearly the same (Sections 3.2 and 7.4). 2. Any relief (or augmentation) of the confining pressure because of this replacement presumably acts over the cross-sectional area of the bottom of the borehole, some 10 cm in diameter, while the torvane is placed at a depth of ~50 cm in the till (Table 4), where the stress relief (or augmentation) is attenuated by redistribution of the unloading (or loading) through the surrounding till, with an attenuation factor probably of the order $(10/50)^2$. 3. Because of the low hydraulic conductivity of the till, the effective pressure will initially react to the change in loading as though undrained, and will then adjust by pore-fluid flow on a time scale $T^* \sim H^2/c_v$ (Section 6) where H is the depth of the torvane in the till, so $T^* \sim 1$ year; this is so much longer than the time-scale of the torvane measurement (a few hours) that the strength measured will differ inappreciably from the in-situ strength.

In the above torvane tests, adhesion of till to the torvane (Figure 6) indicated that the instrument penetrated the till to the depths indicated in the last column of Table 5. The recorded strengths were measured with the vane at these depths. (In the case of Ice Stream D the depth given is the depth in the core from which the tested sample was taken.) These data do not suggest a dependence of strength on depth in the till. In several tests in Ice Streams B and C an effort was made to measure with the torvane a profile of strength up through the till. An example of such a profile from Ice Stream C is given in Figure 17. These results, though somewhat problematical, seem to indicate that the

Table 4. In-situ Strength of Till Measured with Torvane in Hole 91-1 as a Function of Shear Strain rate $\dot{\gamma}$ (represented by vane rotation rate)

Test No.	Till strength		Vane rotation rate	
	at low $\dot{\gamma}$ (kPa)	at high $\dot{\gamma}$ (kPa)	at low $\dot{\gamma}$ (r.p.hr.)	at high $\dot{\gamma}$ (r.p.hr.)
1	2.0 ± 1.5	2.1 ± .5	1	16
2	1.5 ± .1	1.4 ± .1	1	16
3	2.3 ± .05	2.1 ± .05	0.5	32
	-	1.9 ± .05	-	32
4	2.1 ± 1.0	2.3 ± .2	0.5	32
	-	2.1 ± .1	-	32
(av)	2.0 ± 0.3	2.0 ± 0.3	~1	~30

Table 5. In-situ Till Strength: Summary

Study site	Borehole	No. of tests	Strength (kPa)	Depth in till (cm)
Up B	91-1	4	2.0 ± .3	59, 66
New B	95-1	7	2.7 ± .5	35
Up C	96-4	7	3.9 ± 1.0	15 - 20
"	96-6	2	5.5 ± .3	70
"	"	20	4.5 ± 1.0	70
Up D	98-2	10	1.0 ± .3*	42

* Direct shear tests on core sample just after core recovery

strength drops by ~50% in the uppermost 0.5 m of the till, suggesting a shear zone at the top of the till.

6.5 Dependence of Strength on Water Content of Till

Triaxial test results shown in Figure 18 verify for the till the empirical relationships found in soil mechanics between pore volume and effective pressure under loading that compresses or consolidates (compacts) the soil, or under unloading, which allows the soil to expand. The empirical relations, shown by the lines in the figure, are described by *Clarke* [1987, p. 9026]. They are linear relations between the logarithm of the effective pressure σ' (total confining pressure minus pore water pressure) and the void ratio $e = V_w/V_s$ where V_w is the water-saturated pore volume in a given mass of till and V_s is the total solid-particle volume in the same mass. The "Normal Consolidation Line" (NCL) is the e-vs.-σ' path followed under increasing effective pressure applied to a "virgin" till sample that has not been previously consolidated under an effective pressure as high as that being currently applied, or whose "virginity" has been restored by reworking ("remolding") under low effective stress. The empirical NCL relationship is

$$e = e_c - C_c \log(\sigma'/\sigma_c) \qquad (1)$$

where σ_c is a reference effective pressure, e_c a reference void ratio, and C_c a constant called the compression index. In Figure 18 the heavy line labeled "NCL(ISO)" represents (1) for isotropic consolidation, which can be effected in the triaxial apparatus, while the dashed line labeled "NCL (UNI)" is for consolidation that is confined laterally so that the lateral strains are zero (uniaxial compression). The latter is the consolidation geometry provided by standard "oedometer" consolidation-testing machines. The fine broken line labeled "URL (5.7)" ("unloading and reloading line") is an example of the e vs. σ' path followed upon unloading from a point on the NCL (in this case the point at $\sigma' = 5.7$ bar) and also upon reloading up to that point.

In Figure 18 the fine continuous line labeled CSL, the "Critical State Line", is the locus of (σ', e) states that are attained under shear deformation superimposed upon consolidation, which is the condition of most interest here. The CSL and URL are empirical relations analogous to (1); for example, for the CSL,

$$e = e_f - C_f \log(\sigma'/\sigma_f) \qquad (2)$$

The subscript "f" here refers to the condition of shear failure. The NCL(ISO) and CSL for the till were determined by *Tulaczyk* [1999, Chapter 4] by least-squares fit to triaxial test data (e.g. the data for the NCL (ISO) in Figure 18), while the NCL(UNI) and URL are from oedometer data obtained by Hermann Engelhardt. Numerical formulation of these lines is given in Figure 18. Note that $C_f = C_c$, as is generally expected.

Equation (2) provides the basis for a quantitative relation between till strength and water content (expressed as void ratio e). If the strength τ_f is assumed to obey the Coulomb failure condition with negligible cohesion:

$$\tau_f = \mu \sigma' \qquad (3)$$

then combining (2) and (3) and rearranging gives

$$\tau_f = \mu \sigma_f \exp[(e_f - e)(\ln 10)/C_f] \qquad (4)$$

This is an inverse relation between shear strength τ_f and void ratio e or porosity $\pi = e/(1+e)$. It plays an important role in Tulaczyk's new theory of ice stream mechanics (Section 9.2).

The inverse relation (4), in qualitative terms, was noted in Sections 6.2 and 6.4 and an example was shown in Figure 12. When this example is evaluated according to (4), the indicated value of the constant C_f is 0.38, quite different from the $C_f = 0.12$ found by the triaxial tests (see formula for CSL in Figure 18). The reason for the discrepancy is not known but may have to do with the fact that in the tests in Figure 12 the pore water pressure could

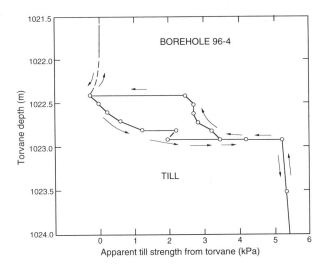

Figure 17. Vertical profile of till strength measured with the torvane in borehole 96-4. The top of the till is at nominal depth 1022.4 m. Arrows indicate the measurement sequence.

not be directly controlled. If we calculate from (4) the till strength that corresponds to the observed till porosities, we obtain a rather large scatter of values. They are shown in Table 6, for porosity ranges and averages found in Ice Streams B, C, and D (Table 3), and for the two versions of equation (4) based on Figures 18 and 12 as discussed above. The latter version gives strength values that seem more reasonable in relation to the till strengths determined by other means (Sections 6.1, 6.2, 6.3, and 6.4).

The inference of basal-till shear strength from till porosity is discussed more fully in a paper by *Tulaczyk et al.* [unpublished]. It concludes that basal effective stresses σ' as low as 2 kPa can be inferred from the data. This is much smaller than the average values 20-60 kPa that would seem to be indicated by the directly measured basal pressures (Section 3.2). From the low inferred value $\sigma' \sim$ 2 kPa, a basal shear strength as low as about 1 kPa is indicated, in general agreement with direct shear measurements on till specimens from near the base of the ice (Sections 6.1 and 6.2).

6.6 Rheological assessment

The results of the mechanical tests on the till in Sections 6.1-6.5 conform fairly well to the attributes of treiboplasticity: a failure strength that is a linearly increasing function of the effective normal stress, that is independent of the strain rate, and that depends on the water-saturated porosity according to a relation that is linear in a semi-log plot. Treiboplasticity therefore seems to provide a good first approximation to the till's mechanical properties. Its internal friction is 0.44 and its apparent cohesion is ~ 1 kPa, values that are typical of soils with ~35% clay content [*Kedzi,* 1974, Table 28; *Terzaghi et al.,* 1996, Fig 19.7; *Lambe and Whitman,* 1969, pp. 307, 313].

In a second approximation the till strength depends slightly on the strain rate. This has significant consequences in the formulation of a rheology for the till. If the till behaves as a perfect treiboplastic material, with strength fully independent of strain rate, its deformation represents a rheologically singular situation in which, as in the perfect plasticity of metals [*Hill,* 1950] and perfectly plastic ice [*Nye,* 1951], the flow is not fully controlled by the stresses and a flow law of conventional type (strain rate as a completely determined function of stress) cannot be written. If however there is an appreciable strain-rate dependence of the strength, the treiboplastic failure condition can be reformulated as a flow law, and the flow mechanics assumes a more familiar form, one that may be more advantageous for numerical modeling than is the perfect-treiboplasticity formulation. It is therefore of practical as well as fundamental interest to consider how much the till rheology deviates from perfect treiboplasticity and to compare it in these terms with the rheology that has been used in ice-stream modeling.

This can be done with a measure of the strain-rate dependence of the strength. As noted in Section 6.2, it is standard practice to take as a measure of this dependence the percentage variation in strength per decade of variation in strain rate, a quantity here designated S. The values of

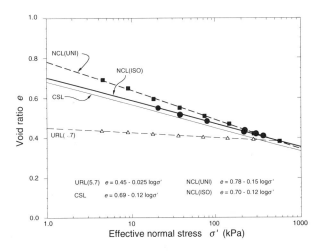

Figure 18. Data from consolidation tests on till. The data for the isotropic NCL and the CSL are from triaxial tests by S. Tulaczyk, and the data for the uniaxial NCL and the URL (for preconsolidation pressure 5.7 bar) are from oedometer tests by H. Engelhardt. See *Tulaczyk et al.,* [2000a, Figure 6A].

Table 6. Calculated Till Shear Strength (kPa) from Equation (4) for Observed Ranges of Porosity (low, average, high) and two Values of the Compression Index C_f

Ice Stream	Porosity π (%)	Calc. 1 τ_f (kPa)	Calc. 2 τ_f (kPa)
B	36	5.1	7.9
"	40	0.7	2.3
"	46	0.02	1.0
C	33	19.5	6.8
"	42	0.2	1.6
"	51	5×10^{-4}	0.2
D	44	0.07	1.0
"	48	0.01	0.5
"	53	10^{-4}	0.1

Calculation 1: $C_f = 0.12$, from Figure 18
Calculation 2: $C_f = 0.38$, from Figure 12

S reported in Sections 6.2, 6.3, and 6.4 from the various sets of tests are 3.4, 4, 5, 6, 4, 11, and 0% per decade (on average $S = 5\% \pm 3\%$ per decade). S values of this order are commonly encountered for soils and fault gouges as reported in the literature [*Berre and Bjerrum*, 1973, p. 6; *Bishop et al.*, 1971, p. 302; *Blanpied et al.*, 1987, Figure 3; *Marone et al.*, 1990, Table 2; *Sheahan et al.*, 1996, Table 1; *Skempton*, 1985, p. 14; *Kamb* 1991, p. 16,586; *Scholtz*, 1998, Figure 1A].

As is shown below, an S-type dependence of strength τ_f on strain rate $\dot{\gamma}$, in which S is constant, independent of τ_f or $\dot{\gamma}$, can be expressed as a nonlinear flow law of the standard form $\dot{\gamma}/\dot{\gamma}_o = (\tau_f/\tau_o)^n$, where $\dot{\gamma}_o$, τ_o, and n are constants, and the exponent n is related to S by $n = (100 \ln 10)/S$. From this relation, the n values that correspond to the S values given above range from 21 to ∞ and cluster around 40. This represents a nonlinearity that is extremely large in relation to typical nonlinearly quasiviscous materials, which generally have $n \leq 5$ [*Garofalo*, 1965, p. 50]. The nonlinearity is so large that the till can be considered to behave rheologically much more nearly as a perfect treiboplastic material (for which $n \to \infty$) than as a normal nonlinear fluid. However, in view of the strain-rate dependence ($S \neq 0$), which is not an attribute of perfect treiboplasticity, we should use some qualified designation like "imperfect treiboplasticity" in referring to the till rheology.

The conclusion that the till rheology is essentially (though imperfectly) treiboplastic is consistent with in-situ till strength measurements in a Swedish glacier [*Hooke et al.*, 1997] and laboratory ring-shear tests on two northern hemisphere tills [*Iverson et al.*, 1998]. The latter tills differ from ice-stream till in having a strength that decreases slightly with strain rate ($S \approx -3.5\%$ per decade, $n \approx -60$). Negative S is sometimes encountered in soil mechanics [*Lambe and Whitman*, 1969, p. 314], and also in fault-gouge mechanics [*Blanpied et al.*, 1987, Figure 3; *Scholtz*, 1998, Figure 1A] where it is called velocity weakening.

In great contrast with $n \sim 40$ (or $n \approx -60$) is the near linearity ($n = 1.33$) of the quasiviscous till flow law put forward by *Boulton and Hindmarsh* [1987, equation 2, p. 9063] on the basis of measurements in a basal tunnel at the terminus of an Icelandic glacier. Reservations about the data interpretation that leads to the value $n = 1.33$ have been expressed by *Iverson et al.* [1997, p. 1058], *Hooke et al.*[1997, p. 173)] and *Iverson et al.* [1998, p. 635].

In formulating a second-approximation till rheology it is important to try to ascertain the cause and significance of the small but non-zero S values. This is a somewhat obscure subject, little discussed in soil-mechanics texts. Some clarity is provided by separately considering tests made under drained conditions, such as the ring-shear tests of *Iverson et al.* [1998], and those made under undrained conditions, such as the triaxial tests in Section 6.3. For undrained tests *Lambe and Whitman* [1969, p. 445] make the following observation: "In all cases where it has been possible to measure the pore pressures during undrained tests at various rates of loading, it has been found that the change in undrained strength results from a difference in induced pore pressure [*Richardson and Whitman*, 1964]. Increasing the rate of strain [causes reductions in] induced pore pressures." A reduction in pore pressure results in increased effective pressure and therefore increased strength. The above observation is confirmed by the triaxial test data for the till: In Figure 16b the ratio of shear strength to effective normal stress for each data point in Figure 14 (see figure caption) is plotted as a function of the corresponding applied strain rate. The result, which includes the effect of the measured pore pressures via their contribution to the effective normal stress value for each data point, shows that the ratio, and hence the strength at a given effective normal stress, does not depend on strain rate. Thus the increase of strength with strain rate shown in Figure 16a is explained as a mechanical effect of reduction in pore pressure induced by increase in strain rate. There is no need for viscous flow elements in the system, although such elements are occasionally alluded to in soil mechanics [e.g. *Lambe and Whitman*, 1969, p. 314; *Mitchell*, 1993, pp. 318, 340].

To use the above conclusion in formulating till rheology, an explanation is needed for the strain-rate induced reduction in pore pressure. It may be attributed to an effect of strain rate on the critical-state porosity (Section 6.5): the CSL in Figure 18 is shifted upward in response to increased strain rate. The consequent tendency for the till to dilate when its strain rate increases has to be counteracted by a reduction in pore pressure so as to maintain a constant pore volume as required by the undrained character of the tests.

Remaining unexplained is the cause of the assumed effect of strain rate on the CSL. Also unexplained is why the S value is so variable from one test to another, and why the internal friction indicated by the ordinate of the horizontal line in Figure 16b is 0.50 rather than 0.44 or 0.45 as determined in Section 6.3.

Iverson et al. [1998, p. 638], following *Tika et al.* [1996], offer the following explanation for the negative S value found in their tests. Because ring-shear tests are inherently drained (if conducted slowly enough for equilibration of pore pressure), the pore pressure is fixed and does not play a role in the response to strain rate. Other factors can therefore make their effect felt. In particular, if increased strain rate raises the CSL, as suggested in the previous paragraph, the deforming till will dilate to follow it. This will result in a less dense framework structure with lessened particle interlocking and with consequent decrease in the angle of internal friction ϕ and therefore in the strength at given effective pressure. Again, a mechanism for coupling the strain rate to the CSL is needed.

It would seem that Iverson et al's. explanation should apply to the direct shear tests in Section 6.2, whereas the tests actually showed variably positive S. Because of the very low hydraulic permeability of the till, it is likely that the pore pressure did not have time to equilibrate fully during the tests, so that the tests had some undrained character and some of the pore-pressure effect resulting in positive S was felt.

"Probably the exact nature of the strain rate effect varies from soil to soil" [*Lambe and Whitman*, 1969, p. 314].

Based on the above discussion of strain-rate dependence of strength there are two ways to proceed in trying to formulate a second-approximation rheology for the till: I. Rheology based on mechanism(s) of strain-rate dependence. II. Rheology on a purely empirical basis. In method I, known or assumed dependences on $\dot\gamma$, such as the CSL and ϕ dependences discussed above, are introduced into the standard Coulomb failure condition (3) (with cohesion = 0). This is illustrated as follows for the undrained situation described above. Represent the effect of $\dot\gamma$ on the CSL by its linearized effect on the ordinate intercept in (2):

$$e = (e_f + g\dot\gamma) - C_f \log(\sigma'/\sigma_f) \qquad (5)$$

in which the coefficient g (unrelated to gravity) is

$$g = \langle de_f/d\dot\gamma \rangle \qquad (6)$$

Here < > represents an averaging from $\dot\gamma = 0$ to the current value of $\dot\gamma$. For undrained conditions, hold e constant at the value e_o (value of e for $\dot\gamma = 0$) and solve (5) with (3) to get τ_f in the same way as (2) was solved with (3) to get (4):

$$\tau_f = \mu\sigma_f \exp[(\ln 10)(e_f - e_o + g\dot\gamma)/C_f] \qquad (7)$$

Inverting for $\dot\gamma$ and introducing (2) with $e = e_o$ and $\sigma' = \sigma_o'$ (the value of σ' applied at $\dot\gamma = 0$) gives

$$\dot\gamma = \frac{C_f}{g} \log\left(\frac{\tau_f}{\mu\sigma_o'}\right) \qquad (8)$$

For g positive, τ_f is seen from (7) or (8) to be an increasing function of $\dot\gamma$. Equation (8) has the form of a rheological flow law for the undrained till. Likewise, a flow law for the till under drained conditions could be obtained by similarly introducing also a dependence of μ on $\dot\gamma$ via e.

In method II one couples with the Coulomb law (3) the best established empirical relation for $\dot\gamma$ dependence. The obvious choice is the S-type dependence of τ_f on $\dot\gamma$, which is found to fit the test data generally. The S dependence is best expressed in differential form:

$$d \ln \tau_f / d \log \dot\gamma = S/100 \qquad (9)$$

where log means $\log_{10}$. From (9) it follows by integration at constant S that

$$\tau_f / \tau_o = (\dot\gamma/\dot\gamma_o)^s \qquad (10)$$

where

$$s = S(100 \ln 10)^{-1} \qquad (11)$$

Here $\tau_o/\dot\gamma_o^s$ represents the integration constant. The S-type law (10) is coupled with the Coulomb failure law (3) by recognizing that τ_o in (10) represents the values of τ_f given by (3) as a function of σ' when $\dot\gamma = \dot\gamma_o$:

$$\tau_o = \mu\sigma' \qquad (12)$$

The result (10) is the inverse of the standard nonlinear quasiviscous flow law

$$\dot\gamma/\dot\gamma_o = (\tau_f/\tau_o)^n \qquad (13)$$

where

$$n = s^{-1} = (100 \ln 10) \, S^{-1} \qquad (14)$$

Hence in the S-based empirical approach to till rheology a power-type flow law (13) emerges, with a very large value of n, corresponding to small s or S in (14) ($S \sim 5 \pm 3\%$ per decade, and $n \sim 40 \pm 20$).

Comparing method II's (13) with method I's (8), the latter's logarithmic form represents a weak dependence of $\dot\gamma$ on τ_f, contrary to the expected strong $\dot\gamma$-on-τ_f dependence that is seen in (13) with large n and that corresponds to the expected weak dependence of τ_f on $\dot\gamma$ represented by (10) with small s. (The foregoing statement re (8) holds in a general way even though the "local n value" $n = d \ln\dot\gamma / d \ln \tau_f = (\ln(\tau_f/\mu\sigma_o'))^{-1}$ becomes very large as $\tau_f \to \mu\sigma_o'$.) This suggests that the method-I approach used in deriving (8) is somehow flawed. At the present stage of understanding of these matters it therefore seems best in formulating the till rheology to pass over the internal till mechanics that generates the strain-rate dependence of the strength and proceed directly to its empirical representation in terms of S, or some other parameterization if anything better can be found.

Note that (13) can be valid as a steady-state flow law only above a certain lower-limit shear stress and corresponding strain rate, below which the flow becomes transient-decreasing in accordance with the Singh-Mitchell creep equation [*Singh and Mitchell*, 1968; *Kamb*, 1991, p. 16,586]. However, this can probably be overlooked in ice-stream modeling.

The full form of the S-based flow law is obtained by combining (12) and (13):

$$\dot\gamma = \dot\gamma_o \left(\frac{\tau_f}{\mu\sigma'}\right)^n \qquad (15)$$

with $n \sim 40 \pm 20$). It differs greatly from normal quasiviscous flow laws (e.g. for ice) not only in the very large value of n but also in the occurrence of $(\mu\sigma')^n$ in the denominator. The latter feature or something like it (usually $(\sigma')^m$ in the denominator with $m \neq n$) occurs in many proposed basal sliding laws [e.g. *Bentley*, 1987, p. 8855]. It also occurs in the till flow laws of *Boulton* and *Hindmarsh* [1987, p. 9063] but with values of m that are very low (1.25, 1.80) though not greatly different from the assigned n values (1.33, 0.625)..

The "Bingham fluid model" of *Boulton and Hindmarsh* [1987, equation 1, p. 9063] has a flow law further modified from (15) by replacing τ_f^n in (15) with $(\tau_f - \tau_o)^n$ for $\tau_f \geq \tau_o$, with $n = 0.625$ (sublinear!), and with $\dot\gamma = 0$ for $\tau_f < \tau_o$. It has been shown that this flow law is not compatible with the triaxial test data [*Tulaczyk, et al.*, 2000a, Figure 5A; *Tulaczyk*, 1999, p. 4-12]. However, a law of this type, but with much larger values of n and m, would seem to be a possible candidate alternative to (15) for representing the till's imperfect treiboplasticity. With $m = n$ it would be essentially the same as the viscoplastic flow law for landslides as formulated by *Iverson* [1985, p. 152], except that in the latter the effect of pore pressure is not included.

A variant on formulation (10) is to express the strain-rate dependence of the strength in the form

$$\tau_f/\tau_o = 1 + s \ln (\dot\gamma/\dot\gamma_o) \qquad (16)$$

[e.g. *Iverson et al.*, 1998, equation 1; *Kamb*, 1991, equation 5; *Scholtz*, 1998, equation 3]. A line of this form is fitted to the strength data in Figure 16a. For small values of $s \ln (\dot\gamma/\dot\gamma_o)$, (16) and (10) are essentially the same, but for a wide range of strain rates (16) becomes only an approximation to (10). The choice between the two is purely empirical, except that the inverse of (16) expresses $\dot\gamma$ as an exponential function of τ_f, which has been given some justification from rate process theory (Mitchell, 1993, p. 352).

7. BASAL WATER SYSTEM

It seems reasonable to assume that water generated under ice streams by basal melting may accumulate in the basal zone or may be transported downstream in this zone, to be released into the sea at or near the (un)grounding line. The system of conduits in which such water storage and transport occurs is called the basal water system or basal hydraulic system. It may bear some resemblance to the basal water systems of temperate glaciers, and has been treated theoretically on this basis [*Weertman and Birchfield*, 1982; *Bindschadler*, 1983, p. 10] or modifications thereof [*Alley*, 1989a]. Here the effort is to bring together the lines of observational evidence bearing on the basal water system of ice streams. Relevant hydraulic observations from Ice Stream B have been discussed extensively by *Engelhardt and Kamb* [1997] (here called "E & K"), and we can limit the treatment here to an update of that discussion. Subsequent to the observations discussed in E & K, additional observations have come from 13 boreholes drilled to the bottom of Ice Stream C and 5 holes drilled to the bottom of Ice Stream D.

7.1 System Existence and Gap Models

In the drilling of a borehole in an ice stream the existence of a basal water system is made seemingly obvious by the rapid disappearance of large quantities of borehole water down the hole starting at the moment when the drill breaks through the basal ice into the till. This has been experienced in all boreholes that reached the base in Ice Streams B, C, and D, with possible exception of holes 95-3 and 96-3 (see below) and also hole 88-2 (Section 7.7). The downrush of water is seen in terms of the rapid drop in borehole water level, typified by the quasi-exponential water-level drop curves (E & K, Figure 3). The water-level drop time ("WLDT", the time for ca. 90% of the total drop) and T, the time constant for a quasi-exponential drop curve, are generally in the range 1-3 minutes in Ice Stream B (see Table 1, last two columns). If the basal water system consists of a partially continuous gap between the bottom of the ice and the top of the till ("gap-conduit model"), then a treatment of the breakthrough drop curve by the method of *Weertman* [1970] indicates that the observed values of T correspond to a basal gap 1.4 to 2.5 mm thick (E & K, Section 9a).

The breakthrough behavior of boreholes in Ice Streams C and D is in a general way similar to that in Ice Stream B, suggesting the presence of a similar basal water system, but with a few differences as discussed below.

The first problem in interpreting the basal water system with the gap-conduit model is that the existence of the gap in the natural state, unmodified by borehole drilling, has been in question for various reasons (E & K, Sections 6, 9b, and 9d; *Tulaczyk*, 1999, Section 5.5.2). One reason is the lack of evidence that more water has ever been produced (pumped out) from the basal system than has been injected into it from the boreholes on breakthrough (E & K, Section 6). Another reason is the low measured propagation speed of the pressure pulse introduced into the till-ice interface when the drill breaks through (E & K, Section 8). This observation is best explained if in the natural state (unmodified by hot-water drilling) there is no continuous macroscopic (≥ 1 mm) water-filled gap between ice and till, and if in the breakthrough process a gap ≥ 1 mm thick is formed by injection of pressurized water along the till-ice interface. An analysis of such a "gap-opening model" seems to indicate that it is able to account for the slow speed of pressure-pulse propagation (E & K, Section 9c). If this is so, then the gap-conduit model of the basal water system, with a ~2 mm natural till-ice gap that pre-exists any gap produced by borehole drilling, is ruled out.

Two new observations change this picture. First, two pressure pulse propagation experiments carried out on Ice Stream C with boreholes 96-4, 96-8, and 96-13 gave a propagation speed much faster than was found in Ice Stream B. The data (in a format like E & K, Figure 19) show a roughly instantaneous propagation, or when scrutinized more closely a propagation speed of about 2.5 m s^{-1}, about 14 times faster that the 0.18 m s^{-1} observed in Ice Stream B (E & K, Section 8). The time resolution of 5 seconds for the pressure data in these figures admits the possibility that the propagation time is essentially 0. To the extent that the pulse propagation speed approaches the speed of sound in water (1400 m s^{-1}), the data permit a pre-existing basal water system with conduits ≥ 1 mm in width or diameter, although the matter is somewhat complicated (E & K, Section 9b). In any case, the large difference between the propagation speeds at the study sites on Ice Streams B and C indicates either a major difference in the nature of a basal water system at the two sites or else the operation of unknown complicating factors at one or the other of the sites. A great peculiarity in the data from Ice Stream C is that in experiment 2, the pulse that arrived at the expected time at borehole 96-4, 50 meters from the source of the pressure pulse (hole 96-13), was a pulse of decreased rather than increased pressure (i.e., a rarefaction).

The second and the most compelling new observation was made in Ice Stream D and is shown in Figure 19. In drilling the first borehole of the season (98-1), before any disturbance of the basal system by connection to a borehole had occurred, the water level in the borehole was pumped down and became constant (with pump running) at a depth of 120 m below the surface. When breakthrough occurred, the borehole water level did not fall as usual, but instead *rose*, to a depth of 107 m. The rise time was 3 minutes. After a few minutes' wait, the pump-out pump was turned off and water was pumped into the hole. The water level

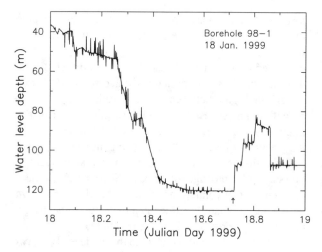

Figure 19. Rise in borehole water level on breakthrough to basal water system in borehole 98-1, Ice Stream D. The rise occurs at J.D. 18.72, marked with an arrow. The gradual drop in water level from depth 37 m at J.D. 18.0 to 120 m at J.D. 18.54 was caused by purposely pumping water out of the hole while drilling proceeded. See text for details.

proceeded to rise, and rose to 97 m depth before the pumping-in was stopped. This rise in water level showed that hydraulic connection with the basal system had been lost, probably by basal till being carried up into the borehole by the inrush of water from below. After a brief wait more water was pumped in, raising the water level again by about 10 m further. Now, however, leakage from the borehole into the basal system was detected in a slow decline of the water level. Then, a few minutes later, the hole broke through abruptly to the basal system and the water level dropped rapidly (in 2 minutes) back to 107 m, the same level reached initially by water upflow from below. The water levels in three subsequent boreholes drilled later nearby (98-2, 98-3, 98-6) dropped to the same 107 m water level after breakthrough.

The experiment in hole 98-1 shows that there exists, prior to any modification by hot-water drilling, a basal water system capable of delivering a substantial quantity of water before any water has been introduced into it from a borehole. The system has, at least locally (over a horizontal distances of at least 21 m), a well defined water pressure, which is maintained on water input or withdrawal. The rise time of 3 minutes corresponds in the gap-conduit model to a gap of 1.8 mm. Although strictly these conclusions apply only to the study site on Ice Stream D, it seems likely that they are valid broadly for the ice streams, which, as noted above, show similar breakthrough-with-water-level-drop behavior in all boreholes studied thus far.

Aside from the anomalous hole 88-2 (E & K, p. 210), the only exceptions to this are two boreholes in which unsuccessful attempts were made to do the experiment that was later successfully carried out in hole 98-1 as described above. The two holes, 95-3 and 96-3, showed no indication or almost no indication of breakthrough. In Section 4.6 this is interpreted for hole 95-3 as an indication that the borehole bottomed in frozen till. An alternative interpretation, applicable also to borehole 96-3, is that when the drill reached the bottom, the borehole water level, which had been drawn down as far as possible in advance, was the same or nearly the same as the water level of the basal system, and for this reason hydraulic connection did not occur. It is possible that a certain overpressure at the base of the borehole water column is necessary to make connection with the basal water system, perhaps by breaking through a hydraulic barrier. Alternatively, because of the draw-down it is possible that a hydraulic connection could have been made between the borehole and the basal system without being detected from a drop or rise in water level. In hole 95-3 there was no change in water level (at 107 m), and in hole 96-3 there was a slow drop of 3 meters (in 36 minutes, to water level 114 m), which may indicate a weak breakthrough to a basal system with water level 114 m. What happened after the possible initial connection is noteworthy. Connection (if any) was lost (in 2 to 20 hours), and by pumping water into the borehole the water level was raised to 57 m in hole 95-3 and to 94 m in hole 96-3. At these levels a substantial overpressure was being applied to the basal zone: expressed as a difference in hydraulic head, the overpressure in hole 95-3 was 107-57 = 50 meters and in hole 96-3 it was 114-94 = 17 meters. The experience assembled in Table 1 shows that breakthrough has occurred with an overpressure as low as 14 meters (borehole 95-2). (Overpressure is the difference between the "pre-breakthrough" and "post-breakthrough" entries in Table 1.) Hence breakthrough is expected, but it did not occur. This could be explained by a frozen bed in both boreholes. But it seems quite unlikely that, out of the many boreholes that have been drilled in the ice streams, a frozen bed would happen to be encountered by just the two boreholes in which during drilling the borehole water level was drawn down to the water level of the basal water system. Instead, the apparent lack of initial breakthrough under these conditions, and/or the lack of later breakthrough when the borehole water level was raised above an overpressure of 14 m, should probably be recognized as a special and peculiar property of the basal water system, perhaps related to the temporary loss of connection that closely followed the initial breakthrough in hole 98-1 as described earlier.

The above conclusion must be qualified by recognition that the two piston cores from borehole 95-3, in which the sediment was clearly sorted and graded, strongly favor the frozen-bed interpretation, because such sediment has otherwise been recovered only from boreholes that bottomed in frozen till (Section 4.5). (See also the discussion in the second-to-last paragraph of Section 6.4.) A small piston core was obtained from borehole 96-3, but because of wash-out by water escaping from the core barrel the sediment was not suitable for comparison with hole 95-3.

The two alternative interpretations of boreholes 95-3 and 96-3, discussed here and in Section 4.5, illustrate the contradictory character of some of the evidence on the basal zone from borehole observations.

7.2 Water-level Oscillations

The water-level drop curves on breakthrough for holes 98-2 and 98-3 show a new feature, a damped oscillation at the end of the drop (Figure 20). The oscillation in hole 98-2 had a period of 1.3 minutes, an initial amplitude of about 4 m (peak to peak), and a decay to the noise level of ~2 m in about 3 minutes. In hole 98-3 the period was 1.6 minutes, initial amplitude 5 m, and decay time about 6 minutes. It seems that what is happening in the oscillation is a seiche-like movement of water back and forth between a pair of closely spaced boreholes open to the atmosphere and connected via a basal water conduit (borehole spacing 2.5 m for pair 98-2 plus 98-1, and the same for pair 98-4 plus 98-3). A dynamical model of such a system [*Kamb*, unpublished] accounts for an oscillation period of about 1.5 minutes if the basal conduit is a till-ice gap of thickness about 2.5 mm.

The occurrence of well-defined waterlevel oscillations in Ice Stream D but not in any of the boreholes in Ice Streams B or C may represent a difference in the nature of the basal water system in these ice streams, or it may alternatively reflect a rather narrow limitation on the range of conditions under which the oscillation can occur, which is what is indicated by the model.

7.3 Water-level Drop Times

Of the eight WLDT's measured in Ice Stream B (Table 1), seven were in the range 2-3 minutes while the remaining one was about 5 minutes (this omits hole 95-4, whose WLDT was only crudely estimated). (Also not included in these WLDT statistics are holes 88-2 and 95-3, which had very anomalous behavior—see Section 7.1 and E & K, p. 210.) For Ice Stream C the corresponding WLDT's are, out of twelve measured, five in the range 2-3 minutes, while the remaining seven are 5 minutes to 30 minutes or even longer. Hole 96-4 is included in the latter group because, although its water level started dropping with a time constant of 2.5 minutes, when it reached a depth of 75 m the water system changed in a way that introduced complications into the drop curve (Figure 21a) and lengthened the WLDT to about 30 minutes.

Water-level drop behavior somewhat similar to 96-4, but even more striking, was shown by borehole 96-12, which was drilled in the middle of the fossil shear margin between Ice Stream C and Ridge B-C. The initial part of the drop curve (Figure 21b) has a quasi-exponential shape with time constant of about 1 *hour*, which is highly abnormal. Below a depth of 90 m the drop was further delayed, with complications somewhat resembling those of borehole 96-4, lengthening the overall WLDT to about 5 hours—extremely abnormal. In terms of this very long WLDT, the basal water system under the shear margin can be recognized as transitional in character between the active ice streams (WLDT 1-3 minutes) and the non-streaming ice sheet, which has no basal water system and in effect an infinite WLDT. The basal water system under the central part of Ice Stream C shows a suggestion of this transitional character in terms of the WLDT statistics discussed above and the anomalous water-level drop curve of hole 96-4.

Further discussion of the significance of the WLDT is in Section 8.2.

7.4 Basal Water Pressures and Effective Pressures

Now that the existence of a basal water system with a locally well defined pressure has been established (Section 7.1), it is necessary to reconsider in this new light the problems posed for concepts of the basal water system by several aspects of the borehole observations of basal water pressures and the effective pressures derived from them (Section 3.2):

(1) The large scatter (± 6 m) in individual water-level depths, corresponding to the scatter of $\pm 0.7 \times 10^5$ Pa in effective pressure values (Section 3.2), implies the presence of large, spatially variable water-pressure gradients that would produce impossibly large, spatially variable local water fluxes in a gap-conduit water system of the type contemplated (E & K, Section 9d).

(2) Negative effective pressures, of which nine are indicated, ranging from -0.1 to -1.0×10^5 Pa (Table 1), are physically impossible in a steady-state gap-conduit system. Negative effective pressure is involved in the gap-opening model considered by E & K (Section 9a) but this represents

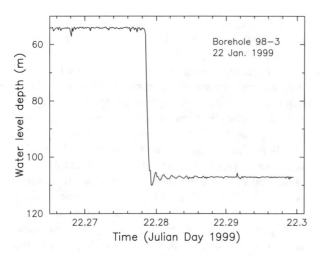

Figure 20. Water-level drop in borehole 98-3, Ice Stream D, showing damped oscillation of the water level about the 107 m depth.

a situation that could not arise in the natural system and if created artificially is a transient that would not long persist. To some extent the negative values may be due to systematic error in the calculation of the ice overburden pressure (E & K, Section 4).

(3) Positive effective pressures range up to $+1.7 \times 10^5$ Pa, which, with the till's internal friction of 0.44 (Section 6.3), would result in till strengths of up to 0.75×10^5 Pa, much larger than the 1-5 kPa from direct measurement (Sections 6.1, 6.2, and 6.4). From theoretical considerations of till-lubricated ice-stream motion (Section 9.2) the till strength should be equal to the basal shear stress, and the basal shear stress should be less than or equal to the driving stress $\rho g h \alpha$. The latter is 0.15×10^5 Pa (for $\alpha = 0.1°$), which is considerably smaller than the 0.75×10^5 Pa till-strength value noted above. Thus there is a contradiction. Perhaps the water-level measurement error is considerably greater than the estimated ±1 m, but that seems unlikely, especially for the sounding-float measurements, to which the pressure-transducer measurements are tied. The above contradiction would be avoided if the ice-stream motion were due to basal sliding sensu stricto, without till lubrication, in which case the till shear strength would have to be larger than the basal shear stress, and a basal sliding mechanism furnishing the observed motion would have to be in operation at the effective pressure level inferred from the observations. A contradiction between the measured till strengths and the strength calculated from the effective pressure would remain.

It has been suggested that the actual ice overburden pressures differ considerably from the calculated values, such that the actual effective pressures are all practically zero. The cause of the required sharp local variations in overburden pressure is not explained and raises mechanical questions. Moreover, with this explanation, the spatially varying water pressures are retained, so that problem (1) remains a problem.

(4) In some cases nearby boreholes show the same or nearly the same water levels, as is expected if the boreholes tap into the same basal conduit (E & K, pp. 209, 215). This seems to be the case for boreholes 96-3, 96-4, 96-8, and 96-13, which were within 50 m of one another and had water level depths 112 ± 2 m, and perhaps also for boreholes 96-9, 96-10, and 96-11, again within 50 m of one another, with water levels of 118 ± 4 m. (The water level depths are indicated in the borehole map, Figure 3.) But the expectation is shattered by holes 96-1 and 96-2, only 2.5 m apart but with water levels of 102 and 117 m, and

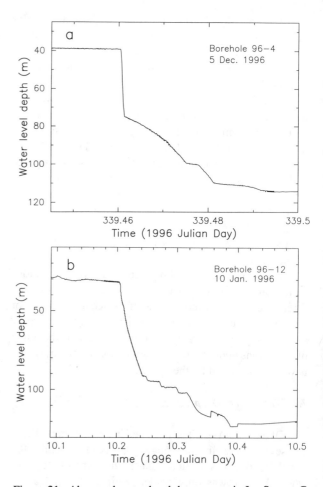

Figure 21. Abnormal water-level drop curves in Ice Stream C: (a) in borehole 96-4; (b) in borehole 96-12, drilled in the center of the fossil marginal shear zone between Ice Stream C and Ridge B-C.

also by holes 96-5 and 96-6, again 2.5 m apart, with water levels of 107 and 96 m.

(5) The cause(s) and significance of the long-term variations in basal water pressure (and/or till pore pressure), which are seen in Ice Stream C (Figure 5) as well as in Ice Stream B (E & K, Section 5), are not known. Likewise for the diurnal variations (Figure 5b; E & K, p. 216).

(6) The problem of explaining the results of the pressure-pulse-propagation experiment in Ice Stream B (E & K, Section 8) is reraised by the newer observations (Section 7.1).

The above problems, and others pointed out here and there throughout this paper, such as the lack of a strong physical distinction between Ice Stream C and Ice Streams B and D (Section 8), seem to be the observationalist's counterpart to the theoretician's "as yet inscrutable property of the bed that escapes our understanding" [*MacAyeal et al.*, 1995, p. 262].

7.5 "Canal" System?

For reasons based on problem (1) above, E & K (Section 9e) concluded that the basal water system must involve localized conduits ("canals") that are ~ 1 m in width and ~ 0.1 m in thickness and that carry most of the water flux of the system. An effort has been made to find such a conduit by searching via boreholes for gradients in basal water pressure and in WLDT and/or time constant T. Holes 95-9, 95-10, and 95-11, for example, were drilled with this objective. However, no observational indication of a canal conduit has been found. This is perhaps not surprising, since basal conduits are difficult to find by means of boreholes in temperate glaciers.

The potential complexity of the basal water conduit system is well illustrated in the flume experiments of *Catania* [1998], which produced a complex variety of braided and unbraided channels and sheet flows at the sediment-"ice" (actually plexiglass) contact, with average channel width dimensions from 0.8 to 18 cm and average depth dimensions from 0.5 to 15 mm. The depth dimensions are roughly comparable to those found here for the gap-conduit model (Sections 7.1, 7.2).

7.6 Basal Melting and/or Freeze-on

A fundamental aspect of the basal water conduit system is the magnitude of the water sources or sinks that feed or deplete it—by basal melting, freeze-on, or seepage of water out of or into the till. A related aspect is the magnitude and time derivative of water storage in conduits or chambers of the water system and in the till. These quantities are of course linked by the water transport flux in the system and by the flux gradients. They are also tied to the thermal conditions at the base of the ice stream, specifically the geothermal gradient, the shear heating due to basal sliding or basal till deformation, and the vertical temperature gradient in the basal ice. Except for the latter, which is discussed in Section 3.1, observational constraints on these quantities are few and are greatly needed to provide a firmly-based concept of the nature and functioning of the basal water system.

Measurement of the transport velocity in the basal water system (E & K, Section 7) has been repeated subsequently once (in boreholes 92-1, 92-2, and 92-4) with similar results, which are compatible with the gap-conduit model with a gap thickness of about 4 mm (E & K, p. 222).

7.7 Detection of Basal Freeze-on

Because in hot-water drilling the breakthrough from ice into unfrozen till seems always to be accompanied by an immediate onset of water-level drop, and because when the drill encounters coarse rock debris in the ice the rate of drill progress is slowed, there is a possibility of detecting a thin layer of frozen till at the bottom of the ice from a time delay between a premonitory slowing of drill advance and the onset of water-level drop. Such a delay appears to have occurred in four boreholes in Ice Stream C and two holes in Ice Stream B. Figure 22 shows an example of the observations. The borehole water-level depth and the drilling-hose load (tension) are plotted as a function of time starting about 80 minutes before breakthrough in hole 95-5. The beginning of the premonitory slowdown is marked with a downward-pointing arrow in the figure. The drill slowdown appears as a progressive decrease in drilling load because the drilling hose was being paid out from the surface at a steady rate, and when the drill advances more slowly than the hose is being paid out, the stretch and hence the tension in the hose decreases. Six minutes after the onset of slowdown, breakthrough began as marked with the upward-pointing arrow in the figure. Breakthrough is indicated by the rapid drop in water level and the sudden increase in drill load caused by the drill being pulled down by the downrushing water.

An alternative interpretation of the premonitory slowdowns, which avoids the possible implication of basal freeze-on under Ice Streams B and C, is to attribute them to delay in the process of hydraulic connection from borehole to basal water system. That such delay can occur seems to be demonstrated by anomalous borehole 88-2, in which the connection was delayed 9 hours after the drill reached the bed as indicated by cessation of drill advance (E & K, p. 210).

8. COMPARISON OF ICE STREAM C WITH ICE STREAMS B AND D

The study of Ice Stream C was undertaken with the idea that, since rapid streaming flow in C stopped about 150 years ago [*Shabtai and Bentley*, 1987; *Alley and Whillans*, 1991; *Retzlaff and Bentley*, 1993; *Smith et al.*, 1999] whereas in B and D (called collectively "B/D" below) the flow continues vigorously, a comparison of C with B/D should reveal which physical factors are responsible for the rapid motion or its cessation. The information in Sections 3-7 has been gathered with this objective in mind. What emerges is that, although C and B/D are quite different in terms of flow velocity (ca. 0.04 m d^{-1} vs. ca 1 m d^{-1}), these ice streams look largely the same in terms of the possibly controlling parameters revealed by the borehole measurements. However, in the data there are a few hints of differences. The similarities are first summarized, and then the hints of differences are considered in some detail.

8.1 Similarities

Ice Stream C has a basal water system generally similar to those of B/D in terms of the breakthrough phenomenon (Sections 7.1, 7.3) and the behavior of the borehole water level in pumping tests (E & K, Section 6). This is somewhat surprising, because drilling in the ice sheet outside the ice streams (including C) has not produced breakthroughs or other indications of a basal water system (Section 4.5; E & K, p. 210). Thus, one might have expected that, as a stopped ice stream, C would be frozen to its bed and would lack a basal water system.

The basal water pressure measured under Ice Stream C corresponds to an average effective pressure of 0.2×10^5 Pa while under B the average is 0.6×10^5 Pa, and under D, 0.4×10^5 Pa (Section 3.2). The difference is not great except as viewed in the light of Section 7.4, item (3), but the difference goes the wrong way, tending to promote more rapid motion in C rather than in B/D. Thus an effective-pressure increase suitable for shutting down the streaming motion of C is not observed. Possibly this is because the large scatter of the individual water-pressure values ($\pm 0.7 \times 10^5$ Pa) hides a small but real difference between the effective pressures under C and B/D.

Ice Stream C is underlain by weak unfrozen till that is similar to the till under B/D in porosity (Section 4.2) and in general lithological and sedimentological characteristics (Section 4.1) including the presence of diatoms. Four of the Ice Stream C cores showed some evidence of having been slightly disturbed by winnowing and resedimentation, as could be expected from the water-jet action of the drill (Section 6.4, fourth paragraph).

The estimated thickness of till based on till core lengths is distinctly smaller for C (≥ 0.3 m) than for B ($\geq$ 3 m at Up B; ≥ 0.7 m at New B), but these estimates cannot be relied upon because drilling-penetration tests indicate much larger thicknesses (≥ 5 m).

The above situation, in which there is a strong difference in streaming velocity between Ice Stream B and C without a strong difference in physical conditions that could be responsible, is a curious counterpart to the situation encountered within Ice Stream B, in which there were rather large time-variations in basal water pressure without a detectable variation in streaming velocity (E & K, p. 217].

8.2 Differences

Although borehole breakthrough behavior in C and B/D is similar in a general way, there is a statistical tendency for the water-level drop time (WLDT) to be longer in C, as discussed in Section 7.3. The tendency manifests itself in two ways: (1) In B/D, water-level drop curves of the normal quasi-exponential form have WLDT's mostly in the range 1-3 minutes, whereas in C more than half of them are in the range from 5 to 30 minutes. (2) In C, two drop curves that begin in quasi-exponential form undergo a change partway down into more complicated, segmented forms that substantially lengthen the drop time (Figure 21). Borehole 96-12 is a particularly special case, first because of its extremely long WLDT of 5 hours (Figure 21b), and second because of its location, in the center of the fossil marginal shear zone that lies between the main mass of Ice Stream C to the north and the B-C Ridge to the south. Because of this location, the breakthrough behavior of borehole 96-12 is probably indicative of the character of a basal water system that is intermediate between streaming and non-streaming conditions.

The significance of the WLDT as a parameter in comparing the basal water systems of Ice Streams C and B/D is that it can be regarded as an inverse measure of the water throughput capability of the system. The related exponential time constant (T in Table 1) is an inverse measure of the gap width in the gap-conduit model (E & K, Section 9a), and consequently is also a measure of

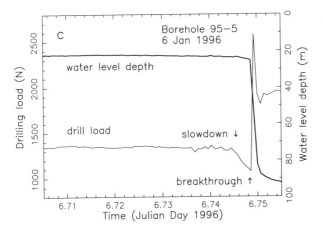

Figure 22. Drilling record showing preliminary slowdown in borehole 95-5 beginning approximately at the downward-pointing arrow. The thin line is the drilling load record, and the heavy line is the borehole water-level record, which confirms the identification of breakthrough in the drill-load record as indicated by the upward-pointing arrow. The abrupt rise in drilling load (hose tension) is caused by downrush of water in the hole at breakthrough.

throughput capability. Why should throughput capability be linked to ice streaming? If ice streaming is associated with enhanced basal melting, then enhanced throughput is necessary to carry the increased basal water flux associated with streaming. Enhanced throughput might be achieved by enlargement of the basal conduit system, perhaps by a type of basal cavitation, or it might be achieved by increased basal water pressure, which in turn might lead to enhanced cavitation or to the type of flow instability considered by Kamb (1991). Why should the indication of throughput capability be statistical? Perhaps this is related to the statistical character of the basal water systems in temperate glaciers, such as Columbia Glacier [Kamb et al., 1994, Section 5].

A related difference between the basal water systems of Ice Streams B and C is the behavior of the pressure-pulse-propagation tests (Section 7.1), which implied low throughput capability for B and relatively high throughput for C, in contradiction to the interpretation of the WLDT's above. This contradiction illustrates the difficulty in reaching a clear and unambiguous understanding of the controls on the ice streaming mechanism.

Torvane data (Section 6.4) indicate that the in-situ strength of the till under Ice Stream C is about 5 kPa, which is about 3 times greater than that under B/D (about 2 kPa/1 kPa). This difference appears to be significant in explaining the slowing of Ice Stream C, according to the theory of ice-stream mechanics discussed in Section 9.2.

8.3 Role of Basal Freeze-on

Although unfrozen till is obtained by piston coring from the base of Ice Stream C, as it is from B/D, there is a possible indication in four Ice-Stream-C boreholes (96-1, 96-6, 96-9, and 96-12) that the drill encountered and drilled through a thin layer of frozen basal till between clean basal ice above and unfrozen till below. This is discussed in Section 7.7. It suggests that a small amount (~ 0.5 m?) of till freeze-on has occurred relatively recently at the base of Ice Stream C, perhaps during the 150 years since the ice stream shut down. The freeze-on interpretation would be more secure if all 10 boreholes with adequate drilling records confirmed it. A similar indication of basal freeze-on has also been encountered in two boreholes in Ice Stream B (holes 95-4 and 95-5), which blurs a possible distinction between Ice Stream B and C as far as the occurrence of basal freeze-on is concerned. However, a statistical distinction can perhaps be recognized, in that the possible freeze-on was encountered in 4 boreholes out of 13 in Ice Stream C but in only 2 out of 29 in Ice Streams B/D. (The 2 in B/D should probably count as only 1 because the two holes are only 15 meters apart.)

If the till layer under Ice Stream C is thin (Section 4.3), a relatively small amount of basal freeze-on might be sufficient to bridge the interval from the top of the till to the bedrock, producing the situation wherein the ice stream becomes frozen to its bed and streaming flow stops. However, this situation is ruled out by the fact that unfrozen till was consistently encountered below the inferred thin basal layer of frozen till. Thus the freeze-on probably is not directly responsible for the stopping of the ice stream by freezing to the bed, but is instead probably a collateral consequence of other changes that are responsible for the stopping (see Section 9.2).

Since the freeze-on process appears to be patchy and discontinuous (apparently detected in 4 boreholes out of 10 in Ice Stream C), one might be tempted to suggest that a scattering of sticky spots at patches where the ice is frozen to its bed is sufficient to arrest the rapid motion of the ice stream while intervening unfrozen-bed areas are able to house an interconnected basal water system. But no evidence supports such a model, because, as noted in the last paragraph, the four detected patches of freeze-on were not frozen to the bed (with possible exception of the patch encountered by borehole 96-1, for which no piston core was obtained).

The amount of freeze-on that could have accumulated in the 150 years since Ice Stream C stopped streaming can be estimated from the temperature gradient in the basal ice (52 C km^{-1}) and the geothermal heat flow (70 mW m^{-2}), assuming that these quantities did not vary appreciably during the 150 years and that basal shear heating was negligible once the ice streaming stopped. (The geothermal heat flow was measured at Siple Dome by *Engelhardt* [2000] and this value is assumed valid under Ice Stream C.) The calculated freeze-on rate is 4.5 mm of ice per year, giving a total of 67 cm in 150 years. This figure should be increased by a factor of about 1.7 to include the frozen till's content of coarser rock particles (excluding clay). Thus the thickness of frozen-on till should be ~ 1 m. It seems possible that the hot water drill could penetrate through a meter of frozen till, provided that the rock clasts encountered are not too large or numerous.

Related to the possible observation of basal freeze-on under Ice Stream C is the observation of a basal temperature 0.35 degrees C below the freezing point (Section 4.6), which contrasts with the near-melting-point condition at the base of Ice Stream B (Section 3.1). According to the interpretation in Section 4.6, the 0.35 C below freezing implies that the melting isotherm lies 7 m below the bottom of the boreholes, but this again is contradicted by the observed breakthrough to a basal water system in all Ice-Stream-C boreholes (except 96-3, discussed in Section 7.1) and by the recovery of unfrozen till cores from the bottom of several of these boreholes (Table 2). At the same time it is difficult to ascribe the sub-freezing observation to measurement error, since the thermistor calibrations show measurement accuracy and long term drift-free precision of ± 0.02 C, and since two independent sets of measurements (in holes 96-2 and 96-12) agree on the 0.35-degree subfreezing basal temperature.

There is a possibility that the 0.35 C discrepancy is due to lowering of the melting point by dissolved impurities (air and salts) in the basal water. For example, if the water were saturated with air at the ambient pressure, the melting point would be depressed to –1.0 C, only 0.1 degree above the measured basal temperature of –1.1 C. Also, an 0.35 degree depression of the melting point would be effected by a salt concentration of 0.56% (compare with sea water salinity of about 3.5%). Basal freeze-on would tend to concentrate impurities in the residual water, because in the freezing process ice rejects impurities. This explanation of the 0.35 C discrepancy fits qualitatively with the theory of ice-stream mechanics discussed in Section 9.2.

9. THEORY

In a broad way the features of the ice-stream basal zone observed in the work summarized here appear to agree with conceptions held and assumptions made in recent and current theoretical treatments and numerical models of the West Antarctic ice streams. But there is a divergence of views on some points. I will not try to make an exhaustive survey of these, but will briefly discuss two that seem particularly important.

9.1 Viscous vs. Treiboplastic Flow

The largest and most fundamental area of difference is in formulation of the basal boundary condition for till lubrication of ice-stream motion. The central feature of this condition is the rheology of the till (Section 6.6). The till has been widely treated as a viscous or quasiviscous fluid with linear or slightly nonlinear flow law of the standard type (13) in which n is 1 or slightly greater than 1. The results in Section 6.6 indicate on the contrary that the till behaves rheologically as a soil or granular solid, with treiboplastic (Coulomb-plastic) failure law (3). Most of the mechanical tests on the till show a small but definite strain-rate dependence of the strength, and if this is incorporated into the rheological formulation the failure law becomes a flow law of power-law type (15), in which the exponent n is a large number ~ 40. Such a rheology is probably more convenient for numerical modeling of ice stream flow than is the perfect treiboplastic failure condition, for which a flow law cannot be written. The large value of n means that there is a great difference in flow-law nonlinearity between the till (highly nonlinear) and the linear viscous or slightly nonlinear quasiviscous rheologies that have been assumed in ice-stream modeling. It seems likely that ice-stream models with a basal boundary condition formulated from the highly nonlinear flow law will be quite different from models based on the linear or only near-linear (13) with n equal to 1 or slightly greater. An example of an ice-stream model based on (13) with large n is the model of Kamb (1991), which shows a type of prompt instability that is an immediate consequence of the flow-law nonlinearity and that has not been recognized in ice-stream models based on linear or slightly nonlinear rheologies.

While the high nonlinearity of the treiboplastic rheology of granular materials seems now well established for ice-stream till, Hindmarsh (1997) has put forward a theory that as the dimensional scale of deformation increases, till rheology undergoes a transition from plastic to viscous at a scale much larger than the grain scale but smaller than the scale of deformation in large ice-sheet models. Above the "cross-over scale" between plastic and viscous deformation the till is apparently supposed to flow as a viscous liquid. The argument for this idea is that "large scale modelling studies using a viscous model of till deformation have been reasonably successful in predicting the geological consequences of ice sheet action". However, I do not see a physical basis for the idea. Hindmarsh goes to some length to build one from a concept of small-scale plastic failure events summing over a larger scale to give viscous flow, but "the key theoretical problem which has yet to be solved is how multiple small scale failure events combine into a viscous type flow" [*Hindmarsh*, 1997, p. 1039].

9.2 The Undrained-treiboplastic-bed Model

On the basis of concepts of the ice-stream basal zone some of which are discussed in the present paper, Slawek Tulaczyk has formulated a new theory of ice-stream mechanics, which shows in a simple way how the physical components of a conceptualized ice-stream mechanism operate and interact to produce a system with well defined stable and unstable states (Tulaczyk, 1999, Chapter 6). He calls the theory the undrained-plastic-bed model (UPB). Only its basis and major features can be sketched here, but a full treatment is available in *Tulaczyk et al.*, (2000b).

The UPB theory describes a till-lubricated model of the ice stream mechanism in which the till behaves rheologically as a treiboplastic material. A straight ice stream of thickness h and width $2w$, with uniform surface slope α and driving stress $\tau_D = \rho g h \alpha$, overlies an unfrozen basal till layer of thickness l, and flows in rectilinear, channel-type flow through an ice sheet of the same h and α. In the spirit of the simplest model of channel flow of glaciers [*Nye*, 1965], all across-slope cross sections of the ice stream are assumed to look the same and behave in the same way (no along-flow variation). The ice stream is supported against the down-slope component of gravity by marginal shear stress τ_M and by uniform basal shear stress τ_B, which is equal to the shear strength τ_f of the till, so that the till layer accommodates an arbitrarily large or small ice motion by shearing at a particular yield stress $\tau_f = \tau_B$. This behavior represents treiboplastic rheology of the bed. The ice-stream flow velocity u is a known function of τ_D, which drives the ice forward, and τ_B, which resists the forward motion. The combined action of u and τ_B generates a frictional heating in the amount $u\tau_B$ (per unit area) within the till. To this is added the geothermal heat flow q_G, less the heat conducted upward into the ice q_I (in response to the vertical temperature profile (Figure 4)), giving a net basal heat budget in the amount $u\tau_B + q_G - q_I$. This heat produces either basal melting at a rate $\dot{m}$ (positive) or basal freeze-on at a rate $\dot{m}$ (negative) at the till-ice interface (Section 7.6). The basal melt water seeps into storage in the till, or else water seeps out of the till to freeze onto the bottom of the ice, as required by $\dot{m}$.

The assumed equality of the basal melting rate and the rate at which water goes into or out of storage as pore water in the till represents the condition of an undrained bed, for which arguments have been given by *Tulaczyk* [1999, p. 5-31]. If the equality did not hold, water would either have to enter or leave macroscopic storage chambers in the basal water system, for which there is no provision in the model, or else water would have to be added to or subtracted from the water in longitudinal transport in the basal water system, violating the assumption of no along-flow variation. The undrained-bed condition seems perhaps questionable in view of the existence proof for the basal water system (Section 7.1), but it should be realized that there is as yet no actual field evidence for the role of the basal water system in ice stream flow. Also it should be appreciated that the model is purposely made as simple as possible, so that the possible complexities involved in water storage in macroscopic chambers (such as a basal gap) and in water transport variably along the length of the ice stream are intentionally avoided at this first step in modeling. In this model the role of the basal water system is that it provides a basis for control of ice-stream motion via the effect of the basal water pressure on the till strength τ_f, which is constant along the length of the ice stream. *Tulaczyk et al.* [2000b] show that the undrained-bed condition can be relaxed, subject to certain assumptions, without a large modification in the formulation or results of the theory.

The change in water storage in the till is a change in the till water content and hence in the till's void ratio e (Section 4.2), at a rate $\dot{e}$ that is proportional to $\dot{m}$. The strength of the till is an exponential function of e (another feature of treiboplasticity, discussed in Section 6.5), so that $\dot{e}$ causes the strength to change at a calculable rate $\dot{t}_f$, and τ_B changes at the same rate $\dot{t}_B = \dot{t}_f$. Thus, starting with a given value of basal shear stress $\tau_B = \tau_f$, we have a corresponding ice-stream flow u, a corresponding shear heating $u\tau_B$, a corresponding net basal heat budget $u\tau_B + q_G - q_I$, a corresponding basal melting rate or freeze-on rate $\pm \dot{m}$, a corresponding rate $\dot{m}$ of water uptake by the till or release from the till, a corresponding time rate of change $\dot{e}$ in the void ratio of the till, and finally, a corresponding rate of change $\dot{t}_f = \dot{t}_B$ via the exponential dependence of τ_f on e.

Formulated explicitly, this chain of correspondences is as follows. It starts with a relation derived by *Raymond* [1996, equation (39)] between τ_B and the ice-stream centerline basal velocity u:

$$u = \frac{1}{2} A (\tau_D - \tau_B)^3 w^4 h^{-3} \quad (17)$$

where A is the constant in the standard $\dot{\gamma} = A\tau^3$ flow law for ice. Because the centerline velocity is fairly representative of the velocity over most of the width of the ice stream [*Echelmeyer et al.*, 1994, Figure 4], (17) is taken to apply to the motion of the ice stream as a whole. The basal melting rate is then

$$\dot{m} = (u\tau_B + q_G - q_I) H^{-1} \quad (18)$$

where H is the latent heat of melting per unit volume. Storage of the melt water in the till requires

$$\dot{e} = (1+e)\dot{m} \, l^{-1} \quad (19)$$

where l is the till thickness. The exponential relation (4) between e and τ_f can be simplified to

$$\tau_f = a \exp(-be) \quad (20)$$

where a and b are aggregations of the constants in (4). To calculate the effect of $\dot{e}$ on τ_f, (20) is differentiated with respect to time:

$$\dot{\tau}_B = \dot{\tau}_f = -\dot{e}ab \exp(-be) = -b\tau_B\dot{e} \quad (21)$$

Equations (17)-(21) provide the basis for understanding how the simple till-lubricated ice stream of the UPB model operates. In general, a given τ_B corresponds to a rate of change $\dot{\tau}_B$ via these relationships. Of particular interest are steady states of the system, which can persist over time. They are states for which $\dot{m}$ (and hence also $\dot{e}$ and $\dot{\tau}_B$) is 0. Such states can exist only if $q_G - q_I$ in (18) is negative, since $u\tau_B$ is inherently positive (or 0). This condition is very likely satisfied in the ice streams, because the temperature gradient in the deep ice is abnormally large (≥ 40 C km^{-1}: Section 3.1). As shown in Figure 23a, for a range of values of $q_G - q_I$ there are two steady states. One of these states is stable and the other is unstable. This is shown by linear stability analysis as follows.

Let τ_B^* be a value of τ_B for which there is a steady state, $\dot{\tau}_B^* = 0$. Consider states for which τ_B differs only slightly from τ_B^*, and let $\Delta\tau_B = \tau_B - \tau_B^*$, a small quantity. Its time derivative $\Delta\dot{\tau}_B$ equals $\dot{\tau}_B$, because $\dot{\tau}_B^* = 0$. It can be shown that $\Delta\dot{\tau}_B$ varies linearly with $\Delta\tau_B$ for small $\Delta\tau_B$, so that one can write

$$\Delta\dot{\tau}_B = \frac{\partial \Delta\tau_B}{\partial t} = p\Delta\tau_B \quad (22)$$

where p is a constant. The solution of (22),

$$\Delta\tau_B = D \exp(pt) \quad (23)$$

(where D is an arbitrary constant), implies that the state is stable if $p < 0$ and unstable if $p > 0$. To determine p in (22), take the derivative of $\dot{\tau}_B$ with respect to τ_B and then set $\tau_B = \tau_B^*$. For differentiation, $\dot{\tau}_B$ as a function of τ_B can be obtained by starting with (21) and introducing successively (19), (18), and (17). One obtains finally

$$p = -b\frac{1+e}{2H}A\frac{w^4}{h^3}\tau_B^* (\tau_D - \tau_B^*)^2 (\tau_D - 4\tau_B^*) \quad (24)$$

Of the factors in (24), the only one that changes sign in the interval $0 < \tau_B^* < \tau_D$ is ($\tau_D - 4\tau_B^*$), so it governs the stability/instability of the system in the state $\tau_B = \tau_B^*$. It is stable if $\tau_B^* < \frac{1}{4}\tau_D$ and unstable if $\tau_B^* > \frac{1}{4}\tau_D$. The dividing line between stable and unstable states, $\tau_B^* = \frac{1}{4}\tau_D$, is at the maximum in the curve of basal melting $\dot{m}$ as a function of τ_B (Figure 23a). The state to the left, at smaller τ_B, and with larger flow velocity u as shown in Figure 23b, is the stable one. It represents a rapidly flowing ice stream, as can be seen from magnitudes of the velocities along the curve $u(\tau_B)$ in Figure 23b. The state to the right, at larger τ_B and smaller u, is unstable. If perturbed toward smaller τ_B the system would move on over to the stable state on the left, as the arrow labeled $\dot{u} > 0$ in Figure 23b indicates. If perturbed toward larger τ_B the system would move on to the right, until it reaches $\tau_B = \tau_D$, at which point ice streaming has become completely shut down ($u = 0$).

9.3 Shut-down of Ice Stream C

The UPB theory provides a possible explanation for the stoppage of streaming motion in Ice Stream C by

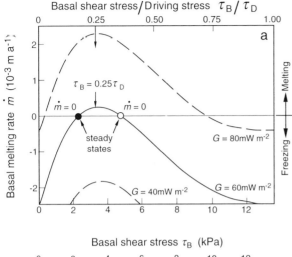

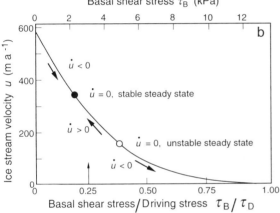

Figure 23. Stable and unstable states of ice-stream motion according to the undrained-treiboplastic-bed theory. (a) Basal melting/freeze-on rate $\dot{m}(\tau_B)$ as a function of basal shear stress τ_B from equations (17) and (18) for three different values of the geothermal heat flux G (called q_G in the text) and fixed basal ice temperature gradient 40 C km^{-1}. Steady states (marked with large dots) occur for $\dot{m} = 0$. The solid dot represent a stable state and the open dot represent an unstable state. (b) Ice-stream flow velocity $u(\tau_B)$ as a function of τ_B, from (17). The two steady states are again shown by dots. An ice-stream system that has been perturbed away from steady state would be represented by a point on the curve $\dot{m}(\tau_B)$ in (a) and the corresponding point on the curve $u(\tau_B)$ in (b). The heavy arrows and the labeling "$\dot{u}>0$" and "$\dot{u}<0$" show the direction in which such a point would move along the curve as time progresses, assuming that finite perturbations behave the same as infinitesimal ones. The sign of $\dot{u}$ is the same as that of $-p\Delta\tau_B$, as follows from (22) and the time derivative of (17). The upper abscissa scale in (a) and the lower abscissa scale in (b) are for τ_B in terms of the ratio τ_B/τ_D, where τ_D is the driving stress, taken to be 13.6 kPa. From *Tulaczyk et al.* [2000b, Figure 6].

linking it to the relatively high shear strength of the till, ~5 kPa, measured with the torvane in Ice Stream C (Section 6.4). The curves in Figure 23 are calculated with a choice of parameters intended to apply the theory to Ice Stream B, but similar results probably apply to Ice Stream C. As shown in the figure, the theory indicates that if during active streaming the basal shear stress is 2 kPa (as it is for Ice Stream B) while the driving stress is 13.6 kPa, with a geothermal heat flux of 60 mW m^{-2} and a basal-ice vertical temperature gradient of 40 C km^{-1}, the streaming flow will be stable at a velocity of about 350 ma^{-1} (to be compared with the actual centerline velocity of 440 ma^{-1} for Ice Stream B). If before shut-down Ice Stream C had τ_B like Ice Stream B, $\tau_B \approx 2$ kPa, and if the basal shear stress were then increased to about 5 kPa or slightly greater, the flow would become unstable and the streaming velocity would progressively decrease, ultimately to zero.

The increase in basal shear stress that would bring on the instability and stoppage could result directly from an increase in shear strength of the till. With its internal friction of 0.44 (Section 6.3), an increase in till strength from 2 kPa to 5 kPa could be produced by a reduction of 7 kPa in basal water pressure, resulting in a 7 kPa increase in effective pressure. This is small compared to the ~100 kPa fluctuations in effective pressure observed in boreholes (Sections 2.3 and 7.4), and thus could be masked by these fluctuations. A possible explanation for the reduction in basal water pressure is the "water piracy" theory [*Alley et al.*, 1994] if there exists a suitably connected basal water system in which the piracy could occur.

Figure 23a indicates that as τ_B increases above about 5 k Pa the basal melting rate $\dot{m}$ becomes negative, so that basal freeze-on takes place as the ice stream slows down and becomes stopped. As noted in Section 8.3, this freeze-on may increase the concentration of impurities in the water immediately below the newly frozen-on ice, which may account for the observations in Ice Stream C of basal temperatures 0.35 C below the pressure melting point for pure water (Sections 3.1, 4.6, 8.3). The frozen-on layer may be the thin basal layer of frozen till apparently encountered in four boreholes in Ice Stream C (Section 8.3).

A way in which observation does not conform to UPB theory outlined above is that the measured till strength of ~ 5 kPa corresponds to the theoretical τ_B at the start of the ice-stream shut-down process rather than at or after completion of the process. The theoretically expected τ_B for complete shut-down is instead τ_D, which is taken to be 13.6 kPa in the calculation leading to Figure 23 and is probably in the range 10-20 kPa for Ice Stream C near Up C. Possibly the measured 5 kPa is low because of mixing of water into the till by the water jet of the drill (Section 6.4).

It appears likely that Ice Stream C is not completely shut down yet, because its flow velocity of 0.04 m d^{-1} at Up C is several times greater than typical non-streaming ice-sheet flow (~ 0.01 m d^{-1}). Thus, its τ_B will not yet have increased fully to τ_D, according to the theory (Figure 23b). This helps to explain why the measured τ_f appears to be less than τ_D. It also explains to some extent why the basal conditions of Ice Stream C are generally similar to those of Ice Streams B and D (Section 8). In particular, it makes understandable the continued existence of a basal water system under Ice Stream C.

9.4 Concluding Remarks

Detailed studies of the ice streams' marginal shear zones have led to the idea that the streaming motion is controlled by the margins and that the basal zone plays only a minor, passive role [e.g. *Echelmeyer et at.*, 1994; *Jackson and Kamb*, 1997, pp. 415 and 423]. Although this is true to an extent, the UPB theory indicates that the basal shear stress has an important influence on the flow, expressed in (17), and determines whether the flow is stable or unstable, as shown in Figure 23. Moreover, one should not lose sight of the fact that the fundamental controlling mechanism that allows an ice stream to attain rapid motion in the first place is the mechanism of basal lubrication, which causes the low basal shear stress. The foregoing paper endeavors to present and interpret observational evidence as to the lubrication mechanism. It shows that while one can claim to understand this mechanism in a general though somewhat vague way, there are a number of detailed aspects for which the observational evidence appears inconclusive or even contradictory and our understanding is inadequate. Thus the ice-stream component of "glaciology's grand unsolved problem"—the West Antarctic Ice Sheet (WAIS) [*Weertman*, 1976]—is as yet only incompletely solved.

Acknowledgments

The results presented here are based on a research program carried out in collaboration with Hermann Engelhardt, whose contributions to it are crucial. I am extremely grateful for his participation and help in countless important ways. Slawek Tulaczyk helped in providing information on the basal till and on his UPB theory. I thank Robin Bolsey for technical assistance of many kinds both in Antarctica and Pasadena. I also thank our colleagues and field assistants, who have helped us so effectively and cheerfully in Antarctica over the years. In rough chronological order: Neil Humphrey, Reed Scherer, Mark Fahnestock, Mark Wumkes, John Chadwick, Jim Berkey, Harold Aschmann, Matthias Blume, Howard Conway, Tom Svitek, Judy Zachariasen, David McKee, Tim Melbourne, Keri Petersen, Patty White, Pavel Svitek, Leanne Allison, Miriam Jackson, Maneesh Sahani, Darius Semmons, Brian Waddington, Sam Webb, Debra Baldwin, Eugene Miya, Katy Quinn, Sabine Schmidt, Keith-Nels Swenson, Marcel Bergman, Sarah Das, Mitch Eaton, Jeff Hashimoto, Rowena Lohman, Beth Pratt, Ginny Catania, Paul Cutler, Elena Hinds, Cheryl Weber, Matt Bachmann, Ben Farrow, Greg Gerbi, Shulamit Gordon, Hans Schwaiger, Erik Mueller, Adam Bucki, Anne Blanchard, Doug Reusch, Sky Colley. Special thanks to Neil Humprhey for building our first piston corer and taking the first till cores, and to Professor Ronald F. Scott (Caltech) and Dr. Less Fruth (Earth Technology, Inc.) for substantial help in carrying out the geotechnical tests of the till. This work was made possible by the support of the U.S. National Science Foundation under grant OPP-9615420 and predecessors.

REFERENCES

Alley, R.B., Water-pressure coupling of sliding and bed deformation: I. Water system, *J. Glaciol.*, 35 (119), 108-118, 1989a.

Alley, R.B., Water-pressure coupling of sliding and bed deformation: II. Velocity-depth profiles, *J. Glaciol.*, 35, 119-129, 1989b.

Alley, R.B., In search of ice-stream sticky spots, *J. Glaciol.*, 39, 447-454, 1993.

Alley, R.B., Progress in ice-stream basal modeling, *Antarctic J. U.S.*, XXIX (5), 61-62, 1994.

Alley, R.B., and D.R. MacAyeal, West Antarctic ice sheet collapse: Chimera or clear danger? *Antarc. J., U.S.*, XXVIII (5), 59-60, 1993.

Alley, R.B., and I.M. Whillans, Changes in the West Antarctic Ice Sheet, *Science*, 254, 953-963, 1991.

Alley, R.B., D.D. Blankenship, C.R. Bentley, and S.T. Rooney, Deformation of till beneath Ice Stream B, West Antarctica, *Nature*, 322, 57-59, 1986.

Alley, R.B., D.D. Blankenship, C.R. Bentley, and S.T. Rooney, Till beneath Ice Stream B: 3. Till deformation: evidence and Implications, *J.Geophys. Res.*, 92, 8921-8929, 1987a.

Alley, R.B., D.D. Blankenship, C.R. Bentley, and S.T. Rooney, Till beneath Ice Stream B: 4. A coupled ice-till flow model, *J.Geophys. Res.*, 92, 8931-8940, 1987b.

Alley, R.B., D.D. Blankenship, S.T. Rooney, and C.R. Bentley, Water-pressure coupling of sliding and bed deformation: III. Application to Ice Stream B, Antarctica, *J. Glaciol*, 35, 130-139. 1989.

Alley, R.B., S. Anandakrishnan, C.R. Bentley, and N. Lord, A water-piracy hypothesis for the stagnation of Ice Stream C, Antarctica, *Ann. Glaciol.*, 20, 187-194, 1994 .

Anandakrishnan, S., and R. B. Alley, Ice Stream C, Antarctica, sticky spots detected by microearthquake monitoring, *Ann. Glaciol.*, 20, 183-186, 1994 .

Anandakrishnan, S., and C.R. Bentley, Micro-earthquakes beneath Ice Stream B and C, West Antarctica: observations and implications, *J. Glaciol., 39*, (133), 455-462, 1993.

Anandakrishnan, S., R.B. Alley, R.W. Jacobel, and H. Conway, The flow regime of Ice Stream C and hypotheses concerning its recent stagnation, Amer. Geophys. Union, this volume, 2000.

Anderson, R.C., Pebble lithology of the Marseilles till sheet in northeastern Illinois, *J. Geol., 63*, 228-243, 1955.

Behrendt, J.C., D.D. Blankenship, C.A. Finn, R.E. Bell, R.E. Sweeney, S.M. Hodge, and J.M. Brozena, CASERTZ aeromagnetic data reveal late Cenozoic flood basalts (?) in the West Antarctic rift system, *Geology 22*, 527-530, 1994.

Bentley, C.R. Antarctic ice streams: a review, *J. Geophys. Res., 92*, 8843-8858, 1987.

Bentley, C.R., Rapid sea-level rise from a West Antarctic ice-sheet collapse: a short-term perspective, *J. Glaciol., 44*, 157-163, 1998a.

Bentley, C.R., Ice on the fast track, *Nature, 354*, 21-22, 1998b.

Bentley, C.R., N. Lord, and C. Liu, Radar reflections reveal a wet bed beneath stagnant Ice Stream C and a frozen bed beneath ridge BC, West Antarctica, *J. Glaciol., 44*, 149-156, 1998.

Berre, T., and L. Bjerrum, Shear strength of normally consolidated clays, *Proc. 8th Int. Conf. on Soil Mech. and Found. Engng., 1*, 39-49, 1973.

Bindschadler, R., The importance of pressurized subglacial water in separation and sliding at the glacier bed, *J. Glaciol., 29*, 3-19, 1983.

Bindschadler R., and T. Scambos, Satellite-image-derived velocity-field of an Antarctic ice stream, *Science, 252*: 242-246, 1991.

Bindschadler R, P. Vornberger, D. Blankenship, T. Scambos, R. Jacobel, Surface velocity and mass balance of Ice Streams D and E. West Antarctica, *J. Glaciol., 42*. 461-475, 1996.

Bindschadler, R.A., R.B. Alley, J. Anderson, S. Skipp, H. Borns, J. Fastook, S. Jacobs, C.F. Raymond, and C.A. Shuman, What is happening to the West Antarctic Ice Sheet? *EOS, 79* (22), 264-265, 1998.

Bishop, W., C.E. Green, V.K. Garga, A. Andersen, and J.D. Brown, A new ring shear apparatus and its application to the measurement of residual strength, *Géotechnique, 12*, 273-328, 1971.

Blake, E.W., U.H. Fischer, and G.K.C. Clarke, Direct measurement of slidng at the glacier bed. *J. Glaciol., 42*, 595-599, 1994.

Blankenship, D.D., C.R. Bentley, S.T. Rooney, and R.B. Alley, Seismic measurements reveal a saturated, porous layer beneath an active Antarctic ice stream, *Nature, 322*, 54-57, 1986.

Blankenship, D.D., C.R. Bentley, S.T. Rooney, and R.B. Alley, Till beneath Ice Stream B: 1. Properties derived from seismic travel times, *J. Geophys. Res., 92*, 8903-8911, 1987.

Blankenship, D.D., R.E. Bell, S.M. Hodge, J.M. Brozena, J.C. Behrendt, and C.A. Finn, Active volcanism beneath the West Antarctic ice sheet and implications for ice-sheet stability, *Nature, 361*, 526-529, 1993.

Blanpied, M.L., T.E. Tullis, and J.D. Weeks, Frictional behavior of granite at low and high sliding velocity, *Geophys. Ret. Lett., 14*, 554-557, 1987.

Boulton, G.S. A paradigm shift in glaciology? *Nature 322*, 18, 1986.

Boulton, G.S., and R.C.A Hindmarsh, Sediment deformation beneath glaciers: rheology and geological consequences: *J. Geophys. Res. 92*, 9059-9082, 1987.

Boulton, G.S., and A.S. Jones, Stability of temperate ice caps and ice sheets resting on beds of deformable sediment. *J. Glaciol., 24*, 29-43, 1979.

Catania, G.A., *A Physical Model of Pressurized Flow Over an Unconsolidated Bed: Implications for Subglacial Braided Channels*, M.S. Thesis, University of Minnesota, Minneapolis, 1998.

Clarke, G.K.C., Subglacial till: A physical framework for its properties and processes, *J. Geophys. Res., 92*, 9023-9036, 1987.

Conway, H., A. Gades, and H. Engelhardt, Ice Stream C, West Antarctica: a sticky problem, submitted for publication, 2000.

Cuffey, K., and R.B. Alley, Is erosion by deforming glacial sediments significant? (Toward till continuity), *Ann. Glaciol, 22*, 17-24, 1996.

Doake, C.S.M., R.M. Frolich, D.R. Mantripp, A.M. Smith, and D.J. Baughan, Glaciological studies on Rutford Ice Stream, Antarctica, *J. Geophys. Res., 92*, 8951-8960, 1987.

Drake, L.D., Mechanisms of clast attrition in basal till, *Geol. Soc. Am. Bull., 83*, 2159-2166, 1972.

Dreimanis, A., Tills: their genetic terminology and classification, in *Genetic Classification of Glacigenic Deposits,* edited by R.P. Goldthwait and C.L. Matsch, A.A. Balkema, Rotterdam, pp. 17-83, 1988.

Dreimanis, A., Formation, deposition, and identification of subglacial and supraglacial tills, in *Glacial Indicator Tracing*, edited by R. Kujansuu and M. Saarnisto, A.A. Balkema, Rotterdam, pp. 35-60, 1990.

Echelmeyer, K.A., W.D. Harrison, C. Larsen, and J.E. Mitchell, The role of the margins in the dynamics of an active ice stream, *J. Glaciol., 40*, 527-538, 1994.

Engelhardt, H., West Antarctic Ice Sheet: high geothermal heat flow and dynamics of ice streams, *Nature,* submitted for publication, 2000.

Engelhardt, H., and Kamb, B., Vertical temperature profile of Ice Stream B, Antarctica, *Antarctic J. U.S., XXVIII* (5), 63-66, 1993.

Engelhardt, H., and B. Kamb, Basal hydraulic system of a West Antarctic ice stream: constraints from borehole observations, *J. Glaciol., 43*, 207–230, 1997.

Engelhardt, H., and B. Kamb, Basal sliding of Ice Stream B, West Antarctica, *J. Glaciol., 44*, 223-230, 1998.

Engelhardt, H., N. Humphrey, B. Kamb, and M. Fahnestock, Physical conditions at the base of a fast moving Antarctic ice stream, *Science, 248*, 57-59, 1990.

Flint, R.F., *Glacial and Quaternary Geology*, Wiley, N.Y., 1971.

Garofalo, F., *Fundamentals of Creep and Creep-Rupture in Metals*, Macmillan, New York, 1965.

Harlan, W.B., K.N. Herod, and D.H. Krinsley, The definition and identification of tills and tillites, *Earth Sci. Rev., 2*, 225-256, 1966.

Hindmarsh, R., Deforming beds: viscous and plastic scales of deformation, *Quat. Sci. Rev., 16*, 1039-1056, 1997.

Hill, R., *The Mathematical Theory of Plasticity*, Oxford, 1950.

Holms, C.D., Drift dispersion in west-central New York, *Geol. Soc. of Am. Bull., 63*, 993-1010, 1952.

Hooke, R.LeB., B. Hanson, N.R. Iverson, P. Janson, and U.H. Fischer, Rheology of till beneath Storglaciären, Sweden, *J. Glaciol., 43*, 172-179, 1997.

Hughes, T.J., West Antarctic ice streams, *Rev. Geophys. Space Phys., 15*, 1-46, 1977.

Hulbe, C.L., *Heat balance of West Antarctic ice streams, investigated with a numerical model of coupled ice sheet, ice stream, and ice shelf flow*, Ph.D. Thesis, University of Chicago, 1998.

Hulbe, C.L., and D.R. MacAyeal, A new numerical model of coupled inland ice sheet, ice stream, and ice shelf flow and its application to the West Antarctic Ice Sheet, *J. Geophys. Res., in press*, 1999.

Hulbe, C.L., and I.M. Whillans, Evaluation of strain rates on Ice Stream B, Antarctica, obtained using GPS phase measurements, *Ann. Glaciol., 20*, 254-262, 1994.

Iverson, N.R., R.W. Baker, and T.S. Mooyer, A ring-shear device for the study of till deformation: tests on tills with contrasting clay contents, *Quat. Sci. Rev., 16*, 1057-1066, 1997.

Iverson, N.R., T.S. Mooyer, and R.W. Baker, Ring-shear studies of till deformation, Coulomb-plastic behavior and distributed strains in glacier beds, *J. Glaciol., 44*, 634-642, 1998.

Iverson, R. M., A constitutive equation for mass movement behavior, *J. Geol., 93*, 143-160, 1985.

Jackson, M.., Dynamics of the Shear Margin of Ice Stream B, West Antarctica, Ph. D. thesis, California Institute of Technology, 1999.

Jackson, M. and B. Kamb, The marginal shear stress of Ice Stream B, West Antarctica: *J. Glaciol., 43*, 415–426, 1997.

Joughin, I., L. Gray, R. Bindschadler, S. Price, D. Morse, C. Hulbe, K. Mattar, and C. Werner, Tributaries of West Antarctic ice streams revealed by RADARSAT interferometry, *Science, 286*, 283-286, 1999.

Kamb, B., Rheological nonlinearity and flow instability in the deforming bed mechanism of ice stream motion: *J. Geophys. Res., 96*, 16,585-16,595, 1991.

Kamb, B., Glacier flow modeling, in *Flow and Creep in the Solar System: Observations, Modeling, and Theory*, edited by D.B.Stone and S.K. Runcorn, NATO ASI Series, vol. 319, pp. 417-506, Kluwer, Dordrecht, 1993.

Kamb, B., M. F. Meier, H. Engelhardt, M. A. Fahnestock, N. Humphrey, and D. Stone, Mechanical and Hydrologic Basis for the Rapid Motion of a Large Tidewater Glacier: 2. Interpretation, *J. Geophys. Res., 99*, 15,231-15,244, 1994.

Kezdi, A., *Handbook of Soil Mechanics*, Elsevier, Amsterdam, 1974.

Lambe, T.W., and R.V. Whitman, *Soil Mechanics*, J.Wiley, New York, 1969.

MacAyeal, D.R., Large-scale ice flow over a viscous basal sediment: theory and application to Ice Stream B, Antarctica, *J. Geophys. Res., 94*, 4071-4087, 1989.

MacAyeal, D.R., Irregular oscillations of the West Antarctic Ice Sheet, *Nature, 359*, 29-32, 1992.

MacAyeal, D.R., R.A. Bindschadler, and T.A. Scambos, Basal friction of Ice Stream E, West Antarctica, *J.Glaciol., 41*, 247-262, 1995.

Marone, C., C.B. Raleigh, and C.H. Scholz, Frictional behavior and constitutive modeling of simulated fault gouge, *J. Geophys. Res., 95*, 7007-7025, 1990.

Meer, J.J.M. van der, Microscopic evidence of subglacial deformation: *Quat. Sci. Rev., 12*, 553-587, 1993.

Meer, J.J.M. van der, Subglacial processes revealed by the microscope: particle and aggregate mobility in till, *Quat. Sci. Rev., 16*, 827-831, 1997.

Mercer, J.H., West Antarctic Ice Sheet and CO_2 greenhouse effect: a threat of disaster, *Nature, 271*, 321-325, 1978.

Mitchell, J.K. *Fundamentals of Soil Behavior* (2nd. ed.), Wiley, N.Y., 1993.

Nye, J.F., The flow of glaciers and ice sheets as a problem in plasticity, *Proc. Roy. Soc. London*, A *207*, 554-572, 1951.

Nye, J.F., The flow of glacier in a channel of rectangular, elliptic, or parabolic cross-section, *J. Glaciol., 5*, 661-690, 1965.

Oppenheimer M., Global warming and the stability of the West Antarctic Ice Sheet, *Nature,393*, 325-332, 1998.

Raymond, C., Shear margins in glaciers and ice sheets, *J. Glaciol., 42*, 90-102, 1996.

Retzlaff, R., and C. R. Bentley, Timing of stagnation of Ice Stream C, West Antarctica, from short-pulse radar studies of buried surface crevasses, *J. Glaciol., 39*, 553-561, 1993.

Richardson, A.M., Jr., and R.V. Whitman, Effect of strain rate upon undrained shear resistance of saturated remolded Fat Clay, *Géotechnique, 13*, 310-346, 1964.

Rooney, S.T., D.D. Blankenship, R.B. Alley, and C.R. Bentley, Till beneath Ice Stream B. 2. Structure and continuity, *J. Geophys. Res., 92*, 8913-8920, 1987.

Rooney, S.T., D.D. Blankenship, R.B. Alley, and C.R. Bentley, Seismic reflection profiling of a sediment-filled graben beneath Ice Stream B, West Antarctica, in *Geological Evolution of Antarctica*, edited by M.R. Thomson et al., British Antarctic Survey, Cambridge, U.K., pp. 261-265, 1991.

Rose, K.E., Characteristics of ice flow in Marie Byrd Land, Antarctica, *J. Glaciol.,24*, 63-75, 1979.

Sheahan, T.C., C.C. Ladd, and J.T. Germaine, Rate-dependent undrained shear behavior of saturated clay, *J. Geotech. Engng., 122*, 99-108, 1996.

Scherer, R.P., Quaternary and Tertiary microfossils from Ice Stream B: evidence for a dynamic West Antarctic Ice Sheet history, *Paleogeography, Paleoclimatology, Paleoceanography, 90*, 395-412, 1991.

Scherer, R.P., A speculative stratigraphic model for the central Ross embayment, *Antarctic. J. U.S., XXIX (5)*, 9-11, 1994.

Scherer, R.P., A. Aldahan, S. Tulaczyk, G. Posnert, H. Engelhardt, and B. Kamb, Pleistocene collapse of the West Antarctic Ice Sheet, *Science, 281*, 82-85, 1998.

Scholz, C. H., Earthquakes and friction laws, *Nature, 391*, 37-42, 1998.

Shabtaie, S. and C.R. Bentley, West Antarctic ice streams draining into the Ross Ice Shelf: configuration and mass balance, *J Geophys. Res., 92*, 1311-1336, 1987.

Singh, A., and J.K. Mitchell, A general stress-strain-time function for soils, *J. Soil Mech. Found. Div. Am. Soc. Civ. Eng., 94*, 21-46, 1968.

Skempton, A.W., Residual strength of clays in landslides, folded strata, and the laboratory, *Géotechnique, 35*, 3-18, 1985.

Smith, A.M., Basal conditions on Rutford Ice Stream, West Antarctica, from seismic observations, *J. Geophys. Res., 102*, 543-552, 1997.

Smith, B., N. Lord, and C.R. Bentley, Radar studies on Ice Stream C: accumulation rates and the time of stagnation (abst.), Agenda and Abstracts, Sixth Annual WAIS Workshop, http://igloo.gsfc.nasa.gov/wais, 1999.

Terzaghi, K., R.B. Peck, and G. Mesri, *Soil Mechanics in Engineering Practice*, Wiley, New York, 1996.

Tika, T.E., P.R. Vaughn, and L.J. Lemos, Fast shearing of pre-existing shear zones in soil, *Géotechnique, 46*, 197-223, 1996.

Truffer, M., *Till Deformation Beneath Black Rapids Glacier, Alaska, and its Implications on Glacier Motion*, Ph.D. Thesis, University of Alaska, Fairbanks, 1999.

Truffer, M., W.D. Harrison, and K.A. Echelmeyer, Glacier motion dominated by processes deep in underlying till, *J. Glaciol., 45* (51), in press, 2000.

Tulaczyk, S., *Basal Mechanics and Geologic Record of Ice Streaming, West Antarctica*, Ph.D. thesis, Calif. Institute of Technology, Pasadena, Calif., 1999.

Tulaczyk, S., B. Kamb, R. Scherer, and H. Engelhardt, Sedimentary processes at the base of a West Antarctic ice stream: constraints from textural and compositional properties of subglacial debris, *J. Sed. Res., 68*, 487–496, 1998.

Tulaczyk, S., B. Kamb,. and H.F. Engelhardt, Basal mechanics of Ice Stream B, West Antarctica 1. Till mechanics, *J.Geophys. Res., 105*, 463-481, 2000a.

Tulaczyk, S., B. Kamb,. and H.F. Engelhardt, Basal mechanics of Ice Stream B, West Antarctica 2. Undrained plastic bed model, *J.Geophys. Res. 105*, 483-494, 2000b.

Tulaczyk, S., B. Kamb, and H.F. Engelhardt, Estimates of effective stress beneath Ice Stream B, West Antarctica, from till preconsolidation and till void ratio, unpublished ms., 2000.

Weertman, J., A method for setting a lower limit on the water layer thickness at the bottom of an ice sheet from the time required for upwelling of water into a borehole, *International Association of Scientific Hydrology, Publication 86*, 69-73, 1970.

Weertman, J., Glaciology's grand unsolved problem, *Nature 260*, 284-286, 1976.

Weertman, J., and G.E. Birchfield, Subglacial water flow under ice streams and West-Antarctic ice-sheet stability, *Ann. Glaciol., 3*, 316-320, 1982.

Whillans, I.M., and C.J. van der Veen, New and improved determinations of velocity of Ice Stream B and Ice Stream C, West Antarctica, *J. Glaciol., 39*, 483-490, 1993.

Whillans, I.M., M. Jackson, and Y.-H. Tseng, Velocity pattern in a transect across Ice Stream B, Antarctica, *J. Glaciol., 39*, 562-572, 1993.

Barclay Kamb, Division of Geological and Planetary Sciences, California Institute of Technology, MC 100-23, Pasadena, CA 91125

THE CONTRIBUTION OF NUMERICAL MODELLING TO OUR UNDERSTANDING OF THE WEST ANTARCTIC ICE SHEET

C. L. Hulbe

NRC, NASA Goddard Space Flight Center, Greenbelt, MD, U.S.A.

A. J. Payne

Department of Geography, University of Southampton, Southampton, U.K.

This paper reviews the contribution of large-scale, numerical modeling to our understanding of the West Antarctic Ice Sheet. We begin with an overview of the rôle numerical modeling plays in West Antarctic research, and introduce the most commonly used methods and terminology. Three sections are devoted to ways in which the three main dynamical compontents of the WAIS, the inland ice sheet, ice streams and ice shelves, have been modeled. Our attention then turns to the results obtained both from the application of single-component and coupled models, including the Quaternary evolution of the ice sheet; its future evolution; the potential for grounding line instability; and the thermal regime of the ice sheet.

INTRODUCTION

Motivation

Mathematical modelling underpins much of our understanding of the behavior of the West Antarctic Ice Sheet (WAIS). Many of the contributions in this collection discuss results from the modelling of particular components of this multifaceted ice-flow system. Examples include work on ice-stream margins [*Raymond*, in press], and beds [*Kamb*, in press], as well oceanic [*Jenkins*, in press] and atmospheric [*Bromwich and Stearns*, in press] interactions. The goal of this paper is to assess the rôle specifically of large-scale numerical modelling of WAIS dynamics. Three distinct types of model have been used: models of grounded ice sheet flow (with no special treatment of ice streams); models of the adjoining ice shelves (in particular the Ronne-Filchner and the Ross); and models of individual ice streams. The latter two are closely related, and are very different from the first type of model. More recently models have been developed which couple some or all of these individual components together.

The principal motive underlying our interest in modeling is to reach beyond the limits of observational studies. The fundamental limit we seek to overcome is time. The dynamical time scales associated with many components of the WAIS are far in excess of the relatively short period for which we have direct observations. Although proxy estimates of past behavior are available from the geological record [*Borns*, in press; *Anderson*, in press], modelling is a vital tool if we are to study ice-flow dynamics on the appropriate time scales. Other important limitations are spatial. Until the advent of remote sensing from satellites [*Bindschadler*, 1998], modelling was also the only way in which the large spatial scales over which ice sheets operate could be addressed. Also, many of the important processes that govern ice-sheet dynamics occur at the basal ice/substrate boundary. Direct observation or indirect, geophysical inference is costly and limited to relatively small areas. Much of our understanding of these key processes therefore comes from the development of theoretical work [*Kamb*, in press]. The characteristic spatial and temporal scales of the problem, as well as the inaccessibility of key processes, have therefore combined to favor a model-based approach to investigating the WAIS.

Mathematical modelling aims toward two goals, which are not entirely complementary. These are the use of models as predictive tools to forecast the future evolution of the WAIS, and heuristic tools to further our understanding of the key processes controlling ice sheet behavior. Concern over anthropogenic climate change, WAIS stability and possible sea-level rise has naturally created a need to forecast future behavior. It should, however, be stressed that these forecasts are only as good as the models on which they are based. In the case of the WAIS, many key processes are not well understood (grounding-line and ice-stream physics to name the two most important) and it may therefore be presumptuous to forecast WAIS behavior over the coming century. The numerous nonlinearities and feedbacks inherent in the WAIS ice-flow system imply that, if they are to be made, forecasts should not be purely deterministic but should be based on a probabilistic approach. Such an approach would use the results from numerous model experiments in which the various model inputs are varied in a consistent fashion (sensitivity analysis). At present, the high computational requirements and technical limitations of most models works against this.

The twins goals of forecasting and understanding are not compatible because they pull the modeller in distinctly different directions. The WAIS ice-flow system consists of many individual subsystems which must all be incorporated if the response of the whole is to be simulated adequately. This inevitably leads to a tendency to produce models of increasing complexity incorporating a growing number of physical processes and parameterizations. The evolution of greater model complexity however makes model behavior increasingly more difficult to interpret. Alternatively, an approach focussed on specific parts of the systems and deliberately limiting potential complexity offers little scope in forecasting but is undoubtedly of greater use in aiding physical understanding of the WAIS.

This paper summarizes the types of model used to study the three individual ice-flow regimes identified above (we will term these ice sheet, shelf and stream). We will discuss the demands of these models in terms of the required boundary and initial condition data, as well as the way in which they are forced. We then consider some of the issues which can be studied either by using these models in isolation or coupled together. Examples include the evolution of the WAIS through a glacial/interglacial cycle; its future evolution on centennial timescales; the potential for internally-generated instability by both thermal and grounding-line mechanisms; and understanding the rôle of ice streams in determining the response of the WAIS to climate change.

Terminology

The proliferation of numerical ice-flow models has led to a wide range of terminology. We therefore wish to clarify some of the basic model properties and establish a standard nomenclature.

The first consideration is spatial. Models embody horizontal space in two principal ways: by studying the dynamics of selected one-dimensional flowlines within the ice sheet or by studying them in the full two-dimensional horizontal plane (either using a prescribed map projection or the more natural latitude/longitude coordinate system). We will term the former type of model flowband and the latter planform. Flowband models have the principal advantage that they vastly reduce the amount of computation involved. Such models have therefore often been used in an exploratory fashion to study the effects of particular physical processes (such as adding subglacial till deformation to ice-sheet stress-balance equations). While the effects of changing flowband width both along a model's domain and through time can be incorporated, changes in flowline orientation and location cannot be addressed. The application of flowband models therefore tends to be limited to either idealized geometries or to situations where flowlines are strongly constrained laterally (e.g., Pine Island Glacier or Rutford Ice Stream). Situations where the direction of ice flow is likely to evolve freely are better studied using two-dimensional planform modelling. A good example is the spatially and temporally variable behavior thought to be exhibited by the Siple Coast ice streams, for which flowband modelling would not be appropriate.

Another way in which models differ is in their incorporation of the vertical dimension. The key aspect is whether processes operating through the vertical extent of the ice mass are incorporated by the use of a vertical average (e.g., isothermal models), or whether the processes are incorporated explicitly. Examples include ice temperature, the stress and velocity components, as well as ice fabric and water content. The latter class of model (when used in planform) has been termed three-dimensional, quasi-three-dimensional, quasi-two-dimensional or 2.5-dimensional. The latter three terms are used to emphasize the fact that a different set of assumptions are often employed in the vertical compared to the horizontal dimensions (a consequence of the great difference between the horizontal and vertical length scales associated with ice masses). Very few models which truly incorporate the vertical dimension have been developed, examples include detailed analyses of the grounding line [*Herterich*, 1987; *Lestringant*, 1994].

Both time-dependent and steady-state models have been used to study the WAIS. At present, the modelling of ice-sheet and ice-shelf flows naturally lend themselves to a time-dependent analysis, whereas that of ice streams tends to be steady state. In the former case, changes in ice-mass geometry arise naturally through divergence in the horizontal flow field and changes in surface mass balance (snow accumulation, as well as subaerial and submarine ice melt). In the case of ice streams, changes in the geometry or speed of the streams can also arise through migration of their lateral margins and upstream onsets, through changes in the properties of subglacial sediments, and perhaps through changes in adjacent streams. The principles governing many ice stream phenomena are not understood and the modelling of the time-dependent evolution of ice streams is therefore still in its infancy. Examples of recent studies addressing this issue are *Marshall and Clarke* [1997] and *Fowler and Johnson* [1996].

Both finite-difference and finite-element techniques have been used in the numerical solution of ice sheet, shelf and stream models. Finite difference models solve a discretized form of the relevant partial differential equations over a regular, structured grid of quadrilaterals. Planform evolution is typically modelled using square grid elements, while models incorporating the vertical do so using a stretched coordinate system leading to quadrilateral elements. Additionally, these models are readily parallelizable and thus capable of performing long-time simulations with reasonable computational resources. Finite elements allow the use of an irregular, unstructured grid (or mesh), often consisting of triangular elements. These models may also use a stretched vertical coordinate system for vertical balance equations. The tolerance of irregular grid geometry allows a great deal of flexibility both in fitting the mesh to highly irregular domains and varying mesh size to focus on key areas. The majority of ice-stream and ice-shelf models employ finite elements because of these advantages. Additionally, the finite-element method solves integral, instead of differential, forms of the governing equations, so boundaries between dynamical regimes may be embedded within the model domain without special parameterizations. However, the changing spatial patterns inherent in systems evolving over time pose a challenge in the design of finite element model domains. Adaptive mesh generation is the best solution to that challenge but is very rare in glaciology. Finite differences have proven the more popular technique for ice-sheet modelling, due principally to their ease of use.

MODELS OF THE INDIVIDUAL COMPONENTS OF THE WAIS FLOW SYSTEM

Ice-sheet models

Numerical modelling of ice-sheet flow stems from work by *Mahaffy* [1976] and *Andrews and Mahaffy* [1976]. These papers develop work by *Nye* [1957] on what has become known as the shallow-ice approximation [*Hutter*, 1983]. The disparity between the vertical and horizontal length scales inherent in ice-sheet flow implies that simple shear dominates and shear stresses balance gravitational driving stresses

$$\tau_{xz}(z) = -\rho g(s-z)\frac{\partial s}{\partial x} \qquad (1)$$

$$\tau_{yz}(z) = -\rho g(s-z)\frac{\partial s}{\partial y}. \qquad (2)$$

The symbols used in (1) and (2) are summarized in Table 1 along with all others used in this paper. As discussed below, this simplification is valid over the vast majority of the WAIS but fails when longitudinal deviatoric or lateral shear stresses are important (in ice streams and ice shelves). The application of the *Glen* [1955] flow law for ice deformation by dislocation creep leads to expressions for horizontal velocity

$$u_x(z) = u_x(h) - 2(\rho g)^n |\nabla s|^{n-1}\frac{\partial s}{\partial x} \times$$
$$\int_h^z A(T)(s-z)^n dz \qquad (3)$$

$$u_y(z) = u_y(h) - 2(\rho g)^n |\nabla s|^{n-1}\frac{\partial s}{\partial y} \times$$
$$\int_h^z A(T)(s-z)^n dz. \qquad (4)$$

Table 1. Symbols used in this paper

Symbol	Definition	Unit
x, y	horizontal dimensions	m
z	vertical dimension [a]	m
t	time	yr
s	ice surface elevation	m
h	bedrock surface elevation	m
z_m	sea-level surface elevation ($z_m \equiv 0$)	m
m	grounding-line position	m
H	ice thickness	m
b	net mass balance [b]	m yr^{-1}
M	basal melt rate	m yr^{-1}
$\tau_{iz}(z)$	shear stress in i-z plane [c]	Pa
$u_i(z)$	horizontal velocity in i direction [c]	m yr^{-1}
$\mathbf{u(z)}$	horizontal velocity vector	m yr^{-1}
$\bar{\mathbf{u}}$	vertically-averaged horizontal velocity vector	m yr^{-1}
$w(z)$	vertical velocity	m yr^{-1}
D	non-linear ice diffusivity	m^2 yr^{-1}
H_*	ice thickness above buoyancy	m
ρ	density of ice	kg m^{-3}
c	specific heat capacity of ice	J kg^{-1} K^{-1}
k	conductivity of ice	W m^{-1} K^{-1}
g	acceleration due to gravity	m s^{-2}
K	multiplier in sliding law	Pa^{-1} m^3 yr^{-1}
A	rate constant in ice deformation law	Pa^{-3} s^{-1}
$\bar{B}$	vertically-averaged rate constant (A)	Pa s$^{1/3}$
β^2	basal resistance to ice flow	Pa s m^{-1}
ν_e	effective ice viscosity	Pa s
n	exponent in ice deformation law	-
T	temperature	K
G	geothermal heat flux	W m^{-2}
Ψ	dissipation or heating due to deformation	W m^{-3}
Φ	subglacial water potential	Pa
ρ_r	density of rock	kg m^{-3}
c_r	specific heat capacity of rock	J kg^{-1} K^{-1}
k_r	conductivity of rock	W m^{-1} K^{-1}
ρ_w	density of fresh water	kg m^{-3}
ρ_m	density of sea water	kg m^{-3}

[a] Positive upwards, origin at sealevel.
[b] Including subaerial, submarine and basal melt.
[c] $i \in (x, y)$, at height z.

The assumption can be made that the parameter A in Glen's flow law is a constant. This allows the integrals in (3) and (4) to be solved and expressions for vertically-averaged horizontal ice velocity ($\bar{\mathbf{u}}$) to be found [*Mahaffy*, 1976; *Oerlemans*, 1982; *Budd et al.*, 1984]. Alternatively, where the variation of A is important, the integrations may be performed numerically [*Huybrechts and Oerlemans*, 1988; *Huybrechts*, 1992; *Herterich*, 1990].

The velocity at the basal boundary of the ice mass ($\mathbf{u(h)}$) must also be specified. This is often referred to as the sliding velocity although this term generally incorporates both classical hard-bed sliding and the deformation of subglacial sediment. There is no universally accepted method of incorporating this into numerical ice-flow models. However, three generalizations are possible. First, the sliding model always determines basal motion as a function of gravitational driving stress (Equation 1 at $z = h$). No account of enhanced longitudinal stress is made (except in the work of *Hulbe* [1998] and *Marshall and Clarke* [1997] see later). In ice-sheet models, basal motion is often restricted to those areas where the ice/bed interface is at the melting temperature. The logic of this is simply that both hard-bed sliding and sediment deformation require the presence of basal meltwater (allowing either cavitation or sediment saturation [*Paterson*, 1994]). Finally, the pressure at which the basal meltwater drainage system operates is also known to be important [*Bindschadler*, 1983]. At present, very few ice-sheet models incorporate basal hydrology and so cannot account for this effect.

The expressions for horizontal velocity (Equations 3 and 4) rely purely on local quantities. This property allows them to be substituted into the continuity equation expressing temporal changes in ice thickness as a function of local mass balance and ice flow divergence

$$\frac{\partial H}{\partial t} = b - \nabla \cdot (\bar{\mathbf{u}} H) \tag{5}$$

$$= b + \nabla \cdot (D \nabla s) - \nabla \cdot (\mathbf{u}(h) H) \tag{6}$$

$$D = 2(\rho g)^n |\nabla s|^{n-1} \int_h^s \int_h^z \times$$
$$A(T)(s-z)^n dz' dz. \tag{7}$$

Equation 6 has become known as the ice-sheet equation and remains the basis of the majority of large-scale ice sheet models. Various approaches to the numerical solution of this non-linear parabolic equation are discussed by *Huybrechts et al.* [1996] and [*Hindmarsh and Payne*, 1996].

Equation 6 can be integrated forward in time to predict changing ice-sheet geometry providing information on surface accumulation is given (this is often provided as a function of mean annual air temperature or ice surface elevation). In the case of ice sheets which lose the majority of their accumulated mass via ablation (e.g., the Greenland Ice Sheet), Equation 6 also provides a natural means of determining the varying horizontal extent of the ice sheet (simply where $H = 0$). The situation is more complicated in Antarctica where grounded ice enters ice shelves and is ultimately lost by calving or submarine melt. These additional factors complicate the application of Equation 6 to determine the extent of grounding, and a pragmatic approach is often necessary. The simplest approach is to specify the exterior boundary of the grounded ice sheet as constant. This assumption is often made if the internal dynamics of ice flow are the principal focus of a study [*Herterich*, 1988; *Payne*, 1999]. Any model which attempts to simulate changes in the grounded extent of the WAIS must couple the equations described in this section with those of the ice-shelf section. The equations can be coupled by parameterization [*Van der Veen*, 1987; *Huybrechts*, 1990a; *Huybrechts*, 1990b], or by continuity and flotation requirements (in the case of at least one finite element scheme, *Hulbe* [1998]) but this alone may be insufficient.

Coupling WAIS flow regimes requires not only technical ability, but a complete understanding of the relevant physics. *Barcilon and MacAyeal* [1993] show that flow perturbations at a stick-slip transition, such as the transition from inland to ice shelf flow, are winnowed within a distance smaller than the ice thickness. However, recent Radarsat-derived observations of ice flow upstream of the WAIS ice stream onsets [*Joughin et al.*, 1999] suggest an extended, rather than abrupt, transition in the flow dynamics. While the notion of an abrupt transition in physics may still apply at the transition from inland to ice shelf flow, it is probably not appropriate at the upstream transition from inland to ice stream flow. To date, very few modelling studies have been made of the coupled sheet, grounding-line, shelf system (for finite difference models [*Muszynski and Birchfield*, 1987; *Huybrechts*, 1992; *Hindmarsh*, 1993], and for a finite element approach [*Hulbe*, 1998]). This topic is discussed further below.

Information on the underlying bedrock topography can also be incorporated into the use of Equation 6. The evolution of bedrock topography can be included using isostasy models of varying sophistication [*Le Muer and Huybrechts*, 1996].

The calculation of the distribution of temperature within ice masses has a long history dating from the work of *Robin* [1955]. Initially, steady-state analytical solutions were used to determine the temperature profile of a column of ice. Developments on this approach included the use of the 'moving-column model' to incorporate the effects of horizontal ice flow [*Budd et al.*, 1971]. *Jenssen* [1977] developed the first time-dependent, numerical model of temperature evolution within ice sheets. This work introduced many innovations which are still in use today, an example being the stretched coordinate system used to fit the computational domain to the irregular form of an ice sheet.

Ice temperature evolves according to the general equation for heat transfer [*Paterson*, 1994]

$$\frac{\partial T}{\partial t} = \frac{1}{\rho c} \frac{\partial}{\partial z} k \frac{\partial T}{\partial z} - \mathbf{u} \cdot \nabla T - w \frac{\partial T}{\partial z} + \frac{\Psi}{\rho c} \tag{8}$$

where the terms on the right represent vertical diffusion, vertical and horizontal advection and the heat produced by ice deformation. Horizontal diffusion can be neglected because of the disparity in the horizontal and vertical length scales of ice sheets. More recent refinements have included the dependence of the thermal properties of ice on temperature [*Huybrechts*, 1992].

The generation of heat by ice deformation (or dissipation) is found as the product of the applied stresses and strain rates, which in the simplified stress regime outlined above reduces to

$$\Psi(z) = -\rho g(s-z)\nabla s \cdot \frac{\partial \mathbf{u}}{\partial z}. \quad (9)$$

Equation (8) requires boundary conditions at the upper and basal ice surfaces. The former is simply the mean annual air temperature (usually given as a function of ice surface elevation using a prescribed lapse rate). The basal boundary condition is

$$\left.\frac{\partial T}{\partial z}\right|_h = \frac{G}{k} + \frac{\tau_{xz}(h) \cdot \mathbf{u}(h)}{k}. \quad (10)$$

This provides two further sources of energy: geothermal heat flux and the frictional heat generated by sliding (depending on how basal motion is treated in the model). The temperatures determined using Equations (8), (9) and (10) are constrained so that they cannot rise above the local pressure-dependent melting point. If the melting point is reached, the excess energy available from dissipation, frictional and geothermal heat sources is used to calculate a melt rate. Melting normally develops first at the ice sheet base. The calculated melt rates may be used to drive a model of basal hydrology but the majority of models simply lose this water.

The vertical velocity field required in Equation (8) is typically calculated from the divergence of the horizontal velocity field found in Equations (3) and (4) on the basis of mass conservation

$$w(z) = -\int_h^z \left(\frac{\partial u_x(z)}{\partial x} + \frac{\partial u_y(z)}{\partial y}\right) dz -$$
$$M + \frac{\partial h}{\partial t} + \mathbf{u}(h) \cdot \nabla h. \quad (11)$$

One of the prime motivations for incorporating temperature evolution within ice-sheet models is the effect of ice temperature on the flow properties of ice and, in particular, the value of A in Glen's flow law (appearing in Equations 3 and 4). While the exponent n in the flow law is thought to vary little from a value of 3 [*Hooke*, 1981], the value of A varies significantly depending on factors such as ice crystal orientation, dust content, water content and temperature. The dependence on temperature is perhaps the strongest effect (or at least the easiest to quantify). *Hooke* [1981] and *Paterson and Budd* [1982] both model this effect using exponential relationships in which A varies by at least three orders of magnitude over the temperature range -50 to 0 °C. This variation is slightly misleading because ice flow is dominated by shearing near the ice-sheet base (note the form of Equations 3 and 4). The temperatures experienced in these areas seldom fall below -20 °C, which reduces the effective range of A to a factor of thirty. It should be noted that temperature relative to melting point should be used in these calculations rather than ice temperature itself.

Models which determine the value of A purely from internal temperature calculations do this using the relationships published by *Paterson and Budd* [1982]. However, an additional flow enhancement factor (typically 5, essentially a constant multiplier applied to the value of A) is often found necessary for numerical models to reproduce WAIS height-to-width ratios correctly. The need for additional enhancement is not fully understood. Smaller factors are often attributed to the larger dust content of deep Wisconsinan-age ice or to the effects of anisotropy [*Li Jun et al.*, 1998; *Li Jun*, 1995] and larger factors are likely related to interactions between ice and till beneath the ice streams.

There are many interactions between ice-sheet form, flow and temperature evolution, which the above set of equations incorporate. Perhaps the most controversial of these is the creep instability identified by *Clarke et al.*, [1977], which is discussed below.

Ritz [1987] questioned the assumption of a constant geothermal heat flux over long time scales. In reality the thermal evolution of ice sheet and underlying bedrock are coupled, and the geothermal heat flux will vary in non-equilibrium conditions. The situation is modelled by coupling Equation (8) with a model of the underlying 2-km thick bedrock slab. Heat transfer in the latter is purely diffusive (groundwater effects are ignored) and involves only the first term on the right-hand side of Equation (8) with thermal properties appropriate to rock

$$\frac{\partial T}{\partial t} = \frac{1}{\rho_r c_r}\frac{\partial}{\partial z}\left(k_r\frac{\partial T}{\partial z}\right). \quad (12)$$

A constant heat flux is then applied to the base of the bedrock slab and geothermal heat flux is determined internally. *Ritz* [1987] found that this effect is important only on long time scales (> 50,000 yr). *Marshall* [1996] has developed a more sophisticated model of heat flow in the underlying bedrock by incorporating permafrost.

The thermodynamics represented by Equation (8) require the material constraint that ice at any depth never warms above its local pressure-dependent melting temperature. However, when warming migrates up from the ice/bed interface, into the ice sheet, a mixture of ice and water at ice-grain boundaries develops. In such a mixture, called polythermal ice, a heat conservation equation must be used to govern the production and advection of water within veins in the ice. A numerical treatment of polythermal conditions has been developed by *Greve* [1995]. Ice-sheet model intercomparison exercises indicate that the incorporation of polythermal ice does not alter overall model behavior significantly [*Huybrechts et al.*, 1996; *Payne et al.*, in press]. This work, along with the inclusion of permafrost in the work of *Marshall* [1996], is part of a on-going theme in the current generation of ice-sheet models: concern about the production and movement of basal meltwater and related mechanisms for enhanced flow.

Ice-shelf models

The calculation of ice shelf flow is complicated by the non-local nature of the problem. While in the ice sheet, mass flux is a function of local thickness and surface gradient, in the ice shelf, spreading at any one location depends on the thickness at all other locations. That is, for the stress balance equations, longitudinal stress gradients are important. Our understanding of ice shelf flow derives from the contributions of many authors, notably: the early work of *Weertman* [1957] on unconfined and confined floating ice and later improvement by *Sanderson and Doake* [1979] who recognized that unlike the situation in grounded ice, vertical shear is negligible; and by *Thomas* [1973] and *Sanderson* [1979] who considered the importance of obstructions to flow (pinning points) within the ice shelf and of bay geometry.

Almost all numerical models of ice shelves solve stress-balance equations [*Morland*, 1987] that depend on the assumption that vertical shear stress is negligible (that is, that horizontal velocity does not vary with depth). Their derivation uses the dynamic boundary conditions of a stress-free top surface and pressure due to sea water applied at the base and along the seaward front of the ice shelf. The seaward condition is simplified by ignoring near-field, three-dimensional effects and applying a depth-integrated sea-water pressure force along the front. The resulting horizontal stress-balance expressions are written:

$$\frac{\partial}{\partial x}\left(2\nu_e H\left(2\frac{\partial \bar{u}_x}{\partial x} + \frac{\partial \bar{u}_y}{\partial y}\right)\right) +$$
$$\frac{\partial}{\partial y}\left(\nu_e H\left(\frac{\partial \bar{u}_x}{\partial y} + \frac{\partial \bar{u}_y}{\partial x}\right)\right) -$$
$$\rho g H \frac{\partial s}{\partial x} = 0 \quad (13)$$

and

$$\frac{\partial}{\partial y}\left(2\nu_e H\left(2\frac{\partial \bar{u}_y}{\partial y} + \frac{\partial \bar{u}_x}{\partial x}\right)\right) +$$
$$\frac{\partial}{\partial x}\left(\nu_e H\left(\frac{\partial \bar{u}_x}{\partial y} + \frac{\partial \bar{u}_y}{\partial x}\right)\right) -$$
$$\rho g H \frac{\partial s}{\partial y} = 0 \quad (14)$$

in which ν_e represents the effective viscosity of ice. The first term in Equations (13) and (14) describes longitudinal strain rates. The second term describes horizontal shear-strain rates. The third term, involving ice thickness and the surface elevation gradient, describes the pressure gradient due to gravity. The ice shelf surface elevation is determined by flotation. Vertical velocity in the shelf may be computed as in Equation (11), but the vertical integral is unnecessary. The equations are valid as long as the characteristic horizontal length scale is large compared to the characteristic vertical scale.

The constitutive relation for ice is applied in the effective viscosity term, ν_e, using the *Glen* [1955] flow law for ice and the incompressibility condition, as in the inland case. It is defined using a rate constant, $\bar{B}$, and the flow-law exponent, n:

$$\nu_e = \frac{\bar{B}}{2}\left[\left(\frac{\partial \bar{u}_x}{\partial x}\right)^2 + \left(\frac{\partial \bar{u}_y}{\partial y}\right)^2 + \frac{1}{4}\left(\frac{\partial \bar{u}_x}{\partial y} + \frac{\partial \bar{u}_y}{\partial x}\right)^2 + \frac{\partial \bar{u}_x}{\partial x}\frac{\partial \bar{u}_y}{\partial y}\right]^{\frac{1-n}{2n}}.(15)$$

The rate constant is a depth-average value

$$\bar{B} = \frac{1}{H}\int_h^s A^{-\frac{1}{n}}dz. \quad (16)$$

Kinematic boundary conditions are required to solve Equations (13) and (14). In general, shelf margins are fixed with a no-slip (zero velocity) condition except where grounded ice discharges into the shelf. Coupling the ice sheet and ice shelf equations at the grounding line poses a challenge that is an ongoing subject of investigation. Most finite-difference models use the parameterization of *Van der Veen* [1987]. Finite-element models do not face the same numerical difficulties at the grounding line as finite-difference schemes and a mass-flux continuity condition is sufficient for such models [*Hulbe*, 1998; *MacAyeal*, 1989]. The coupling issue may be avoided by assuming that the grounding line passes no dynamic information between grounded and floating ice [*Hindmarsh*, 1993; *Hindmarsh*, 1996] and thereby eliminating ice shelves entirely. Unfortunately, such an assumption is unlikely valid at ice stream/ice shelf junctions.

Ice-stream models

One of the most notable differences among ice sheet models in use today is in the embodiment of ice streams. Ice streams flow over weak, water-saturated till with the result that basal shear stress and horizontal velocity is nearly constant with depth. Resistance to flow is provided in large part by ice-stream shear margins, secondarily by resistance at the bed and perhaps by backpressure from the ice shelf into which the West Antarctic ice streams flow. The effect is a stress regime similar to that of ice shelves. Ice stream speed is much larger than would be predicted by gravitational driving stress alone. Two approaches are taken to simulating ice stream flow, one which treats them using a special case of ice sliding and one which treats them as semi-grounded ice shelves (sometimes termed "shelfy-streams").

At present, the more common approach to account for ice stream flow in the WAIS is to design a special case of inland ice sliding. The special sliding parameterizations assume that as the streams flow toward the grounding line, the ice is increasingly supported by basal water pressure (that is, the ice goes nearly afloat). In this way, the observed inverse relationship between gravitational driving stress and ice velocity *Rose* [1979] is reproduced. For example, ice stream flow in the x-direction is embodied as

$$u_x(h) = -K\frac{\tau_{xz}(h)}{H_*^2} \quad (17)$$

$$H_* = s - 20 - \frac{H}{9} \quad (18)$$

where H_* is a measure of ice thickness above buoyancy (in m, and based on work by *Crary* [1962]) and K is a constant. This relationship was found to reproduce observed balance fluxes over the WAIS well. The contribution of basal sliding increases nearer the coast because an increasing proportion of the ice column is supported by buoyancy. This more than compensates any reduction in ice flow by internal deformation caused by falling gravitational driving stress. This and a similar procedure stemming from the work of *Weertman* [1964] and *Bindschadler* [1983] has, until very recently, been the only method by which the radically different style of ice-stream flow has been incorporated into large-scale models of the WAIS. One slight improvement on this situation is to allow sliding only where basal ice is at its melting point.

An alternative is to treat ice streams as a special case of ice shelf flow, with the recognition that flow over the bed shape, and perhaps till deformation or water content, provides some resistance to ice flow. *MacAyeal* [1989;1992] modifies the equations for ice-shelf flow using a linear-viscous till rheology. For example, x-direction stress balance equation for ice-stream flow is written:

$$\frac{\partial}{\partial x}\left(2\nu_e H\left(2\frac{\partial \bar{u}_x}{\partial x} + \frac{\partial \bar{u}_y}{\partial y}\right)\right) + \\ \frac{\partial}{\partial y}\left(\nu_e H\left(\frac{\partial \bar{u}_x}{\partial y} + \frac{\partial \bar{u}_y}{\partial x}\right)\right) - \\ \rho g H \frac{\partial s}{\partial x} - \bar{u}_x \beta^2 = 0 \quad (19)$$

where β^2 represents basal resistance to ice flow. Other ice-till coupling models are possible and may be preferable [*Kamb*, 1991; *Tulaczyk*, 1998]. Model experiments show that, with the stress balance described in Equation (19), some form of basal resistance, in addition to resistance from the margins and from the ice shelf, is required to slow stream flow [*Hulbe*, 1998]. This embodiment of ice streams has been used with success to reproduce unique features of the present-day WAIS ice streams, such as shear margins and 'U-shaped' transverse speed profiles [*Hulbe*, 1998]; to generate oscillatory behavior in WAIS flow [*MacAyeal*, 1992]; and in the continuum-mixture model of *Marshall and Clarke* [1997].

Inverse models

Inverse models have been applied in ice stream and ice shelf modelling in order to deduce poorly-known variables such as basal resistance and ice softness

[*MacAyeal*, 1993; *MacAyeal et al.*, 1995; *Rommelaere*, 1997]. Under-determined boundary conditions or physical parameters are estimated by inverting the model dynamics to fit a set of observational data, using prescribed uncertainties. The process may also provide insight into the physical processes involved.

Control methods are particularly well-suited to explore the force budget of ice streams, a topic of special interest in West Antarctic glaciology. As discussed above, the stress balance of ice streams is similar to that of ice shelves, except for some interaction between the base of the stream and the bed over which it flows. Borehole observation of that interaction and analysis of till samples retrieved from beneath Ice Streams B and C suggest that the mechanical coupling afforded by till is too weak to influence significantly the overall stress balance of the ice streams [*Kamb*, 1991]. Instead, it is supposed that distributed basal-stress asperities must provide the resistance needed to balance ice stream flow [*Kamb*, 1991]. Such 'sticky spots' have been difficult to discover by surface measurement of strain rates on one active ice stream [*Hulbe and Whillans*, 1994]. However, given observations of ice stream velocity and geometry, estimates of their associated uncertainties, and a numerical model of ice stream flow, it is possible to invert for the unknown basal-coupling parameter.

MacAyeal et al. [1995] use the control method to estimate basal friction (coupling between ice and bed) of Ice Stream E. That ice stream is a good target for such an investigation because satellite-derived observations of ice velocity afford the spatial coverage needed to perform a meaningful inverse calculation. Surface undulations observed by satellite but not present in the available ice surface elevation and thickness data are accommodated by specifying some error for those quantities. That uncertainty is important, because it allows the model to generate the bed roughness elements that may balance ice stream flow. Indeed, the ice stream surface topography derived from the *MacAyeal et al.* [1995] inversion more closely resembles the satellite-observed topography than does the starting data set. The basal-stress field computed by the inverse model is characterized by an irregular pattern of locally large stress ($> 5 \times 10^4$ Pa) and broader regions of small stress. The implication is that sticky spots are important to ice-stream stress balance although the model's physical description of sticky spots is incomplete. Moreover, model results suggest a level of detail in the interaction between ice and bed that is below the resolution of whole-ice sheet models, forcing models to rely on simpler basal sliding laws that reproduce the statistics of inverse model results.

MODELS OF WHOLE ICE-SHEET EVOLUTION

Numerical modelling of the WAIS was pioneered by *Budd et al.* [1971] in their Derived Physical Characteristics. This landmark work introduced many concepts and techniques which are still used in glaciology today. The study assumed that the ice sheet is in steady state and that ice always flows downhill (a consequence of the shallow-ice approximation outlined above). The measured ice-surface topography can therefore be used to identify flowlines (shown in Figure 1). The flux of ice (in m^2 yr^{-1}) is then calculated by integrating snow accumulation along these flowlines. This so-called balance flux can then be used to determine vertically-averaged horizontal ice velocity ($\bar{u}$), if ice thickness is known.

The study then goes on to determine steady-state ice temperature distribution given the velocity field derived from the balance fluxes. Figure 2 shows the results of this study for basal temperature. The basic pattern has been reproduced by many subsequent studies. Cold ice is predicted under the divide areas of Marie Byrd and Ellsworth Land. Nearer the coast, basal melt predominates, in particular there is extensive melt along the Siple Coast. The interpretation of this basic pattern is that the vertical advection of cold ice from the surface dominates under the divide. Moving away from this area, increased ice flow leads to increased energy dissipation and frictional heating (in the presence of sliding). A host of additional characteristics are also computed by *Budd et al.* [1971], including ice residence times, vertical and horizontal strain rates, and gravitational driving stresses.

Rose [1979] uses similar techniques in his study of ice flow in Marie Byrd Land. This was the first numerical study which made explicit reference to the Siple Coast ice streams. A steady-state analysis of basal temperatures was performed using the concept of balance flux. Frictional heat generation from the basal motion of ice was included and the primary result was that the ice streams are likely to experience basal melting, while the inter-stream ridges are likely to be frozen to the bed. These results were only slightly affected by the use of a range of geothermal heat fluxes (0.045 to 0.075 W m^{-2}). *Rose* [1979] also draws attention to the defining dynamical feature of ice streams, the inverse relationship between gravitational driving stress and balance velocity.

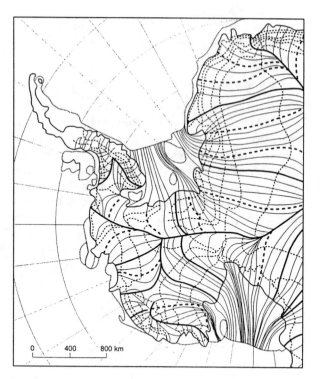

Fig. 1. Selected flowlines across the present-day Antarctic Ice Sheet. The trajectories shown are calculated by assuming that ice always flows along the path of steepest downhill gradient and would only be followed in reality if the ice sheet were in long-term equilibrium. The dashed and continuous heavy lines refer to divides in ice drainage; the thin, continuous lines are ice-flow trajectories; and the thin, dashed lines are ice-surface elevation contours (500 m interval). Based on *Budd et al.* [1971].

Throughout the 1980's Budd and co-workers developed the time-dependent numerical models of the WAIS [*Budd and Smith*, 1982; *Budd et al.*, 1984; *McInnes and Budd*, 1984; *Budd et al.*, 1985; *Budd and Jenssen*, 1987]. The basic shallow-ice model discussed above was employed in both planform and flowband (Pine Island to Ice Stream B) forms to study behavior over glacial-interglacial timescales (100 kyr). Innovations included parameterizations of the effect of changing ice thickness and surface air temperature on accumulation rate and isostatic depression.

Budd et al. [1985] and *Budd and Jenssen* [1987] perform similar analyses to *Rose* [1979]. However, their work implies that the area under Ice Stream B and the downstream portion of Ice Stream C is frozen. These papers are also amongst the first to attempt to model the flow of basal meltwater. This is done using the thin-film model of *Weertman* [1966], in which meltwater flows down the gradient of the surface defined by the water potential (assuming that basal water pressure is equal to ice overburden pressure)

$$-\nabla \Phi = -g(\rho \nabla s + (\rho - \rho_w)\nabla h). \quad (20)$$

The only rôle of the thermal regime of the basal ice appears to be as a source of meltwater, which can then drain freely irrespective of the temperature of the overlying ice. Three principal routes of meltwater drainage are identified: under Ice Streams B and D (both carrying approximately 1×10^4 m^2 yr^{-1}), and under Ice Stream C (carrying approximately 1×10^3 m^2 yr^{-1}). Model-derived water fluxes should at present be considered with caution, as the magnitude of the result is sensitive to the surface snow accumulation rate, a poorly-parameterize quantity [*Fastook*

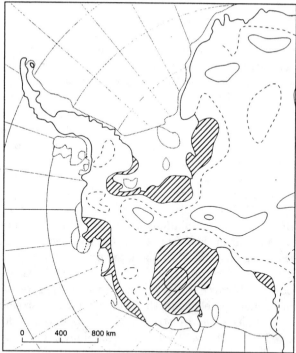

Fig. 2. Calculated basal ice temperatures for the present-day Antarctic Ice Sheet. The technique assumes steady state and uses velocities derived from the flowlines shown in Figure 1. The contours are -10 °C (dashed) and -20 °C (continuous), and the stippled area is at melting point. Based on *Budd et al.* [1971].

and Prentice, 1994]; to spatial variations in geothermal flux, which are nearly entirely unknown [Budd and Jenssen, 1987]; and to errors in basal frictional heating, which are model-dependent [Hulbe, 1998].

More complete theories of ice sheet basal water flow that pertain to the basal environment of the WAIS have been developed [Walder and Fowler, 1994; Fowler and Schiavi, 1998], as have new theories of interaction between meltwater and till [Tulaczyk, 1998]. These descriptions have yet to be incorporated into whole-ice sheet simulations.

Oerlemans [1982] uses a vertically-integrated planform model of ice flow similar to that of Budd and Smith [1982] to study the time-dependent response of the ice sheet to sea level and air temperature forcing. The ice-flow model is coupled with simplified models of isostasy and ice-shelf formation. Despite a relatively coarse spatial resolution of 100 km, the present-day distribution of ice thickness is reasonably well reproduced. The model also highlights the sensitivity of the WAIS to environmental change.

A major advance in the numerical modelling of ice masses came with the development of models which coupled the temporal evolution of ice flow and temperature [Huybrechts and Oerlemans, 1988; Herterich, 1988]. This saw the development of what has become the 'standard' model used throughout much of the 1980's and 1990's, the basis of which is described above.

A further advance was made by coupling models of ice sheet evolution with those of ice-shelf evolution [Huybrechts, 1990a; Huybrechts, 1990b; Huybrechts and Oerlemans, 1990; Huybrechts, 1992; Böhmer and Herterich, 1990; MacAyeal, 1992]. This has allowed us to address the long-term evolution of the WAIS without recourse to overly restrictive assumptions. Such models incorporate many components:

1. ice deformation within the grounded ice sheet (see above);

2. temperature evolution with the grounded ice sheet (see above);

3. a parameterization of the effect changing air temperatures on accumulation (and ablation) [Fortuin and Oerlemans, 1990];

4. a model of the isostatic effect of changing ice loads [Le Muer and Huybrechts, 1996];

5. a parameterization of ice sliding;

6. a specific model of the complex stress regime at the grounding line (see relevant section below); and

7. the deformation of ice within floating ice shelves (see above).

These models are usually driven by time series of regional air temperature change (available from ice-core studies) and the eustatic component of sea-level change. They have been used to address two main questions: the expansion and contraction of the WAIS during a complete glacial-interglacial cycle; and the effects of future greenhouse-induced polar warming. The answers to both of these questions rely on the complex interactions between grounding-line migration, changing interior ice thickness and varying accumulation rates, as well as isostatic response, ice temperature and ice viscosity.

Quaternary evolution

The studies by Huybrechts [1990a;1990b;1992] provide the only models of coupled ice sheet evolution over a glacial-interglacial cycle. Figure 3 shows the evolution of ice volume and representative East and West Antarctic surface elevations from a typical run. Changing ice volume is principally a consequence of the areal expansion and contraction of the grounded ice sheet. Regional changes in ice thickness arise from these fluctuations in the location of the grounding line. Other potential contributions include changing snow accumulation rates (reduced in the cooler climate of a glacial) and ice temperatures (cooler temperatures during a glacial result in slower flowing, thicker ice). However, both effects prove to be relatively minor.

The modelled expansion of the WAIS during a glacial fills both the Ross and Ronne-Filchner basins, although the latter zone appears to be more prone to grounding. WAIS retreat during deglaciation lags the eustatic forcing in the model by some 10 kyr. The cause of this behavior is difficult to interpret and possible effects include late-glacial warming (leading to enhanced accumulation rates) and delayed isostatic response. The EAIS, in contrast, expands little and is constrained by the proximity of the present-day grounding line to the continental break.

The predicted basal thermal regime of the WAIS during glacial and interglacial conditions share the general properties discussed by Budd et al. [1971]. Basal melting is predicted in a broad zone adjacent to to 0.1 m yr^{-1} predominate over the WAIS, only 0.006 m yr^{-1} of which can be explained by ice temperature effects [Huybrechts, 1992]. Any attempt at forecasting future WAIS response to anthropogenic climate change must therefore incorporate these inheritance effects and not treat the present-day ice thickness distribution as an equilibrium initial condition.

Future evolution

Huybrechts and Oerlemans [1990] use the same model to investigate the consequences of future, anthropogenic climate change of the ice sheet. Low and high scenarios are employed with warmings of 4.2 and 8.4 °C over pre-industrial, mean annual temperatures, respectively. These temperature increases lead to enhanced accumulation rates and introduce surface melting to the WAIS. Ablation occurs in low-lying coastal areas and is modelled using a day-degree technique [*Braithwaite and Olesen*, 1989]. In the model, the effects of increased ablation dominate those of increased accumulation only after a 8.3 °C warming. Two experiments are contrasted: static ones where ice flow is ignored, and dynamic ones where the effects of changing ice flow are incorporated. The latter experiments indicate that the low scenario would lead to a sea-level fall of approximately 0.2 m by 2350. In contrast, the dynamic response in the high scenario is a sea-level rise of approximately 0.5 m. The vast majority of this rise is brought about by slight reductions of ice thickness at the grounding line which cause grounding-line retreat and the enhanced flow of ice from the grounded ice mass. Grounding-line retreat is centered on the Antarctic Peninsula ice shelves, with little predicted to occur around the Ross and Ronne-Filchner Ice Shelves. These values equate to mean rates of sea-level change of -0.67 and +1.67 mm yr^{-1} respectively over the coming centuries.

Fastook and Prentice [1994] use similar accumulation and ablation parameterizations to evaluate the effects of a range of perturbations to mean annual air temperature. The ice flow model is, however, limited to grounded, ice-sheet flow and does not model the physics of the grounding line explicitly. The results correspond well with those of *Huybrechts and Oerlemans* [1990] in that only temperature increases in excess of 9 °C lead to sea-level rise, any smaller perturbation leading to net ice-volume increase.

Verbitsky and Oglesby [1995] use a linearized, isothermal ice-flow model, which is forced by the *net* snow accumulation rates predicted by an at-

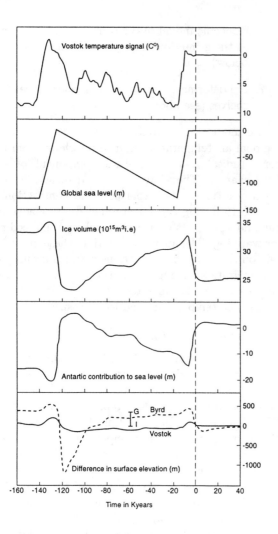

Fig. 3. Forcing (mean annual air temperature and eustatic sealevel) and predicted evolution of key glaciological variables (volume, contribution to sealevel and local ice thickness at Byrd and Vostok stations) over the last glacial-interglacial cycle. Based on *Huybrechts* [1990b].

the Siple Coast, as well as in smaller areas surrounding the Rutford Ice Stream and the Thwaites and Pine Island Glaciers. The outward migration of the grounding line during a glacial, and the associated ice thickenning, leads to more extensive basal melt. In particular, a broad zone of melting develops which connects the Siple Coast with the Thwaites/Pine Island area.

An important conclusion from these studies is that the WAIS (see the Byrd line in Figure 3) is still undergoing thinning in response to grounding-line retreat during the late glacial. Predicted values of up mospheric global circulation model. An experiment employing doubled atmospheric CO_2 content is described however the results are discussed on 10 to 150 kyr timescales and it is therefore difficult to compare them with the work of *Huybrechts and Oerlemans* [1990] and *Fastook and Prentice* [1994].

Thompson and Pollard [1997] also employ a full atmospheric global circulation model to predict net snow accumulation rates over Antarctica and the likely changes associated with a doubling of atmospheric CO_2. They do not employ an ice sheet model and so their estimates can only be compared to the static experiments of *Huybrechts and Oerlemans* [1990]. Mean Antarctic accumulation rates are estimated to increase from 0.18 to 0.21 m yr^{-1}. This equates approximately to a -1.0 mm yr^{-1} rate of sea-level change, compared to rates of -1.3 and -0.7 mm yr^{-1} (from the static low and high scenarios) from *Huybrechts and Oerlemans* [1990]. It should, however, be stressed that predictions over these centennial timescales should be based on dynamic rather than static simulations. Recently, *Huybrechts and de Wolde* [1999] have updated this work with similar results.

POTENTIAL SOURCES OF INSTABILITY

The possible threat to global sea level on human time scales has directed one course of WAIS model development toward the search for instability in the coupled ice sheet/ ice shelf system. Initially, the concern was over rapid migration of the WAIS grounding line triggered by the ongoing rise in sea level. The discovery of ice-rafted debris pulses in the North Atlantic [*Bond and Lotti*, 1995] sparked interest in oscillatory behavior of ice sheets independent of climate change. It should be noted that the term 'instability' is often used colloquially, to mean rapid change in ice sheet volume. The work of [*Hindmarsh*, 1996] is a notable exception. Here, we take instability to mean a large change in ice sheet flow and volume due to internal dynamical or thermodynamical processes.

Two main sources of instability have been identified in the literature. They are associated with the behavior of the grounding line and with the interaction of ice flow and ice temperature. The implications of these sources of instability for the WAIS have concerned numerical modellers over much of the last two decades. More recently, with the appreciation of the importance of ice streams in draining the WAIS, the two topics have become linked with the rôle of ice streams and the deformation of subglacial sediment [*Hindmarsh*, 1993; *MacAyeal*, 1992].

Although little has been written about other sources of WAIS instability, it should be stressed that many such works exist in the wider glaciological literature. Examples include valley-glacier surging and the rapid retreat of tidewater glaciers. In particular, surging theory lays great emphasis on the rôles of subglacial hydrology and geology in switching between different modes of glacier flow. Current models of the WAIS (as discussed above) essentially ignore these processes and, by doing so, may be omitting important sources of potential instability. It is to hoped that future models will address this shortcoming.

Here we introduce the theoretical reasoning underlying the proposed grounding-line and thermomechanical instabilities and then go on to review more recent applications of this theory to the WAIS. It should be stressed that, in both cases, the coupled models described above have the potential to show these instabilities, in that the basic physics on which the instabilities rely is embodied in the models. However, these models tend to react in a well-defined way to changes in external forcing over glacial-interglacial cycles. This implies that the many additional feedbacks inherent in these models damp the potential for instability.

Grounding-line instability

Early in the history of WAIS studies, the unique geographic setting of the ice sheet led naturally to concern over the stability of its grounding line. Clearly, grounding lines are capable of rapid migration under certain physical circumstances. It can be shown [*Hindmarsh*, 1993] that the local nature of inland ice sheet stress balance permits migration of the grounding line to be computed according to conservation of mass in the ice sheet and a flotation condition. The expression for the grounding line migration rate in one direction, $\dot{m}$, thus derived is:

$$\dot{m} = -\frac{\frac{\partial s_m}{\partial t} - \frac{\rho_m}{\rho_i}\dot{z}_m - \left(\frac{\rho_m}{\rho_i} - 1\right)\frac{\partial h_m}{\partial t}}{\frac{\partial s_m}{\partial x} + \left(\frac{\rho_m}{\rho_i} - 1\right)\frac{\partial h_m}{\partial x}} \quad (21)$$

in which the subscript m refers to values at the grounding line, and z_m is the elevation of the sea surface. As the denominator in Equation (21) approaches zero, that is, where the ice sheet is very lightly grounded, $\dot{m}$ must grow quite large. Rapid grounding line migration may occur but this is distinct from an instability in the underlying physics.

The origin of the marine ice sheet instability hypothesis lies in the classic *Weertman* [1957] analysis of ice shelf spreading. In that work, an expression for unidirectional spreading of confined floating ice is derived

$$\frac{\partial u}{\partial x} = 3^{(1-2)/n} \left[\frac{\rho g H}{4\bar{B}}\left(1 - \frac{\rho_i}{\rho_m}\right)\right]^n \quad (22)$$

using the standard assumptions discussed above. The recognition that in the floating ice, and thus near the grounding line, $\partial u/\partial x \propto H^3$ sparked the instability hypothesis. Flow at the grounding line should therefore increase rapidly as the grounding line retreats exposing progressively thicker grounded ice. Subsequent modelling studies of grounding line migration and ice sheet stability [*Weertman*, 1974; *Hughes*, 1975; *Thomas and Bentley*, 1978] in essence transmitted that fast stretching across the grounding line into the ice sheet to produce rapid grounding line retreat into the inland-deepening basin of West Antarctica. The pin upon which the hypothesis rests is the transmission of stress from floating to grounded ice. *Hindmarsh* [1993] argues that the transition zone between grounded ice sheet and floating ice shelf is of such limited longitudinal extent that it is unlikely for such transmission to take place. Numerical simulations [*Herterich*, 1987; *Lestringant*, 1994] and an analytical analysis of stick-slip transitions [*Barcilon and MacAyeal*, 1993] support this view.

A semi-analytical analysis of equilibrium states for marine ice sheets [*Hindmarsh*, 1993] finds an infinity of equilibrium profiles for a range of ice sheet grounding line positions and bed geometries, including beds with negative slopes in the direction of the ice divide. The implication is neutral equilibrium of the inland ice sheet/ice shelf grounding line. If the equilibria are found to be stable, the marine ice sheet instability hypothesis must be rejected. Evaluating the stability of the marine ice sheet equilibrium property is troublesome because numerical representations of the grounding line can themselves be unstable [*Hindmarsh*, 1993; *Ritz*, 1992]. Building on earlier work, *Hindmarsh* [1996] undertakes a normal-mode stability analysis of marine ice sheets uncoupled from their ice shelves and finding no instability, concluded that the grounding line is Lyapunov stable (a less restrictive test of stability than asymptotic stability). However, the situation in West Antarctica is complicated by the presence of ice streams, which broaden the transition from ice sheet to ice shelf flow. *Hindmarsh* [1993] suggests that the introduction of ice stream dynamics into the transition zone may have a stabilizing effect on the coupled system. Contrarily, MacAyeal's [1992] addition of deforming till into the ice stream dynamics produces oscillatory behavior. Thus, while the original marine ice sheet instability hypothesis is cast in doubt, the stability properties of the WAIS itself are not completely resolved.

Thermomechanical instability

Three types of thermomechanical instability are identified in ice sheet models. These relate to creep instability; the downstream transition from frozen to melting basal conditions; and the occurence of warm-based ice encircled by frozen bed conditions. The first two processes are related to the temperature-dependence of the flow law for ice. The third relies on the geometry of the basal temperature field. These instabilities are internal, in that they depend on flow-dependent thermal evolution of the ice sheet.

The basic thermomechanical ice-flow model introduced above contains the potential for positive feedback between predicted ice velocity and temperature fields (Equations 3 and 8 respectively). The essence of the feedback lies in the dependence of the flow-law rate factor (A) on temperature, and the dissipation term in the temperature-evolution equation (Equation 8). *Clarke et al.*, [1977] introduced the term 'creep instability' for the process whereby an initial temperature anomaly leads to enhanced ice flow, increased dissipation and further warming. *Yuen and Schubert* [1979] suggest that the process could lead to large-scale surging of the Antarctic Ice Sheet.

Initial numerical studies of creep instability employed vertical, one-dimensional models. Recent investigations have been conducted within the larger context of model development and validation, as discussed below. *Huybrechts and Oerlemans* [1988] use a thermomechanical flowband model to assess the effects of changing ice-sheet geometry (affecting gravitational driving stress) and horizontal temperature advection. No indications of runaway warming are found and the modelled ice sheet responds to imposed climatic change in a well-behaved fashion. *Hulbe* [1998] discusses the positive feedback in the context of a quasi-three-dimensional thermomechanical finite element model that includes both horizontal and vertical temperature advection and diffusion. In that analysis, the tendency toward excessive heating in deep ice is mitigated by corresponding large vertical strain rates, which thin the ice and thus increase upward diffusion of temperature, and secondarily by enhanced downstream advection.

Payne [1995], *Pattyn* [1996] and *Greve and MacAyeal* [1996] study a related form of thermomechanical instability. In these models, an instability arises because of an assumed abrupt increase in sliding velocity with the onset of basal melting. The sudden transition leads to a pronounced step in the ice-surface profile above the warm-cold ice transition. The steep surface slope in turn increases the gravitational driving stress and deformational velocity, and thus viscous dissipation also increases dramatically (Equations 1 and 9). *Payne* [1995] estimates a sixteen-fold increase in dissipation for a doubling of ice surface slope. The location of the warm-cold ice transition point can migrate rapidly upstream as a consequence of this localized heating and associated enhanced flow, causing a surge. Eventually, reduced ice thicknesses and enhanced cold-ice advection lead to stagnation. The validity of this instability mechanism depends on the abruptness of the transition from frozen, immobile to warm-based, sliding basal conditions. This is, in turn, determined by subglacial hydrology and the deformation mechanism of subglacial sediments [*Fowler and Johnson*, 1996].

A third form of internal instability is discussed by *Oerlemans* [1983] and *MacAyeal* [1992], in studies that seek cyclic behavior in ice sheets. Both employ thermomechanical models that associate the presence of basal meltwater with enhanced basal sliding. The latter uses ice-stream specific stress-balance equations but ignores horizontal temperature advection. *Oerlemans* [1983] uses a constant climate forcing while *MacAyeal* [1992] specifies a climate cycle according to the Vostok ice core record. The result in both cases is a cycle of slow ice sheet growth and rapid discharge. Interestingly, MacAyeal's [1992] inclusion of subglacial till dynamics leads to periodic ice-sheet fluctuations that are out of phase with the climate forcing. The cyclic behavior relies on the development of basal ice at its melting point in the interior of the ice sheet, where ice is thick, while ice nearer the margins remains frozen to the bed. Eventually, the pool of warm-based ice breaks through the encircling cold-based ice, leading to a large, rapid surge. The thin, post-surge ice sheet refreezes to its bed, thickens over time, and the cycle repeats.

The spatial pattern of warm basal temperature in the interior and cold basal temperature near the margins, required by this instability, is the opposite to that predicted by the majority of WAIS thermomechanical models [*Budd et al.*, 1971; *Huybrechts*, 1992; *Payne*, 1999]. The main process favouring warm-based divides is the increased thermal insulation afforded by thick ice. Processes favouring cold-based divides and warm-based margins are enhanced cold-ice advection at the divide and increased dissipation as ice discharge increases towards the margin. The models which predict cold-based interiors are physically more realistic because horizontal temperature advection is fully incorporated. However, MacAyeal's [1992] model treats ice-stream dynamics explicitly, with the physically realistic effect of decreasing dissipation near the margin.

Observations of the present-day WAIS and glacial geomorphology of the former Laurentide Ice Sheet (LIS), which may also have discharged in some areas via fast-flowing ice streams, also suggest the cold-interior pattern. However, the present-day WAIS and the imprint left by the former LIS are late in the surge stage of the cycle described by *Oerlemans* [1983] and *MacAyeal*, [1992]. Moreover, the present-day WAIS poses a difficult test for these models because its observed spatially heterogenous pattern of basal melting and freezing is likely to be due, in part, to spatial variations in bed topography and geology.

CONCLUSIONS

This review has introduced some of the key models in use today to study various aspects of WAIS behavior and their application in understanding WAIS response to past and future climate change, as well as in identifying sources of internal instability. Although much progress has been made in understanding the controls on this behavior, several lines of future investigation are apparent. These can be roughly divided into two groups: improved boundary condition and test data; and incorporation of more appropriate physics.

Climate and mass-balance related boundary conditions are vital to correct simulation of ice sheet processes and evolution. Present-day atmospheric boundary conditions such as mean annual air temperature and snow accumulation are known to a level of accuracy commensurate with that required by ice sheet models although the simple climate parameterizations favored by ice sheet modellers fail to capture some important peculiarities of West Antarctic precipitation [*Fastook and Prentice*, 1994; *Hulbe*, 1998]. Even more troublesome is the melt rate from the underside of ice shelves, which has been shown to influence WAIS models [*Huybrechts and Oerlemans*, 1990; *MacAyeal and Thomas*, 1986] but for which we have very limited data. The same is also true of the geothermal heat warming the underside of the grounded WAIS, which has been shown to influence the spatial extent of basal melting enormously [*Budd and Jenssen*, 1987; *Hansen and Greve*, 1996] but for which there is virtually no data. In addition, the process of iceberg calving and, more generally the disintegration of ice shelves, is generally not included in WAIS models or is included using extremely crude parameterizations of the process [*Payne et al.*, 1989]. A notable attempt to improve on numerical simulation of iceberg calving can be found in *Fastook and Schmidt* [1982].

Model validation and testing is an important, though underattended issue in ice sheet modelling. The modelling community has developed a series of validation experiments through the European Ice Sheet Modelling Initiative (EISMINT), which is supported by the European Science Foundation. In addition to providing tests for ice shelf and ice sheet model numerics, workshops and short courses sponsored by EISMINT have aided the investigation of thermomechanical instabilities, grounding line parameterizations, and other special topics. The tests may prompt the discovery of errors or numerical instabilities in ice sheet models. Unfortunately, sound numerics are not sufficient. Performance tests are also necessary before model predictions can be given any credibility. Model predictions are normally tested on a qualitative basis by comparing computed and measured ice surface topography [*Huybrechts*, 1992] or the locations of concentrations of ice flow [*Payne*, 1998]. This is particularly worrying given the number of free variables which can be tuned to produce a good comparison with the field data. In the case of surface topography, these free variables include ice rheology (in particular the effect of temperature), bedrock topography, isostatic rebound, snow accumulation and inheritance effects from the last glacial.

The lack of adequate testing is in part due to the paucity of field data at the appropriate spatial and temporal scales. In particular, models of basal thermal regime (identified as being crucial in determining the likelihood of thermal instability) remain untested except at a very limited number of ice-core locations and by broad comparison with radio-echo sounding maps of brightness at the ice/bed interface [*Hulbe*, 1998]. Although new data are continually collected and old data are reworked to produce more accurate or complete observational data sets, it is unlikely that the density of field data will ever be sufficient to test some modelled variables. Indeed, one of the benefits of numerical modelling is the ability to bridge gaps between observations [*Licht and Fastook*, submitted]. Thus, it is the responsibility of numerical modellers to express clearly the ambiguities involved in their work and the responsibility of readers to be careful in their interpretation of model-based predictions.

The realization of the importance of ice streams in draining the WAIS has given much impetus to recent model development work. Ways in which the ice-streams models can incorporated into models of the larger WAIS ice-flow system are currently being sought. The existing generation of ice sheet models incorporate the effects of ice streaming using localized sliding laws, which detailed studies have shown to be entirely inappropriate. This theme is also linked to an increasing concern with the dynamics of basal ice and the ice/substrate interface. Examples include the inclusion into models of temperate ice layers [*Greve*, 1995; *Hansen and Greve*, 1996], basal hydrology [*Fowler and Johnson*, 1996], subglacial sediment deformation [*MacAyeal*, 1992] and basal thermal regime [*Hulbe*, 1998].

OUTLOOK

The present-day WAIS appears to be late in the cycle of ice sheet growth and retreat. The thin ice cover (approximately 1000 km) in the ice stream region is not conducive to warming basal ice. How far into the future can the WAIS sustain ice stream flow? Whole ice-sheet models suggest that the answer lies in understanding the heat balance at the ice/bed interface. Another important factor, porosity (and thus deformability and water content) of sub-ice stream till, has only recently been suggested [*Tulaczyk*, 1998].

Model sensitivity experiments have been conducted to evaluate the importance of various aspects of ice flow and heat sources to basal melting and, in turn, to the WAIS's unique flow style [*Fastook and Prentice*, 1994; *Payne*, 1998; *Hulbe*, 1998]. The most significant contributors to present-day basal melting are suggested to be the small snow accumulation rates over the Siple Coast and the latent heat of meltwater stored at or near the ice/bed interface [*Fastook and Prentice*, 1994; *Hulbe*, 1998]. The detailed effects of basal topography, basal water drainage patterns, and the most recent theories of till mechanics are yet to be explored in the context of planform models.

Speculation about the future course of WAIS flow is complicated by uncertainty about future climate change. Ongoing warming, observed elsewhere in the southern hemisphere, may bring warmer surface temperatures, larger snow accumulation rates, or both to West Antarctica [Budd and Simmonds, 1991]. If the effect of future warming is melting of ice shelves alone, the grounding line stability analysis of Hindmarsh [1993] suggests that there will be little effect on grounded ice flow. The dynamic connection between ice shelves and Siple Coast ice streams may modify that outcome. Simulations of WAIS response to CO_2-doubling induced climate change, conducted with broad-scale thermodynamic finite difference models predict no significant change in grounded ice volume over the next several centuries [Huybrechts and Oerlemans, 1990], though the questions raised by the Hindmarsh [1993] grounding-line stability analysis must lend caution to the interpretation of those results. Additionally, those models did not seek to represent the details of ice stream flow.

From this review, it is clear that the state of the art in numerical ice sheet modelling is only beginning to rise to the unique challenges of the WAIS. Models which incorporate the intricacies of longitudinal and transverse stresses are urgently needed to aid in the study of ice-stream margins, onset areas and sticky spots, as well as grounding lines (all of which likely fall between the well constrained sheet and stream flow regimes discussed above). Beyond this, the glaring omission of subglacial processes from large-scale ice sheet models must be addressed before modellers can start to address the wealth of data becoming available from field and remote-sensing studies of the area. Clearly, there is much room for improvement in numerical models of the WAIS.

Acknowledgments. C.L.H. is supported by a National Research Council Research Associateship. We would like to thank Steve Price for his helpful comments on an early manuscript. We also appreciate the many useful suggestions made by Shawn Marshall and an anonymous referee, and gratefully acknowledge the work of editors Bob Bindschadler and Richard Alley.

References

Anderson, J. B., Marine history, *this volume*, in press.

Andrews, J. T., and M. A. W. Mahaffy, Growth rate of the Laurentide Ice Sheet and sea level lowering, *Quat. Res.*, *6*, 167–183, 1976.

Barcilon, V., and D. R. MacAyeal, Steady flow of a viscous ice stream across a no-slip/slip-free transition at the bed, *J. Glaciol.*, *39*, 167–185, 1993.

Bindschadler, R., The importance of pressurized subglacial water in separation and sliding at the glacier bed, *J. Glaciol.*, *29*, 3–19, 1983.

Bindschadler, R., Monitoring ice sheet behavior from space, *Rev. Geophys.*, *36*, 79–104, 1998.

Böhmer, W. J., and K. Herterich, A simplified three-dimensional ice-sheet model including ice shelves, *Ann. Glaciol.*, *14*, 17–19, 1990.

Bond, G. C., and R. Lotti, Iceberg discharges into the North Atlantic on millennial time scales during the last glaciation, *Science*, *267*, 1005–1010, 1995.

Borns, H. W., Terrestial history, *this volume*, in press.

Braithwaite, R., and O. B. Olesen, Calculation of glacier ablation from air temperature, West Greenland, in *Glacier fluctuations and climatic change*, edited by J. Oerlemans, pp. 219–233, D. Reidel, Dordrecht, 1989.

Bromwich, D., and C. Stearns, Atmospheric environment and mass delivery, *this volume*, in press.

Budd, W. F., and D. Jenssen, Numerical modelling of the large-scale basal water flux under the West Antarctic Ice Sheet, in *Dynamics of the West Antarctic Ice Sheet*, edited by C. J. van der Veen, and J. Oerlemans, pp. 293–320, D. Reidel, Dordrecht, 1987.

Budd, W. F., and I. Simmonds, The impact of global warming on the Antarctic mass balance and global sea level, in *The Role of Polar Regions in Global Change*, edited by G. Weller, pp. 489–494, University of Alaska, Fairbanks, 1991.

Budd, W. F., and I. N. Smith, Large-scale numerical modelling of the Antarctic Ice Sheet, *Ann. Glaciol.*, *3*, 42–49, 1982.

Budd, W. F., D. Jenssen, and U. Radok, *Derived physical characteristics of the Antarctic Ice Sheet*, Publication No. 18, University of Melbourne, 1971.

Budd, W. F., D. Jenssen, and I. N. Smith, A three-dimensional time-dependent model of the West Antarctic Ice Sheet, *Ann. Glaciol.*, *5*, 29–36, 1984.

Budd, W. F., D. Jenssen, and B. J. McInnes, Numerical modelling of ice stream flow with sliding, in *Research Notes 28*, edited by T. H. Jacka, pp. 130–137, ANARE, Tasmania, 1985.

Clarke, G. K. C., U. Nitsan, and W. S. B. Paterson, Strain heating and creep instability in glaciers and ice sheets, *Rev. Geophys.*, *15*, 235–247, 1977.

Crary, A. P., *Glaciological studies at Little America Station, Antarctica, 1957 and 1958*, IGY Glaciological Report 5, 1962.

Fastook, J. L., and M. Prentice, A finite-element model of Antarctica: sensitivity test for meteorological mass-balance relationship, *J. Glaciol.*, *40*, 167–175, 1994.

Fastook, J. L., and W. Schmidt, Finite-element analysis of calving from ice fronts, *Ann. Glaciol.*, *12*, 103–106, 1982.

Fortuin, J. P. F., and J. Oerlemans, Parameterization of the annual surface temperature and mass balance of Antarctica, *Ann. Glaciol.*, *5*, 78–84, 1990.

Fowler, A. C., and C. Johnson, Ice sheet surging and ice stream formation, *Ann. Glaciol.*, *23*, 68–73, 1996.

Fowler, A. C., and E. Schiavi, A theory of ice-sheet surges, *J. Glaciol.*, *44*, 104–118, 1998.

Glen, J. W., The creep of polycrystalline ice, *Proc. R. Soc. London, Ser. A*, *228*, 519–538, 1955.

Greve, R., Thermomechanisches verhalten polythermer eisschilde – theorie, analytik, numerik, Ph.D. thesis, Technische Hochschule, Darmstadt, 1995.

Greve, R., and D. R. MacAyeal, Dynamic/thermodynamic simulations of Laurentide ice sheet instability, *Ann. Glaciol.*, *23*, 328–335, 1996.

Hansen, I., and R. Greve, Polythermal modelling of steady states of the Antarctic Ice Sheet in comparison with the real world, *Ann. Glaciol.*, *23*, 382–387, 1996.

Herterich, K., On the flow within the transition zone between ice sheet and ice shelf, in *Dynamics of the West Antarctic Ice Sheet*, edited by C. J. van der Veen, and J. Oerlemans, pp. 185–202, D. Reidel, Dordrecht, 1987.

Herterich, K., A three-dimensional model of the Antarctic Ice Sheet, *Ann. Glaciol.*, *11*, 32–35, 1988.

Herterich, K., A simplified three-dimensional ice-sheet model including ice shelves, *Ann. Glaciol.*, *14*, 17–19, 1990.

Hindmarsh, R. C. A., Qualitative dynamics of marine ice sheets, in *Ice in the Climate System*, edited by W. R. Peltier, NATO ASI Ser., Ser. I, 12, pp. 67–99, 1993.

Hindmarsh, R. C. A., Stability of ice rises and uncoupled marine ice sheets, *Ann. Glaciol.*, *23*, 105–115, 1996.

Hindmarsh, R. C. A., and A. J. Payne, Time step limits for stable solutions of the ice sheet equation, *Ann. Glaciol.*, *23*, 74–85, 1996.

Hooke, R. L., Flow law for polycrystalline ice in glaciers, *Rev. Geophys.*, *19*, 664–672, 1981.

Hughes, T. J., The West Antarctic Ice Sheet: instability, disintegration, and initiation of Ice Ages, *Rev. Geophys.*, *13*, 502–526, 1975.

Hulbe, C. L., Heat balance of West Antarctic ice streams, investigated with a numerical model of coupled ice sheet, ice stream and ice shelf flow, Ph.D. thesis, University of Chicago, 1998.

Hulbe, C. L., and I. Whillans, Evaluation of strain rates on Ice Stream B, Antarctica, obtained using differential GPS, *Ann. Glaciol.*, *20*, 254–262, 1994.

Hutter, K., *Theoretical glaciology*, D. Reidel, Dordrecht, 1983.

Huybrechts, P., A 3-D model for the Antarctic Ice Sheet: A sensitivity study on the glacial-interglacial contrast, *Climate Dyn.*, *5*, 79–92, 1990a.

Huybrechts, P., The Antarctic Ice Sheet during the last glacial-interglacial cycle: a three-dimensional experiment, *Ann. Glaciol.*, *14*, 115–119, 1990b.

Huybrechts, P., *The Antarctic Ice Sheet and environmental change: a three-dimensional modeling study*, Berichte zur Polarforschung, 99, 1992.

Huybrechts, P., and J. de Wolde, The dynamic response of the Greenland and Antarctic ice sheets to multiple-century climatic warming, *J. Climate*, *12*, 2169–2188, 1999.

Huybrechts, P., and J. Oerlemans, Evolution of the East Antarctic Ice Sheet: A numerical study of thermomechanical response patterns with changing climate, *Ann. Glaciol.*, *11*, 52–59, 1988.

Huybrechts, P., and J. Oerlemans, Response of the Antarctic Ice Sheet to future greenhouse warming, *Climate Dyn.*, *5*, 93–102, 1990.

Huybrechts, P., A. J. Payne, and EISMINT Intercomparison Group, The EISMINT benchmarks for testing ice-sheet models, *Ann. Glaciol.*, *23*, 1–12, 1996.

Jenkins, A., Ocean environment and mass removal, *this volume*, in press.

Jenssen, D., A three-dimensional polar ice-sheet model, *J. Glaciol.*, *18*, 373–389, 1977.

Joughin, I., L. Gray, R. Bindschadler, S. Price, D. Morse, C. Hulbe, K. Mattar, and C. Werner, Tributaries of West Antarctic ice streams revealed by RADARSAT interferometry, *Science*, *286*, 283–286, 1999.

Kamb, B., Rheological nonlinearity and flow instability in the deforming bed mechanism of ice stream motion, *J. Geophys. Res.*, *96*, 16585–16595, 1991.

Kamb, B., Ice-stream beds, *this volume*, in press.

Le Muer, E., and P. Huybrechts, A comparison of different ways of dealing with isostasy: examples from modelling the Antarctic Ice Sheet during the last glacial cycle, *Ann. Glaciol.*, *23*, 309–317, 1996.

Lestringant, R., A two-dimensional finite-element study of flow in the transition zone between and ice sheet and an ice shelf, *Ann. Glaciol.*, *20*, 67–72, 1994.

Li Jun, Interrelation between flow properties and crystal structure of snow and ice, Ph.D. thesis, University of Melbourne, 1995.

Li Jun, T. H. Jacka, and V. Morgan, Crystal size and microparticle record in the ice core from Dome Summit South, Law Dome, East Antarctica, *Ann. Glaciol.*, *27*, 343–348, 1998.

Licht, K. J., and J. L. Fastook, Model simulation of Late Quaternary ice advance and retreat, Ross Sea, Antarctica, *Quat. Sci. R.*, submitted.

MacAyeal, D. R., Large-scale flow over a viscous basal sediment: Theory and application to Ice Stream B, Antarctica, *J. Geophys. Res.*, *94*(B4), 4071–4087, 1989.

MacAyeal, D. R., Irregular oscillations of the West Antarctic Ice Sheet, *Nature*, *359*, 29–32, 1992.

MacAyeal, D. R., A tutorial on the use of control methods in ice-sheet modeling, *J. Glaciol.*, *39*, 91–98, 1993.

MacAyeal, D. R., and R. H. Thomas, The effects of basal melting on the present flow of Ross Ice Shelf, Antarctica, *J. Glaciol.*, *32*, 72–86, 1986.

MacAyeal, D. R., R. A. Bindschadler, and T. A. Scambos, Basal friction of Ice Stream E, West Antarctica, *J. Glaciol.*, *41*, 247–262, 1995.

Mahaffy, M. A. W., A three-dimensional numerical model of ice sheets: Tests on the Barnes Ice Cap, Northwest Territories, *J. Geophys. Res.*, *81*, 1059–1066, 1976.

Marshall, S. J., Modelling Laurentide Ice Sheet thermodynamics, Ph.D. thesis, University of British Columbia, 1996.

Marshall, S. J., and G. K. C. Clarke, A continuum mixture model of ice stream thermodynamics in the Laurentide Ice Sheet 1. Theory, *J. Geophys. Res.*, *102*, 20599–20613, 1997.

McInnes, B. J., and W. F. Budd, A cross-sectional model for West Antarctica, *Ann. Glaciol.*, *5*, 95–99, 1984.

Morland, L. S., Unconfined ice-shelf flow, in *Dynamics*

of the West Antarctic Ice Sheet, edited by C. J. van der Veen, and J. Oerlemans, pp. 99–116, D. Reidel, Dordrecht, 1987.

Muszynski, I., and G. E. Birchfield, A coupled marine ice-stream ice-shelf model, *J. Glaciol.*, *33*, 3–15, 1987.

Nye, J. F., The distribution of stress and velocity in glaciers and ice sheets, *Proc. R. Soc. London, Ser. A*, *239*, 113–133, 1957.

Oerlemans, J., A model of the Antarctic Ice Sheet, *Nature*, *297*, 550–553, 1982.

Oerlemans, J., A numerical study on cyclic behaviour of polar ice sheets, *Tellus*, *35*, 81–87, 1983.

Paterson, W. S. B., *The Physics of Glaciers*, Pergamon, New York, 3rd edn., 1994.

Paterson, W. S. B., and W. F. Budd, Flow parameters for ice sheet modelling, *Cold Reg. Sci. Technol.*, *6*, 175–177, 1982.

Pattyn, F., Numerical modelling of a fast flowing outlet glacier: Experiments with different basal conditions, *Ann. Glaciol.*, *23*, 237–246, 1996.

Payne, A. J., Limit cycles in the basal thermal regime of ice sheets, *J. Geophys. Res.*, *100*(B3), 4249–4263, 1995.

Payne, A. J., Dynamics of the Siple Coast ice streams, West Antarctica: results from a thermomechanical ice sheet model, *Geophys. Res. Lett.*, *25*, 3173–3176, 1998.

Payne, A. J., A thermomechanical model of ice flow in West Antarctica, *Climate Dyn.*, *15*, 115–125, 1999.

Payne, A. J., et al., Results from the EISMINT Phase 2 Simplified Geometry Experiments: the effects of thermomechanical coupling, *J. Glaciol.*, in press.

Payne, A. J., D. Sugden, and C. Clapperton, Modeling the growth and decay of the Antarctic Peninsula ice sheet, *Quat. Res.*, *31*, 119–134, 1989.

Raymond, C. F., Ice-stream margins, *this volume*, in press.

Ritz, C., Time dependent boundary conditions for calculation of temperature fields in ice sheets, *IAHS Publ.*, *170*, 207–216, 1987.

Ritz, C., Un modele thermo-mechanique d'evolution pour le bassin antarctique Vostok-Glacier Byrd: sensibilite aux valeus des parametres mal connus, Ph.D. thesis, Laboratoire de Glaciologie et Geophysique de L'Environment, Saint Martin d'Heres Cedex, 1992.

Robin, G. Q., Ice movement and temperature distribution in glaciers and ice sheets, *J. Glaciol.*, *2*, 523–532, 1955.

Rommelaere, V., Trois problemes inverses en glaciologie, Ph.D. thesis, l'Universite Joseph Fourier, Grenoble, 1997.

Rose, K. E., Characteristics of ice flow in Marie Byrd Land, Antarctica, *J. Glaciol.*, *24*, 63–75, 1979.

Sanderson, T. J. O., Equilibrium profile of ice shelves, *J. Glaciol.*, *22*, 435–460, 1979.

Sanderson, T. J. O., and C. S. M. Doake, Is vertical shear in an ice shelf negligible?, *J. Glaciol.*, *22*, 285–292, 1979.

Thomas, R. H., The creep of of ice shelves: interpretation of observed behaviour, *J. Glaciol.*, *12*, 55–70, 1973.

Thomas, R. H., and C. R. Bentley, A model for Holocene retreat of the West Antarctic Ice Sheet, *Quat. Res.*, *10*, 150–170, 1978.

Thompson, S. L., and D. Pollard, Greenland and Antarctic mass balances for present and doubled atmospheric CO_2 from the GENESIS version-2 global climate model, *J. Climate*, *11*, 871–900, 1997.

Tulaczyk, S. M., Basal mechanics and geological record of ice streaming, West Antarctica, Ph.D. thesis, California Institute of Technology, Pasadena, 1998.

Van der Veen, C. J., Longitudinal stresses and basal sliding: a comparative study, in *Dynamics of the West Antarctic Ice Sheet*, edited by C. J. van der Veen, and J. Oerlemans, pp. 223–248, D. Reidel, Dordrecht, 1987.

Verbitsky, M. Y., and R. J. Oglesby, The CO_2-induced thickening/thinning of the Greenland and Antarctic Ice Sheets as simulated by a GCM (CCM1) and an ice-sheet model, *Climate Dyn.*, *11*, 247–253, 1995.

Walder, J. S., and A. Fowler, Channelized subglacial drainage over a deformable bed, *J. Glaciol.*, *40*, 3–15, 1994.

Weertman, J., The deformation of floating ice shelves, *J. Glaciol.*, *3*, 38–42, 1957.

Weertman, J., The theory of glacier sliding, *J. Glaciol.*, *5*, 287–303, 1964.

Weertman, J., Effect of meltwater layers on the dimensions of glaciers, *J. Glaciol.*, *6*, 191–207, 1966.

Weertman, J., The stability of the junction of an ice sheet and an ice shelf, *J. Glaciol.*, *13*, 3–11, 1974.

Yuen, D. A., and G. Schubert, The role of shear heating in the dynamics of large ice masses, *J. Glaciol.*, *24*, 195–212, 1979.

C. L. Hulbe, Code 971, Oceans and Ice Branch, Laboratory for Hydrospheric Sciences, NASA Goddard Space Flight Center, Greenbelt, MD 20771, U.S.A. (e-mail: chulbe@ice.gsfc.nasa.gov)

A. J. Payne, Department of Geography, University of Southampton, Highfield, Southampton SO17 1BJ, U.K. (e-mail: A.J.Payne@soton.ac.uk)

RUTFORD ICE STREAM, ANTARCTICA

C. S. M. Doake, H. F. J. Corr, A. Jenkins, K. Makinson, K. W. Nicholls, C. Nath, A. M. Smith, and D. G. Vaughan

British Antarctic Survey, Natural Environment Research Council, Cambridge, UK

Rutford Ice Stream is in many ways a typical Antarctic outlet glacier. Constrained by a subglacial-bed trough to the east of the Ellsworth Mountains, it drains an area of 49,000 km² of the West Antarctic Ice Sheet. Varying in width from 20 to 30 km, flowing fast (up to 400 m/a) for more than 150 km before it starts to float, and over 2000 m thick along most of its length, it discharges 18.5±2 Gt of ice per year across its grounding line. It has an average driving stress of 40 kPa, which is resisted by lateral shear stresses at the margins in boundary layers up to about 10 km wide, and by basal shear stress in the middle third of the ice stream. Seismic studies of the base reveal varied conditions, with soft deformable till and more competent sediments. Stresses in the margins of up to 160 kPa lead to fracturing and crevassing, highlighted as bright bands in satellite synthetic aperture radar (SAR) images. Shallow seismic refraction and radar measurements indicate that fracture is initiated at depths around 10-20 m, consistent with the SAR penetration depths. Indications of change come from SAR interferometry of the upstream shear margin, where decadal fluctuations in the velocity profile suggest the effective width of the ice stream is varying. The limit of tidal flexing has been accurately located with SAR interferometry and shows no change in position between 1992 and 1996. Downstream of the grounding line there is a strong pattern of ice thickness variation advecting with the flow. We do not have a good explanation for the pattern, but it could have been caused by fluctuations in the position of the grounding line as a consequence of changes in ice thickness advecting downstream. The extent of the pattern suggests that the changes were occurring between 100 and 400 years ago.

INTRODUCTION AND BACKGROUND

Fast flowing outlet glaciers and ice streams form the dominant drainage system for the Antarctic Ice Sheet (e.g. Figure 1 in *Budd and Warner*, 1996). It has been estimated that although only 13% of the coastline consists of outlet glaciers and ice streams [*Drewry et al.*, 1982], they are responsible for discharging about 90% of the accumulation falling inland of the coastal zone [*McIntyre*, 1985]. By acting as transition regions between ice sheet flow and ice shelf flow they play a role, not yet properly understood, in controlling the state of equilibrium of the ice sheet [*Hindmarsh*, 1993; 1996]. Changes in ice stream flow would alter the mass balance of the ice sheet, affecting directly global sea level. The conditions for ice streams to form are not fully determined, although factors such as the topography and nature of the underlying bed and the climatic regime, as typified by accumulation and temperature, are important. Recent experiments with numerical models [*Payne et al.*, in press] show that thermomechanical coupling can generate instabilities in the flow, giving patterns resembling ice streams even on a flat bed. Modest bed topography can be sufficient to anchor ice streams, although there may be oscillatory behaviour if adjacent ice streams can capture neighbouring catchment areas [*Payne and Donglemans*, 1997]. The Siple Coast ice streams are examples of ice streams formed over a bed with very subdued topography, while most outlet glaciers elsewhere in Antarctica are constrained by bed troughs of varying size and depth.

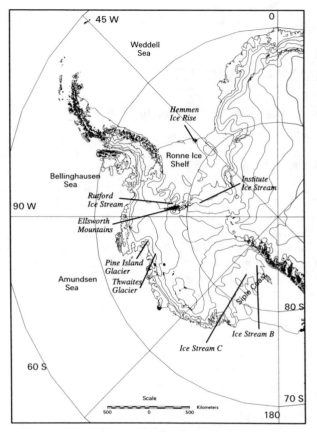

Fig. 1. Location map.

leads to a general absence of crevassing there, which was one of the primary attractions of the area when a study of grounding line dynamics was initiated in 1978. The other reason for selecting this area was that surveys could be tied to rock stations in the Ellsworth Mountains, allowing absolute velocities to be obtained [*Stephenson and Doake*, 1982]. Before satellite fixing was available, this was an important consideration.

Most of the ice stream area falls within the satellite orbit limits of 82 °S for Landsat and 79.5 °S for ERS SAR imagery, providing good high resolution coverage at both visible and radar wavelengths (Figure 3). Identification of surface features related to basal topography and the flow patterns have helped in the interpretation of ground data and in understanding the dynamics. Since the launch of ERS-1 in 1991, SAR images have revealed other features. Most prominent are bright shear margins where crevassing, not always apparent on the surface, increases the radar backscatter coefficient. Similar radar-bright shear bands have been identified on other ice streams and glaciers and around ice rises in ice shelves [*Sievers et al.*, 1993]. The (sinuous) location of the grounding line has been most accurately identified by SAR interferometry [*Goldstein et al.*, 1993; *Rignot*, 1998a], confirming earlier estimates using tiltmeters [*Stephenson et al.*, 1979; *Stephenson*, 1984].

Comprehensive ground surveys of Rutford Ice Stream were carried out between 1978 and 1986 to measure strain, velocity and elevation [*Doake et al.*, 1987]. A

Rutford Ice Stream provides an example of a fast flowing glacier constrained for most of its course by a deep bedrock trough, bounded by the Ellsworth Mountains on its west flank and Fletcher Promontory to its east (Figures 1 and 2). The rugged nature of the Ellsworth Mountains, which include the Vinson Massif, at 4897 m the highest point in Antarctica, contrasts with the much smoother outline of Fletcher Promontory, resting on a gently inclined bed up to a few hundred meters below sea level. The bed of Rutford Ice Stream reaches depths of more than 2000 m below sea level, creating a vertical relief of 7 km in a horizontal distance of less than 40 km. The ice stream occupies a major tectonic feature which has probably given it a stable position for millions of years [*Storey et al.*, 1988].

Today the major part of the ice stream flows southeast into Ronne Ice Shelf. The velocity decreases downstream from the grounding line for 190 km, turning north-east before starting a monotonic increase to the ice front [*Jenkins and Doake*, 1991]. The resulting longitudinal compressive flow in the grounding zone area

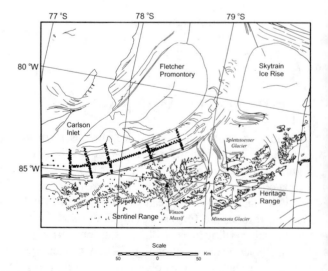

Fig. 2. Stake positions of survey carried out between 1978-1986 and of GPS measurements made between 1994-1996. Line features are derived from Landsat 5 MSS imagery acquired in January-March 1986. Grounding line at mouth of ice stream is taken from SAR interferometry (see Figure 8).

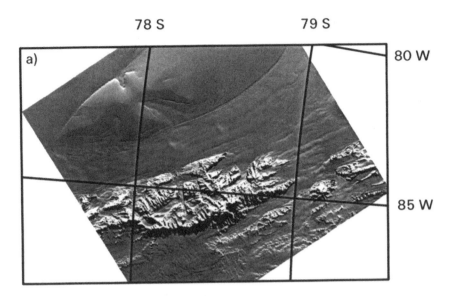

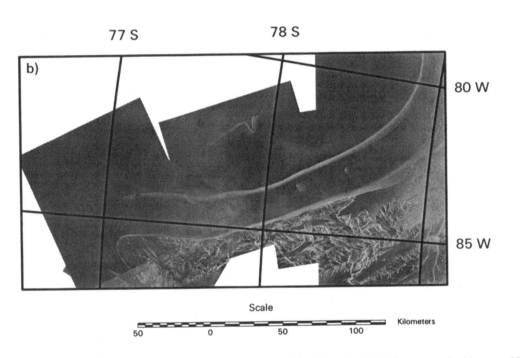

Fig. 3. a) Landsat image acquired on 3 February 1974. b) Mosaic of SAR images acquired between 10 and 12 January 1992 on descending orbits (look direction is approximately to the north-west). Bright bands of high backscatter delineate the shear margins of the ice stream.

triangulation network consisted of a 140-km line along the center and across the grounding line, and five transverse arms, one on floating ice and the rest on grounded ice (Figure 2). Surveying was done with theodolite and tellurometers, controlled by stations fixed with the TRANSIT satellite doppler positioning system. More detailed surveys across shear margins, along profiles near the grounding line and at individual stakes further upstream, have since been carried out with GPS receivers. Ice sounding radars have been used to measure ice thickness over the ice stream and to study near-surface conditions in a shear margin. Seismic techniques have been used to examine the nature of the bed at several locations.

GLACIOLOGICAL REGIME

Surface Topography

Surface elevations have been measured by several methods. Along the stake network they have been calculated from measurements of vertical angles tied to a TRANSIT satellite datum [*Frolich et al.*, 1987]. On a basin scale, airborne geophysical surveys have used pressure altimetry and, more recently, differential carrier phase GPS [*Corr and Doake*, 1998]. High floating Tropical Wind Energy conversion and Reference Level Experiment (TWERLE) balloons crossed the continent during the 1970s measuring surface elevation with a radar altimeter [*Levanon*, 1982]. But the most consistent data set has been obtained by radar altimetry from the ERS satellites [*Bamber and Bindschadler*, 1997]. Away from slopes that exceed about one degree, satellite altimetry gives comprehensive coverage, although the footprint of about 20 km precludes a useful resolution of much less than 5 km over rougher terrain.

A clear relationship between surface slope and brightness in high resolution satellite (SPOT) imagery was shown by *Vaughan et al.* [1988] which formed the basis of a 'shape from shading' technique to recover topography from satellite images [*Cooper*, 1994].

Drainage Basin

Using a 1 km gridded data set from the Byrd Polar Research Center, the drainage basins of Antarctica have been defined using the hydrological GIS tools in ARC/Info. Results from this analysis show that Rutford Ice Stream drains an area of about 49,000 km^2, a tenth of the West Antarctic Ice Sheet. Figure 4 shows surface elevations for the drainage basin.

For the purposes of this paper, we consider Rutford Ice Stream to extend into Ronne Ice Shelf as far east as about 80 °W, to a line joining Fletcher Promontory and Skytrain Ice Rise (Figure 2). While some glaciers cut through the mountains, notably Newcomer Glacier at the northern end and Nimitz, Minnesota and Splettstoesser glaciers between the Sentinel and Heritage ranges, most of the ice is discharged through the main ice stream (Figure 2). The catchment area has an ice divide bordering the drainage basins of Institute Ice Stream, Pine Island Glacier, Carlson Inlet and Evans Ice Stream (Figure 4). The watershed to the west of the Ellsworth Mountains is not well defined and a portion of the upper basin of Pine Island Glacier may drain into Rutford Ice Stream. Ice flowing around the Sentinel Range, the northern and highest part of the Ellsworth Mountains, constitutes the main drainage pattern of Rutford Ice Stream. The majority of the ice feeding the ice stream begins west of the Ellsworth Mountains at elevations up to 2000 m above sea level, flows northward then turns sharply eastward around the north end of the Sentinel Range before flowing southward into the fast moving part of the ice stream. The bed beneath the ice stream dips to more than 2000 m below sea level.

Ice Thickness and bed Elevation

Ice sounding radars have successfully measured thickness over most of the ice stream and catchment area. Numerous profiles have been flown both along and across the ice stream [*Corr et al.*, 1996; *Corr and Doake*, 1998], giving good resolution not only of the fast flowing part of the ice stream but also of the upstream area and drainage basin. At the upper end on the ice stream, thicknesses reach 3100 m. The main ice stream flows in a trough where adjacent to the Ellsworth Mountains thicknesses of more than 2300 m are found and the bed is at about 2000 m below sea-level [*Frolich and Doake*, 1988]. In a few areas there have been extensive seismic shooting experiments, mainly for investigating basal properties, but giving thickness as well. [*Smith*, 1997a; 1997b]. The two methods agree well, where there is overlap.

Flow

Strain-rate and velocity results have been presented by *Doake et al.*, [1987], *Frolich et al.*, [1987, 1989] and *Frolich and Doake* [1988]. The pattern of strain-rate trajectories shows the isotropic points expected from the lateral variations in width and bed topography [*Frolich et al.*, 1989]. The onset region of fast flow [*Bindschadler et al.*, this volume] occurs at the northern end of the Sentinel Range near (77 °S, 86 °W) (Figure 3). Velocities rise to about 200 m/a soon after the onset and have reached 300 m/a by the time the ice stream flows past the junction with Carlson Inlet. On the Carlson Inlet side of the shear margin, velocities are less than 10 m/a. Maximum velocities of around 400 m/a are reached about 40 km upstream of the grounding line on the Ellsworth Mountains side of the center-line.

SAR interferometry has been used by *Stenoien* [1998] to calculate velocity in the drainage basin to the north and west of the Ellsworth Mountains. Overlapping ascending and descending passes allowed two components of the velocity vector to be measured which, when integrated from a point with known velocity, gave estimates of the

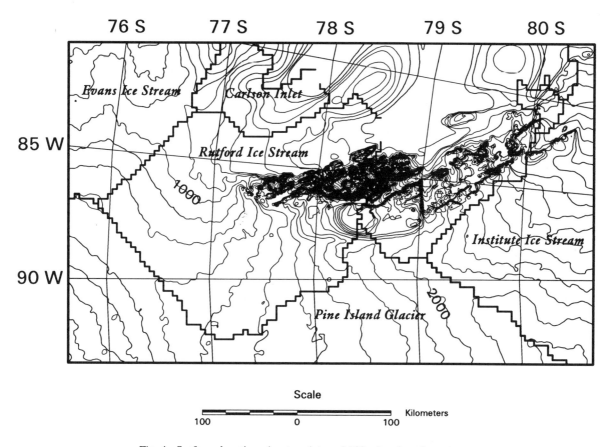

Fig. 4. Surface elevations (contour interval 100 m) and catchment area.

full velocity field. A maximum of 150 m/a was found for the area near (77 °S, 86 °W), close to the onset region.

Accumulation and Temperature Measurements.

Accumulation measurements are sparse. Two shallow cores to 10 m depth obtained at the lower end of the ice stream yielded values of annual accumulation of 0.3 ± 0.03 Mg m^{-2} a^{-1}. Additional accumulation results come from stake measurements. The pattern is of an increase in accumulation going upstream [*Doake et al.*, 1987], which is part of a regional gradient stretching across Ronne Ice Shelf and Ellsworth Land. The main source area for the precipitation is probably in the Bellingshausen and Amundsen seas, transported by the prevailing depression tracks coming off the South Pacific. Using a revised data base for accumulation over the ice sheet [*Vaughan et al.*, 1999] and delineating the catchment basin from satellite radar altimetry (Figure 4), gives a balance flux for the Rutford Ice Stream of 18.7±1 Gt/a. An estimate of the discharge across the grounding line is 18.5 ± 2 Gt/a [*Crabtree and Doake*, 1982], suggesting that the ice stream is in overall equilibrium.

Temperatures measured at 10 m depth range from -26 °C at the junction with Carlson Inlet to -28 °C near the grounding zone of the ice stream. This kind of inversion is common on ice shelves and their immediate hinterland and is caused by the thermal structure of the atmospheric boundary layer [*King et al.*, 1998].

SHEAR MARGINS

The shear margins on Rutford Ice Stream show a variety of features. On the western side, tributary glaciers flowing from the Ellsworth Mountains create surface undulations which show up as lineations on satellite visible imagery and aerial photography. Just downstream of the grounding line, the Minnesota, Nimitz and Splettstoesser glaciers join the Rutford between the Sentinel and Heritage ranges and create a zone of large crevasses as well as surface undulations. On the eastern side of the ice stream, abutting Fletcher Promontory, the shear margin has a more subdued topographic signature reflecting the smaller amount of ice flowing into the margin. However, there are characteristic patterns, such as the 'wave' nature of undulations on the floating portion around Fletcher Promontory, where large open crevasses are observed. In general, there are fewer open crevasses on the eastern side compared with the western. Upstream, at the junction with Carlson Inlet, there is little change in bed topography across the margin, making it similar to those found on the Siple Coast ice streams. GPR measurements show no sign of buried crevasses either in the margins of Carlson Inlet or in its main body, unlike on Ice Stream C [*Retzlaff and Bentley*, 1993], suggesting it has been inactive for at least 300 years (the estimated age of the deepest ice sounded by the GPR).

SAR images show ice stream margins well, highlighting them as bright bands, up to about 5 km wide, of increased backscatter compared with surrounding areas. Similar bright regions have been identified around ice rises, for example Hemmen Ice Rise in the Ronne Ice Shelf, [*Sievers et al.*, 1993]. Fracturing is believed to be responsible for the increased scattering, occurring down to the penetration depth of tens of meters in dry snow of the C-band ERS and Radarsat radars [*Partington*, 1998]. The longer wavelength of the L-band radar on the JERS satellite gives even better discrimination of the shear margins, especially in regions where the ice may be warmer and contain more icy, near-surface layers.

There is a strong correlation between backscatter coefficient and strain-rate across the shear margin. A relationship between the backscatter coefficient and the strain rate has been determined such that a critical value of strain rate is required for the onset of the margin signature, and is related to the critical stress needed for fracture and crevassing [*Vaughan*, 1993; *Vaughan et al.*, 1994]. The backscatter coefficient is dependent on the orientation of the radar beam, giving the qualitative result that the fractures or crevasses are aligned in the direction expected from theory, i.e. pointing upstream towards the middle of the ice stream. Comprehensive measurements using shallow ground probing radar (GPR) on a rectangular grid pattern across the shear margin with Carlson Inlet reveal a host of reflectors at depths between 5 and 40 m aligned in the expected direction. A consistent picture emerges of fracturing generated where the stress exceeds a critical value of about 160 kPa. This critical stress need not necessarily be attained at the surface but may be reached at shallow depths, down to a few tens of meters, as revealed by SAR backscatter. In places fractures reach the surface and form open crevasses, which may in turn be covered by snow fall. In effect, shear margins are defined by a critical value of the stress or equivalently, by a change in the SAR backscatter coefficient. How the position of shear margins might be related to subglacial boundaries of temperature, topography, lubrication etc is discussed by *Raymond et al.* (this volume).

Flow over two prominent knolls produces features similar in some respects to those found in the margins (Figure 3). SAR imagery reveals increased backscatter, beginning upstream of the topographic expression. Shallow seismic refraction soundings undertaken about 2 km upstream of the upstream knoll shows anomalous results, consistent with a fractured zone at a few tens of meters depth scattering seismic waves. GPR sounding confirms that there are buried crevasses, whose tops gradually rise and break the surface as the knoll is approached.

Detailed velocity measurements have been carried out across the shear margin with Carlson Inlet. Two lines of stakes near the upper transverse arm of the 1978-1986 stake scheme have been measured with GPS in 1994 and 1996, giving 3-D movements accurate to a few mm. Continuous profiling with kinematic GPS across one of the lines tested the achievable accuracy with a faster technique. These results have been used to calibrate SAR interferograms for 1992 and 1994 and compared with survey velocities measured in 1984-86 [*Frolich and Doake*, 1998]. The conclusion is that there has been a significant 'fluttering' of the margin velocity, such that on a decadal time scale the velocity can fluctuate by tens of meters per year in the central part of the shear zone. An alternate interpretation is of the ice stream width changing.

Modelling by R.M. Frolich (personal communication, 1998) using an approximation that treated the ice stream as an ice shelf with basal friction [*MacAyeal*, 1989] showed that a softening in the margins was necessary to explain the observed flow pattern. Either the development of a fabric or an increase in temperature are possible explanations [*Echelmeyer et al.*, 1994]. Several holes were drilled in 1993 across the shear margin with Carlson Inlet, close to where the GPS velocity profiles were obtained (Figure 2), using hot water to penetrate down to 300 m. Four holes were instrumented with thermistors and three remained the following year for remeasurement. Unfortunately the hole closest to the center of the shear margin was lost (the suspicion is that it was very close to a subsurface fracture). Results from the remaining three (Figure 5) show that the shear margin was about 0.1 °C warmer at 300 m depth than on either side, similar to the pattern seen in Ice Stream B [*Harrison et al.*, 1998]. All three profiles showed a minimum temperature at about 100 m depth, which can be modelled by assuming a 0.5 °C warming in the last 100 -150 years.

R.M. Frolich (personal communication, 1998) has developed a reduced model of induced anisotropy in ice

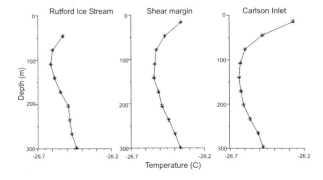

Fig. 5. Temperature profiles in the shear margin measured at three sites, one near Rutford Ice Stream, one near the middle of the shear margin and the other near Carlson Inlet. The spacing between the Rutford Ice Stream and middle sites is 5.7 km and between the middle and Carlson Inlet sites is 2 km.

streams and confined ice shelves where the absence of leading order vertical shear allows semi-analytic modelling of anisotropic effects [*Morland and Staroszczyk*, 1998]. 'Enhancement factors' have been used in the conventional flow law to simulate anisotropy, and are needed to explain velocity profiles on Ice Stream B, where a combined temperature/fabric enhancement of up to 10 has been required. Perhaps surprisingly, ice samples from 300 m depth in the shear margin of Ice Stream B do not show a strongly developed fabric [*Jackson and Kamb*, 1997], so there is still uncertainty about how to account for the localized softening in shear margins [*Raymond et al.*, this volume].

BASAL REGIME

The nature of the glacier bed and the way in which the basal conditions control the ice stream flow have been investigated using seismic techniques. Movement can be by three mechanisms: ice creep, basal sliding or deformation of mobile sediment. The relative proportions of these mechanisms will affect the way the flow regime develops - or remains unchanged - with time, under the regional climatic, geological and oceanographic conditions.

High resolution seismic reflection surveys were carried out on Rutford Ice Stream between 1991 and 1993 [*Smith*, 1994]. The surveys covered a region between about 40 and 90 km upstream from the grounding line, straddling the central part of the flow. The data acquired were of sufficiently high quality to allow the reflection coefficient of the ice bed interface to be calculated and hence the acoustic impedance of the bed material. The results show a pattern of varying basal conditions. Regions of soft deforming till and others of more competent sediment occur adjacent to one another [*Smith*, 1997a]. This pattern varies both along the ice stream (i.e. parallel to ice flow) and in places across the ice stream too [*Smith*, 1997b]. The proportion of the ice stream width which appears to be underlain by deforming till increases with distance downstream. Sub-parallel stripes of the two different bed types are aligned roughly in-line with the ice flow direction. A similar pattern has been found beneath the downstream part of Ice Stream B [*Atre and Bentley*, 1994]. By three-quarters of the way downstream (about 40 km from the grounding line), only a quarter of the ice stream width appears to consist of non-deforming sediments and this region is concentrated roughly in the middle of the flow.

The effect of basal variability on the ice stream flow will depend on the friction associated with the different bed types. This, in turn, will be strongly influenced by the sub-glacial hydrology, about which very little is known beneath Rutford Ice Stream. The downstream increase in bed deformation may be associated with a decreasing effective pressure (the ice column getting closer to floating on the sub-glacial water) leading to dilation and weakening of the sediments, allowing them to deform more easily. Alternatively, the ice could be so close to flotation further upstream, that it is lifted off the bed, dislocating it from the sediments and hindering deformation. Whichever the case, the non-deforming parts of the bed are probably characterised by greater amounts of basal sliding. What cannot be interpreted from the seismic reflection data are the relative degrees of restraint to ice flow associated with the two types of bed. In addition, the degree of motion accommodated by enhanced creep within the ice is still unknown. However, the relatively low basal shear stress (around 40 kPa on average), suggests that it is unlikely to be the major component of ice flow.

The seismic reflection surveys were extended in 1997-8 and supplemented with passive seismic monitoring to investigate the distribution of friction at the ice stream bed [e.g. *Blankenship et al.*, 1987; *Anandakrishnan and Bentley*, 1993]. The hypothesis was that non-deforming sediments were associated with higher friction and greater amounts of seismicity. Preliminary results appear to confirm this [*Smith*, 1998]. That these friction variations are not reflected directly in the surface velocity field is almost certainly due to the thickness of the ice - considerably more than 2 km over all the seismic lines. Deep variations in ice flow would almost certainly be smoothed out over this thickness of ice, though they may still influence the overall ice stream flow and possible future variations in the ice stream dynamics [*MacAyeal et al.*, 1995].

Basal Conditions and the State of Equilibrium of the ice Stream

Interpretation of the seismic data suggests that the ice stream bed consists of sediments. In some places the sediments are undergoing significant deformation but elsewhere basal sliding is a more important flow mechanism. Bed deformation necessarily involves downstream transport of the sedimentary material, which must be replenished for the ice stream flow to continue for a long period of time. As there are no indications of significant non-steady behaviour in Rutford Ice Stream, we examine whether the suggested basal regime is consistent with this evidence.

Smith [1997c] has considered a simple model in which the bed is taken to be a slab of sediment 200 km long, 25 km wide and 20 m thick. Only the most downstream third is assumed to be dilated and thus identified as deforming. Assuming that the velocity within the deforming sediment decreases linearly from the ice velocity (350 m/a) at the top surface to zero at the bottom, the sediment flux discharged at the grounding line is approximately 0.1 km^3/a. This flux must be matched by sediment supply from further upstream for long term equilibrium. Smith calculates the supply of sediment by three mechanisms. One mechanism is melting of sediment-rich basal layers in the ice stream. A basal melt rate of 47 mm/a in ice with 10% sediment by volume in the bottom 5 m as in Byrd core [*Gow et al.*, 1979] would account for 4% of the discharge. Another mechanism is downstream transport from the non-dilated area. Assuming that shearing in the bed can occur without dilation at speeds of up to 75 m/a [*Alley et al.*, 1987], then this mechanism could account for 40% of the discharge. The third mechanism is erosion of the rock substrate beneath the dilated part of the sediment slab. A rate of 10 mm/a as suggested for Ice Stream B [*Alley et al.*, 1987] would be enough for 20% of the discharge. However, the whole sediment slab contains only enough material for 1000 years discharge for the assumed grounding line flux of 0.1 km^3/a.

These calculations suggest that transport and erosion could supply up to 60% of the material required to maintain the basal sediment system. Estimated factors could easily be inaccurate by a factor of two or more and error bars are probably at least 100% of resulting values. Hence, the supply, transport and deposition of sediment beneath the ice stream may or may not be in equilibrium. Although this appears to be an inconclusive result, the possibility that the sediment system *could* be in equilibrium is important. There is no evidence for unsteady flow of Rutford Ice Stream upstream of the grounding line and on a time scale of a thousand years it does not appear to have undergone any dramatic change. This is consistent with a basal system which is able to retain its basic characteristics for that length of time. However, it raises questions of how much the flow depends on deforming sediments, what determines the supply of sediment to the ice stream bed, and how the flow in the region could change if all the sediments were eroded.

ONSET OF STREAMING

The onset of fast flow can be defined as a shift in the relationship between driving stress and velocity. In ice streams the velocity increases although the driving stress decreases. Upstream of the onset the velocity increases as the driving stress increases [*Bindschadler et al.*, this volume].

An airborne campaign of geophysical surveying with ice sounding radar and aeromagnetics around the north end of the Ellsworth Mountains in 1997/98 covered the important area where the onset of streaming occurs. Preliminary analysis of the data shows that there is a topographic low in the bed at (77 °S, 86 °W) (Figure 6), around the position where the ice stream velocity increases. This might be interpreted as scour of overlying sediment. SAR interferometry shows that along the 'center-line' the velocity increases over a distance of around 50 km from a typical value of tens of meters per year to hundreds of meters per year (Figure 7). Care should be taken in interpreting the interferogram as it shows the velocity component in only one direction (south-east north-west), so it is not an accurate representation of the velocity in the direction of flow.

McIntyre [1985] has pointed out a correlation between a velocity increase and a step in the bed for other Antarctic ice streams where a transition from converging to streaming flow takes place. The transition is assumed to be associated with a change from motion due to internal deformation to basal sliding. It is usually accompanied by a change in surface slope, allowing satellite imagery in the visible wavebands to be used to delineate the extent of ice streams. McIntyre suggested that the transition to fast flow was not due solely to the basal ice temperature being at the pressure melting point, but that the topography played a crucial triggering role. In his model, a topographic step

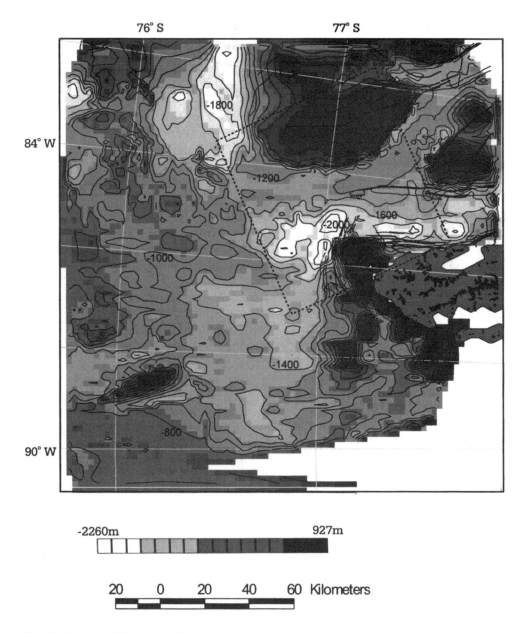

Fig. 6. Topographic low in bed at head of Rutford Ice Stream. Contour lines of bed elevation are at 200 m intervals. The dotted box outlines the area covered by Figure 7.

controls the onset of sliding and thus induces a form of stability in the ice sheet. Erosion of the sub-glacial headwall of the step at rates of say 4 mm/year, which is the value deduced for Byrd Glacier, would allow a slow development of the glacial trough and upstream migration of the ice stream.

Rutford Ice Stream conforms broadly to this general model. There is converging flow in most of the catchment area from the divide down to the onset of streaming. In the onset area, there is a step in the bed of around 100 m. Sliding is inferred from the rapid downstream increase in surface velocity but slow decrease in surface slope. Ice thicknesses vary around a mean of 2500 m.

GROUNDING ZONE AND TIDAL FLEXING

Determining the precise position of the grounding line is complicated by the fact that most methods measure the

230 THE WEST ANTARCTIC ICE SHEET: BEHAVIOR AND ENVIRONMENT

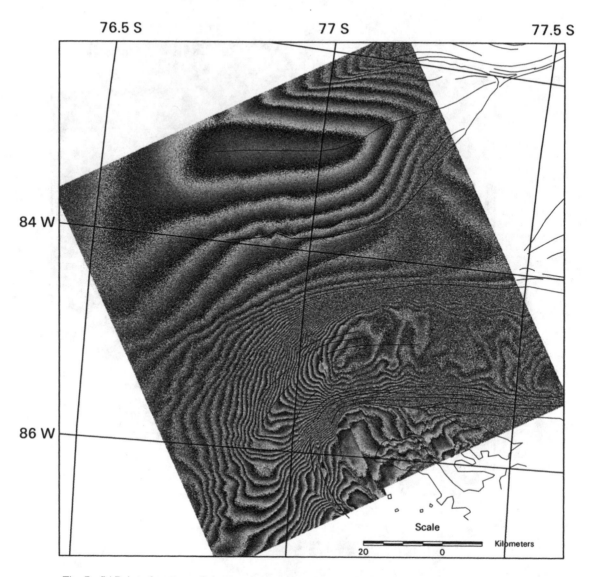

Fig. 7. SAR interferogram of onset region constructed from ERS-1 images acquired on 25 and 28 January 1994 (orbits 13223 and 13266, frame 5265). Fringe pattern illustrates increase in velocity downstream. Contour interval of fringes is about 8 m/yr in satellite look direction (approximately northwest), but varies with incidence angle. Line features are taken from Landsat 5 MSS imagery acquired between January-March 1986

limit of tidal flexing, sometimes called the hinge line. The difference between the grounding line, flexing limit, and position of hydrostatic equilibrium has been discussed by *Smith* [1991] and *Vaughan* [1994]. Tilt meters [*Stephenson et al.*, 1979; *Stephenson*, 1984], kinematic GPS profiling [*Vaughan*, 1995], and SAR interferometry [*Rignot*, 1998a] can only measure the upstream limit of surface deflection or vertical movement. Radar and seismic sounding have been unable to identify reliably a grounding line on the ice stream. There is little obvious change in character of echos from ice that is definitely grounded and definitely afloat [*Smith*, 1991]. Strand cracks have been noticed by field parties working in the grounding zone, and occur between the hydrostatic point and the limit of flexing.

Best estimates of the hinge line position come from SAR interferometry because of the technique's high spatial resolution, wide area coverage, and low noise [*Goldstein et al.*, 1993; *Rignot*, 1998a]. The position shown in Fig 8 has been taken from an interferogram computed from SAR data acquired on 7 and 13 February 1992. The estimated error in position is around 200 m,

Stephenson, 1984] are close to the interferometry position. However, the new position places the hinge line on the crest of the knoll (Figure 8b), while the old position was closer to its base. Along the shear margin with Fletcher Promontory, the hinge line is inland of the topographic break in slope seen on the Landsat image (Figure 8b) and also inland of the band of high backscatter shown in the SAR image (Figure 8a). The limit of flexing found by *Vaughan* [1995] by kinematic GPS surveys agrees well with the interferometrically derived hinge line.

A simple elastic bending model is adequate to explain tiltmeter, kinematic GPS profile and SAR interferometry results [*Smith*, 1991; *Vaughan*, 1995; *Rignot*, 1998a]. Fitting the model to both the GPS and the SAR data gave similar results, showing that an elastic modulus of 0.88 ± 0.35 GPa is appropriate for ice flexing at tidal frequencies [*Vaughan*, 1995; *Rignot*, 1998a]. However, a similar analysis using profiles derived from SAR interferometry for Petermann Gletscher, Greenland yielded a rather different value for the elastic modulus of 3.0 ± 0.2 GPa. [*Rignot*, 1996]. The discrepancy is unexplained. The tidal amplitude (several meters) over the floating portion has been measured by gravimeter [*Doake*, 1992], confirmed by GPS profiling [*Vaughan*, 1995] and agrees with an ocean tidal model [*Robertson et al.*, 1998].

The sinuous nature of the hinge line where it curves around a surface knoll, reflecting a step in the bed, shows that the sub-glacial topography plays an important role in determining its position. Interferometry has shown similar pinning of hinge lines on knolls for Carlson Inlet [*Rignot*, 1998a] and Pine Island Glacier [*Rignot*, 1998b]. A relatively steep bed slope such as that around the knoll would tend to stabilise the position (to small perturbations) while the flatter slopes on either side would allow more rapid migration, perhaps in response to changes in ice thickness propagating down the glacier.

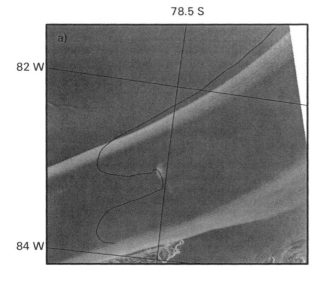

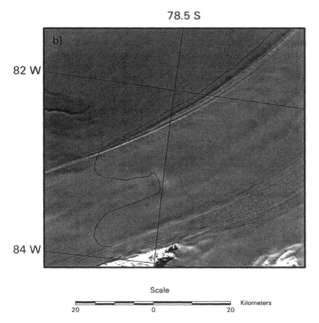

Fig. 8. Hinge line position inferred from interferometry displayed on: a) ERS-1 SAR backscatter image, showing tidal flexing inland of band of high backscatter in shear margin; b) Landsat image, showing flexing limit on crest of knoll at grounding line, and inland of topographic break in slope along shear margin with Fletcher Promontory.

which includes both the uncertainty in selecting where the fringe pattern changes, representing the limit of flexing, and also geo-referencing the interferogram. No change in hinge line position has been observed between 1992 and 1996 [*Rignot*, 1998a]. Earlier estimates of position around the downstream knoll where it was inferred from tiltmeter measurements of tidal flexure [*Stephenson et al.*, 1979;

FLOATING PORTION

Estimates of surface velocity derived from comparison of Landsat (1974) and SPOT (1987) images [*Vaughan et al.*, 1988] relied on there being a sufficiently distinctive pattern that was stable enough to be recognised after an interval of 13 years. The pattern was provided by an undulating surface topography of up to 30 m relief reflecting the underlying thickness variations. The fact that this pattern cannot be reproduced by flow across the present day grounding line position [*Smith and Doake*, 1994] illustrates that there have been considerable fluctuations in the last few hundred years. It is uncertain whether or not the thickness pattern has been produced by a change in position of the grounding line or by variations in flow from the side glaciers such as Minnesota Glacier. It is unlikely that variations in melt rate are responsible because of the short spatial scale of the pattern.

The present-day thickness profile at the grounding line has two thicker branches on either side of a thinner central portion which has flowed over the knoll. In the 1974 Landsat image, the first significant departure from this profile occurs about 35 km downstream of the grounding line position (see Figure 4 in *Smith and Doake*, 1994). A velocity of around 350 m/a [*Jenkins and Doake*, 1991] gives a time of 100 years (or a date about 1874 AD) since the last major fluctuation, although there are indications of more recent disturbances. The pattern of undulations is visible for about another 100 km downstream before it is smoothed away, so the onset of the process creating the pattern must have started at least 300 years earlier, i.e. before 1574 AD.

First attempts to explain this pattern [*Doake et al.*, 1987] were misled by the character of the radar echo from the base of the thinner ice areas. The contrast between flat, smooth and strong basal echoes from the thicker ice and the extended, more intermittent echoes from areas with thin ice suggested that the thinner ice was lightly grounded. Comparison between surface elevation and ice thickness tended to support this conclusion because although the surface elevations in the thin ice areas were lower compared with the surrounding ice, they were apparently too high to be floating in hydrostatic equilibrium. However, the realisation from the SPOT 1987 image that the pattern was moving with the ice shelf [*Vaughan et al.*, 1988] meant that a reappraisal was necessary with the conclusion that the whole ice stream was floating in this area.

Jenkins [1988] used profiles of surface elevation and ice thickness to plot a map of isostatic anomaly (measured surface elevation minus surface elevation required for hydrostatic equilibrium). Positive anomalies coincided with surface depressions and thinner ice. The size of the anomalies, mostly between ±10 m but rising to +20 m in places, suggests that limited vertical shear stresses can be supported in floating ice and are of the same magnitude found, for example, in ice rumples. However, marine ice could be accreted in the thin ice areas by the same mechanism as operates in other areas on the Ronne Ice Shelf - upwelling of water to a depth where its temperature is at or colder than the in situ pressure freezing point, so forming frazil ice crystals [*Bombosch and Jenkins*, 1995]. If the radar reflections were from an internal horizon, such as that found in the center of Ronne Ice Shelf at the boundary between meteoric and marine ice, then a greater actual ice thickness than that measured could explain most of the positive anomalies. An argument against this is that the reflection coefficients found were higher than those calculated for the marine ice areas in the Ronne Ice Shelf. Also, it is difficult to conceive of a mechanism whereby the actual ice thickness is thinner than that measured, needed to explain negative anomalies. Surface features on ice streams, created by bed undulations with wavelengths of the order of the ice thickness, can last for several hundred years [*Gudmundsson et al.*, 1998]. The character and extent of the pattern here are compatible with theories of persistence of surface features with short spatial scales and require no special properties of ice rheology.

Rignot [1998a] has revived the idea that the ice shelf may be intermittently grounded at low tides, based on his interpretation of an area of closely spaced fringes in SAR interferograms, about 20 km downstream from the grounding line knoll. However, no signs of tidal flexing were found from tiltmeters installed further downstream [*Jenkins*, 1988]. Seismic data show that water column thicknesses in the area are generally more than 100 m, although a very localised shoaling may have been missed by the seismic sites being on a grid spacing of about 10 km [*Smith and Doake*, 1994]. The sea bed topography consists of two deep troughs separated by a 500 m high ridge. It is possible that the interferometry is picking up complex tidal interactions caused by the sea bed topography, similar to the 'bump and dimple' feature identified in SAR interferometry along the Ronne Ice Front [*Rignot et al.*, submitted].

Melt rates at a number of points 25 km apart along a flowline have been calculated from surface measurements of thickness, velocity and their gradients [*Jenkins and Doake*, 1991]. Values decrease from about 2 to 4 m/a near the grounding line to less than 1 m/a 100 km downstream, but with considerable variability and uncertainty caused mainly by the uneven nature of the thickness pattern. *Smith* [1996] used the variation in depth of internal reflectors in seismic profiles across the grounding line at the knoll to deduce melt rates in the first 4 km of floating ice. His values ranged up to 7 m/a, with a mean of about 3 m/a in the downstream half of the line. *Corr et al.* [1996] tried to calculate average melt rates between a series of 'gates' or cross-sections about 10 km apart using airborne radar ice thickness measurements. In the area 70 km immediately downstream of the grounding line, the average melt rate was 2.7 ± 0.5 m/a. Errors in calculating the velocity profiles were too large to be able to deduce meaningful melt-rate values between individual gates.

STATE OF EQUILIBRIUM

The force balance of the ice stream has been calculated by *Frolich et al.* [1987] and *Frolich and Doake* [1988]. On average, the driving force of 40 kPa is equally partitioned between restive forces at the base and shear stresses from the margins. Steps in the bed generate 'bending' or 'bridging' stresses, which are important locally. Detailed analysis shows that most of the restraint from the margins is felt only in boundary layers about 10 km wide. In the middle portion of the ice stream, the lateral shear restraint is close to zero and thus the driving force must be balanced mainly by shear stress at the base [*Frolich and Doake*, 1988]. This conclusion is supported by the seismic evidence that the non-deforming sediments are concentrated in the middle of the flow.

There are not many indications of change in Rutford Ice Stream and its drainage basin. Perhaps the most obvious is the pattern of surface undulations that persists for 150 km downstream from the grounding line. However, without contemporary evidence of changes at the grounding line or in the tributary glaciers such as Minnesota Glacier, a convincing and reasonable explanation for the pattern is not offered here. A plausible explanation is that fluctuations in ice thickness, caused by medium to long term (10 to 100 years) variations in accumulation rate in the drainage basin, make the grounding line oscillate around a mean position that is pinned by the knoll. Support for this hypothesis is given by the observations using SAR interferometry of small fluctuations in the shear margin with Carlson Inlet. Lack of any significant change in the grounding line position between 1992 and 1996 from SAR interferometry is compatible with the unchanging cross sectional thickness profile immediately downstream of the grounding line.

There appears to be a tongue of thicker ice about 20 km downstream from the grounding line on the west side which may indicate an influx sometime early this century. To decipher the pattern in terms of the history of the grounding line position would be an inverse problem where it is necessary to allow for spreading, melting, relaxation etc. This has not been attempted.

DISCUSSION

Rutford Ice Stream appears to be in balance at present between input over its catchment basin and output across the grounding line. Because its character is typical of many Antarctic ice streams and there is already extensive knowledge of its regime, it offers an excellent natural laboratory for further study of critical processes operating under near equilibrium conditions.

It is interesting to speculate on the varied observations made on ice streams such as Rutford Ice Stream (fluctuations in margin velocity/position), Pine Island Glacier (retreat in grounding line position), Thwaites Glacier (increase in velocity at grounding line), and the Siple Coast ice streams (changes in flow regime, recent grounding line advances and retreats). Are we seeing responses representing the natural variability of ice streams which are basically in harmony with their environment, or is each ice stream independently revealing an aspect of change in response to a forcing which may be external, such as climate change, or internal, such as the basal thermal regime?

Marine ice sheets have been modelled as land-based ice sheets but with a bed below sea level and discharging at the margins into an infinite sink [*Weertman*, 1974; *Hindmarsh*, 1993]. While ice rises embedded in ice shelves may conform to this ideal [*Hindmarsh*, 1996], discharge from the West Antarctic Ice Sheet is dominated by ice stream flow. If we are to understand the nature of a marine ice sheet like the West Antarctic Ice Sheet and to predict its future behaviour, then we need to improve our understanding both of the fundamental processes which control the flow of ice streams and of the role played by ice streams in determining the stability of the ice sheet.

Acknowledgments. We would like to thank all those who have helped us to gather data in the field and to Matthew Lythe who helped process data for some of the figures. Bob Bindschadler and an anonymous referee made useful comments improving the manuscript.

REFERENCES

Alley, R.B., D.D. Blankenship, S.T. Rooney, and C.R. Bentley, Till beneath Ice Stream B. 4. A coupled ice-till flow model, *Journal of Geophysical Research*, 92, 8931-8940, 1987.

Anandakrishnan, S. and C.R. Bentley. Micro-earthquakes beneath ice streams B and C, West Antarctica: observations and implications, *Journal of Glaciology*, 39, 455-462, 1993.

Atre, S.R. and C.R. Bentley. Indication of a dilatant bed near Downstream B camp, Ice Stream B, Antarctica. *Annals of Glaciology*, 20, 177-182, 1994.

Bamber, J.L. and R.A. Bindschadler, An improved elevation dataset for climate and ice-sheet modelling: validation with satellite imagery, *Annals of Glaciology*, 25, 439-444, 1997.

Bindschadler, R.A., J.L. Bamber, and S. Anandakrishnan, Onset of Streaming Flow in the Siple Coast region, West Antarctica, this volume.

Blankenship, D.D., S. Anandakrishnan, J. Kempf and C.R. Bentley, Microearthquakes under and alongside Ice Stream B, detected by a new passive seismic array, *Annals of Glaciology*, 9, 30-34, 1987.

Bombosch, A. and A. Jenkins, Modelling the formation and deposition of frazil ice beneath Filchner-Ronne Ice Shelf, *Journal of Geophysical Research*, 100, 6983-6992, 1995.

Budd, W.F. and R.C. Warner, A computer scheme for rapid calculations of balance-flux distributions, *Annals of Glaciology*, 23, 21-27, 1996.

Cooper, A.P.R., A simple shape-from-shading algorithm applied to images of ice-covered terrain, *IEEE Transactions on Geoscience and Remote Sensing*, 32, No. 6, 1196-1198, 1994.

Corr, H., M. Walden, D.G. Vaughan, C.S.M. Doake, A. Bombosch, A. Jenkins, and R.M. Frolich, Basal melt rates along the Rutford Ice Stream., in *Filchner-Ronne Ice Shelf Programme, Report No. 10*, edited by H. Oerter, pp 11-15, Alfred-Wegener Inst. for Polar and Mar. Res., Bremerhaven, Germany, 1996.

Corr, H.F.J. and C.S.M. Doake, Rutford Ice Stream reservoir, in *Filchner-Ronne Ice Shelf Programme, Report No 12*, edited by H. Oerter, pp 19-23, Alfred-Wegener Inst. for Polar and Mar. Res., Bremerhaven, Germany, 1998.

Crabtree, R.D. and C.S.M. Doake, Pine Island Glacier and its drainage basin: results from radio echo-sounding. *Annals of Glaciology*, 3, 65-70, 1982.

Doake, C.S.M., Gravimetric tidal measurements on Filchner Ronne Ice Shelf, in *Filchner-Ronne Ice Shelf Programme, Report No. 6*, edited by H. Oerter, pp 34-39, Alfred-Wegener Inst. for Polar and Mar. Res., Bremerhaven, Germany, 1992.

Doake, C.S.M., R.M. Frolich, D.R. Mantripp, A.M. Smith, and D.G. Vaughan, Glaciological studies on Rutford Ice Stream, Antarctica, *Journal of Geophysical Research*, 92, 8951-8960, 1987.

Drewry, D.J., S.R. Jordan, and E. Jankowski, Measured properties of the Antarctic ice sheet: surface configuration, ice thickness, volume and bedrock characteristics, *Annals of Glaciology*, 3, 83-91, 1982.

Echelmeyer, K.A., W.D. Harrison, C. Larsen, and J.E. Mitchell, The role of the margins in the dynamics of an active ice stream, *Journal of Glaciology*, 40, 527-538, 1994.

Frolich, R.M., D.R. Mantripp, D.G. Vaughan, and C.S.M. Doake, Force balance of Rutford Ice Stream, Antarctica, In *The Physical Basis of Ice Sheet Modelling. IAHS Publication No. 170*, edited by E.D. Waddington and J.S. Walder, 323-331, 1987.

Frolich, R.M. and C.S.M. Doake, Relative importance of lateral and vertical shear on Rutford Ice Stream, Antarctica, *Annals of Glaciology*, 11, 19-22, 1988.

Frolich, R.M., D.G. Vaughan, and C.S.M. Doake, Flow of Rutford Ice Stream and comparison with Carlson Inlet, Antarctica, *Annals of Glaciology*, 12, 51-56, 1989.

Frolich, R.M and C.S.M. Doake, SAR interferometry over Rutford Ice Stream and Carlson Inlet, Antarctica, *Journal of Glaciology*, 44, 77-92, 1998.

Goldstein, R., H. Engelhardt, B. Kamb, and R.M. Frolich, Satellite radar interferometry for monitoring ice sheet motion: application to an Antarctic ice stream, *Science*, 262, 1525-1530, 1993.

Gow, A.J., S. Epstein and W. Sheehy, On the origin of stratified deposits in ice cores from the bottom of the Antarctic ice sheet, *Journal of Glaciology*, 23, 185-192, 1979.

Gudmundsson, G.H., C.F. Raymond, and R. Bindschadler, The origin and longevity of flow stripes on Antarctic ice streams, *Annals of Glaciology*, 27, 145-152, 1998.

Harrison, W.D., K.A. Echelmeyer, and C.F. Larsen, Measurement of temperature within a margin of Ice Stream B, Antarctica: implications for margin migration and lateral drag, *Journal of Glaciology*, 44, 615-624, 1998.

Hindmarsh, R. C. A., Qualitative Dynamics of Marine Ice Sheets, In: *NATO ASI Series Ice in the Climate System. Vol. V.* edited by W.R. Peltier, 67-99, 1993.

Hindmarsh, R. C. A., Stability of ice rises and uncoupled marine ice sheets, *Annals of Glaciology*, 23, 94-104, 1996.

Jackson, M. and B. Kamb, The marginal shear stress of Ice Stream B, West Antarctica, *Journal of Glaciology*, 43, 415-426, 1997.

Jenkins, A., Recent investigations of surface undulations where Rutford Ice Stream enters Ronne Ice Shelf, in *Filchner Ronne Ice Shelf Programme, Report No. 4*, edited by H. Oerter, pp 12-17, Alfred-Wegener Inst. for Polar and Mar. Res., Bremerhaven, Germany, 1988.

Jenkins, A. and C.S.M. Doake, Ice-ocean interaction on Ronne Ice Shelf, Antarctica, *Journal of Geophysical Research*, 96, 791-813, 1991.

King, J.C., M.J. Varley, and T.A. Lachlan-Cope, Using satellite thermal infrared imagery to study boundary layer structure in an Antarctic katabatic wind region, *International Journal of Remote Sensing*, 19, 3335-3348, 1998.

Levanon, N., Antarctic ice elevation maps from balloon altimetry, *Annals of Glaciology*, 3, 184-186, 1982.

MacAyeal, D.R., Large-scale ice flow over a viscous basal sediment - theory and application to Ice Stream-B, Antarctica, *Journal of Geophysical Research*, 94, 4071-4087, 1989.

MacAyeal, D.R., R.A. Bindschadler, and T.A. Scambos, Basal friction of Ice Stream E, West Antarctica, *Journal of Glaciology*, 41, 247-262, 1995.

McIntyre, N.F., The dynamics of ice-sheet outlets, *Journal of Glaciology*, 31, 99-107, 1985.

Morland, L.W. and R. Staroszczyk, Viscous response of polar ice with evolving fabric. *Continuum Mechanics and Thermodynamics.* 10, 135-152, 1998.

Partington, K.C., Discrimination of glacier facies using multi-temporal SAR data, *Journal of Glaciology*, 44, 42-53, 1998.

Payne, A.J. and P.W. Donglemans, Self organization in the thermomechanical flow of ice sheets, *Journal of Geophysical Research*, 102, 12219-12234, 1997.

Payne, A.J. and 10 others, Results from the EISMINT Phase 2 simplified geometry experiments: the effects of thermomechanical coupling, *Journal of Glaciology*, in press.

Raymond, C.F., K.A. Echelmeyer, I.M. Whillans, and C.S.M. Doake, Ice stream shear margins, this volume.

Retzlaff, R. and C.R. Bentley, Timing of stagnation of Ice Stream C, West Antarctica, from short-pulse radar studies of buried surface crevasses, *Journal of Glaciology*, 39, 553-561, 1993.

Rignot, E., Tidal motion, ice velocity and melt rate of Petermann Gletscher, Greenland, measured from radar interferometry, *Journal of Glaciology*, 42, 476-487, 1996.

Rignot, E., Radar interferometry detection of hinge-line migration on Rutford Ice Stream and Carlson Inlet, Antarctica, *Annals of Glaciology*, 27, 25-32, 1998a.

Rignot, E., Fast recession of a West Antarctic glacier, *Science*, 281, 549-551, 1998b.

Rignot, E., L. Padman, D.R. MacAyeal, and M. Schmeltz, Analysis of sub-ice-shelf tides in the Weddell Sea using SAR interferometry, *Journal of Geophysical Research*, submitted.

Robertson, R., L. Padman, and G.D. Egbert, Tides in the Weddell Sea, in *Ocean, Ice and Atmosphere: Interactions at the Antarctic Continental Margin, Antarctic Research Series*, vol. 75, edited by S. Jacobs and R. Weiss, pp 341-369, AGU, Washington, D.C., 1998.

Sievers, J., R. Hartmann, D. Kosmann, A. Reinhold, and K-H. Thiel, Utilisation of ERS-1 data for mapping of Antarctica, in *Space at the service of our environment. Proceedings of the First ERS-1 Symposium, ESA SP-359*, edited by B. Kaldeich, pp 247-251, ESA, The Netherlands,1993.

Smith, A.M., The use of tiltmeters to study the dynamics of Antarctic ice-shelf grounding lines, *Journal of Glaciology*, 37, 51-58, 1991.

Smith, A.M., Introduction to high resolution seismic surveys on Rutford Ice Stream, in *Filchner-Ronne Ice Shelf Programme Report No. 7*, edited by H. Oerter, pp 39-41, Alfred-Wegener Inst. for Polar and Mar. Res., Bremerhaven, Germany, 1994.

Smith, A. M., Ice shelf basal melting at the grounding line, measured from seismic observations, *Journal of Geophysical Research*, 101, 22,749-55, 1996.

Smith, A.M., Basal conditions on Rutford Ice Stream, West Antarctica, from seismic observations, *Journal of Geophysical Research,* 102, 543-552, 1997a.

Smith, A.M., Variations in basal conditions on Rutford Ice Stream, West Antarctica, *Journal of Glaciology,* 43, 245-255, 1997b.

Smith, A.M., Seismic investigations on Rutford Ice Stream, West Antarctica. *PhD thesis*, Open University, Milton Keynes, UK, pp 207, 1997c.

Smith, A.M., Glaciological-Geophysical Investigations on Rutford Ice Stream and Carlson Inlet, 1997-98, in *Filchner-Ronne Ice Shelf Programme Report No. 12*, edited by H. Oerter, pp 77-84, Alfred-Wegener Inst. for Polar and Mar. Res., Bremerhaven, Germany, 1998.

Smith, A.M. and C.S.M. Doake, Seabed depths at the mouth of the Rutford Ice Stream, Antarctica, *Annals of Glaciology*, 20, 353-356, 1994.

Stenoien, M.D., Interferometric SAR observations of the Pine Island Glacier catchment area. *Ph.D. thesis*, University of Wisconsin-Madison, 1998.

Stephenson, S.N., C.S.M. Doake, and J.A.C. Horsfall, Tidal flexure of ice shelves measured by tiltmeter, *Nature*, 282, 496-97, 1979.

Stephenson, S.N. and C.S.M. Doake, Dynamic behaviour of Rutford Ice Stream, *Annals of Glaciology*, 3, 295-99, 1982.

Stephenson, S.N., Glacier flexure and the position of grounding lines: measurement by tiltmeters on Rutford Ice Stream, Antarctica, *Annals of Glaciology*, 5, 165-169, 1984.

Storey, B.C. I.W.D. Dalziel, S.W. Garrett A.M. Grunow, R.J. Pankhurst and W.R. Vennum, West Antarctica in Gondwanaland: Crustal blocks, reconstruction and break-up processes, *Tectonophysics* 155, 381-390, 1988.

Vaughan, D.G., Relating the occurrence of crevasses to surface strain rate, *Journal of Glaciology*, 39, 255-266, 1993.

Vaughan, D.G., Investigating tidal flexure on an ice-shelf using kinematic GPS, *Annals of Glaciology*, 20, 372-376, 1994.

Vaughan, D.G., Tidal flexure at ice shelf margins, *Journal of Geophysical Research*, 100, 6213-6224, 1995.

Vaughan, D.G., C.S.M. Doake, and D.R. Mantripp, Topography of an Antarctic ice stream. *SPOT 1 - Image utilization, assessment, results. Centre National d'Etudes Spatiale*, 167-174, 1988.

Vaughan, D.G., R.M. Frolich, and C.S.M. Doake, ERS-1 SAR: stress indicator on Antarctic ice streams, in *Space at the service of our environment. Proceedings of the Second ERS-1 Symposium. ESA SP-361*, edited by B. Kaldeich, pp 183-186, ESA, The Netherlands, 1994.

Vaughan, D.G., J.L. Bamber, M. Giovinetto, J. Russell, and A.P.R. Cooper, Reassessment of net surface mass balance in Antarctica, *Journal of Climate*, 12, 933-946, 1999.

Weertman, J., Stability of the junction of an ice sheet and an ice shelf, *Journal of Glaciology*, 13, 3-11, 1974.

Hugh Corr, British Antarctic Survey, Madingley Road, CAMBRIDGE CB3 0ET, UK. (e-mail: hfjc@bas.ac.uk)

Christopher Doake, British Antarctic Survey, Madingley Road, CAMBRIDGE CB3 0ET, UK. (e-mail: csmd@bas.ac.uk)

Adrian Jenkins, British Antarctic Survey, Madingley Road, CAMBRIDGE CB3 0ET, UK. (e-mail: a.jenkins@bas.ac.uk)

Keith Makinson, British Antarctic Survey, Madingley Road, CAMBRIDGE CB3 0ET, UK. (e-mail: kmak@bas.ac.uk)

Keith Nicholls, British Antarctic Survey, Madingley Road, CAMBRIDGE CB3 0ET, UK. (e-mail: kwn@bas.ac.uk)

Chandrika Nath, British Antarctic Survey, Madingley Road, CAMBRIDGE CB3 0ET, UK. (e-mail: chna@bas.ac.uk)

Andrew Smith, British Antarctic Survey, Madingley Road, CAMBRIDGE CB3 0ET, UK. (e-mail: ams@bas.ac.uk)

David Vaughan, British Antarctic Survey, Madingley Road, CAMBRIDGE CB3 0ET, UK. (e-mail: dgv@bas.ac.uk)

A REVIEW OF PINE ISLAND GLACIER, WEST ANTARCTICA: HYPOTHESES OF INSTABILITY VS. OBSERVATIONS OF CHANGE

David G. Vaughan[1], Andrew M. Smith[1], Hugh F. J. Corr[1], Adrian Jenkins[1], Charles R. Bentley[2], Mark D. Stenoien[2], Stanley S. Jacobs[3], Thomas B. Kellogg[4], Eric Rignot[5] and Baerbel K. Lucchitta[6]

The Pine Island Glacier ice-drainage basin has been cited as the part of the West Antarctic ice sheet most prone to substantial retreat on human time-scales. Here we review the literature and present new analyses showing that this ice-drainage basin is glaciologically unusual. Due to high precipitation rates near the coast, Pine Island Glacier basin has the second highest balance flux of any extant ice stream or glacier. Well-defined tributaries flow at intermediate velocities through the interior of the basin and have no regions of rapid velocity increase. The tributaries coalesce to form Pine Island Glacier which has characteristics of outlet glaciers (e.g. high driving stress) and of ice streams (e.g. shear margins bordering slow-moving ice). The glacier flows across a complex grounding zone into an ice shelf. There, it comes into contact with warm Circumpolar Deep Water which fuels the highest basal melt-rates yet measured beneath an ice shelf. The ice front position may have retreated within the past few millennia but during the last few decades it appears to have shifted around a mean position. Mass balance calculations of the ice-drainage basin as a whole show that there is currently no measurable imbalance, although there is evidence that some specific areas within the basin are significantly out of balance. The grounding line has been shown to have retreated in recent years. The Pine Island Glacier basin is clearly important in the context of the future evolution of the West Antarctic ice sheet because theoretically, it has a high potential for change and because observations already show change occurring. There is, however, no clear evidence to indicate sustained retreat or collapse over the last few decades.

1. INTRODUCTION

The West Antarctic ice sheet (Figure 1) drains into the Southern Ocean by three main routes; through the ice streams on the Siple Coast into the Ross Ice Shelf, through the glaciers and ice streams feeding Ronne Ice Shelf, and through the glaciers which debouch, either directly or through small ice shelves, into the Bellingshausen and Amundsen seas. While the dynamics of the Siple Coast ice streams have been studied under the West Antarctic Ice Sheet Initiative (WAIS), and those feeding Ronne Ice Shelf have been studied under the auspices of the Filchner-Ronne Ice Shelf Programme (FRISP), there has been no coordinated effort to understand the dynamics of glaciers feeding the Bellingshausen and Amundsen seas. Consequently, this area is seldom visited and its glaciology is poorly understood.

The largest glaciers in this sector are Pine Island Glacier and Thwaites Glacier. Both transport ice from the interior of the West Antarctic ice sheet to the Amundsen Sea. In terms of the mass of snow accumulating in their

[1]British Antarctic Survey, Natural Environment Research Council, Cambridge, U.K.
[2]Geophysical and Polar Research Center, University of Wisconsin, Madison, Wisconsin
[3]Lamont-Doherty Earth Observatory, University of Columbia, New York
[4]Institute for Quaternary Studies, University of Maine, Orono, Maine
[5]Jet Propulsion Laboratories, Pasadena, California
[6]U.S. Geological Survey, Flagstaff, Arizona

Copyright 2001 by the American Geophysical Union

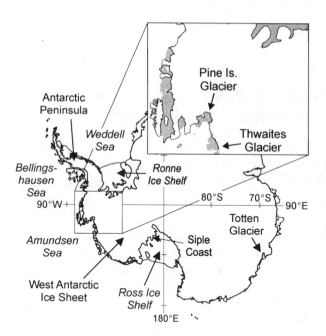

Fig. 1. Location map. Shaded region shows area covered by Figures 2-7.

catchment basins, Pine Island and Thwaites glaciers are respectively, the second and fifth most active basins in Antarctica [*Vaughan and Bamber*, 1998]. Pine Island Glacier alone accounts for around 4% of the outflow from the entire Antarctic Ice Sheet. The ice-drainage basins that feed these glaciers rest on beds as much as 2500 m below sea level, perhaps the deepest in Antarctica, and some authors have suggested that this in itself implies a great potential for rapid collapse [e.g. *Fastook*, 1984; *Thomas*, 1984].

Together, Pine Island and Thwaites glaciers may be key to the future evolution of the West Antarctic ice sheet, but in this review, we concentrate on the Pine Island Glacier basin alone. We do this because, in addition to theories of instability, there is a growing body of observations of change and unsteady flow there. After some introductory notes we consider each of the component parts of the basin in turn. We then consider the interactions between the basin and sea into which it flows. We assess the evidence for both long-term and recent changes in the ice cover of the region. Finally, we consider how relevant those observations may be to models which have predicted that Pine Island Glacier might be particularly prone to collapse.

1.1 Introductory notes on nomenclature

1.1.1 *Glaciers, ice streams and outlet glaciers.* The terms *glacier*, *ice stream* and *outlet glacier* are often loosely applied in the scientific literature as well as the non-specialist press. Here we use widely accepted definitions; *ice streams*, being areas of fast-moving ice sheet bounded by slower moving ice; *outlet glaciers*, being fast-moving ice bounded by nunataks or mountain ranges [*Bentley*, 1987; *Swithinbank*, 1954]; and *glaciers*, being a generic term for a "mass of snow and ice continuously moving from higher to lower ground, or if afloat, spreading continuously" [*Armstrong et al.*, 1973]. The distinction between ice streams and outlet glaciers "becomes rather hazy in practice" [*Bentley*, 1987], and is particularly acute in this case, as Pine Island and Thwaites *glaciers* share most of the dynamical characteristics of pure ice streams and need not be considered as inherently different.

1.1.2 *Floating portion of Pine Island Glacier*. *Hughes* [1980] stated that neither Pine Island Glacier nor Thwaites Glacier are "buttressed by a confined and pinned ice shelf". *Stuiver et al.* [1981] also stated that they "are unimpeded by an ice shelf". At that time, the position of the grounding line of Pine Island Glacier was poorly mapped, and Pine Island Glacier was assumed to calve directly into Pine Island Bay. Airborne radar sounding soon revealed that the seaward ~80 km of the glacier was indeed floating [*Crabtree and Doake*, 1982]. The original idea that these glaciers are dynamically different from others in West Antarctica has, however, persisted. Subsequent authors support the original notion, that Pine Island Glacier does not debouch through "an ice shelf" [*Kellogg and Kellogg*, 1987], "a substantial ice shelf" [*Jenkins et al.*, 1997], or "a large ice shelf" [*Rignot*, 1998]. Taking the widely accepted definition of an ice shelf, a "*floating ice sheet of considerable thickness attached to a coast*" [*Armstrong et al.*, 1973], it is clear that the floating portion of Pine Island Glacier, together with the adjacent floating areas, do constitute an *ice shelf*. Whether the ice shelf is a significant dynamic control on the glacier is, however, still an open question.

1.1.3 *Pine Island Bay*. Strictly, *Pine Island Bay* is the bay (approximately 75 x 55 km) into which flows the ice from Pine Island Glacier [*Alberts*, 1981], although the term is sometimes applied to a rather wider area.

1.1.4 *West Antarctic ice sheet*. The West Antarctic ice sheet is not an officially-recognised placename. We follow the accepted usage, meaning the term to refer to the ice sheet that covers West Antarctica, but excluding the Antarctic Peninsula.

1.2 Introductory note on meteorology

Since the 1960s, it has been widely recognised that the coastal portions of West Antarctica bordering on the

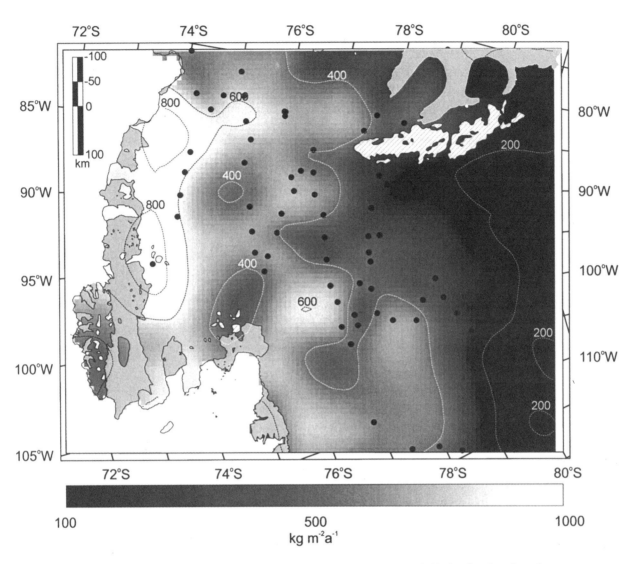

Fig. 2. Map of the area around the ice-drainage basin of Pine Island Glacier showing sites of measured surface mass balance (black dots), and interpreted grid of surface mass balance derived from field measurements and passive microwave satellite data from *Vaughan et al.* [1999].

Amundsen and Bellingshausen seas experience high precipitation rates compared with the rest of Antarctica [*Shimizu*, 1964]. These rates are matched only on the Antarctic Peninsula and around the coast of Wilkes Land [*Giovinetto*, 1964]. A recent compilation of net surface mass balance derived from *in situ* measurements and satellite data [*Vaughan et al.*, 1999] is shown in Figure 2. It agrees broadly with earlier estimates [eg. *Giovinetto and Bentley*, 1985], but shows an increased level of detail.

Meteorologically, the high precipitation rate in this sector results from synoptic-scale cyclones which travel around the Antarctic in the circumpolar trough. The trough is deepest over the Amundsen Sea, and many synoptic-scale cyclones come ashore here, producing considerable precipitation in the coastal zone. The effect is seen in moisture transport calculations [*Bromwich*, 1988] and in precipitation fields derived from General Circulation Models [*e.g., Connolley and King*, 1996]. Precipitation may be particularly high during winter months when the circumpolar trough moves south and more cyclones track across the coastal region [*Jones and Simmonds*, 1993].

There are no meteorological stations in the Bellingshausen/Amundsen Sea sector of West Antarctica. Consequently, direct measurements of decadal climate change (or stasis) have yet to be reported from either the Pine Island Glacier or Thwaites Glacier basins, although it

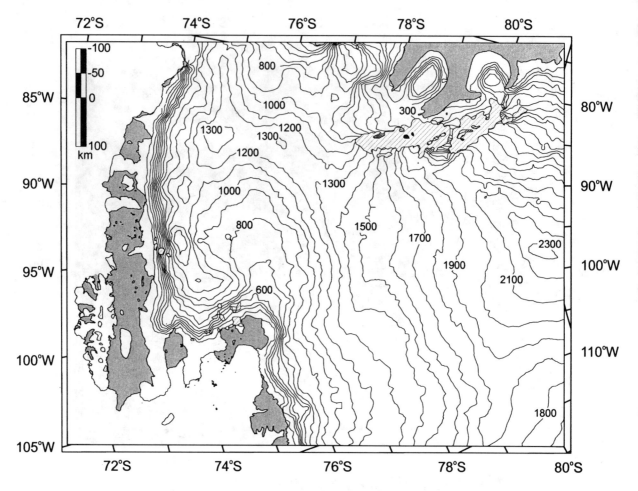

Fig. 3. Map of the area around Pine Island Glacier showing surface-elevation contours (meters above sea level) derived from Geodetic Mission of the ERS-1 satellite on a 5-km grid [*Bamber and Bindschadler*, 1997]. Ice shelves are shaded.

is possible that surface elevation changes (see Section 2.7) do reflect recent anomalous precipitation rates [*Wingham et al.*, 1998]. Offshore, a reduction in sea-ice extent in both the Amundsen and Bellingshausen seas has been noted in all seasons over the two decades prior to 1995 [*Jacobs and Comiso*, 1997]. This perhaps reflects a change in surface temperatures.

2. THE INTERIOR ICE-DRAINAGE BASIN

The interior of ice-drainage basins are sometimes viewed as cisterns, which passively accumulate ice and then supply it to the glacier (or ice stream) at whatever rate the glacier can transport it away. In this section, however, we present evidence that flow in the Pine Island Glacier basin is far from homogeneous. There is no clear distinction between ice-sheet and glacier flow and flow in the basin may have a strong influence on the overall configuration and glacier activity.

2.1 Delineation of the ice-drainage basin

Field-measurements of surface elevation in the Pine Island Glacier basin are few, but altimetry from the ERS-1 satellite is available to 81.9°S, which includes the entire basin. This altimetry has been used to create several high-resolution Digital Elevation Models (DEMs) of Antarctica [e.g., *Bamber and Bindschadler*, 1997; *Legresy and Remy*, 1997, *Stenoien*, 1998; *Liu et al.*, 1999].

Here we use a 5 km-resolution ERS-1-derived DEM [*Bamber and Bindschadler*, 1997] (Figure 3) to delineate the Pine Island Glacier basin and its neighbours. For comparison, a 200-m resolution DEM [*Liu et al.*, 1999] was also used to delineate the Pine Island Glacier basin

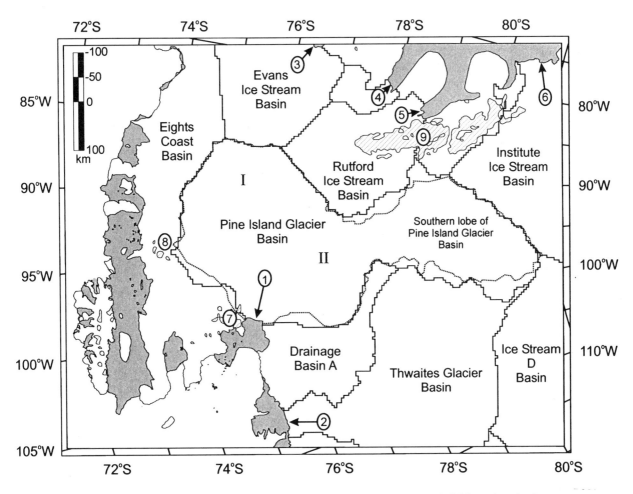

Fig. 4. Map of ice-drainage basins in the vicinity of Pine Island Glacier. Solid lines show basin boundaries derived from *Bamber and Bindschadler* [1997]. Dotted line shows the Pine Island Glacier basin derived from *Liu et al.* [1999]. Major glaciers are numbered and their direction of flow indicated by arrows; 1. Pine Island Glacier, 2. Thwaites Glacier, 3. Evans Ice Stream, 4. Carlson Inlet, 5. Rutford Ice Stream, 6. Institute Ice Stream. Rock outcrops are hatched with the major mountain ranges numbered (7. Hudson Mountains, 8. Jones Mountains, 9. Ellsworth Mountains). I and II indicate regions for which *Stenoien* [1998] presented separate mass-balance calculations. The coastline is derived from the Antarctic Digital Database [SCAR, 1993]. Ice shelves are shaded.

alone (Figure 4). The method used to produce the DEM is described in detail by *Vaughan et al.* [1999]. In summary, we identified segments of grounding line and then delineated the basins that feed them by tracing the line of steepest ascent inland as far as the ice divide. This procedure was limited to the grounded ice sheet as it assumes that ice-flow is parallel to the surface slope.

Table 1 shows the area of the Pine Island Glacier drainage basin measured from the above delineation, using an equal area projection, together with earlier estimates for comparison. There was considerable disagreement between the early estimates. While the more recent ones that rely on ERS-1 data have reduced the uncertainty, some still remains, presumably resulting from the different methods of analysis. For this review, we use an average of the three most recent values (165,000±7000 km^2) as a reasonable estimate of the basin area, although we accept that this may be improved in future by further analysis.

2.2 Shape of the catchment basin

The shape of the Pine Island Glacier basin (Figure 4) is similar to earlier delineations. It consists of two lobes, one immediately upstream of Pine Island Glacier and another,

Table 1. Estimates of ice-drainage basin area and balance flux for the Pine Island Glacier basin.

Basin Area (1000 km^2)	Balance Flux (Gt a^{-1})	Source
159	63.4	From DEM by *Liu et al.* [1999] and surface balance compilation of *Vaughan et al.* [1999]
175	69	From DEM by *Bamber and Bindschadler* [1997] and surface balance compilation of *Vaughan et al.* [1999]
159 ± 1	63.9 ± 6	[*Rignot*, 1998]
	76	[*Bentley and Giovinetto*, 1991] (Arithmetic mean of estimates from *Crabtree and Doake* [1982] and *Lindstrom and Hughes* [1984])
182	65.9 ± 5	[*Lindstrom and Hughes*, 1984]
214 ± 20	86 ± 30	[*Crabtree and Doake*, 1982]

the *southern lobe*, feeding the first through a neck less than 100 km across. A delineation of the catchment basins of the 70 largest glaciers in Antarctica, by a similar method, showed this configuration is unusual [*Vaughan and Bamber*, 1998]. Generally, ice-drainage basins which drain through glaciers or ice streams, are uniformly convergent. Only the Ice Stream C basin has a similar "necked" shape [*Joughin et al*, 1999]. It is possible that the existence of the *southern lobe* of the Pine Island Glacier basin indicates unsteady conditions in the basin, with this lobe currently being transferred between catchment basins. Alternatively, it may simply reflect unusual bed morphology.

2.3 Mass input and balance flux

Overlaying the basins for Pine Island Glacier derived in Section 2.2, on a grid representing the mean surface balance over Antarctica [*Vaughan et al.*, 1999], we have estimated the total rate of snow accumulation in the Pine Island Glacier basin (Table 1). This is the amount of ice that must leave the basin for mass balance to be maintained and is termed the *balance flux*. The aggregate of the three most recent estimates gives a balance flux for Pine Island Glacier of (66 ± 4) Gt a^{-1}. The uncertainty is derived from the spread of the results, but is consistent with the uncertainties in area (± 4%) and accumulation (± 5%) [*Vaughan et al.*, 1999].

A similar process has been used to find the balance fluxes of the other major glaciers of Antarctica, and while many glaciers are fed by basins with larger areas, only one balance flux exceeds that of Pine Island Glacier. Totten Glacier, East Antarctica has a balance flux of around 75 Gt a^{-1} [*Vaughan and Bamber*, 1998]. Outside Antarctica, the most active glacier is Jacobshavns Isbræ, Greenland which supports about half this flux [*Bindschadler*, 1984].

2.4 Glacier tributaries

Two techniques employing satellite data and yielding wide coverage allow us to identify areas where ice-flow is concentrated within the interior of the ice-drainage basin. The pattern that emerges is one of great complexity, with many tributaries coalescing to form the main glacier.

2.4.1 Satellite Altimetry.
We can derive some understanding of the distribution of ice flow in the interior of the basin using the DEMs discussed in Section 2.1. The method first calculates the flow-direction for each cell. It then assigns to each cell a numerical value corresponding to the number of other cells whose accumulation will eventually flow through it. The technique is known as *flow-accumulation* and a grey-scale representation of this *flow-accumulation* grid (Figure 5) gives an indication of where flow is more convergent within the basin.

Figure 5 shows a system of tributaries which merge about 100 km above the grounding line to form the single unit of flow which is Pine Island Glacier. These tributaries are identifiable several hundreds of kilometers inland, much further inland than the point where the ice enters a more confined channel, previously suggested to be the start of channelized ice flow [*Lucchitta et al.*, 1995; *Lucchitta et*

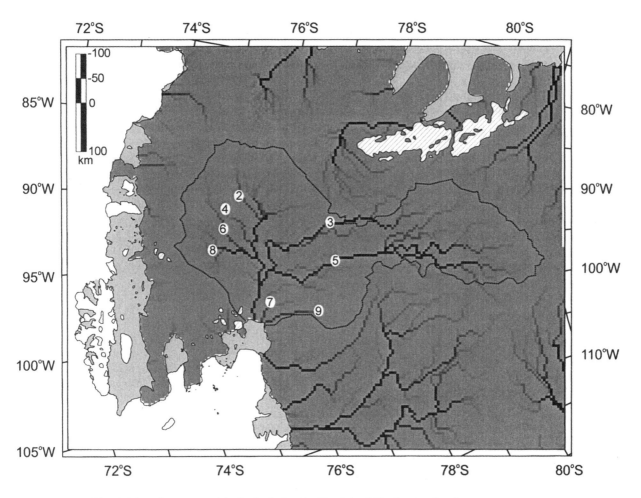

Fig. 5. Map of area around the basin drained by Pine Island Glacier showing flow-accumulation derived from 5-km resolution surface elevation grid. Grid cells are shaded such that cells fed by many others are darker than those fed by only a few. The darker areas thus represent areas into which the flow is channelled. The numbered features are the tributaries as identified by *Stenoien* [1998]. Stenoien's tributaries 1 and 10 are not resolved on this representation.

al., 1994]. This set of tributaries is almost the same as that determined by *Stenoien* [1998] from interferometric Synthetic Aperture Radar (SAR) images (see figure 6.14 of *Stenoien* [1998]) and in Figure 5 they are numbered using Stenoien's designation. One tributary (5) drains much of the southern lobe of the catchment basin described in Section 2.2 Its presence is perhaps not surprising given the narrow neck where the southern lobe joins the rest of the drainage basin.

Stenoien [1998] suggested that similar patterns of tributaries have not been seen elsewhere, and that they are perhaps unique to Pine Island Glacier. There is, however, evidence elsewhere in Antarctica for tributary systems in other basins. Bright margins in ERS-1 SAR data have shown that Evans Ice Stream forms at the confluence of at least five tributaries [*Jonas and Vaughan*, 1996]. Radarsat data shows that Recovery Glacier has two tributaries which extend hundreds of kilometers inland [*Jezek, et al.*, 1998]. Flowlines in Landsat imagery show that Institute Ice Stream also has several tributaries [*Mantripp et al.*, 1996]. Finally, *Joughin et al* [1999] show a tributary system feeding ice streams on the Siple Coast.

The presence of several tributaries in the interior drainage system may imply that Pine Island Glacier is unlikely to respond dramatically to changes in one locality. For example, if "water-piracy" [*Alley et al.*, 1994] or a reduced supply of basal till were to shut-off one of the tributaries, the others would probably be unaffected and the flux in Pine Island Glacier may suffer little change.

2.4.2 *Interferometric SAR.* Goldstein et al. [1993] used SAR data from the ERS-1 satellite to construct interferometric SAR (InSAR) images of ice flow in

Antarctica. These showed ice movement only along the line of sight to the satellite, but the method has since been refined to produce a 2-dimensional velocity-field for parts of the interior of the Pine Island Glacier basin [*Stenoien*,1998]. The procedure yields an understanding that is more quantitative than that from the satellite altimetry presented above, although coverage of the Pine Island Glacier basin is less complete. *Stenoien's* [1998] 2-dimensional velocity field covers much of the northern part of the drainage basin (see figures 5.14 and 5.15 of *Stenoien* [1998]). The data cover the upstream regions of many of the tributaries (2, 4 and 6 in Figure 5) and the middle regions of two which originate in the southern lobe of the basin (3 and 5 in Figure 5). There are no points of ground control in this area and Stenoien derived an absolute velocity field by assuming that a saddle on the ice divide had zero velocity. Ice speed away from the tributaries is low (0-50 m a^{-1}) but increases within them to more than 150 m a^{-1} upstream of the confluence of tributaries 2, 3 and 5 (Figure 5).

Of interest is Stenoien's observation that none of the tributaries for which data are available shows a rapid increase in ice speed, but rather a gradual increase down the length of each tributary. A similar pattern has also been observed on other West Antarctic ice streams *(Joughin et al.*, 1999) The lack of a sudden velocity increase, suggests notions of a bi-stable state of glacier-flow i.e. fast or slow, may be unrealistic, but rather that a progressive response to changing boundary conditions is possible.

Taken together, InSAR and flow-accumulation show that the interior of the Pine Island Glacier basin is complex with around 10 tributary ice streams coalescing to form a single glacier. None of these tributaries appears to have a well-defined region of rapid velocity increase and what controls their location and longevity remains to be determined. Radar data presented in Section 2.6 suggest that the control may be through basal conditions. However, at present even sophisticated thermomechanical models of the area fail to reproduce this complex flow pattern [e.g. *Payne*, 1999].

2.5 *Subglacial topography*

The bed topography of the West Antarctic ice sheet was first mapped in detail using a combination of traverse data, airborne sounding data and TWERLE balloon altimetry [*Jankowski and Drewry*, 1981]. While this study clearly delineated the major subglacial features of the area, the availability of new data prompts us to repeat the exercise.

Figure 6 shows a new compilation of bed topography beneath the grounded portion of the Pine Island Glacier basin. To create a grid of ice thickness we used (Figure 6a) traverse data [*Bentley and Ostenso*, 1961; *Behrendt*, 1964; *Bentley and Chang*, 1971], airborne radar data [*Jankowski and Drewry*, 1981; British Antarctic Survey unpublished data] and rock outcrops, which were used as an isopleth of zero ice thickness. The resulting grid of ice thickness was subtracted from the ERS-1 derived DEM of surface elevation described in Section 2.1 to produce bed topography (Figure 6b).

The bed topography (Figure 6b) shows clearly the main features identified in earlier compilations: the Bentley Subglacial Trench; the Byrd Subglacial Basin, which here reaches almost 2000 m below sea level; and between these depressions, the "sinuous ridge" described by *Jankowski and Drewry* [1981]. The Ellsworth Subglacial Highlands, are also well-defined. Despite significantly improved data coverage in the Pine Island Glacier basin, Figure 6b shows no new substantial features except a trough 1000 m below sea level, in which Pine Island Glacier and its main tributaries flow.

2.6 *Driving stress*

The *driving stress* in an ice sheet is calculated from the surface slope and ice thickness according to a simple relation [*Paterson*, 1994; page 241]. Here we have calculated the driving stress for the region (Figure 7) using the ERS-1-derived DEM and the ice thickness grid described above. The calculated driving stress is negligible near the ice divide where the surface slopes are low; intermediate (50-110 kPa) on the slow-moving areas between the ice divides and the tributary glaciers; and low (<50 kPa) on the tributary glaciers, but rises to >110 kPa along the main trunk of Pine Island Glacier.

An airborne sortie was flown from the inactive Siple Station (75° 54' S 84° 30' W) to the ice front of Pine Island Glacier in 1998 (the flight-track is shown in Figure 6a). It covered much of the main tributary of Pine Island Glacier (that formed by numbers 2, 4, 6 and 8 in Figure 5). Ice-penetrating radar data from this sortie show that the margin of this main tributary is marked by a downward step in the bed elevation and a change to a smoother ice-base reflection which has an "ice shelf-like" character (Figure 8). This change in reflection character is believed to indicate a transition from a frozen bed (rough) to one which is at the pressure-melting point (smooth). Although the flight track does not follow an ice flow line exactly, the driving stresses derived from the along-track data compare well with those derived from the gridded datasets shown in Figure 7. The driving stress calculated from the along-track topography has four distinct zones (Figure 8); the interior of the basin (50-75 kPa), the main tributary glacier (around 30 kPa), the main trunk of Pine Island Glacier

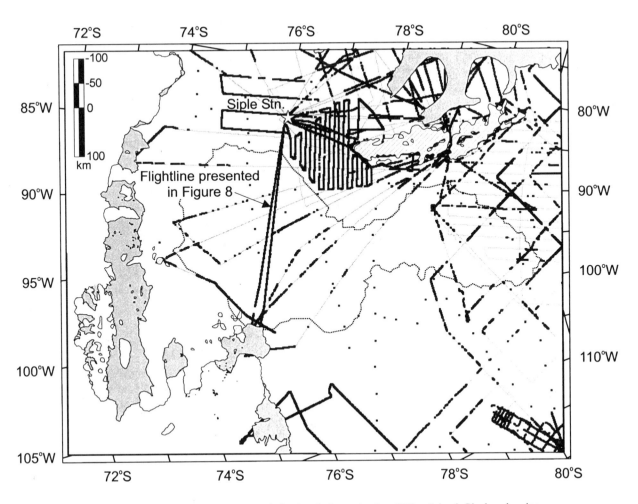

Fig. 6a. Map of the area around the ice-drainage basin of Pine Island Glacier showing measurements of ice thickness from airborne survey and oversnow traverses. Thin grey lines indicate unsuccessful sounding by airborne survey.

(over 100 kPa) and the floating ice (less than 10 kPa). The marked change in bed roughness across the tributary margin may also indicate that the location of the tributary and underlying geological constraints are closely related. The low driving stress suggests that the tributary flows over a well-lubricated bed.

The pattern of driving stresses suggests that the Pine Island Glacier basin is dynamically different from the idealised ice-stream basin. Much of the basin comprises a slow-moving ice sheet which may be cold-based, as suggested by the character of the radar reflection (Figure 8). This ice sheet feeds a number of wet-based (Figure 8), lubricated tributaries with relatively low driving stresses (around 30 kPa). These merge to form Pine Island Glacier, which has a much higher driving stress (>100 kPa), more akin to an East Antarctic outlet glacier, than some of the West Antarctic ice streams [*Bentley*, 1987].

2.7 *Surface elevation change*

ERS-1 satellite altimetry data for the period 1992-1996 were analysed for evidence of surface-elevation change [*Wingham et al.*,1998]. The data covered most of the interior of the Antarctic Ice Sheet north of 82°S. The analysis showed only one region of spatially-coherent surface-elevation change. Thinning at a mean rate of 11.7 ± 1.0 cm per year was indicated in the Pine Island Glacier-Thwaites Glacier basin. *Wingham et al.* [1998] indicated that the change was centered and most significant over the Thwaites Glacier basin (see their Figure 2), rather than the Pine Island Glacier basin, but the trend did appear to extend across both. The simplest interpretation is that the surface lowering resulted from a change in surface mass balance. Alternatively, a change in the glacier flux due to increased discharge or grounding-line retreat might also be

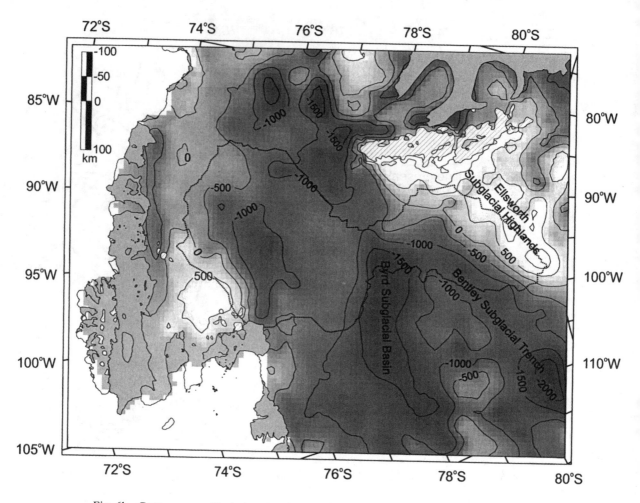

Fig. 6b. Contour map of bed elevation (meters above sea level). A grid of ice thickness was calculated using the measurements shown in Figure 6a. This grid was subtracted from a surface-elevation grid [*Bamber and Bindschadler*, 1997] to produce a grid of bed elevation.

the cause. However, in either case, the shortness of the observation period gives little indication of future behavior. It is hoped that more detailed analysis of the ERS-1 altimetry will refine the pattern of change. NASA's Geoscience Laser Altimeter System (GLAS), scheduled for launch in 2001, will allow similar measurements to be made even in the coastal margin of Antarctica.

3. THE GLACIER

Until the discovery of the network of ice tributaries in the basin [*Stenoien*, 1998; Section 2.4], Pine Island Glacier was generally considered to extend only around 70 km above the grounding line to where the ice is first channelled into parallel flow (Figure 9). This may still be still a useful definition, since it draws some distinction between the tributaries and the main trunk of the glacier and approximately marks the increase in driving stress mentioned in Section 2.6. Thus defined, Pine Island Glacier is bounded to the north by nunataks in the Hudson Mountains and to the south by slow-moving ice sheet.

3.1 Surface features

Surface features on Pine Island Glacier revealed by Landsat and SAR imagery have been shown by various authors and are reproduced in Figure 9. Flowlines of the type discussed by *Whillans and Merry* [1993] show considerable convergence at the head of the glacier, around the zone of arcuate "crevasses" revealed by ERS-SAR images (shown in Figure 9 and in greater detail by *Lucchitta et al.* [1995]). These presumably mark a zone of longitudinal extension. *Lucchitta et al.* [1995] noted that these "crevasses" had not been previously described and

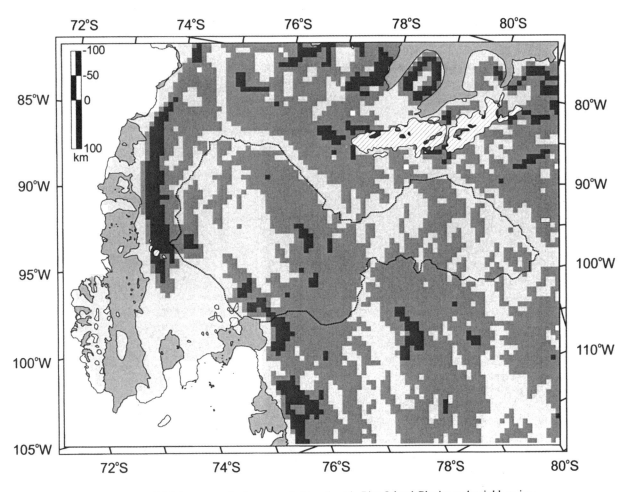

Fig. 7. Map of driving stress for the grounded ice sheet in Pine Island Glacier and neighbouring ice-drainage basins derived from surface elevation (Figure 3) and bed elevation (Figure 6b). Light-grey shading denotes driving stresses in the range 0-50 kPa, mid-grey 50-110 kPa, and dark-grey greater than 110 kPa.

are not shown by visible images, and concluded that they may be covered by a layer of snow.

Below the zone of convergence and arcuate crevasses, flowlines on the glacier are roughly parallel. The surface has smooth, long-wavelength (a few kilometers) undulations that are typical of fast-moving glaciers.

3.2 Velocity

Lucchitta and others [*Lucchitta et al.*, 1995; *Lucchitta et al.*, 1994; *Ferrigno et al.*, 1993] have measured the ice velocity on Pine Island Glacier using sequential SAR images acquired by the ERS-1 satellite. They found that on the main trunk of the grounded portion, the center-line speed ranged from about 1 km a^{-1} near the arcuate crevasses, to 1.5 km a^{-1} at the grounding line identified by *Crabtree and Doake* [1982]. The flow speed then rose rapidly to 2.5 km a^{-1} between that grounding line and the one identified by *Rignot* [1998], and then remained approximately constant to the ice front. Their velocity measurements were generally higher than earlier estimates [*Kellogg and Kellogg*, 1987; *Lindstrom and Tyler*, 1984; *Crabtree and Doake*, 1982; *Williams et al.*, 1982] although the data are not adequate to determine if there has been any acceleration.

3.3 Grounding line

The grounding line of Pine Island Glacier is not easily discernable on either the Landsat or ERS-1 SAR imagery in Figure 9, but its position was determined by hydrostatic calculations based on airborne data [*Crabtree and Doake*, 1982] (Figure 9). This hydrostatic condition downstream of this grounding line was, however, not entirely clear.

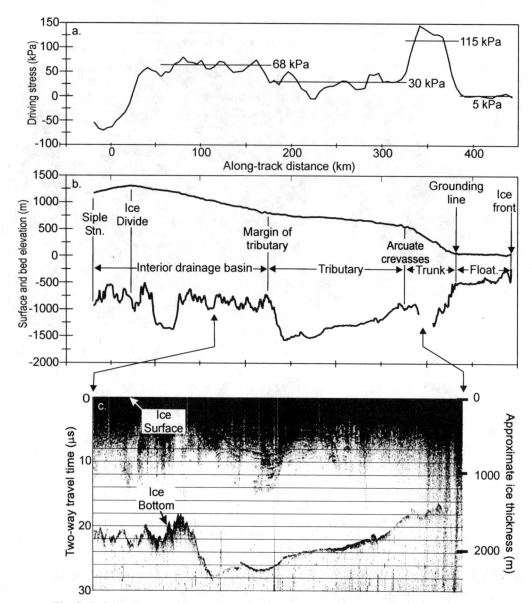

Fig. 8. Driving stress, topography and radar data collected during airborne survey flight-track identified in Figure 6a. a) Driving stress calculated using 10-km smoothing along the flight-track. The driving stress is calculated to be positive in the direction of Pine Island Glacier ice front and is negative where the ice flows into a different basin. b) Ice-surface and bottom topography along the flight-track with ice-flow features marked. c) Section of radar data, showing the difference in character between the ice-bottom return from the interior drainage basin (rough) and from the tributary glacier (smooth). This difference probably reflects the difference between a frozen bed and one that is at the pressure-melting point.

Thomas [1984] argued that a zone of partial grounding might exist for 30 km downstream. This downstream position has now been confirmed, using InSAR to detect the limit of tidal flexing [*Rignot*, 1998]. That analysis also showed that the limit of flexure extends seaward (around 15 km) near the centre of the glacier with a re-entrant on either side (Figure 9). This pattern is possibly an indication of a bedrock obstruction similar to that known to underlie Rutford Ice Stream [*Stephenson*, 1984; *Doake et al*, this volume] or it might be solely due to the glacier being thicker close to its center line.

The position of the limit of flexure was measured at five epochs between 1992 and 1996, which showed that it moved during this period [*Rignot*, 1998]. The simplest

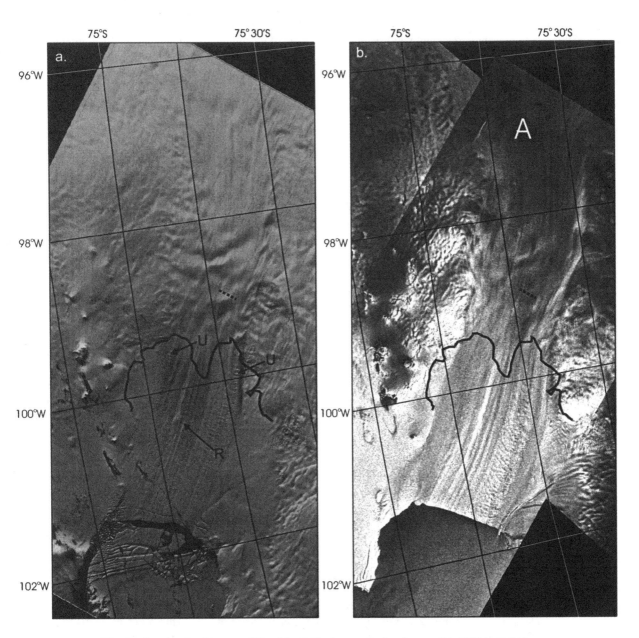

Fig. 9a. Landsat 1 sub-scene of Pine Island Glacier acquired on January 24, 1973 (path 246, row 114). "R" marks the position of the raft of ice discussed in Section 4.1. "U" indicates plumes of ice thickness undulations formed close to the grounding line and dissipating towards the ice front (discussed in Section 4.1). The dotted line marks the grounding line identified by *Crabtree and Doake* [1982] and the black line indicates the limit of tidal flexing determined by *Rignot* [1998] for 21 January 1996.

Fig. 9b. Mosaic of two ERS-1 SAR images (orbit 3174, frames 5193 and 5211) acquired December 4, 1992 showing the same area. "A" marks the set of arcuate crevasses identified by *Lucchitta et al.* [1995].

interpretation is that between 1992 and 1994 there was a retreat in the position of the grounding line in the center of the glacier that averaged 1.2 ± 0.3 km a^{-1}. The pattern of change within the re-entrant parts is not so clear, though Rignot calculated that it could be caused by a thinning rate of 3.5 ± 0.9 m a^{-1}. This is a small fraction of the basal melt rates in this area (see Section 4.2) and thus could result from a relatively small change in the oceanographic conditions. Alternatively, it could also be caused by a thinning of the glacier upstream of the grounding line.

4. THE ICE SHELF

Pine Island Glacier debouches into an ice shelf comprising the ~80 km floating portion of the glacier plus the slow-moving floating ice sheet that surrounds it. The floating portion of Pine Island Glacier is easily identified by plentiful flowlines generated near the grounding line, which continue to the ice front showing little divergence.

4.1 Surface features

Kellogg et al. [1985] found that dense (650 kg m^{-3}) "well-sintered" firn predominated at the surface of the floating portion of Pine Island Glacier, close to the ice front. They interpreted this as resulting from strong katabatic winds, producing net sublimation from the ice surface. The extent, persistence or magnitude of this negative net surface mass balance is, however, not well-established.

Landsat and SAR images of the floating portion of Pine Island Glacier (Figure 9) show crevasses formed at the grounding line, moving in plumes to the ice front. Much of the southern side of the floating portion of the glacier is covered by periodic transverse surface undulations visible on the Landsat images. These also form in plumes emanating from the grounding line and dissipating towards the ice front. Airborne radar-sounding data show that these undulations have an amplitude of around 20 m, a horizontal wavelength of around 2.5 km and are hydrostatically compensated by ice thickness changes of ~190 m. The ice flow velocity measured on this section of ice shelf was 2.3 - 2.6 km a^{-1} [*Lucchitta et al.*, 1997] which suggests that one undulation is produced each year, although the mechanism that causes them is uncertain.

A former feature of the floating portion of Pine Island Glacier, not previously discussed but clearly visible in Figure 9, was a raft of thicker ice embedded in the ice shelf. In later images, the raft was seen to have been advected downstream at approximately the same speed as the ice-shelf flow. The origin and significance of this raft is unclear and it is no longer available for investigation as it would have calved from the ice shelf in the late-1980s. Similar features have been noted in grounded ice streams [e.g., *Whillans et al.*, 1993] and possibly in floating ice shelves [*Cassasa et al.*, 1991].

4.2 Basal melting

In early 1994, the research ship *Nathaniel B. Palmer* entered Pine Island Bay and conducted an oceanographic survey of the area which has led to three studies revealing a most unusual oceanographic regime:

- *Jacobs et al.* [1996] used oceanographic measurements and a "salt-box" calculation to show that the mean basal melt rate beneath the floating portion of Pine Island Glacier was 10-12 m a^{-1}, five times the highest rate previously published (on George VI Ice Shelf [*Bishop and Walton*, 1981]). They concluded that these high melt-rates were driven by relatively warm Circumpolar Deep Water (CDW) flooding this portion of the continental shelf, combined with the deep draft of Pine Island Glacier.

- *Jenkins et al.* [1997] determined that the position of the ice-shelf front showed no persistent trend in the period 1973-1994 (in Figure 10 we extend this series to 1966-1998). Combining this with a flux calculation across the grounding line and with the assumption that it is a steady-state system, they calculated a mean basal melt rate over the ice shelf of 12 ± 3 m a^{-1}.

- *Hellmer et al.* [1998] used analyses of dissolved oxygen and oxygen isotopes to confirm the strong melt-water signal in the outflow and applied a thermohaline model. This suggested that in some areas, the basal melt-rate is twice the mean value and that temporal variations in the temperature of the inflowing CDW could cause substantial changes in the basal melt-rate.

Rapid melting from beneath the floating portion of Pine Island Glacier was later confirmed using satellite measurements of mass balance [*Rignot*, 1998] giving a mean melt-rate of (24 ± 4) m a^{-1} increasing to (50 ± 10) m a^{-1} close to the grounding line. These melt-rates are larger than those calculated by *Jenkins et al.* [1997] due to a new position for the grounding line constrained by InSAR observations of tidal flexing. Using this grounding line position, we increase the estimate of mean melt-rate produced by Jenkins et al. to around 17 m a^{-1}, and that from Jacobs et al. by a similar amount.

These observations of a thick ice shelf coming in contact with relatively warm Circumpolar Deep Water has led to speculation that ice-ocean interactions in Pine Island Bay may be similar to those prevalent during the Last

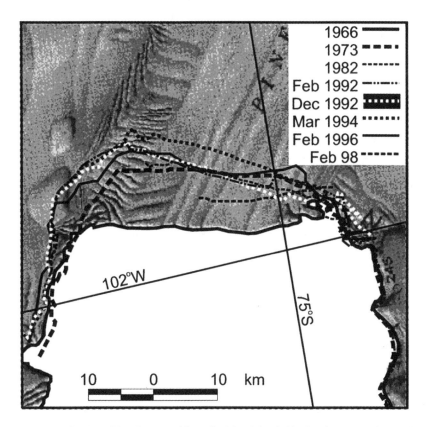

Fig. 10. Map of selected ice-front positions for Pine Island Glacier, between 1966 and 1998 overlaid on an excerpt of a sketch-map drawn from aerial photography collected in 1966 (USGS, 1993; original map prepared in 1967). Sources: 1966, Aerial photography (USGS, 1993); January 24,1973, Landsat 1 (path 246, row 114); January 15, 1982, Landsat image; February 9, 1992, ERS-1 SAR image; December 4, 1992, ERS-1 SAR images (orbit 3174, frames 5193 and 5211); 15 March 1994, ERS-1 SAR image; February, 1996, ERS-1 SAR image; February 13, 1998, flight-track of BAS airborne survey aircraft which flew along the ice front.

Glacial Maximum when much of the West Antarctic ice sheet probably extended out to the edge of the continental shelf and came into contact with similarly warm water [e.g. *Jenkins et al.*, 1997]. In Pine Island Bay we now know that such conditions generate high sub-ice shelf melt-rates and we infer that these rates may have also been prevalent during glacial periods, severely limiting the size of any ice shelves. In addition, the results highlight the importance of oceanographic conditions as a significant control on the present and future configuration of the West Antarctic ice sheet.

4.3 *Ice front stability*

The ice-front position of Pine Island Glacier has been reconstructed by several authors using several sources of data and covering various periods. Figure 10 shows the position of the ice front sporadically since 1966. The pattern suggests a retreat of around 10 km between 1966 and 1973, followed by a period of general stability, and readvance in recent years. However, this simple interpretation should be qualified on two counts. a) The presence of an iceberg just off the ice front on the sketch map drawn from 1966 aerial photography. This indicates that a significant calving had recently occurred, and that prior to this the ice front was even further advanced than the most extreme position shown. The same is true for the 1973 image. b) The ice velocity at the front (>2.5 km a^{-1}) allows for fluctuations within the intervals between the observations, large enough to exceed the extreme positions shown in Figure 10. However, it is interesting that the ice front position *appears* to be relatively stable, despite the high ice velocity. Further monitoring would be required to confirm whether or not this is the case.

Kellogg and Kellogg [1987] have suggested that the Thwaites Iceberg Tongue was not actually formed from calving of Thwaites Glacier Tongue but might have calved from Pine Island Glacier. This is, however, unlikely, as a profusion of surface transverse lineations on the iceberg tongue matched well with the transverse lineations on Thwaites Glacier, but did not match the longitudinal lineations that predominate on Pine Island Glacier [*Ferrigno et al.*, 1993].

In summary, there are insufficient data available to discern a decadal trend in the ice-front position of Pine Island Glacier, although it is probable that some retreat (~10 km) has occurred in the last 30 years. (Longer-term retreat of the ice shelf front is discussed in Section 5).

5. THE MARINE ENVIRONMENT

5.1 *Retreat of ice in Pine Island Bay*

Seabed sediments provide information on glacier retreat in Pine Island Bay. Results from four cores within Pine Island Bay and from 19 cores on the outer continental shelf and eastern Amundsen Sea have been presented [*Anderson and Myers*, 1981; *Kellogg and Kellogg*, 1987; *Kellogg et al.*, 1987], but their value is limited because they contain very little material suitable for radiocarbon dating. The radiocarbon dates that do exist, only poorly constrain the retreat of the ice sheet in Pine Island Bay to the last few millennia. Prior to this, the ice sheet may have occupied the entire Pine Island Bay, and perhaps butted against the Thwaites Glacier Tongue to form an extensive ice shelf or ice sheet.

Kellogg and Kellogg [1987] suggested the retreat was very recent, with Pine Island Bay being filled with grounded ice only 100 years ago, but this interpretation relied heavily on an ongoing retreat rate of ~0.8 km per year inferred from aerial photography acquired in 1966 and Landsat imagery acquired in 1973. The variable position of the ice front shown in Figure 10 now casts doubt on such an extrapolation.

5.2 *Ice Extent at the Last Glacial Maximum*

The sediment cores collected on the outer continental shelf, near 110°W [*Anderson and Myers*, 1981], and north of Thurston Island between 100°W and 102°W [*Kellogg and Kellogg* 1987; *Kellogg et al.*, 1987] show a thin (0-15 cm) upper layer of sandy mud, probably of Holocene age, containing common planktonic and calcareous benthic foraminifera. Diatoms are relatively rare in this layer despite high abundances in the surface water. A compact, poorly-sorted diamicton underlies the sandy mud. This diamicton is generally more than 2.3 m thick, and probably represents deposition beneath grounded ice that extended to the continental shelf break.

If we assume that the diamicton is a remnant of subglacial basal till, then grounded ice probably remained over most of the Amundsen Sea continental shelf until relatively late in the Holocene. The postglacial sediments on the outer shelf are much thinner in Pine Island Bay than on the outer shelf in the Ross Sea [*Kellogg et al.*, 1979; *Domack et al.*, 1999; *Licht et al.*, 1999; *Shipp et al.*, 1999] which, given similar deposition rates, might suggest an earlier deglaciation in the Ross Sea than in the Amundsen Sea. Finally, if we assume that post-glacial deposition rates were similar to current measured rates (e.g., 10-35 cm a^{-1} beyond coastal Alaskan glaciers [*Molnia and Carlson*, 1999] and ~10 cm a^{-1} in Antarctic Peninsula fjords [*Domack and McClennen*, 1996]), then we can conclude that deglaciation occurred, at most, a few thousand years ago.

6. MASS BALANCE

Comparisons of balance flux (Table 1) and grounding line flux (Table 2) for Pine Island Glacier show a progression toward reduced uncertainty, but significant uncertainty still remains. Our preferred estimate of overall mass balance is -2.4 ± 4 Gt a^{-1} (the difference between the balance flux determined in this study and the grounding line flux calculated by *Rignot* [1998]). This value is not significantly different from zero. Given a catchment basin area of ~170 000 km^2, this would be equivalent to a lowering of surface elevation in the range 1.5 - 3 cm a^{-1}, depending on the density of the layers being lost.

Rignot [1998] determined the ice thickness by inverting the ice-surface elevation at the limit of flexing using a hydrostatic condition but there is evidence that the limit of flexing is often many kilometers upstream of the hydrostatic limit. One example is on Rutford Ice Stream where along the center-line the limit of flexing is 2 km upstream of the hydrostatic point and the surface is 50 m above the hydrostatic condition [*Vaughan*, 1994; *Smith*, 1991]. If a similar situation applies on Pine Island Glacier then the ice flux across the grounding line may have been over-estimated by Rignot. However, as at present we cannot assess the likelihood of this possible error, we use Rignot's grounding line flux for our preferred mass balance estimate.

The mass balance within two regions of the drainage basin has also been calculated (see figure 6.8 and table 6.1 of *Stenoien*, [1998]). The north-eastern part of the basin has a positive mass balance (6.4±3.7 Gt a^{-1}), whereas in the region just north of the narrow neck it is negative

Table 2. Estimates of grounding-line flux for Pine Island Glacier.

Mass (Gt a^{-1})	Source
68.4 ± 2	[Rignot, 1998]
> 56 ± 6	[Jenkins et al., 1997]
70	[Lucchitta et al., 1995]
25.5 ± 5	[Lindstrom and Hughes, 1984]

(-7.7±4.7 Gt a^{-1}). These values were calculated using a hand-drawn map of mean surface mass balance, but repeating the calculation using an updated map of surface mass balance [Vaughan et al.,1999], gave the same answer. Distributed evenly across the areas, these imbalances represent changes in surface elevation of (23±14) cm a^{-1} and (-51±31) cm a^{-1}, respectively.

There is thus a contradiction between mass balance calculations and measurements of change in surface elevation [Wingham et al.,1998; see Section 2.7] which seems to imply one of three possibilities; a) substantial changes in the density-depth relation in the snow in the period 1992-1996, b) unusually low precipitation accumulation in the period 1992-1996 or, c) one, or both of the analyses is substantially in error.

7. DISCUSSION AND CONCLUSIONS

The Pine Island Glacier ice-drainage basin is an area of particular interest and in some ways, may be unique. The shape of the basin is unusual. It comprises two lobes joined by a narrow neck less than 100 m across. The southern lobe has a higher ice-surface elevation and does not appear to correlate with any significant bed feature. Surface elevation in the ice-drainage basin dropped over four years [Wingham et al., 1998], though the reasons for this are uncertain. Precipitation rates are high and the basin has the second-highest balance flux of any extant ice stream or glacier (66±4 Gt a^{-1}). Although there are indications that the ice sheet may be out of balance locally [Stenoien, 1998], mass balance calculations of the basin as a whole show that there is currently no measurable, significant imbalance. The tributaries which drain the basin flow at intermediate speeds (50-150 m a^{-1}) and show no regions of marked velocity increase. They coalesce to form Pine Island Glacier which has high driving stresses (>100 kPa) similar to East Antarctic outlet glaciers, but also shear margins bordering slow-moving ice, often typical of West Antarctic ice streams. The bed of the slow-moving ice in the drainage basin appears to be frozen. In the tributaries, the bed is at the pressure-melting point and is well-lubricated, presumably associated with the reduction observed in the driving stress. This thawed bed continues into Pine Island Glacier. The glacier flows over a complex grounding zone where Rignot [1998] measured a grounding line retreat along the center of the glacier of a few kilometers over a period of a few years. Once afloat, the base of the glacier comes into contact with warm Cicumpolar Deep Water which generates basal melt-rates of 10-20 m a^{-1} or more [Jacobs et al., 1996; Jenkins et al., 1997; Hellmer et al., 1998; Rignot, 1998], the highest rates yet measured beneath any ice shelf. The front of the ice sheet or shelf in Pine Island Bay may have retreated significantly in the last few millennia [Kellog and Kellog, 1987], but over the last few decades it appears to have been more stable, shifting back and forth no more than a few kilometers.

These observations summarize what is known about Pine Island Glacier and its drainage basin. Several of them indicate evidence of change in this part of the West Antarctic ice sheet on time scales varying from millennia to a few years. However, at present we are cautious as to the long-term significance of these changes. Although we note that the unusual basin shape could indicate ongoing transfer of the southern lobe between catchments, there is no evidence to support this and the shape may simply be reflecting the bed topography in some way. We are uncertain as to the whether the observed surface-elevation change is the result of changing precipitation or else changing ice flow. In addition, the apparent local imbalances do not seem to concur with the changes in surface elevation. We do not understand yet if the recent grounding line retreat has been associated with a significant change in the force-balance of the glacier. Although the ice-ocean interaction in Pine Island Bay appears to be unusually dynamic, we cannot be sure of the past or future nature of the intrusion of warm water which is responsible. Finally, more radio-carbon dating and better discrimination between ice-rafted, sub-ice shelf and sub-ice sheet sediments are required to improve our confidence in the indications of ice-front retreat over the past millennia. Hence, as yet, none of the observed changes make a strong case for ongoing basin-scale ice-sheet change or readjustment and certainly, none suggest that the ice sheet in this area has entered a phase of significant collapse or retreat.

At a time when the paradigm of marine-ice-sheet instability is being questioned, our observations of change, perhaps appear ambiguous and inconclusive. Interpreting them as precursors of collapse would clearly be unjustified. They are certainly not yet sufficient to rigorously test the various theoretical models and ideas of ice sheet stability and collapse which have been discussed over the years [e.g. *Weertman*, 1974; *Mercer*, 1978; *Hughes*, 1980; *Fastook*, 1984; *MacAyeal*, 1992; *Hindmarsh*, 1993; *Bentley*, 1998]. This is the case, even though Pine Island Glacier has occupied a central position in discussions regarding the stability of the West Antarctic ice sheet. It can be argued that research has made relatively poor progress in the Pine Island Glacier region, particularly when compared to the Siple Coast ice streams or those which drain into the Ronne Ice Shelf. This shortcoming suggests that we should prepare field and remote sensing experiments that will allow us to determine the causes of change in the ice sheet. Only such an understanding will lead us to a sound foundation for predicting future behaviour.

Acknowledgments. We thank Howard Conway and Robert Bindschadler for constructive reviews. SSJ acknowledges support of the NASA Polar Research Program for the LDEO contribution (#6034).

REFERENCES

Alberts, F.G., *Geographic names of the Antarctic*. National Science Foundation, Washington. 1981

Alley, R.B., S. Anandakrishnan, C.R. Bentley and N. Lord, A water-piracy hypothesis for the stagnation of Ice Stream C, Antarctica, *Ann. Glaciol.*, 20, 187-194, 1994.

Anderson, J.B. and M.C. Myers, USGSC Glacier Deep Freeze 81 Expedition to the Amundsen Sea and Bransfield Strait, *Antarct. J. U.S.*, 16, 5, 1981

Armstrong, T., B. Roberts and C.W.M. Swithinbank, *Illustrated glossary of snow and ice*, Scott Polar Research Institute, Cambridge, 1973.

Bamber, J.L. and R.A. Bindschadler, An improved elevation dataset for climate and ice-sheet modelling: validation with satellite imagery, *Ann. Glaciol.*, 25, 439-444, 1997.

Behrendt, J.C., Distribution of narrow-width magnetic anomalies in Antarctica, *Science*, 144, 3621, 993-994, 1964.

Bentley, C.R., Antarctic ice streams: a review, *J. Geophys. Res.*, 92, 8843-8858, 1987.

Bentley, C.R., Rapid sea-level rise from a West Antarctic ice-sheet collapse: a short-term perspective, *J. Glaciol.*, 44, 157-163, 1998.

Bentley, C.R. and N.A. Ostenso, Glacial and subglacial topography of West Antarctica. *J. Glaciol.*, 3, 882-911. 1961.

Bentley, C.R. and F.K. Chang, Geophysical exploration in Marie Byrd Land, Antarctica, in *Antarctic Snow and Ice Studies 2, Antarct. Res. Ser.* vol 16, edited by A.P. Crary, pp 1-38, AGU, Washington, D.C., 1971.

Bindschadler, R., Jakobshavn Glacier drainage basin: a balance assessment, *J. Geophys. Res.*, 89, 2066-2072, 1984.

Bishop, J.F. and J.L.W. Walton, Bottom melting under George VI Ice Sheet, Antarctica, *J. Glaciol.*, 27, 429-447, 1981.

Bromwich, D. Snowfall in high southern latitudes, *Rev. Geophys.*, 26, 149-168, 1988.

Cassasa, G., K.C. Jezek,, J. Turner and I.M. Whillans, Relict flow stripes on the Ross Ice Shelf. *Ann. Glaciol.*, 15, 132-138, 1991

Connolley, W.M. and J.C. King, A modelling and observational study of East Antarctic surface mass balance, *J. Geophys. Res.*, 101, 1335-1343, 1996.

Crabtree, R.D. and C.S.M. Doake, Pine Island Glacier and its drainage basin: Results from radio-echo sounding, *Ann. Glaciol.*, 3, 65-70, 1982.

Doake, C.S.M., H.F.J. Corr, A. Jenkins, K. Makinson, K.W. Nicholls, C. Nath, A.M. Smith and D.G. Vaughan, Rutford Ice Stream, Antarctica, *this volume*.

Domack, E. W. and C. E. McClennen, Accumulation of glacial marine sediments in fjords of the Antarctic Peninsula and their use as late Holocene palaeoenvironmental indicators, in *Foundations for Ecological Research West of the Antarctic Peninsula, Antarct. Res. Ser.* vol 70, edited by R.M. Ross et al., pp 135-154, AGU, Washington, D.C., 1996.

Domack, E.W., E.A. Jacobson, S. Shipp and J.B. Anderson, Late Pleistocene-Holocene retreat of the West Antarctic Ice-Sheet system in the Ross Sea: Part 2 - Sedimentologic and Stratigraphic Signature, *Geological Society of America Bulletin*, 111, 1517-1536, 1999.

Fastook, J.L., West Antarctica, the sea-level controlled marine instability: past and future, in *Climate Processes and Climate Sensitivity, Geophys. Mono.* vol 29, edited by J.E. Hansen and T. Takahashi, pp 275-287, AGU, Washington, D.C., 1984.

Ferrigno, J.G., B.K. Lucchitta, K.F. Mullins, A.L. Allison, R.J. Allen, and W.G. Gould, Velocity measurements and changes in the position of Thwaites Glacier/iceberg tongue from aerial photography, Landsat images and NOAA AVHRR data, *Ann. Glaciol.*, 17, 239-244, 1993.

Giovinetto, M.B., The drainage systems of Antarctica: Accumulation, in *Antarctic Snow and Ice Studies, Antarct. Res. Ser.* vol 2, edited by M. Mellor, pp 127-155, AGU, Washington, D.C., 1964.

Giovinetto, M.B. and C.R. Bentley, Surface balance in ice drainage systems of Antarctica, *Antarct. J. US.*, 20, 6-13, 1985.

Goldstein, M., H. Engelhardt, B. Kamb, and R.M. Frolich, Satellite radar interferometry for monitoring ice sheet motion: application to an Antarctic ice stream, *Science*, 262, 1525-1530, 1993.

Hellmer, H.H., S.S. Jacobs, and A. Jenkins, Ocean erosion of a floating Antarctic Glacier in the Amundsen Sea, in *Ocean, Ice and Atmosphere: Interactions at the Antarctic Continental Margin, Antarct. Res. Ser.* vol 75, edited by S.S. Jacobs and R.F. Weiss, pp 83-100, AGU, Washington, D.C., 1998.

Hindmarsh, R.C.A., Qualitative dynamics of marine ice sheets, in *Ice in the Climate System, NATO ASI.Series* vol I 12, edited by W.R. Peltier, pp 68-99, Springer-Verlag, Berlin Heidelberg, 1993.

Hughes, T.J., The weak underbelly of the West Antarctic Ice Sheet, *J. Glaciol.*, 27, 518-525, 1980.

Jacobs, S.S., H.H. Hellmer, and A. Jenkins, Antarctic ice sheet melting in the Southeast Pacific, *Geophys. Res. Lett.*, 23, 957-960, 1996.

Jacobs, S.S. and J.C. Comiso, Climate variability in the Amundsen and Bellingshausen Seas, *J. Clim.*, 10, 697-709, 1997.

Jankowski, E.J. and D.J. Drewry, The structure of West Antarctica from geophysical studies, *Nature*, 291, 17-21, 1981.

Jenkins, A., D.G. Vaughan, S.S. Jacobs, H.H. Hellmer, and J.R. Keys, Glaciological and oceanographic evidence of high melt rates beneath Pine island glacier, West, Antarctica, *J. Glaciol.*, 43, 114-121, 1997.

Jezek, K.C., H.G. Sohn and K.F. Noltimier, The Radarsat Antarctic Mapping Project, *IGARSS '98 - 1998 International Geoscience and Remote Sensing Symposium, Proceedings*, Vol 1-5, Chapter 888, 2462-2464, 1998.

Joughin, I., L. Gray, R. Bindschadler, S. Price, D. Morse, C. Hulbe, K. Matter and C. Werner, Tributaries of West Antarctic Ice Streams Revealed by RADARSAT Interferometry, *Science*, 286, 283-286, 1999.

Jonas, M. and D.G. Vaughan, ERS-1 SAR mosaic of Filchner-Ronne-Schelfeis, *Filchner Ronne Ice Shelf Programme Reports*, 10, edited by H. Oerter, 47-49, AWI, Bremerhaven, Germany, 1996.

Jones D.A. and I. Simmonds, A climatology of Southern Hemisphere extratropical cyclones, *Climate Dynamics*, 9, 135-145, 1993.

Kellogg, T.B., R.S. Truesdale, and L.E. Osterman, Late quaternary extent of the West Antarctic Ice Sheet: new evidence from Ross Sea cores, *Geology*, 7, 249-253, 1979.

Kellogg, T.B., D.E. Kellogg, and T.J. Hughes, Amundsen Sea sediment coring, *Antarct. J. U.S.*, 20, 79-81, 1985.

Kellogg, T.B. and D.E. Kellogg, Recent glacial history and rapid ice stream retreat in the Amundsen Sea., *J. Geophys. Res.*, 92, 8859-8864, 1987.

Kellogg T.B., D.E. Kellogg, E.D. Waddington, and J.S. Walder., Late Quaternary deglaciation of the Amundsen Sea: implications for ice sheet modelling, in *The physical basis of ice sheet modelling, IAHS Pub. 170*, edited by E.D. Waddington and J.S. Walder, pp 349-357,Int. Assoc. of Hydrol. Sci., Wallingford, England, 1987.

Legresy, B. and F. Remy, Altimetric observations of surface characteristics of the Antarctic Ice Sheet, *J. Glaciol.*, 43, 265-275, 1997.

Licht, K.J., N.W. Dunbar, J.T. Andrews, and A.E. Jennings, Distinguishing subglacial till and glacial marine diamictons in the western Ross Sea, Antarctica: implications for a last glacial maximum grounding line, *Bulletin of the Geological Society of America*, 111, 91-103, 1999.

Lindstrom, D. and D. Tyler, Preliminary results of Pine Island and Thwaites Glaciers Study, *Antarct. J. U.S.*, 19, 53-55, 1984.

Liu, H., K. Jezek, and B. Li, Development of an Antarctic digital elevation model by integrating cartographic and remotely sensed data: A geographic information system based approach., *J. Geophys. Res.*, 104, 99-23,213, 1999.

Lucchitta, B., C. Rosanova, and K. Mullins, Velocities of Pine Island and Thwaites Glacier, West Antarctica, from ERS-1 SAR images, *Ann. Glaciol.*, 21, 277-283, 1995.

Lucchitta B.K., C.E. Smith, J. Bowell, and K.F. Mullins, Velocities and mass balance of Pine Island Glacier, West Antarctica, derived from ERS1-SAR. *ESA Publication SP-361*, 147-151, 1994.

Lucchitta B.K., and C.E. Rosanova, Velocities of Pine Island and Thwaites Glaciers, West Antarctica, from ERS1-SAR images. *ESA Publication SP-414*, 819-824, 1997.

MacAyeal, D.R., Irregular oscillations on the West Antarctic Ice Sheet, *Nature*, 359, 29-32, 1992.

Mantripp, D.R., J. Sievers, H. Bennat, C.S.M. Doake, K. Heidland, J. Idhe, M. Jonas, B. Reidel, A.V. Robinson, R. Scharroo, H.W. Shenke, U. Shirmer, F. Stefani, D.G. Vaughan and D.J. Wingham, *Topographic map (satellite image map), Filchner-Ronne-Shelfeis (2nd Edition). Map at 1 : 2 000 000*, Institut für Angewandte Geodäsie, Frankfurt am Main, Germany, 1996.

Mercer, J.H., West Antarctic Ice Sheet and CO_2 greenhouse effect: A threat of disaster, *Nature*, 271, 321-325, 1978.

Molnia, B.F. and P.R. Carlson, Surface sedimentary units of northern Gulf of Alaska continental shelf, *Bulletin of the American Association of Petroleum Geologists*, 62, 633-643, 1999.

Paterson, W.S.B., *The physics of glaciers*, Elsevier Science, Oxford, 480pp, 1994.

Payne, A.J., A thermomechanical model of ice flow in West Antarctica. *Climate Dynamics*, 15, 115-125, 1999.

Rignot, E.J., Fast recession of a West Antarctic Glacier, *Science*, 281, 549-551, 1998.

SCAR, *Antarctic digital database user's guide and reference manual*, Scientific Committee on Antarctic Research, Cambridge, xi+156pp, 1993.

Shimizu, H., Glaciological studies in West Antarctica, in *Antarctic Snow and Ice Studies, Antarct. Res. Ser.* vol 2, edited by M. Mellor, pp 37-64, AGU, Washington, D.C., 1964.

Shipp, S., J. Anderson and E. Domack, Late Pleistocene-Holocene retreat of the West Antarctic Ice-Sheet system in the Ross Sea: Part 1 - Geophysical Results, *Geological Society of America Bulletin*, 111, 1486-1516, 1999.

Smith, A.M., The use of tiltmeters to study the dynamics of Antarctic ice-shelf grounding lines, *J. Glaciol.*, 37, 51-58, 1991.

Stenoien M.D., *Interferometric SAR observations of the Pine Island Glacier catchment area*, Unpublished Ph.D. thesis, University of Wisconsin-Madison, 127pp, 1998.

Stuiver, M., G.H. Denton, T.J. Hughes and J.L. Fastook, History of the marine ice sheet in West Antarctica during the last glaciation: a working hypothesis, in *The Last Great Ice Sheets*, edited by G.H. Denton and T.J. Hughes, Wiley-Interscience, New York, 1981.

Stephenson, S.N., Glacier flexure and the position of grounding lines: Measurements by tiltmeter on Rutford Ice Stream, Antarctica, *Ann. Glaciol.*, 5, 165-169, 1984.

Swithinbank C.W.M., Ice streams, *Polar Record*, 7, 185-186, 1954.

Thomas, R.H., Ice sheet margins and ice shelves. In: J.E. Hansen and T. Takahashi (Eds), Climate Processes and Climate Sensitivity, in *Climate Processes and Climate Sensitivity, Geophys. Mono.* vol 29, edited by J.E. Hansen and T. Takahashi, pp 265-274, AGU, Washington, D.C., 1984.

USGS, *Antarctica Sketch Map, Thurston Island - Jones Mountains, 1:500,000*, USGS, Washington, D.C., 1993.

Vaughan, D.G., Investigating tidal flexure on an ice shelf using kinematic GPS, *Ann. Glaciol*, 20, 372-376, 1994.

Vaughan, D.G., J.L. Bamber, M. Giovinetto, J. Russell, and A.P.R. Cooper, Reassessment of net surface mass balance in Antarctica, *J. Clim.*, 12, 933-946, 1999.

Vaughan D.G. and J.L. Bamber, Drainage basin analysis and improved calculation of balance fluxes for West Antarctic ice streams and glaciers, in *Abstracts, AGU Chapman Conference on the West Antarctic Ice Sheet, Orono, Sept.1998*, 1998.

Weertman, J., Stability of the junction of an ice sheet and an ice shelf, *J. Glaciol.*, 13, 3-11, 1974.

Whillans, I.M. and C.J. Merry, Ice-flow features on Ice Stream B, Antarctica, revealed by SPOT HRV imagery, *J. Glaciol.*, 39, 515-527, 1993.

Whillans, I.M., M. Jackson and Y-H Tseng, Velocity pattern in a transect across Ice Stream B, Antarctica, *J. Glaciol.*, 39, 562-572, 1993.

Williams, R.S., J.G. Ferrigno, T.M. Kent, and J.W. Schoonmaker Landsat images and mosaics of Antarctica for mapping and glaciological studies, *Ann. Glaciol.*, 3, 321-326, 1982.

Wingham, D.J., A.J. Ridout, R. Scharroo, R.J. Arthern, and C.K. Schum, Antarctic elevation change from 1992 to 1996 *Science*, 282, 456-458, 1998.

C. R. Bentley and M.D. Stenoien, Geophysical and Polar Research Center, University of Wisconsin - Madison, Madison, Wisconsin 53706, USA. (email: bentley@geology.wisc.edu)

S. S. Jacobs, Lamont-Doherty Earth Observatory, University of Columbia, Palisades, New York 10964, USA. (email: sjacobs@ldeo.columbia.edu)

T. B. Kellogg, Institute for Quaternary Studies, University of Maine, Orono, Maine 04469, USA. (email: tomk@iceage.umeqs.maine.edu)

B.K. Lucchitta, U.S. Geological Survey, Flagstaff, Arizona 86001, USA. (email: blucchitta@flagmail.wr.usgs.gov)

E. Rignot, Jet Propulsion Laboratory, Pasadena, California 91109, USA. (email: eric@adelie.jpl.nasa.gov)

D. G. Vaughan, A.M. Smith, H. F. J. Corr, A. Jenkins, British Antarctic Survey, Natural Environment Research Council, Cambridge CB3 0ET, U.K. (email: d.vaughan@bas.ac.uk)

ICE STREAMS B AND C

I. M. Whillans

Byrd Polar Research Center and Department of Geological Sciences, Ohio State University, Columbus, Ohio

C. R. Bentley

Geophysical and Polar Research Center, University of Wisconsin, Madison, Wisconsin

C. J. van der Veen

Byrd Polar Research Center and Department of Geography, Ohio State University, Columbus, Ohio

The mapping and description of the kinematics and dynamics of ice streams B and C are reviewed. The discussion centers around the themes of why ice streams are fast despite small driving stress and why and how ice streams change with time. The mapping has described the limits to the ice streams, their surface and bed features and crevasses. For ice stream B, velocities are in excess of those needed to evacuate current snow accumulation, so the catchment of ice stream B is thinning. Also, ice stream B is widening and slowing. In contrast, the lower and middle reaches of ice stream C have mainly stopped and are thickening. The upper portion of ice stream C is active, and there must be some special ongoing activity at the region joining active and stopped ice. The bed under ice stream B has a layer of soft sediment. This sediment has probably been in traction from the ice above and the process is likely active now. Debris has collected in a delta-like feature under the mouth of ice stream B. An analysis of the budget of forces shows that gravitational action on ice stream B is opposed mainly from the sides, meaning that the bed is nearly perfectly lubricated. A calculation along the flowline shows where the ice stream begins and how some of the characteristics change along-glacier. For ice stream C the reaction to the driving stress is from the bed. Various suggested hypotheses for the controls on ice stream behavior are reviewed in the light of measurement programs that were targeted to test them. The analysis does not favor dominance of the hypotheses of basal heat feedback, global warming, piracy, height above buoyancy, deforming bed, ongoing surge, or active volcanism in ice stream behavior. Rather an active ice stream has a very weak bed, probably because it is soft and moldable, and frictional drag comes from the sides. The cause of switches in time of basal drag are not yet known.

INTRODUCTION

Ice stream B (Figure 1) is the archetypal example of an ice stream. An ice stream contrasts with other outlet glaciers in that these other glaciers are fast but have very steep surface slopes, meaning that the gravitational driving stress is large. An example of such an outlet glacier is Byrd Glacier, which passes through Transantarctic Mountains. It achieves velocities of 0.8 km a^{-1} under a mean driving stress of 220 kPa (calculated from elevation decrease from 550 m to 150 m over 40 km and ice thickness 2500 m [*Whillans et al.,* 1989]), yet as discussed next, ice stream B achieves similar speeds under a driving stress of 15 kPa, only 7% of that of Byrd Glacier. The other distinguishing feature of ice streams is the lack of close bed-topographic control on the routes. These characteristics lead to the ice streams catching special scientific attention.

This first anomaly of the flow of ice stream B is displayed in Figure 2. Driving stress is only about 15 kPa (3rd panel), yet speeds reach 800 m a^{-1} (4th panel). Moreover, over the span from –200 km to +50 km speed becomes larger as the gravitational forcing becomes smaller. Such a relation is counter to the concepts of classical glaciology. This mysterious inverse relation between forcing and response calls for an explanation.

Next to ice stream B is ice stream C (Figure 3). Its presence is evident on imagery, but its middle and lower reaches have small speeds (Figure 3; [*Whillans and Van*

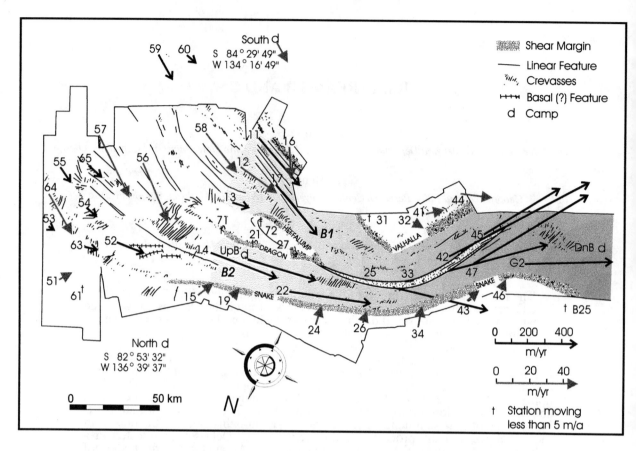

Fig. 1. Ice stream B showing surface features and velocities. The features derive from interpretations of aerial photos and satellite images, the velocities from ground-based surveys [*Whillans and Van der Veen*, 1997]. Reproduced from the Journal of Glaciology with permission from the International Glaciological Society and the authors.

der Veen, 1993]) and buried crevasses indicating former activity [*Bentley et al.*, 1985]). The characteristics of ice stream C and the history of its discovery are discussed in the companion paper by *Anandakrishnan et al.* (this volume) and raise a second mystery: how and why the ice streams change.

In this report some of the evidence relating to these two mysteries (fast flow under small driving stress for ice stream B and to stoppage of ice stream C) is reviewed and the current status of the problems of ice stream flow is summarized.

MAPPING THE ICE STREAMS

A combination of aerial photography, satellite imagery, and radio-echo sounding has been used to map the ice streams. A compilation of much of the data is presented in Figures 1 and 3. The first mapping of the surface and bed elevations and ice thickness was carried out as part of the NSF-SPRI-TUD radio-sounding program of the 1960's and 1970's [*Robin et al.*, 1970; *Rose*, 1979]. Newer, more detailed maps stem principally from extensive airborne radar sounding carried out in 1984-85 [*Shabtaie and Bentley*, 1987, 1988; *Shabtaie et al.*, 1987] and 1988-89 [*Retzlaff et al.*, 1993] with ground control by satellite tracking [*Whillans and Van der Veen*, 1993]. The 1984-85 set of flights were reconnaissance in nature and employed analog recording of the data, whereas the 1988-89 flights were on regularly spaced grids with 5- or 10-km line spacings and data were recorded digitally.

Part of a satellite image is shown in Figure 4a. It reveals crevasses, drift mounds, flow traces and ridges and troughs. A chaotic zone of crevasses and outboard arcuate crevasses marks each lateral boundary (Figure 4b; also discussed in *Raymond et al.* [this volume]). The interstream ridges (except ridge A/B) have very simple smooth surfaces that slope toward the ice streams. The along-flow beginning and end to ice stream flow are not clearly evident in the images or photos.

The surface of ice stream B exhibits an irregular topography of uncertain origin. Some small features are migrating with time. Most features are too small to depict in Figures 1 and 3. An image trackable feature 'a', near the mouth of ice stream B, is traveling down-

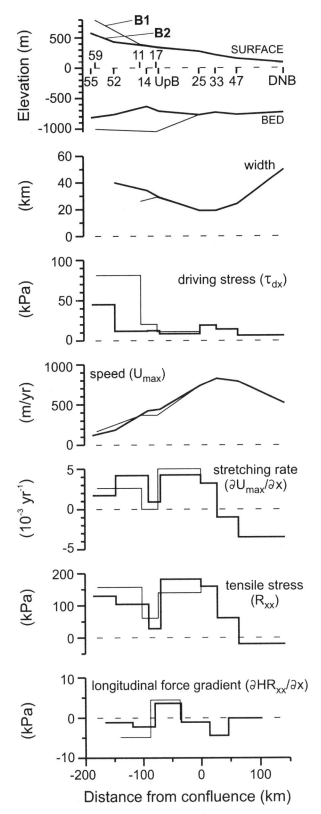

Fig. 2. Data and steps in computation of budget of forces. Origin of coordinate system is at station 25 (near the junction of tributaries B1 and B2). Elevations and station names (top panel) and speeds (U_{max}, 4th panel) from *Whillans and Van der Veen* [1993]. Bed elevation (top panel) from *Retzlaff et al.* [1993]. Width of ice stream, W, from distance between zones of severe crevasses as evident on photographs and satellite images (at the upper end of the ice stream this width differs from that of *Shabtaie et al.* [1987], who based it on correlation of zones of intense clutter between ice-sounding radar flight lines). Driving stress (3rd panel), stretching rate, longitudinal tensile stress, and longitudinal force gradient are computed from data in the panels above. Resistive stresses involve the rate factor, $B = 540$ kPa a$^{1/3}$, corresponding to the depth-weighted mean using the temperature profile measured at the UpB camp [*Engelhardt et al.*, 1990] and the temperature dependence given in *Hooke* [1981].

glacier [*Bindschadler and Vornberger*, 1998]. Near the head of ice stream B, at UpB, many topographic features are migrating up-glacier [*Hulbe and Whillans*, 1997]. In contrast, the boundary between ice streams B1 and B2, downstream of the Unicorn and containing stations 25, 33, 42, and 45 (Figure 1), persists as a suture zone for 250 km down-glacier from their merger.

Most shear margins to the ice streams B and C are not distinct in surface topography. Exceptions to this remark are a few margins that contain surface valleys, such as the down-glacial portion of the northern shear margin (the Snake) of ice stream B [*Shabtaie et al.*, 1987] and the northern margin of ice stream A. Northern ice streams D and E are different from ice streams B and C, being bounded by especially steep slopes over a zone about 1 km wide just outboard of the shear margins [*Stephenson and Bindschadler*, 1990]. For ice streams B and C, the lack of major relief in most of the margins demonstrates that surface relief is not a necessary characteristic.

Also, there is no clear association of ice-stream location with subglacial topography. Much of ice stream B1 overlies a deep subglacial trough, but part also overlies a subglacial ridge. The upper parts of ice stream B1 are associated with a deep-lying bed, but so are the neighboring inter-stream ridges. There is a shallower trough beneath ice stream B2 near station UpB, but it trends diagonally across the ice stream, not along its axis. From a contour map of subglacial topography alone, it would be very difficult to predict where ice streams might form. There is some indication that the association between subglacial troughs and ice-stream axes may become closer up-glacier where the subglacial relief is greater [*Shabtaie and Bentley*, 1988], but it is not possible to test this hypothesis definitively because the positions of some of the ice-stream boundaries are not well known in that region.

Close to the Ross Ice Shelf, the combined trunk of ice stream B widens into a nearly flat area known as an "ice plain." There the mean surface slope is only 3.5×10^{-4} and surface elevations are only a few tens of meters above hydrostatic equilibrium in the ocean [*Shabtaie and Bentley*, 1987].

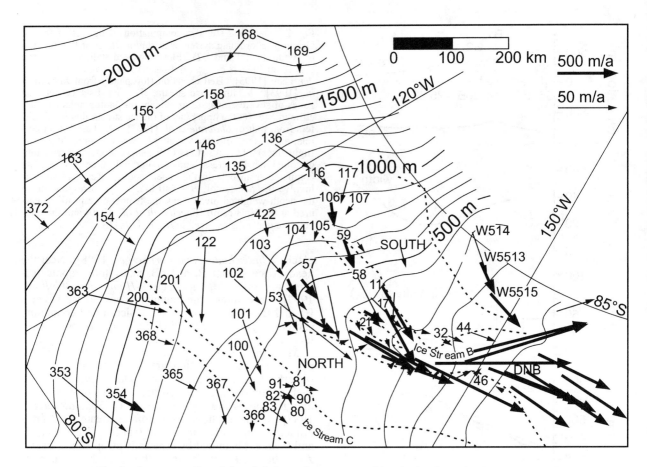

Fig. 3. Ice streams B and C and their catchment areas with velocity determinations superposed. Velocities are *from Price and Whillans* [1998]. Shear margins are from *Shabtaie et al.* [1987].

The surfaces of ice streams are lower than those of the interstream ridges to either side. For ice stream B the slope from interstream ridge into ice stream drives ice flow across the shear margin into the ice stream from the sides. Speeds on the interstream ridges are about that expected to balance the measured surface accumulation rate [*Whillans and Van der Veen*, 1993].

At the head of the small-activity ice stream C there is a bulge in the surface [*Joughin et al*, 1999; *Spikes et al*, 2000]. This form is believed to be linked to the future reactivation of ice stream C but the precise location of the future active ice stream is not clear.

In short, the locations of the ice streams are not definitively associated with basal topography. Their locations must be at least partially controlled by basal or internal conditions, particularly in their more down-glacier portions. Inland, the lateral boundary between inland-ice flow and ice-stream flow is not clearly understood.

Exposed Crevasses

Broad-scale maps of location and type of exposed crevasses have been made using photos and satellite imagery. The first maps are based on aerial photography [*Vornberger and Whillans*, 1990]. Later maps include information from SPOT imagery (Figure 1). Other imagery that can be useful are DISP [*Bindschadler and Vornberger*, 1998] and AVHRR [*Bindschadler and Vornberger*, 1990]. LANDSAT is not helpful because ice stream B lies well beyond its latitudinal range. Figure 1 includes simplified tracings of crevasse patterns on ice stream B.

A characteristic feature of ice streams is the crevassed shear margins. These margins separate the active ice streams from the slowly moving interstream ridges. The pattern of crevasses within a shear margin is shared with other glaciers with large strain rates at the sides, but the pattern is on a very grand scale in the case of ice streams. An image of the Dragon is shown in Figure 4*b*. *Raymond et al.* [this issue] discuss shear margins more thoroughly. As observable with visible imagery, a shear margin begins with a series of irregular crevasses or buckles in ice flow. The Dragon begins with a fan shaped set of crevasses (Figure 4*a*). Within a well-developed shear margin the crevasses are arcuate on the outboard side and, with drift mounds, form a chaotic zone on the inboard side of the shear margin

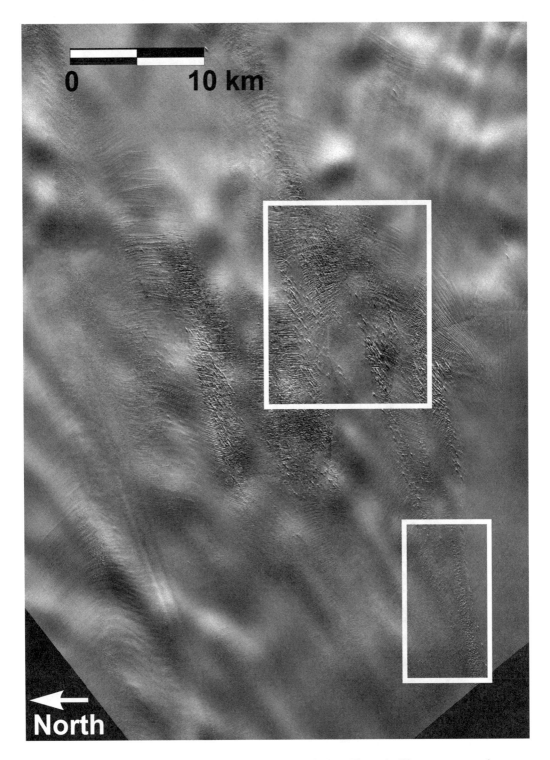

Fig. 4a. Head of ice stream B, between stations 56 and 14 on Figure 1. Flow traces cross the entire image from top to bottom. Most crevasses are transverse to ice motion (which is top to bottom). Drift mounds are aligned diagonally upper-left to lower-right. [SPOT High Resolution Visible image, K092 J569, December 15, 1989, 15:29 GMT; solar azimuth: 83.1°; solar elevation: 24.1°].

Fig. 4b. Subscene from Figure 4a showing the Dragon shear margin. The active ice stream moves from top to bottom of the image on the left-hand side. The uncrevassed region to the right is the nearly stagnant interstream ridge (Unicorn). The rightmost crevasses are arcuate and hooked down-glacier.

(Figure 4b). Ice from the interstream ridge flows laterally into the shear margin where it is taken up as part of ice stream flow. The shear margins can retain their simple forms over distances in excess of 250 km (in the case of the Snake).

Most observable crevasses in the body of the ice stream are transverse crevasses (for example in Figure 4a). Measurements of velocity, reported on below, show that the flow pattern in the ice stream is, to a first approximation, simple lateral shear. Crevasses formed under such a regime are diagonal to flow, but with age and down-glacial transport existing crevasses continue to open and rotate. In such a simple shear regime these crevasses are widest when they are transverse to flow. It is these that are most readily observed. During rotation the crevasses experience strike-slip shearing along their lengths. This leads to complex shapes and folding of intercrevasse slivers. Further rotation causes the crevasses to close.

There is a much smaller crevasse density near the kinematic centerline of the ice stream (left-hand portion of Figure 4a). This is a site of small lateral shear stress, meaning that there is less along-crevasse shearing. Existing crevasse bridges are less distorted and less evident in visual inspections near the centerline. The small crevasse density and the less distortion of bridges at the centerline mean that there are many sites where the Twin Otter aircraft can make safe landings. This is the reason that most point determinations of ice-stream velocity are near the kinematic centerline (cf. Figure 1).

Most of the crevasses in the ice stream are bridged and most bridges sag. Some bridge sections fall, leaving a hole. Such cavities disrupt the surface wind pattern. Snow drift collects downwind of the cavity to form drift mounds [*Vornberger and Whillans*, 1990]. Often the hole is rebridged but the drift mound remains visible. The drift mounds in Figures 4a and 4b and seen in photos and overflight are believed to form in this way. It is sagging bridges and drift mounds that can be mapped and tracked with time to determine motion [*Scambos and Bindschadler*, 1993; *Bindschadler et al.*, 1996; *Whillans and Tseng*, 1995; *Whillans et al.*, 1993; *Whillans and Van der Veen*, 1997].

Crevasses Intersect at Oblique Angles

The simple viewpoint is that considering that the void within a crevasse cannot support tensile stress perpendicular to crevasse orientation, a new crevasse formed in ice with existing crevasses should be parallel or perpendicular to earlier crevasses [*Van der Veen*, 1998]. Some glaciers exhibit this pattern (e.g. Mulock Glacier in the Transantarctic Mountains [*Swithinbank*, 1988, page B31], also Skeiðarárjökull, Iceland, during a surge [*Björnsson*, 1998, cover photo], arctic glaciers [*Herzfeld*, 1998, figures 168 and 190]). On the ice streams, the observation of frequent oblique crevasse intersections (Figure 4c) indicates that new crevasses must form in unfractured ice, beneath the depth of penetration of preexisting crevasses. Old crevasses must heal at depth because otherwise the older crevasse would open wider instead of a new crevasse forming. A speculation on why pre-existing crevasses are not foci for new crevasses is that the deep extensions of crevasses are not fractures but are recrystallization fronts that heal and strengthen with strain [*Whillans et al.*, 1993]. Other models for such behavior are possible. For an arctic glacier in Canada, *Hambrey and Müller* [1978, p. 59] make similar observations but suggest upward water migration as the physical process, a process not feasible on the ice streams. The common interpretation is that deep parts of crevasses can heal and new crevasses form at other angles.

The speed of upward growth of new crevasses in ice streams is not known, it could be nearly instantaneous or very slow, perhaps with crevasses not always reaching the surface. The lack of resolution of this issue means that crevasse age estimates based on crevasse depth requires some line of argument about past depth of the upper portion of crevasses.

There are some special patterns to crevasses observed on ice stream B. Splaying crevasses just upglacier of the DnB camp (Figure 1) indicate lateral spreading and crevasse advection as the ice stream fans out toward the ice shelf. The Dragon shear margin (Figure 4a) and the suture between tributaries B2a and B2b originate with patterns of reversed splaying crevasses (between stations 54, 55 and 65 in Figure 1). Such a pattern suggests lateral spreading around an obstruction to flow. Other shear margins begin with special crevasse patterns: packages of crevasses called the chromosomes in the case of the Heffalump, warps or buckles in the ice surface in the case of the Snake, and crossing crevasses in a pattern reminiscent of simple sketches of flying seagulls where flow splits into tributaries B1b and B2a. These patterns, and the reason for crevasses to appear in groups, are discussed by *Merry and Whillans* [1993]. Other special crevasse patterns have uncertain genesis. A discontinuous line of crevasses diagonally upglacier from the onset of the Dragon (starting at the right-hand edge of Figure 4a and angled diagonally up-glacier at about 45° to the flow traces) was mistaken as an extension of the Dragon (on the line between stations 71 and 65, Figure 1) by *Shabtaie and Bentley* [1988]. Its location with respect to bed topography [*Retzlaff et al.*, 1993] indicates that it relates to flow over a basal disturbance.

Subsurface Crevasses

A survey of the strain grid near UpB by short-pulse radar reveals two areas of crevasses that do not reach the surface [*Clarke and Bentley*, 1994]. A question is whether some of these crevasses are buried or whether they ever reached the surface.

One set of crevasses (which we call group I) very near the surface, is limited to the glacier-left part (viewing with glacier flow) nearest the shear margin (lines U, V, and X of the OSU strain grid, the southern portion of study in *Hulbe and Whillans* [1994]). This set appears to comprise buried examples of the type of crevasse that are visible at the surface up-glacier. The

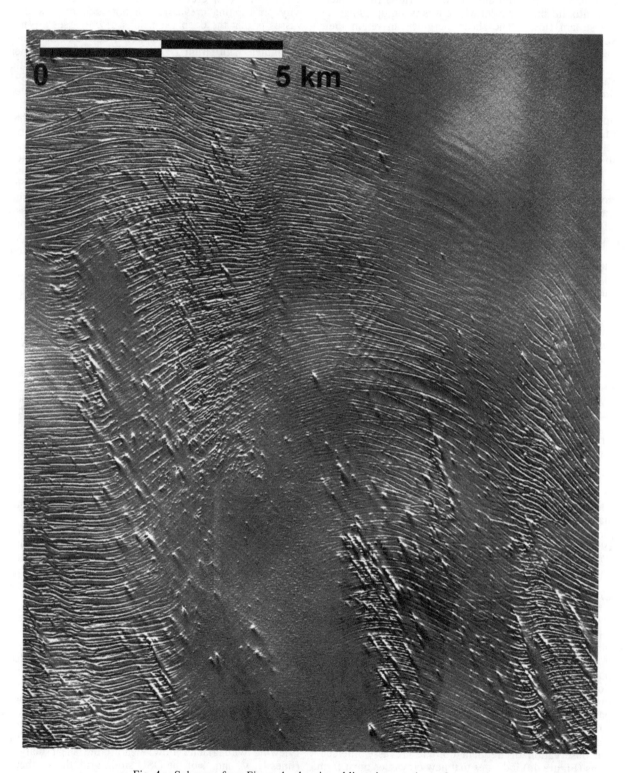

Fig. 4c. Subscene from Figure 4a showing oblique intersections of crevasses.

net rotation of these buried crevasses is greater farther from the glacier margin, as expected for steady flow because that ice has been in the ice stream longer and experienced more net rotation.

A second (group II) and deeper set of crevasses (deeper than 30 m) occurs beneath the first set and is also closer to the ice-stream centerline. The depth of the top of crevasses in this group is in snow strata about 200 years old (this is an age estimate if the crevasses reached the snow surface at formation). The down-glacial portion of this group (group IIdn) contains crevasses that are rotated by at least 45° out of alignment with present-day principal strain rates. Ice rotation rates vary within the region [*Whillans et al.*, 1993; *Hulbe and Whillans*, 1997], but a representative value is 1×10^{-3} a^{-1} (from the C line [*Hulbe and Whillans*, 1997]). With this rotation rate, the crevasse misalignment is achieved in 200 years (an accurate calculation would need to consider the crevasse-perpendicular rotation rate, but the result would be of similar magnitude). These calculations of crevasse depth and orientation are both consistent with the crevasses in this down-glacier group being some centuries old. Using present-day velocities, the nascent site for these crevasses must lie about 40 km up-glacier of the strain grid (typical speed of 200 m a^{-1}) – possibly at the very large crevasses just up-glacier of station 14 (Figure 1). The up-glacier members of this group (group IIup) are aligned perpendicular to the principal extending strain rates, suggesting a much younger age or ongoing formation – possibly at the crevasses closer to station 14.

The third region (group III) is at the kinematic centerline and no crevasses are detected by radar, consistent with typically smaller stress levels at the centerline. More important is the history of that ice. That ice must have passed glacier-right of the large crevasse field up-glacier of station 14.

Flow Traces

Active fast flow is marked by longitudinal ridges and furrows, called flow traces or flow stripes. They originate at disturbances to flow and the scar is carried off with the ice flow. The "suture" joining tributaries B1 and B2 is a prominent flow trace passing through stations 25 and 33 in Figure 1. Lesser flow traces originate within shear margins, such as the Heffalump [*Merry and Whillans*, 1993].

Flow traces have been used to define ice streams [*Hodge and Dopplehammer*, 1996; *Bell et al.*, 1998; *Stephenson and Bindschadler*, 1990]. Ice streams do contain flow traces, but flow traces also occur on outlet glaciers through the Transantarctic Mountains (e.g. figure 36 in *Swithinbank* [1988]) as well as on medium-speed ice in West Antarctica [*Bindschadler et al.*, 1996]. Flow traces originate far up-glacier [*Joughin et al.*, 1999] and can disappear down-glacier (e.g. at S80°20', W133°, in figure 4 of *Bindschadler et al.* [1996]; also noted by *Stephenson and Bindschadler* [1990]), presumably where velocity fluctuations within the inland ice or ice stream (figure 2 in *Bindschadler* [1993]) disrupt simple flow lines. The presence of flow traces can not be taken as a general diagnostic of ice stream flow, rather ice streams are defined here as grounded glaciers that are fast despite small driving stress (following *Bentley* [1987]).

Possibly flow traces are caused by strength variations in the ice, as found to be the explanation for the topographic features at UpB [*Hulbe and Whillans*, 1997]. Other suggestions are made by *Stephenson and Bindschadler* [1990], *Merry and Whillans* [1993] and *Gudmundsson et al.* [1998].

VELOCITIES

The most direct measurements of velocity have been obtained by leaving markers in the ice surface and surveying at least twice with ground-based satellite tracking. These velocities are depicted in Figures 1 and 3.

Speeds increase from about 300 m a^{-1} at the onset of ice stream flow to 865 m a^{-1} at the narrowest section and then decrease as the ice stream spreads onto the ice plain [*Whillans and Van der Veen*, 1993; *Bindschadler et al.*, 1993]. Here the 'onset' is taken to be at the site of major along-glacier decrease in driving stress despite fast speed. Distinct crevassed shear margins begin at about the same sites, where a speed contrast of about 100 m a^{-1} between potential ice stream and potential interstream ridge occurs over a distance of about 10 km or less (e.g. near stations 13 and 63 in Figure 1). The two main tributaries (B1 and B2) have very similar speeds, and join forming a suture across which there is very little shear. Sub-tributary B1b is slower and there is a shear zone where sub-tributary B1a fuses with it.

Spatially more dense measurements of velocity are obtained from feature tracking with repeat photography. A detailed map of speeds in the 10-block of tributary B2 (near the UpB camp) shows local longitudinal fluctuations of about 1% over 1000 m, and rather larger variability in the transverse component of velocity (figures 3b and 3d of *Whillans et al.* [1993]). *Whillans et al.* [1993] suggest that some of the fluctuations are due to rafts of stiff ice being advected along flow. A very precise survey of strains and vertical motions of a strain grid around the UpB camp leads to a further model for the fluctuations [*Hulbe and Whillans*, 1997]. Topographic features and the strain rate pattern in the ice stream are evolving with time. This result is interpreted as being due to tipped bands of crystallographically soft ice being advected along flow. This soft ice acts rather like soft fault gouge favoring lateral compression and growing topographic thrust-like structures (Figure 5). This discovery of strength heterogeneity accounts for the anomalous computation of reverse basal drag at certain sites [*Whillans and Van der Veen*, 1993]. That calculation was made under the assumption that there were no major horizontal gradients in ice strength. This advecting marbling or foliation of the glacier means that the flow of the ice stream is not steady on the time scale of advection of these features.

Two further blocks of feature tracking have been studied. Transects within each block have been stacked

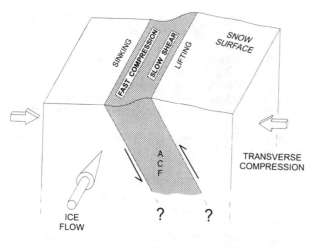

Fig. 5. Interpretation of the process leading to surface topographic variations on ice stream B. There is an inclined band of ice with a more intense crystal orientation fabric. This leads to zones of faster deformation and differential vertical motion [*Hulbe and Whillans, 1997*]. Reproduced from the Journal of Glaciology with permission from the International Glaciological Society and the authors.

to produce single width-scaled transverse profiles of ice speed (Figure 6, top panel). Speed is fastest near the centerline and decreases towards each margin. As developed in the appendix, the theoretical pattern for speed within the ice stream is a decrease as the 4th power of distance from the center (taking basal drag, thickness, and surface slope constant across the section and using the usual constitutive relation). This pattern is followed in good approximation, notable exceptions being that the speed maximum in the 40-block is glacier right of the geometric center, and there are local fluctuations in speed.

Anisotropy

There is strong evidence for crystalline anisotropy in the ice from studies of seismic- and electromagnetic-wave velocities. Oblique-angle seismic reflection studies near UpB find that, through most of the ice thickness, the ice-crystal c-axes lie in or near the vertical plane normal to the flow direction of the ice stream and are distributed randomly within that plane [*Blankenship, 1989*]. This is corroborated by detailed radar-polarization experiments, which also find that the c-axes near UpB lie near the transverse plane [*Liu et al., 1994*]. This fabric is one that might be produced by protracted simple lateral shearing.

The study of radar polarization also discovered a transverse change in the fabric over a horizontal distance of 100 to 200 m along a line some 15 km up-glacier of UpB. The discontinuity lies within a few hundred meters of the crevasse-free zone near the kinematic centerline (as discussed under Subsurface Crevasses, above). This suggests that nearly all the ice of the ice stream, except that very close to the centerline,

has developed a preferred crystal orientation fabric and that there are variations in the intensity of fabric.

The patch of 'soft' ice reported by *Hulbe and Whillans* [1997] (Figure 5) was not studied with radar polarization measurements so it is not proved that the two studies are finding the same type of feature. However, it seems reasonable to suppose that both found the same sort of important horizontal gradients in ice strength due to preferred crystal orientations within the ice stream.

Speed of Deep Versus Surface Ice

The net rate of vertical shear through the ice stream at three sites near the UpB camp has been measured by tracking basal radar diffraction patterns with time. The diffractions originate from irregularities at the base of the ice; shifts in the interference or fading pattern are tracked with repeated mapping at the surface [*Liu et al., in press*]. The results are significantly different at the three sites. The deduced speed contrast between surface and bed (with one standard error uncertainties) are 1.0 (0.1), 0.1 (0.1), and -0.5 (0.1) m a^{-1} (the third site suggests faster deep ice). The down-glacier surface slopes are 0.0044, 0.0007 and 0.0031 and driving stresses are 40, 6 and 29 kPa, respectively. The variation may be due to flow in response to the local slope (and driving stress) or to shearing within 'weak bands' [*Hulbe and Whillans, 1997*]. The third site lies over such a weak band.

For the first site, the differential speed can be explained by lamellar ('laminar') flow in the vertical and longitudinal plane if basal drag equals two thirds of the local driving stress, that is 26 kPa. However, in light of the effects of other stresses and of strength inhomogeneities [*Hulbe and Whillans, 1997*], precise estimates of basal stress from such differential motion may not be possible.

MASS BALANCE

Early results of mass balance calculations indicate thinning of ice stream B [*Shabtaie and Bentley, 1987; Shabtaie et al., 1988; Whillans and Bindschadler, 1988*]. Of these, the most data rich calculation finds that input in the catchment of the ice streams is 27% less than output from the mouth of ice stream B [*Whillans and Bindschadler, 1988*]. This translates to a mean thinning in the catchment at 0.06 m a^{-1}. Assessments of uncertainties indicate that the inherent spatial variability of accumulation rate limits mass balance determinations to about 7% confidence (standard error, divided by discharge) [*Venteris and Whillans, 1998*], and that the limits to uncertainties in catchment area are about 9% [*Price and Whillans, 1998*]. The southern boundary to the catchment of ice stream B is very poorly determined, so the real uncertainty in catchment area is larger, say 12%. The uncertainty in discharge is relatively small. Assuming that the errors in accumulation rate and in area combine statistically, as is common in mass balance calculations, leads to a standard error of

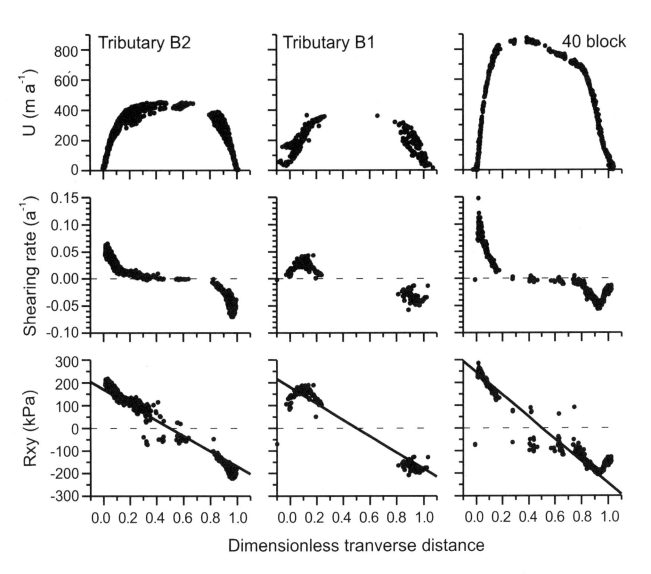

Fig. 6. For three cross-transects of ice stream B, longitudinal speed (top panels), lateral shearing rate (middle) and lateral shear stress (bottom). The lines in the bottom panels represent theoretical values if lateral drag opposes all of the action of the ice stream. Motion is toward the viewer. Transverse distance is scaled to ice stream width. [*Whillans and Van der Veen, 1997*]. Reproduced from the Journal of Glaciology with permission from the International Glaciological Society and the authors.

$(7^2 + 12^2)^{1/2}$ = 14% on the estimate of mass input to the ice stream system. Ice stream B together with its catchment is thinning at a rate that is two standard errors different from zero.

Shabtaie et al. [1988] compute a pattern of mass imbalance within the ice stream. The uncertainties allow for mass balances of zero. An uncertainty of 5% was assigned of the velocity profiles. This uncertainty is valid for the single velocity profile available at that time, at the mouth of the ice stream. Local fluctuations in ice speed are now known to be larger up-glacier (c.f. Figure 6). Modern uncertainties associated with mass balance calculations find that mass balances within ice stream B that are not different from zero.

Ice stream C is thickening in its lower and middle reaches (see *Anandakrishnan et al.* [this volume], for more detail). This is known because there is net accumulation of snow [*Whillans and Bindschadler, 1988*] and yet almost no discharge [*Whillans and Van der Veen, 1993*]. The feeding into the upper reach is moving at typical speeds (Figure 3). There is no major lateral ice divergence between these sites, so there must be a region of thickening between the upper and middle reaches. This pattern is confirmed by more recent work [*Joughin et al., 1999*].

The coffee can method has been applied just outboard of the dragon at station 21 (Figure 1). It shows thinning at 0.096 (0.044) m a^{-1} [*Hamilton et al., 1998*].

CHANGES WITH TIME

Non-traditional methods for detecting changes have demonstrated very important recent events. The evidence includes direct measurement of speed change with time on the ten-year scale, buried former ice streams and migration of shear margins. These results show that the ice streams are changing much more rapidly than classical glaciology, as used by *Whillans* [1982] for example, could permit.

Direct measurements of velocity and margin position show widening and slowing of ice stream B. Based on repeat velocity determination [*Stephenson and Bindschadler*, 1988] and on repeat satellite imagery [*Bindschadler and Vornberger*, 1998], the mouth of ice stream B is found to have been widening at 137 m a^{-1} and slowing by 2.4% a^{-1} since at least 1963. The upper reach of ice stream B is slowing by 0.7% a^{-1} [*Hulbe and Whillans*, 1997], and widening at 17 (6) m a^{-1}, based on crevasse shape with error estimate [*Hamilton et al.*, 1998], and at 7 m a^{-1}, based on temperature measurements [*Harrison et al*, 1998], and at 9.7 (1.1) m a^{-1}, based on repeat velocity determination [*Echelmeyer and Harrison*, 1999]. If these two sites are a reliable sampling, then the entire ice stream is slowing and widening.

Past changes in the ice streams are evident from peculiar surface forms and from buried features. The surface of interstream ridge A/B is lumpy, quite unlike the smooth surface of interstream ridge B/C (Figure 7, also *Bindschadler and Vornberger* [1990]). Ice speed is 12.6 m a^{-1} (Figure 1) on this interstream ridge, a typical value for an interstream ridge. The region has not been properly investigated, but the tentative interpretation is that it was formerly flowing faster, but has slowed down to form irregular highs and lows.

There are stranded shear margins. The interstream ridge between tributaries B1 and B2 (the Unicorn) contains a hook-shaped ridge (approximately joining stations 72 and 27 in Figure 1). This ridge is probably due to a buried old shear margin. The depth to subsurface crevasses in the northern part of the Unicorn indicates that the shear margin migrated gradually from the hook-shaped ridge, where it stood about 190 years before present, to the present position of the Dragon, which it reached about 130 years before present [*Clarke et al.*, in press]. There is an apparently similar stranded margin at the mouth of ice stream B [*Bindschadler and Vornberger*, 1990]. These two samples suggest that the ice stream was wider in the past and became narrower. The measurements noted above indicate that the sense of width change is now reversed.

Ice stream C, the next ice stream north, was also active, but its middle and lower reaches are now nearly inactive (Figure 3). Proof of former activity is the existence of subsurface crevasses [*Robin et al.*, 1970; *Rose*, 1979; *Bentley et al.*, 1985]. These crevasses probably once reached the surface, especially near the lateral margins, so the depth of burial indicates a time of about 140 years (corrected for the decade that has elapsed since the measurements) [*Retzlaff and Bentley*, 1993; *Bentley et al.*, in press] since the activity ceased. The up-glacial portion is traveling at 'normal' speeds for ice feeding into an ice stream (Figure 3). There must be a growing bulge near longitude W135°.

BED MOBILITY

Seismic wide-angle reflection experiments at UpB in 1983-84 obtained both compressional-wave (P-wave) and shear-wave (S-wave) reflections not only from the base of the ice, but also from a second reflector a few meters deeper [*Blankenship et al.*, 1986, 1987]. Such dual reflections from each of two layers have not been obtained elsewhere on the ice streams, despite attempts at DnB and UpC and in later years at UpB. For data from 1983-84, inversion of reflection times yields determinations of the P- and S-wave velocities in the subglacial sedimentary layer, showing that the porosity of the sediments is about 0.4 and that the effective pressure is only about 50 kPa (the ice overburden is 9000 kPa). These numbers indicate that the sediment, presumed to be glacial till, is dilated, water-saturated, and very weak, characteristics that are confirmed by drilling to the bed [*Kamb*, this volume; *Engelhardt and Kamb*, 1998].

Extensive seismic reflection profiling reveals that the subglacial sediment layer is widely distributed near the UpB camp. The average thickness is 6 to 7 m [*Rooney et al.*, 1987]. The upper surface of the layer is smooth, but the lower surface is fluted parallel to ice-stream movement. These sediment-filled flutes are as much as 13 m deep and 1000 m across. In one or two locations the sediment is no thicker than the resolution of the seismic experiments (about 2 m), so it may pinch out entirely. Nowhere on the reflection profiles is any feature discerned to penetrate more than a few meters into the ice from the bed. The bed is smooth as well as soft.

The presence of soft, deformable, dilated subglacial sediment (probably till) opens up the possibility that debris has been, and presumably still is being, transported with the glacier or with subglacial water motion. The best evidence for till deposition is the discovery of a feature similar to a delta beneath the mouth of ice stream B [*Shabtaie and Bentley*, 1987] and seismic evidence for foreset beds within it [*Rooney*, 1987].

There is little information on the time scale for deposition of this delta-like feature. It could be forming at present or it could be a relict from a time of adjustment of the ice sheet, for example to the rise in global sea level 8000 years ago. A change in configuration of the ice would lead to excavation of many soft deposits that had accumulated in up-glacial basins. That is, debris transport could be continuous or episodic and it is not known what the current phase of activity may be. However, some debris must be in traction today, if only due to the effect of roughness elements (on the order of 1 m in size, as deduced from radar diffraction studies) that must be plowing the bed.

Other Features of the Bed

There are glacier-parallel stripes in bed character as deduced by mapping acoustic impedance using the phase of P-wave reflections under ice streams B and C

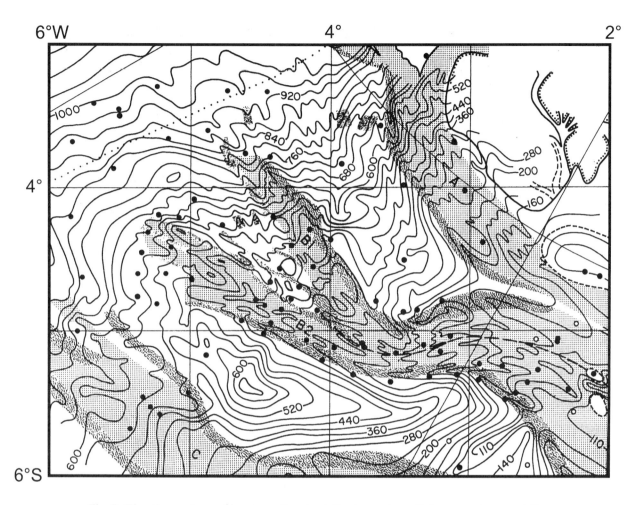

Fig. 7. Elevation contours of ice stream B and neighboring regions. Dots represent survey stations used for elevation control, many of which appear in Figures 1, 2 and 3. Squares are about 110 km on a side. [*Shabtaie et al.*, 1987].

[*Atre and Bentley*, 1993, 1994]. Because estimated impedances in a dilated bed (porosity 0.4) and in the lowermost ice are very nearly the same, minor differences in the nature of the sediments composing the bed, or the physical state of the bed (e.g. the porosity) can change the impedance contrast. Lateral variations of this kind provide an explanation for the relatively high impedances found in some places beneath ice stream B. Under ice stream C, the bed beneath the faster flowing (10 m a^{-1} at station 90 in Figure 3) ice with surface flow traces (in satellite imagery, [*USGS*, 1992]) has a low-impedance bed (consistent with higher-porosity subglacial sediments), whereas that under the very slow (2 m a^{-1} at stations 80 and 82 in Figure 3) ice showing a mottled surface in satellite images has a relative higher-impedance bed (consistent with more bed strength).

Basal stripes are evident in radar as well. Near UpB the basal echo strength varies slowly in the direction of ice motion and rather more in the transverse direction (Figure 8). The transverse scale of variation is some 10 to 20 m. There is a longer 150 m scale as well in transverse variation [*Novick et al.*, 1994]. The bed contains flow-aligned stripes of differing bed type.

Microearthquake studies have yielded evidence of sticky spots (sites of concentrated basal drag) beneath both ice streams [*Anandakrishnan et al.*, this volume]. However, the analysis of the pattern in ice velocity across the full width of ice stream B (next section, Figure 6) finds no dynamically important concentrations of friction, meaning that the earthquake-determined sticky spots do not provide a significant restraint to the flow of ice stream B.

Folding of the internal structure of ice streams has been sought with detailed radar work. In most places, radar-reflecting internal layers are continuous. At UpB they are parallel to the surface in the flow direction but show divergence from the surface in the transverse direction [*Schultz et al.*, 1987]. Near the mouth of ice stream B and within ice stream C there are transverse folds in the internal layers, and the axial planes are tilted

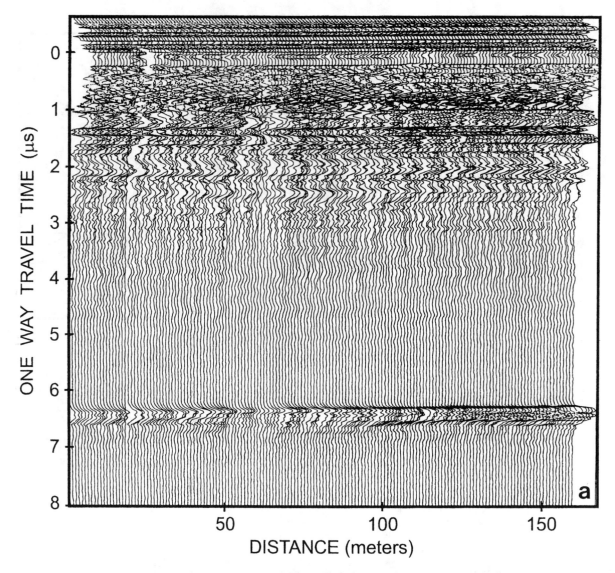

Fig. 8. Stacked radio-echo traces from near the UpB camp. Part a is a transverse transect showing that bed returns fluctuate over horizontal distances of 10 to 20 m. Part b is a longitudinal transect showing that bed variations have a much longer scale length. [*Schultz et al.*, 1987]. Reproduced from the Annals of Glaciology with permission from the International Glaciological Society and the authors.

variously up- and down-glacier [*Jacobel et al.*, 1993; *Jacobel and Grommes*, 1994]. These must indicate a temporally and spatially varying basal drag somewhere up-glacier. It may be a similar phenomenon to the thrusting discovered near the UpB camp [*Hulbe and Whillans*, 1997].

Conditions under the inactive ice stream C have been investigated by several techniques that are discussed by *Anandakrishnan et al.* [this volume]. Most of the bed is soft and wet; in some places the water layer may be at least several centimeters thick.

The simple extension of the finding at both UpB and UpC that basal water is nearly at overburden pressure [*Kamb*, this volume] is that water should flow at the ice-bed interface and collect according to the relief on the bed and pressure gradients due to ice thickness gradients. On the ice streams, the basal hydrologic gradient is dominated by surface slope, bed slope being secondary. Water flow according to the surface topography leads to divergence and convergence in the expected subglacial water flow lines and so to regions without a water supply, presumably a sticky spot to ice sliding, and to subglacial ponds. However, the maps of basal radar echo strength do not correlate with the likely distribution of subglacial water [*Novick et al.*, 1994]. Other considerations affect basal reflection strength; possibly basins filled with sediment affect the radar in the same way as water-filled ponds. However, the

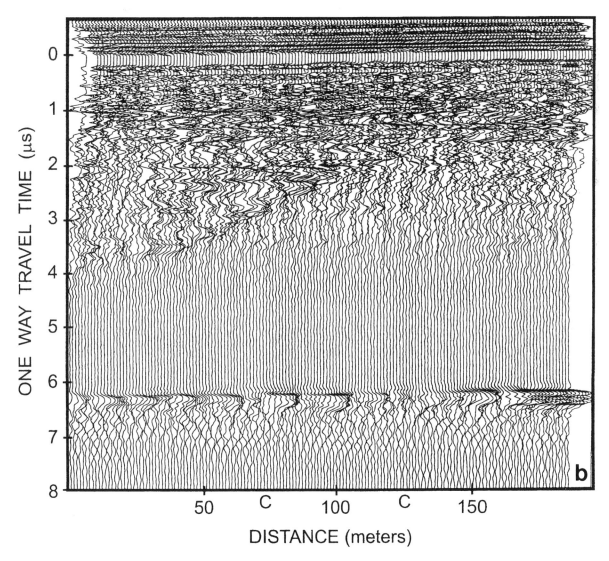

Figure 8. Continued.

straightforward interpretation is that water collection does not seem to be a factor in bed character.

The contrast between the beds of ice streams B and C and the intervening interstream ridge is addressed with an analysis of the strength of airborne-radar echoes from the downstream ends of ice streams B and C [*Bentley et al.,* 1998]. Returns from beneath interstream ridge B/C are weaker than those from the base of the ice streams, most probably because the interstream ridge is frozen to its bed. Returns from the bed of ice stream C are bright, in many places brighter than those from ice stream B. The beds of both ice streams are probably wet. The cause for brightness variations within ice stream C is not known, but suggestions are that there is trapped seawater, or that some other property of the bed affects reflection strength, or that there are variations in the amount of energy lost by scattering from subsurface crevasses.

FORCE BUDGET

The gross-scale mechanics controlling glacier flow are assessed by evaluating the action effect of gravity, the driving stress, and the potential reactions. The reactions or resistance can come from the bed, from the sides or from the ends of the glacier section being considered. Driving stress is calculated from the geometry of the glacier. Net resistances from along the glacier or from the sides are assessed from measured strain rates using the constitutive relation for ice to obtain stresses. Basal resistance is deduced as the residual in the calculation needed to balance forces.

The general formula for force budget may be expressed for a unit of map area [*Van der Veen and Whillans,* 1989, equation 14]:

$$\tau_{dx} = \tau_{bx} - \frac{\partial}{\partial x}HR_{xx} - \frac{\partial}{\partial y}HR_{xy} \quad (1)$$

In this case the forces in the down-glacier x-direction are considered. The action or driving stress is on the left-hand side and the terms on the right hand side are, respectively, basal drag, differential longitudinal tension and differential lateral drag. Resistive stresses, represented by R, are thickness means and they are estimated from strain rates.

Action of Gravity

Glacier motion is driven by gravity as described by the driving stress, τ_{dx}, being the horizontal action per unit map area. It is calculated from the product of surface slope, $\partial h/\partial x$, and ice thickness, H:

$$\tau_{dx} = -\rho g H \frac{\partial h}{\partial x} \quad (2)$$

in which, ρ and g represent ice density and acceleration due to gravity, respectively. In practice, the most critical quantity is surface slope because it varies by large factors. The driving stress along ice stream B is depicted in Figure 2, 3rd panel.

Longitudinal Tension / Compression

The longitudinal term, $\partial HR_{xx}/\partial x$, is evaluated from measurements or assessments of longitudinal and transverse stretching rates together with the constitutive relation for ice. Longitudinal stretching is computed from measurements of velocity along the ice stream centerlines:

$$\dot{\varepsilon}_{xx} = \frac{\partial U_{max}}{\partial x} \quad (3)$$

It is depicted in Figure 2, 5th panel. Transverse strain rate is evaluated from longitudinal variations in ice-stream width, W, as shown in Figure 2, 2nd panel:

$$\dot{\varepsilon}_{yy} \equiv \frac{\partial u_y}{\partial y} \cong \frac{U_{max}}{W}\frac{\partial W}{\partial x} \quad (4)$$

in which the simplification is made that the centerline speed can be substituted for the width mean. This is not a necessary simplification, but one learns *a posteriori* that the accuracy of transverse strain rate is not critical to the analysis of force budget. Longitudinal tension is calculated from these strain rates using the constitutive relation [*Hooke*, 1981]:

$$R_{xx} = B\dot{\varepsilon}_e^{-2/3}[2\dot{\varepsilon}_{xx} + \dot{\varepsilon}_{yy}] \quad (5)$$

The rate factor, $B= 540$ kPa a$^{1/3}$ is obtained from the values presented by *Hooke* [1981] and the depth-temperature profile at UpB [*Engelhardt et al.*, 1990]. As is usual, the exponent in the flow law is taken to be 3. In evaluating the effective strain rate, $\dot{\varepsilon}_e$, vertical shearing is neglected. This is appropriate in the analysis of the budget of forces because most vertical shearing occurs in deep ice. This warm ice is not strong and so does not carry large horizontal stresses, R_{xx} and R_{yy}. Vertical normal strain rate is obtained from horizontal strain rates by invoking incompressibility. The expression used for effective strain rate is

$$\dot{\varepsilon}_e = [\dot{\varepsilon}_{xx}^2 + \dot{\varepsilon}_{yy}^2 + \dot{\varepsilon}_{xx}\dot{\varepsilon}_{yy} + \dot{\varepsilon}_{xy}^2]^{1/2} \quad (6)$$

Lateral shear strain, $\dot{\varepsilon}_{xy}$, varies across the ice stream. At the kinematic centerline its value is zero. The tensile stress so computed for the centerline is displayed in the 6th panel of Figure 2. The pattern in R_{xx} is nearly the same as the pattern in longitudinal stretching, $\partial U_{max}/\partial x$, from which it is mainly derived.

Tensile stress, R_{xx}, is large (about 150 kPa) along most of the ice stream (6th panel of Figure 2). It evokes the vision that ice streams are pulling ice out of the interior ice reservoir [*Hughes*, 1998]. However, it is the longitudinal gradient in tension that is important to the budget of forces. This is displayed in the last panel of Figure 2, as $\partial HR_{xx}/\partial x$. Compared to driving stress, τ_{dx}, which is 10 kPa or more, longitudinal tensile gradients are small (5 kPa or nearer zero). Differential longitudinal tension is a minor player in the balance of forces all along ice stream B.

There is a short reach within tributary B2, just up-glacier of the UpB camp, where tension is smaller than for most of the rest of the ice stream. This reduced tension would mean fewer exposed crevasses, and indeed the UpB camp area is the only place on the middle reach of ice stream B where large aircraft could land safely. The small tension is a consequence of the more nearly constant speed in this reach. That in turn is due to the more nearly parallel sides and nearly constant thickness of the ice stream. The gradient from small tension at UpB to larger tension down-glacier leads to the biggest value in longitudinal force gradient for the ice stream (Figure 2, last panel).

Near the ice shelf, tension drops to near zero, then becomes slightly negative. This small value of R_{xx} means that there is very little stress transmission between the ice shelf and ice stream.

Buttressing Force From the Ice Shelf

It was once a common view that the ice shelf may be holding back the grounded ice sheet [*Jenssen et al.*, 1985, page 2; *Bindschadler and Vornberger.*, 1990; *MacAyeal*, 1989]. Measurements at the mouth of ice stream B show that the longitudinal resistive stress is nearly zero (6th panel in Figure 2 at $x = 100$ km). The contribution of the ice shelf to the budget of forces for the ice stream is negligible because the ice shelf is thin and stresses are small. Should sea level or ice thickness near the flotation point change then there would be a change in the area and maybe value of basal drag for the ice stream, most likely an along-glacier migration in flotation position. Some scientists suggested other patterns in reaction, including strong longitudinal stresses and longitudinal stress gradients. The simpler model would retain the current pattern in force transmission.

A change in any one part of the system, such as reduction of ice shelf, must eventually propagate along glacier to affect other parts, but is as yet undemonstrated with observation. There is evidence for down-glacier propagation of disturbances in the Ross Ice Shelf [*Casassa et al.*, 1991]. None is definitively reported as of yet for up-glacier propagation, although *Retzlaff and Bentley* [1993] suggest that a wave of stagnation moved up-glacier from the mouth of ice stream C. The Holocene rise in sea level must have affected the ice sheet in such a sense.

Tension From the Inland Ice

Another suggestion has been that the ice streams may be pulling away from the inland ice ("pulling power" [*Hughes*, 1998, pages 51 and 110]). Speeds do increase down-glacier at the onset to ice-stream flow (Figure 2, 5th panel), implying longitudinal tension (6th panel). However, the gradient in this tension is very small (bottom panel), indicating that it plays a small role in ice stream dynamics. *Hughes* [1998, pages 51 and 166] suggests that there is an arc of "primary transverse crevasses across the head of ice streams." There is no such arc for ice stream B (Figure 4a). As shown in the lowest panel of Figure 2, the tendency for ice streams to pull ice out of the inland ice reservoir is minor.

However, as with ice-shelf buttressing, the onset region is part of a linked system. Should for some reason the ice streams be removed there would be up-glacially propagating effects.

Lateral Drag

Drag from the sides is estimated from measurements of lateral shear strain rate and used with the constitutive relation to compute lateral shear stress. The first results come from a line of poles surveyed across the Dragon, near UpB [*Echelmeyer et al.*, 1994; *Harrison et al.*, 1998]. Further data come from the repeat photogrammetry of three sections across the ice stream. Figure 6 shows lateral shearing rate and the computed shear stress. Lateral drag varies continuously and nearly linearly from one side of the ice stream to the other, meaning that there are no major sticky spots. The line in Figure 6 (3rd row of panels) represents the stress necessary to oppose the action of gravity (as developed in the appendix). The inferred stresses do balance, or more than balance the gravitational driving force.

Jackson and Kamb [1997] provide an explanation for the apparent overbalance. Working with ice samples recovered from the ice stream they determine that ice from the shear margin is weaker than isotropic ice because of a special crystal orientation fabric. Allowing for this effect brings the calculated stresses in Figure 6 close to the theoretical solid line, removing the discrepancy. Moreover, the stress required to deform the ice samples at the measured rate accords with the inference from the full transect.

The dominance of lateral drag applies at the mouth of ice stream B as well [*Bindschadler et al.*, 1987]. There, 83% (standard error: 18%) of the driving force is opposed by lateral drag [*Whillans and Van der Veen*, 1997].

Differential lateral drag, $\partial HR_{xy}/\partial y$, is linked with centerline speed, U_{max}, and ice stream width, W (appendix). The result is displayed in Figure 9 (middle panel). It is about 10 kPa, and that is enough to oppose most of the driving stress within the ice stream. Up-glacier of about $x = -100$ km, lateral drag plays a much smaller role.

Basal Drag: Variation Across the Ice Stream

Basal drag is computed as the residual of the sum of driving stress, longitudinal force gradients and lateral drag. The residual computed in this way is very small, being some 0 to 7 kPa, for each of the sections of ice stream B that have been studied (Figure 6). Basal drag is small right to the edge of the ice stream (match of points to line in Figure 6; also *Echelmeyer et al.* [1994]). This raises the difficult question of what is responsible for such small friction despite ever decreasing speeds toward the margin (and large basal friction at the outboard margins).

Basal Drag: Variation Along the Ice Stream

Supposing that the three transects of Figure 6 are representative of the entire ice stream, the procedure in the appendix is used to compute basal drag all along ice stream B (Figure 9, bottom panel).

Up-glacier of $x = -100$ km, basal drag is large and about equal to the driving stress. Lateral drag and longitudinal force gradients are not important. This is much as has been found elsewhere on inland ice when averages are taken over spatial scales of 20 km or more (*Whillans and Johnsen* [1983] for Byrd Station, Antarctica; *Whillans and Jezek* [1987] for Dye3, Greenland).

Down-glacier of $x = -100$ km, basal drag is much smaller, in fact nearly zero. Lateral drag takes up the dominant role, becoming important where basal drag decreases.

A concern has been that the findings on basal water flow and storage and of ice friction near UpB may not be representative for the ice stream as a whole. There are fewer exposed crevasses near UpB than elsewhere on the ice stream. Maybe the reasons for fewer crevasses are also reasons to suspect that basal conditions are unusual. The reason for the paucity of exposed crevasses is now known to be mainly because longitudinal stretching is smaller than elsewhere, being about 2×10^{-3} a^{-1} at UpB [*Hulbe and Whillans*, 1994] as opposed to 4×10^{-3} a^{-1} over the long reach between the UpB camp and the narrows at station 33 (Figure 1). The smaller rate of stretching means that crevasse bridges do not fail at UpB. The reason for smaller

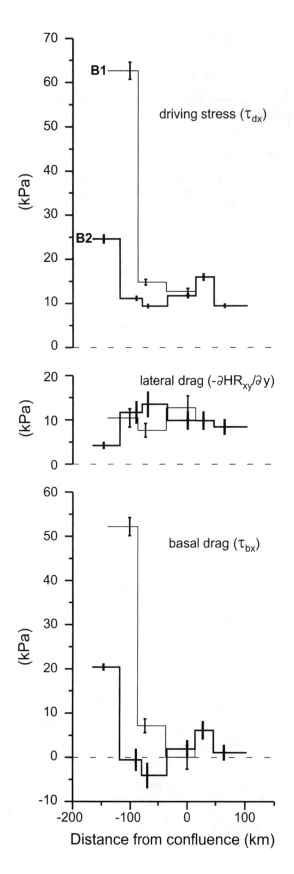

Fig. 9. Components in the budget of forces along ice stream B. Driving stress is repeated from Figure 2, with uncertainties that trace approximately equally from surface elevation (standard error: 2 m) and thickness (50 m). The uncertainty in lateral drag, $\partial H R_{xy}/\partial y$, derives mainly from uncertainty in the value of B, with values in the range 400 to 750 kPa a$^{1/3}$ considered. Basal drag is computed from the sum of the terms above, with a small contribution from longitudinal force gradient from Figure 2.

stretching is that the ice thickness increases downglacially [*Novick et al.,* 1994] and there is only a minor narrowing of ice stream width [*Whillans et al.,* 1993]. Nevertheless, crevasse density is large [*Clarke and Bentley,* 1994], it is just that fewer crevasses reach the surface. The measured patterns in velocity for other studied sections of ice stream B differ from each other as much as from the section at UpB, so there does not appear to be anything especially unusual about the UpB camp region. This means that the results of studies near the UpB camp are probably representative of most of the ice stream.

Velocity gradients on the nearly inactive reaches of ice stream C are too small (Figure 3) for horizontal force transmission to be significant beyond the scale of about 5 km. At that scale basal drag balances driving stress.

ONSET

Using the definition of the onset to be where driving stress decreases and speed increase along glacier, the onset is at x ≈ -100 km, the empirical finding is that

up-glacier of the onset:

$$\tau_{dx} = \tau_{bx} \approx 20 \text{ to } 70 \text{ kPa; speed} < 300 \text{ m a}^{-1}.$$

down-glacier of the onset:

$$\tau_{dx} = \partial H \overline{R}_{xy}/\partial y \approx 10 \text{ kPa; speed: } 300 \text{ to } 800 \text{ m a}^{-1}.$$

A very simple picture emerges from this study. For the inland ice, basal drag is at normal levels for an ice sheet [*Drewry,* 1983], in the range 20 to 70 kPa, and the large-scale balance of forces is between gravity (as described by the driving stress) and the bed. At the ice-stream onset, basal drag decreases to values of 0 to 5 kPa, and the large-scale balance of forces is between gravity (which generates a driving stress of about 10 kPa) and the ice-stream sides. Basically, basal drag is large under inland ice, but is nearly zero under the ice stream and mechanical control changes from the bed, under inland ice, to the sides, within the ice stream.

There are no distinct topographical steps that delimit the transition from inland-ice flow to ice-stream flow at the head of the ice streams. For example, the beginning of ice stream B2 and its shear margins, the snake and dragon, start at a nearly level bed, and the beginning of ice stream B1b, and its shear margin, the Heffalump, begins with a small upward step.

Other definitions of the onset lead to:

x = -400 km: steeper slope [*Bell et al.*, 1998].

x = -400 km: lateral shearing [*Anandakrishnan et al.*, 1998].

x = -100 km: first crevasses in a shear margin (Figure 1).

x = -310 km: first flow traces [*Hodge and Dopplehammer*, 1996; *Scambos and Bindschadler*, 1997; *Bindschadler and Vornberger*, 1990].

x = -200 km: most up-glacial crevasse [*Bindschadler*, 1997].

Most of these other definitions imply longer ice streams than the definition used here.

HYPOTHETICAL CONTROLS

A central need is to understand the physical reasons for ice streams and their changes with time. From their first discovery, hypotheses have been advanced to account for the ice streams and their changes. Not all the hypothetical explanations are mutually exclusive and competitive with one another. It is possible that several hypothesized processes can operate at once. As with other branches of science, the glaciological community has devised tests to seek to disprove various hypotheses. The endeavor has been very successful, and the range of feasible processes has been considerably reduced. A large number of hypotheses have been advanced for the behavior of ice streams. Here we review each of them.

Basal Heat Feedback

One of the first models for the behavior of ice streams is that the faster the ice moves the more work is expended as basal heat and the more heating or basal melting is produced. Basal heat can raise the temperature to melting and basal water can allow sliding and lubricate the flow. This model is still attractive.

Calculations of basal heating support this model [*Rose*, 1979]. The motion of the ice stream is taken to be opposed by friction at the bed, and although the driving stress of an ice stream is small, the speed is so fast that heat production (the product of driving stress and speed) is much more than for inland ice. This heat input may keep the ice streams active.

Using this model *Payne* [1998] is able to numerically reproduce fast ice streams on wet beds separated by nearly stagnant interstream ridges that are basally frozen. The location of ice streams corresponds with actual ice streams. There are two major observational discrepancies with the model. First, the interior of the ice sheet at Byrd Station is predicted to be basally frozen, yet observations find it and the region around it to be wet based [*Ueda and Garfield*, 1970; *Whillans and Johnsen*, 1983]. Second, the lower and middle reaches of ice stream C are predicted to be nearly stopped on account of being basally frozen, yet the bed is largely wet [*Shabtaie et al.*, 1987; *Bentley et al.*, 1998]. One explanation for the discrepancies is that Payne selected a very small value for geothermal heat flow. Should a more common value [e.g. *Sclater et al.*, 1980] be selected, then the wet-based conditions would be calculated to be more pervasive, as observed, and not so narrowly confined to ice streams. Payne uses the model to describe time changes in ice streams. It is not demonstrated that such changes would occur with a more common value of geothermal heat flow. Possibly this model can explain frozen beds under interstream ridges, but further processes must be invoked to explain the positions of ice stream onsets.

Basal heating may explain abrupt change in speed at the sides of the ice streams (e.g. Figure 6). *Jacobson and Raymond* [1998] introduce the concept of an "edge shield" in which heat dissipation in the ice of the shear margin raises its temperature and reduces the vertical temperature gradient and upward heat conduction. With this edge shield, geothermal and frictional heat cannot be conducted upward as rapidly as otherwise. This leads to wet-based conditions under the shear margin, but not under the interstream ridge. There is a possible feedback between the generation of an edge shield, or warmed margin, and the time evolution of the position of the shear margin.

A second version of this model argues that the water produced lubricates the motion. The more water production the more potential lubrication. Ice stream C may have shut down because the lubricating water stopped being produced in sufficient quantity. That is, the ice stream did not necessarily freeze to the bed, it is just that the lubrication ceased to be sufficiently thick.

A third version of the model emphasizes water flowing in from up-glacier. The critical question concerns how this water flows, as a film, in wide shallow canals, or in narrow channels and whether the flow style can change. More widely distributed lubrication leads to smaller basal drag and faster ice speeds. Drawing on models for automatically surging valley glaciers, most particularly Variegated Glacier, Alaska, U.S.A., raises the suggestion that the behavior of the ice stream system is linked with instabilities in basal lubrication, although perhaps without generation of a full surge cycle. *Retzlaff and Bentley* (1993) suggest that the stagnation of ice stream C involved a drop in water pressure at the bed due to a conversion from water flow as a sheet to channelized water flow.

All three versions of the model of basal heat balance deserve fresh visits. The discovery that basal drag under ice streams is very small means that there is less heat production than earlier supposed. Following a flowline from inland ice to the ocean, one would find a maximum in basal heat production near the ice stream onset, where driving stress is large and opposed by friction at the bed and ice speed is large. Within the ice stream, speed is very large but basal drag is very small, but the combination can still lead to significant heat dissipation at the bed [*Van der Veen and Whillans*, 1996]. However, the way that basal drag and speed can change together with ice-stream evolution is not clear. Perhaps water can freeze on and consume available basal water under ice streams. This process could eliminate the lubrication and eventually cause an increase in basal drag under the ice stream and reduce speed.

Piracy

Rose [1979] proposes that ice stream B is capturing the catchment area of ice stream C. Inland ice feeding into ice streams B and C separates in a region with very small surface slope. The direction of driving stress is not clear, and there could be an ongoing shift in flow line direction. That is, flowlines that formerly were directed down ice stream C are now shifted toward ice stream B. However, this vision of the time evolution of the pair of ice streams is not compatible with measured velocities at the head of the ice streams.

The ice feeding into ice stream C is behaving normally (Figure 3; *Anandakrishnan and Alley* [1997]). In fact, velocities are fast for such a small driving stress and are directed down ice stream C. This reach is either an active ice stream or very nearly so. Such velocities disprove the ongoing ice-piracy model.

A variant on the piracy model is that it is rerouting of subglacial water that initiates the switches in behavior. Gradients in basal hydrologic potential are so small at the head of ice stream C that it is not clear which route is taken by the subglacial water from under much of the inland ice. *Alley et al.* [1994] and *Anandakrishnan and Alley* [1997] propose that there has been a switch, that water once descending ice stream C now crosses beneath interstream ridge B/C and then descends ice stream B. The result is that water supply to the middle and lower reaches of ice stream C has been cut off, reducing lubrication and slowing ice stream speed.

This model depends on the validity of the hypothesis that an up-glacier supply of basal water is needed to maintain fast flow. It also requires that the rerouted water cross beneath an interstream ridge without causing fast ice motion of the ice above. Perhaps the water passes beneath the interstream ridge in a narrow channel. A further idea is that the piracy model applied in the past and that since then the upper reach to ice stream C has thickened. Such a thickening may make such past piracy difficult to demonstrate.

Height Above Buoyancy

The fast sliding of ice streams is aided by large basal water pressures. An early suggestion is that these pressures are linked to the depth below sea level [*Budd et al.*, 1979]. The implication is that there is a free water connection between the ice sheet base and the ocean so that the effective pressure at the bed is equal to the height of the ice above ocean buoyancy. There is evidence for ponded water far in the interior at Byrd Station [*Whillans and Johnsen*, 1983] and under the ice streams [*Shabtaie et al.*, 1987; *Bentley et al.*, 1998]; such ponding implies pressures vastly in excess of that expected from the depth below sea level. In the case of UpB, the height-above-ocean-buoyancy model leads to a calculated difference between overburden and water pressures that is a factor of 50 too large [*Bentley*, 1987]. This model underestimates basal water pressure.

Possibly basal water pressure plays a role, but it cannot be the dominant parameter. For example, basal pressures are close to the glacial-overburden pressure near Byrd Station, yet motion is slow. At the other extreme, basal pressures are at buoyancy for the ice shelf where motion is fast. Something else, such as the roughness and strength of the bed must dominate the determination of basal drag.

Deforming Bed

A series of papers describes the view that a laterally continuous viscous stratum may exist between the glacier and the static bedrock beneath [*Alley et al.*, 1986, 1987; *MacAyeal*, 1994]. Based on seismic imagery, the stratum is some 0 to 13 m thick. The hypothesis is that shearing through this stratum of till accounts for much of the motion of the ice stream and is a process for horizontal transport of debris. A prediction of this model, shared with other models, is the presence of a till delta at the mouth of the ice stream. As noted above, such a till delta may have been found.

This model for control of ice stream motion seemed particularily attractive when basal drag was believed to be significant at the bed and the bed to consist continuously of soft sediment of appropriate viscosity. However, it is now known that basal drag is very small under active ice stream B, the ongoing mechanical control instead coming from the ice stream sides. Also samples recovered from the bed have large porosity and extreme weakness (2 kPa; *Kamb* [1991]). The water content of the debris is similar to that of mudflows which have viscosities (0.03×10^{-12} to 3×10^{-12} kPa-a, Table 5 in *Iverson* [1997]) many orders of magnitude weaker than that proposed for beneath ice streams (0.2 kPa-a [*Alley et al.*, 1986; *Jenson et al.*, 1996]). Most basal deformation must be plastic very near the ice-bed contact [*Kamb*, 1991] and have little direct control on ice stream motion.

In places, the ice stream could be dragging along some debris [*Tulaczyk et al.*, 1998]. Moreover, there is the possibility that this debris is collected at potential sticky spots under the ice stream, keeping the bed consistently smooth and drag resistant.

Moldable Bed

It seems probable that the bed can deform, although perhaps not continuously in either space or time [*Hindmarsh*, 1997]. In fact, considering that ice velocity and thickness change with time in at least minor ways on the ice stream, the bed of the ice stream must be continuously reformed to maintain small friction. This can happen if the bed is moldable such that sites that tend to oppose flow are removed.

This model has some of the weaknesses of other models. For example, it does not explain the abrupt change in speed at lateral margins nor why the bed under interstream ridges is strong. Maybe if erosion rate is linked with some power-law product of slippage rate and drag then there may be an accounting for these characteristics. Also, the model does not explain why the middle and lower reaches of ice stream C stopped.

A variation in emphasis on this model (or the deforming bed model) is that a layer of sedimentary rock beneath the ice is needed for fast flow. *Bell et al.* [1998] and *Anandakrishnan et al.* [1998] use remote sensing methods to infer that there are sedimentary rocks under part of the catchment to ice stream C. Observed ice speed is somewhat faster over some of the sedimentary rock than over igneous rock and other sedimentary rock. They interpret the faster speed as being ice-stream behavior, although it does not fit the definition of ice stream used here. *Bell et al.* [1998] propose further that very thick (much more than 100 m) sedimentary rock is needed for fast flow, but the physical reason for this is not clear. Perhaps the issue is that a certain moldable type of sedimentary rock can be formed into a more streamlined bed that offers less friction.

Ongoing Surge

A suggestion is that the ice stream systems could be in various stages of surge cycles. Surges are known for certain valley glaciers and are associated with ice speed increases of 10 to 100 fold with time. Part of the upper portion of the glacier is depleted during a surge and extra ice is transported to the ablation zone. After the surge phase, the glacier nearly stops and the depleted zone is rebuilt while the surge lobe decays. Often a new surge occurs after recovery. The ice streams do have speeds that are about 100 times faster than the neighboring interstream ridge ice and mass balance measurements do indicate thinning in the catchment of ice stream B. The suggestion is that the lower and middle reaches of ice stream C are at a post-surge phase and the surge lobes of extra ice are in the ice shelf [*Casassa et al.*, 1991] where they dissipate.

By comparing early mass balance calculations along ice stream B [*Shabtaie et al.*, 1988] with the results of a numerical model *Bindschadler* [1997] suggests that ice streams are the result of a sudden weakening of ice strength and flow resistance. These calculations of mass imbalance are suspect, as explained above. Bindschadler further suggests that there is an ongoing headward migration of ice stream onsets, although the kinematic wave analysis is based on a model of basal sliding that does not accord with observations.

A convincing demonstration of surge activity would require evidence of the surge cycle. Perhaps the remnants of formerly active ice stream segments (c.f. Changes with Time, above) are surge-depleted zones that are now in a build up stage. The one such site checked (the Unicorn, ridge B1/B2) is currently thickening [*Hamilton et al.*, 1998]. Maybe the fast ice streams are in full surge stage. The growing bulge in the upper reach of ice stream C, between the active and inactive portions of ice stream C, could be called a surge front. As yet, the existence of surges in Antarctica is an unproved possibility.

Whether the ice streams surge may not be a question distinct from others addressed here. Surging is a descriptive term for cyclic thickening and thinning of glaciers; it does not describe a physical process. For a cause of the changes one must invoke some other hypothesis.

Active Volcanoes

A subglacial structure having the shape of a composite volcano has been found under the inland ice upglacier of ice stream B/ice stream C. *Blankenship et al.* [1993] propose that the surface moat or depression to the side is not the result of ice flow around the volcano, but is due to basal melting associated with volcanism. Active subglacial volcanism or focused geothermal heat flow is a possibility in this region of thinned and stretched crust. However, a link between volcanoes and ice streams in position and time of activity has not been established. In Iceland, volcanic activity and fast glacier motion are temporally independent [*Björnsson*, 1998].

Global Warming

A common public perception is that the ice sheets are sensitively vulnerable to a global warming. This view is advanced in the title and header to the review by *Oppenheimer* [1998]. The implication is that there is evidence of a tendency toward century scale dramatic change. We are not aware of any such evidence, and none is presented by Oppenheimer.

CONCLUSIONS

The first mystery of why the ice streams are so fast despite small gravitational driving stress is partly resolved. There is a switch at the onset to ice stream flow where the bed becomes extremely weak, offering almost no friction. Thereafter, ice stream motion is restrained by friction from the sides and maybe from some rough areas within the ice stream. Basically, there is a change from control at the bed to control at the sides.

Several important physical processes involved in the first mystery await resolution. For example, the reason why the bed is so weak and why it remains weak is not established. Also, the processes operating at the lateral margins are not even fully identified, let alone properly modeled. For example, *Van der Veen and Whillans* [1996] model that a widening ice stream should become faster, yet the observations show the reverse, slowing during widening. Important as the remaining issues may be, the community has efficiently sorted out the broad issue involved in the first mystery: the ice streams are fast because their bases are nearly frictionless.

The second mystery is the discovery of rapid changes in the ice sheet. An explanation for the changes is not now available, but the explanation must be associated with the process responsible for weak beds, this process must lead to instabilities. A proper resolution of this second mystery requires an understanding of what keeps the beds frictionless and what changes friction.

These mysteries have implications beyond that of the cause of ice streams and the stability of the West Antarctic Ice Sheet. Their solutions will be important to deducing the physical reasons for the major fluctuations of other ice sheets, such as the Laurentide Ice Sheet and Scandinavian Ice Sheet, which also operated on soft beds and which also underwent rapid time fluctuations.

APPENDIX

Transverse Variation in Stress and Velocity

The distribution of stress and velocity across an ice stream is developed from theoretical principles. Here the theory behind the comparison with observation in Figure 6 is developed.

Balance of forces is considered for a column of unit map area. Let the coordinate x represent the along-flow axis and y across flow. Let τ_{bx} represent drag at the bed. The motive effect of gravity is described by τ_{dx}, the driving stress. Depth-averaged stresses acting on vertical faces of the column are R_{xx} (longitudinal normal) and R_{xy} (lateral drag). The stresses associated with longitudinal tension or compression and with lateral drag are large in the upper, cold portion of the glacier, and smaller at depth. In the present analysis, mean values through the thickness, H, are used. The balance of the various forces acting on the column is described by [Van der Veen and Whillans, 1989]:

$$\tau_{dx} = \tau_{bx} - \frac{\partial}{\partial x} HR_{xx} - \frac{\partial}{\partial y} HR_{xy} \qquad (A1)$$

The stresses are represented by the letter, R, to emphasize that they are stresses that normally Resist, or oppose the motion of the glacier.

To derive a transverse profile of speed, equation (A1) is integrated with respect to distance across the ice stream from the kinematic centerline (where $y = 0$, $R_{xy} = 0$)

$$\int_0^{HR_{xy}} d(HR_{xy}) = \int_0^y \left[\tau_{bx} - \tau_{dx} - \frac{\partial}{\partial x}(HR_{xx}) \right] d\bar{y} \qquad (A2)$$

Now define:

$$C = -\left[\tau_{bx} - \tau_{dx} - \frac{\partial}{\partial x}(HR_{xx}) \right] \qquad (A3)$$

and assume that C is constant across the ice stream. In fact, longitudinal force gradients (per unit width), $\partial HR_{xx}/\partial x$, are everywhere negligibly small, so they do not contribute meaningfully to any transverse variations in C. Driving stress, τ_{dx}, does vary somewhat, mainly because of transverse variations in surface slope (Figure 7). The transverse variation in basal drag, τ_{bx}, is not *a priori* known. Completing the integration of Equation (A2):

$$HR_{xy} = -Cy \qquad (A4)$$

It is this function that is drawn as a straight line in Figure 6, for τ_{bx} and $\partial HR_{xx}/\partial x$ both set to zero. The measurements follow the line reasonable well, and so support the assumption that C is nearly constant across the ice stream and that basal drag, τ_{bx}, is indeed nearly zero.

An expression for velocity is obtained by utilizing the constitutive relation [Hooke, 1981]

$$R_{xy} = B \left(\frac{1}{2} \frac{\partial u_x}{\partial y} \right)^{1/3} \qquad (A5)$$

for the case of simple shear, which is a good approximation for most of the width of the ice streams (Figure 6). A depth-mean value for the rate factor, B, is used.

Combining Equations (A4) and (A5) to eliminate R_{xy}, and solving for u_x:

$$\int_{U_{max}}^{u_x} du_x = -2 \int_0^y \left(\frac{Cy}{HB} \right)^3 d\bar{y} \qquad (A6)$$

which leads to glacier speed varying as the 4th power with respect to distance from the kinematic centerline.

The integral of Equation (A6) is linked with ice-stream width by invoking the criterion that the along-glacier component of velocity is zero at the shear margins $u_x(y = \pm W) = 0$, in which W represents the ice-stream halfwidth. Integrating:

$$u_x = \frac{C^3(W^4 - y^4)}{2(HB)^3} \qquad (A7)$$

This expression is a theoretical profile of the longitudinal component of velocity across the ice stream. Most measured profiles are well approximated by this.

Values of C are needed for the computations leading to Figure 2. These are computed from centerline speeds by rearranging Equation (A7):

$$C = \frac{HB}{W} \left(\frac{2U_{max}}{W} \right)^{1/3} \qquad (A8)$$

This expression is used to evaluate the role of lateral drag along the ice stream (wherever the simple relation of Equation (A3) is valid). Lateral drag in the budget of forces is

$$\frac{\partial HR_{xy}}{\partial y} = -C = -\frac{HB}{W} \left(\frac{2U_{max}}{W} \right)^{1/3} \qquad (A9)$$

This expression allows the role of lateral drag to be estimated from ice stream geometry (H, W), centerline speed, U_{max}, and the flow law for ice. It is used to compute lateral drag (2nd panel in Figure 9).

Acknowledgements. We thank Sridhar Anandakrishnan for comment and Mike Willis for comment and for figure preparation. Supported by grants from the US NSF-OPP. Byrd Polar Research Center contribution number 1171. Geophysical and Polar Research Center, University of Wisconsin-Madison, contribution 587.

REFERENCES

Alley, R. B., D. D. Blankenship, C. R. Bentley, and S. T. Rooney, Deformation of till beneath ice stream B, West Antarctica, *Nature, 322,* 57-59, 1986.

Alley, R. B., D. D. Blankenship, C. R. Bentley, and S. T. Rooney, Till beneath ice stream B. 3. Till deformation: evidence and implications, *Journ. Geophys. Res., 92,* 8921-8929, 1987.

Alley, R. B., and I. M. Whillans, Changes in the West Antarctic Ice Sheet, *Science, 254,* 959-963, 1991.

Alley, R. B., S. Anandakrishnan, C. R. Bentley, N. Lord, and S. Shabtaie, A water-piracy hypothesis for the stagnation of ice stream C, *Ann. Glaciol., 20,* 187-194, 1994.

Anandakrishnan, S., and R. B. Alley, Stagnation of ice stream C, West Antarctica by water piracy, *Geophys. Res. Ltrs., 24,* 265-268, 1997.

Anandakrishnan, S., D. D. Blankenship, R. B. Alley, and P. L. Stoffa, Influence of subglacial geology on the position of a West Antarctic ice stream from seismic observations, *Nature, 394,* 62-65, 1998.

Anandakrishnan, S., and C. R. Bentley, Micro-earthquakes beneath ice streams B and C, West Antarctica: observations and implications, *Journ. Glaciol., 39,* 455-462, 1993.

Anandakrishnan, S., R. B. Alley, R. W. Jacobel, and H. Conway, Ice Stream C slowdown is not stabilizing West Antarctic Ice Sheet, this volume.

Atre, S. R., and C. R. Bentley, Laterally varying basal conditions under ice streams B and C, West Antarctica, *Journ. Glaciol., 39,* 507-514, 1993.

Atre, S. R., and C. R. Bentley, Indication of a dilatant bed near downstream B camp, ice stream B, *Ann. Glaciol., 20,* 177-182, 1994.

Bell, R. E., D. D. Blankenship, C. A. Finn, D. L. Morse, T. A. Scambos, J. M. Brozena, and S. M. Hodge, Influence of subglacial geology on the onset of a West Antarctic ice stream from aerogeophysical observations, *Nature, 394,* 58-62, 1998.

Bentley, C. R., Antarctic ice streams: a review, *Journ. Geophys. Res., 92,* 8843-8858, 1987.

Bentley, C. R., S. Shabtaie, D. G. Schultz, and S. T. Rooney, Continuation of glaciogeophysical survey of the interior Ross embayment (GSIRE): summary of 1984-85 field work, *Antarctic Journ. U.S., XX,* 63-64, 1985.

Bentley, C. R., N. Lord, C. Liu, Radar reflections reveal a wet bed beneath stagnant Ice Stream C and a frozen bed beneath Ridge BC, West Antarctica, *Journ. Glaciol., 44,* 157-164, 1998.

Bentley, C. R., B. E. Smith, and N. E. Lord, Radar studies on Roosevelt Island and Ice Stream C, *Antarctic Journ. U.S.,* in press.

Bindschadler, R., Siple Coast Project research of Crary Ice Rise and the mouths of Ice Streams B and C, West Antarctica: review and perspectives, *Journ. Glaciol., 39,* 539-552, 1993.

Bindschadler, R., Actively surging West Antarctic ice streams and their response characteristics, *Ann. Glaciol., 24,* 409-414, 1997.

Bindschadler, R. A., S. N. Stephenson, D. R. MacAyeal, and S. Shabtaie, Ice dynamics at the mouth of ice stream B, Antarctica, *Journ. Geophys. Res., 92,* 8885-8894, 1987.

Bindschadler, R., and P. Vornberger, AVHRR imagery reveals Antarctic ice dynamics, *EOS, 71,* 741-742, 1990.

Bindschadler, R., P.L. Vornberger, and S. Shabtaie, The detailed net mass balance of the ice plain on Ice Streams B, Antarctica: a geographic information system approach, *Journ. Glaciol., 39,* 471-482, 1993.

Bindschadler, R., P. Vornberger, D. Blankenship, T. Scambos, and R. Jacobel, Surface velocity and mass balance of Ice Streams D and E, West Antarctica, *Journ. Glaciol., 42,* 461-475, 1996.

Bindschadler, R., and P. Vornberger, Changes in the West Antarctic ice sheet since 1963 from declassified satellite photography, *Science, 279,* 689-692, 1998.

Björnsson, H., Hydrological characteristics of the drainage system beneath a surging glacier, *Nature, 395,* 771-774, 1998.

Blankenship. D. D., Seismological investigations of a West Antarctic ice stream, Ph.D. thesis, University of Wisconsin-Madison, 1989.

Blankenship, D. D., C. R. Bentley, S. T. Rooney, and R. B. Alley, Seismic measurements reveal a saturated, porous layer beneath an active Antarctic ice stream, *Nature, 322,* 54-57, 1986.

Blankenship, D. D., C. R. Bentley, S. T. Rooney, and R. B. Alley, Till beneath ice stream B. 1. Properties derived from seismic travel times, *Journ. Geophys. Res., 92,* 8903-8911, 1987.

Blankenship, D. D., R. E. Bell, S. M. Hodge, J. M. Brozena, J. C. Behrendt, and C. A. Finn, Active volcanism beneath the West Antarctic ice sheet and implications for ice-sheet stability, *Nature, 361,* 526-529, 1993.

Budd, W. F., P. L. Keage, and N. A. Blundy, Empirical studies of ice sliding, *Journ. Glaciol., 23,* 157-170, 1979.

Casassa, G., K. C. Jezek, J. Turner, and I. M. Whillans, Relict flow stripes on the Ross Ice Shelf, *Ann. Glaciol., 15,* 132-138, 1991.

Clarke, T. S., and C. R. Bentley, High-resolution radar on ice stream B2, Antarctica: measurements of electromagnetic wave speed in firn and strain history from buried crevasses, *Ann. Glaciol., 20,* 153-159, 1994.

Clarke, T. S., C. Liu, N. E. Lord, and C. R. Bentley, Evidence for a recently abandoned shear margin adjacent to Ice Stream B2, Antarctica, from ice-penetrating radar measurements, *Journ. Geophys. Res.,* in press.

Drewry, D. J., *Antarctica: glaciological and geophysical folio,* Cambridge, University of Cambridge, Scott Polar Research Institute, 1983.

Echelmeyer, K. A., W. D. Harrison, C. Larson, and J. E. Mitchell, The role of the margins in the dynamics of an active ice stream, *Journ. Glaciol., 40,* 527-538, 1994.

Echelmeyer, K. A. and W. D. Harrison, Ongoing margin migration of Ice Stream B, *Journ. Glaciol., 45,* 361-369, 1999.

Engelhardt, H., and B. Kamb, Basal sliding of ice stream B, West Antarctica, *Journ. Glaciol., 44,* 223-230, 1998.

Engelhardt, H., N. Humphrey, B. Kamb, and M. Fahnestock, Physical conditions at the base of a fast moving Antarctic ice stream, *Science, 248,* 57-59, 1990.

Gudmundsson, G. H., C. F. Raymond, and R. Bindschadler, The origin and longevity of flow stripes on Antarctic ice streams, *Ann. Glaciol., 27*, 145-152, 1998.

Hambrey, M. J., and F. Müller, Structures and ice deformation in the White Glacier, Axel Heiberg Island, Northwest Territories, Canada, *Journ. Glaciol., 20*, 41-66, 1978.

Hamilton, G. S., I. M. Whillans, and P. J. Morgan, First point measurements of ice-sheet thickness change in Antarctica, *Ann. Glaciol., 27*, 125-129, 1998.

Harrison, W. D., K. A. Echelmeyer, and C. F. Larsen, Measurement of temperature in a margin of ice stream B, Antarctica: implications for margin migration and lateral drag, *Journ. Glaciol., 44*, 615-624, 1998.

Herzfeld, U. C., The 1993-1995 surge of Bering Glacier (Alaska) – a photographic documentation of crevasse patterns and environmental changes, *Trierer Geographische Studien, Heft 17*, 1998.

Hindmarsh, R. C. A., Deforming beds: viscous and plastic scales of deformation, *Quatern. Sci. Rev., 16*, 1039-1056, 1997.

Hodge, S. M., and S. K. Dopplehammer, Satellite images of the onset of streaming flow of ice streams C and D, West Antarctica, *Journ. Geophys. Res., 101*, 6669-6677, 1996.

Hooke, R. LeB., Flow law for polycrystalline ice in glaciers: comparison of theoretical predictions, laboratory data, and field measurements, *Rev. Geophys. Space Phys., 19*, 664-672, 1981.

Hughes, T.J., *Ice Sheets,* Oxford, U.K., Oxford University Press, 343 pp., 1998.

Hulbe, C. L., and I. M. Whillans, Evaluation of strain rates on ice stream B, Antarctica, obtained using GPS phase measurements, *Ann. Glaciol., 20*, 254-262, 1994.

Hulbe, C. L., and I. M. Whillans, Weak bands within Ice Stream B, West Antarctica, *Journ. Glaciol., 43*, 377-386, 1997.

Iverson, R. M., The physics of debris flows, *Rev. Geophys., 35*, 245-296, 1997.

Jackson, M., and B. Kamb, The marginal shear stress of ice stream B, West Antarctica, *Journ. Glaciol., 43*, 415-426, 1997.

Jacobel, R. W., A. M. Gades, D. L. Gottschling, S. M. Hodge, and D. L. Wright, Interpretation of radar-detected internal layer folding in West Antarctic ice streams, *Journ. Glaciol., 39*, 528-537, 1993.

Jacobel, R. W., and B. J. Gommes, Internal layer folding patterns from radar studies of ice streams B and C, *Antarctic Journ. U.S., 29*, 66-68, 1994.

Jacobson, H. P., and C. F. Raymond, Thermal effects on the location of ice stream margins, *Journ. Geophys. Res., 103*, 12111-12122, 1998.

Jenson, J. W., D. R. MacAyeal, P. U. Clark, C. L. Ho, and J.C. Vela, Numerical modeling of subglacial sediment deformation: implications for the behavior of the Lake Michigan Lobe, Laurentide Ice Sheet, *Journ. Geophys. Res., 101*, 8717-8728, 1996.

Jenssen, D., W. F. Budd, I. N. Smith, and U. Radok, On the surging potential of polar ice streams, Part II: ice streams and physical characteristics of the Ross Sea drainage basin, West Antarctica. Cooperative Institute for Research in Environmental Sciences, University of Colorado at Boulder, report DE/ER/60197-3, 1985.

Joughin, I., L. Gray, R. Bindschadler, S. Price, D. Morse, C. Hulbe, M. Karim, and C. Werner, Tributaries of West Antarctic ice streams by RADARSAT interferometry, *Science, 286*, 283-286, 1999.

Kamb, B., Rheological nonlinearity and flow instability in the deforming bed mechanism of ice stream motion, *Journ. Geophys. Res., 96*, 16585-16595, 1991.

Kamb, B., Basal zone of the West Antarctic ice streams and its role in their streaming motions, this volume.

Liu, C., C. R. Bentley, and N. Lord, C axes from radar depolarization experiments at UpB in 1991-92, Antarctica, *Ann. Glaciol., 20*, 169-176, 1994.

Liu, C., C. R. Bentley, and N. Lord, Velocity difference between the surface and the base of Ice Stream B, West Antarctica, from radar-fading pattern experiments, *Ann. Glaciol., 29*, in press.

MacAyeal, D. R., Large-scale ice flow over a viscous basal sediment: theory and application to ice stream B, Antarctica, *Journ. Geophys. Res., 94*, 4071-4087, 1989.

Meier, S., Portrait of an Antarctic outlet glacier, *Hydrological Sciences – Journal des Science Hydrologiques, 28*, 430-416, 1983.

Merry, C. J., and I. M. Whillans, Ice-flow features on ice stream B, Antarctica, revealed by SPOT HRV imagery, *Journ. Glaciol., 39*, 515-527, 1993.

Novick, A. N., C. R. Bentley, and N. Lord, Ice thickness, bed topography, and basal reflection coefficients from radar sounding, Upstream B, West Antarctica, *Ann. Glaciol., 20*, 148-152, 1994.

Oppenheimer, M., Global warming and the stability of the West Antarctic ice sheet, *Nature, 393*, 325-332, 1998.

Payne, A. J., Dynamics of the Siple Coast ice streams, West Antarctica: results from a thermomechanical ice sheet model, *Geophys. Res. Ltrs, 25*, 3173-3176, 1998.

Price, S. F., and I. M. Whillans, Delineation of a catchment boundary using velocity and elevation measurements, *Ann. Glaciol., 27*, 140-144, 1998.

Raymond, C. F., K. A. Echelmeyer, I. M. Whillans and C. S. M. Doake, Ice stream shear margins, this volume.

Retzlaff, R., N. Lord, and C. R. Bentley, Airborne radar studies: ice streams A, B, and C, West Antarctica, *Journ. Glaciol., 39*, 495-506, 1993.

Retzlaff, R., and C. R. Bentley, Timing of stagnation of ice stream C, West Antarctica, from short-pulse-radar studies of buried surface crevasses, *Journ. Glaciol., 39*, 553-561, 1993.

Robin, G. de Q., S. Evans, D. J. Drewry, C. H. Harrison, and D. L. Petrie, Radio-echo sounding of the Antarctic ice sheet, *Antarctic Journ. U.S., V*, 229-232, 1970.

Rooney, S. T., Subglacial geology of ice stream B, West Antarctica, Ph.D. Thesis, University of Wisconsin-Madison, 1987.

Rooney, S. T., D. D. Blankenship, R. B. Alley, and C. R. Bentley, Till beneath ice stream B. 2. Structure and continuity, *Journ. Geophys. Res., 92*, 8913-8920, 1987.

Rose, K. E., Radio echo sounding studies of Marie Byrd Land, Antarctica, Ph.D. thesis, University of Cambridge, 1978.

Rose, K. E., Characteristics of ice flow in Marie Byrd Land, Antarctica, *Journ. Glaciol., 24*, 63-74, 1979.

Scambos, T. A., and R. Bindschadler, Complex ice stream flow revealed by sequential satellite imagery, *Ann. Glaciol., 17*, 177-182, 1993.

Schultz, D. G., L. A. Powell, and C. R. Bentley, A digital recording system for echo studies on ice sheets, *Ann. Glaciol., 9*, 206-210, 1987.

Sclater, J. G., C. Jaupart, and D. Galson, The heat flow through oceanic and continental crust and the heat loss of the Earth, *Rev. Geophys. Space Phys., 18*, 269-311, 1980.

Shabtaie, S., and C. R. Bentley, West Antarctic ice streams draining into the Ross Ice Shelf: configuration and mass balance, *Journ. Geophys. Res., 92*, 1311-1336, 1987.

Shabtaie, S., I. M. Whillans, and C. R. Bentley, The morphology of ice streams A, B, and C, West Antarctica, and their environs, *Journ. Geophys. Res., 92*, 8865-8883, 1987.

Shabtaie, S., and C. R. Bentley, Ice-thickness map of the West Antarctic ice streams by radar sounding, *Ann. Glaciol., 11*, 126-136, 1988.

Shabtaie, S., C. R. Bentley, R. A. Bindschadler, and D. R. MacAyeal, Mass balance studies of ice streams A, B, and C and possible surging behavior of ice stream B, *Ann. Glaciol., 11*, 137-149, 1988.

Spikes, B., B. Csatho, and I. Whillans, Airborne laser profiling of an Antarctic ice stream for change detection, *Int. Arch. Photogram. Remote Sensing, 32 (part 3-W14)*, 169-175, 2000.

Stephenson, S. N. and R. A. Bindschadler, Observed velocity fluctuations on a major Antarctic ice stream, *Nature, 334*, 695-697, 1988.

Stephenson, S. N., and R. A. Bindschadler, Is ice-stream evolution revealed by satellite imagery? *Ann. Glaciol., 14*, 273-277, 1990.

Swithinbank, C., Antarctica. U.S. Geological Survey Professional Paper 1386-B, 278 pp., 1988.

Tulaczyk, S., H. Engelhardt, B. Kamb, and R. P. Scherer, Sedimentary processes at the base of a West Antarctic ice stream, constraints from textural and compositional properties of subglacial debris, *Journ. Sedimentary Res., 68*, 487-496, 1998.

Ueda, H. T., and D. E. Garfield, Deep core drilling at Byrd Station, Antarctica, *IAHS Publ. No 86*, 53-62, 1970.

USGS, Experimental Edition maps, Antarctica Satellite Image Maps, U.S. Geological Survey, Reston VA 22092 U.S.A., 1992.

Van der Veen, C. J., and I. M. Whillans, Force budget: I. Theory and numerical methods, *Journ. Glaciol., 35*, 53-60, 1989.

Van der Veen, C. J., and I. M. Whillans, Model experiments on the evolution and stability of ice streams, *Ann. Glaciol., 23*, 129-137, 1996.

Van der Veen, C. J., Fracture mechanics approach to penetration of surface crevasses on glaciers, *Cold Regions Sci. Technol., 27*, 31-47, 1998.

Venteris, E. R., and I. M. Whillans, Variability of accumulation rate in the catchments of ice streams B, C, D and E, Antarctica, *Ann. Glaciol., 27*, 227-230, 1998.

Vornberger, P. L., and I. M. Whillans, Crevasse deformation and examples from ice stream B, Antarctica, *Journ. Glaciol., 36*, 3-9, 1990.

Whillans, I. M., Reaction of the accumulation zone portions of glaciers to climatic change, *Journ. Geophys. Res., 86*, 4274-4292, 1982.

Whillans, I. M., and R. Bindschadler, Mass balance of ice stream B, *Ann. Glaciol., 11*, 187-193, 1988.

Whillans, I. M., and S. J. Johnsen, Longitudinal variations in glacial flow: theory and test using data from the Byrd Station Strain Network, *Journ. Glaciol., 29*, 79-97, 1983.

Whillans, I. M., and K. C. Jezek, Folding in the Greenland Ice Sheet, *Journ. Geophys. Res., 92*. 485-493, 1987.

Whillans, I. M., Y. H. Chen, C. J. van der Veen, and T. J. Hughes, Force budget: III. Application to three-dimensional flow of Byrd Glacier, Antarctica, *Journ. Glaciol., 35*, 68-80, 1989.

Whillans, I. M., and C. J. van der Veen, Patterns of calculated basal drag on ice streams B and C, Antarctica, *Journ. Glaciol., 39*, 437-446, 1993.

Whillans, I. M., and C. J. van der Veen, New and improved determinations of velocity of ice streams B and C, West Antarctica, *Journ. Glaciol., 39*, 483-490, 1993.

Whillans, I. M., M. Jackson, and Y-H. Tseng, Velocity pattern in a transect across ice stream B, Antarctica, *Journ. Glaciol., 39*, 562-572, 1993.

Whillans, I. M., and Y-H. Tseng, Automatic tracking of crevasses on satellite images, *Cold Regions Sci. Technol., 23*, 201-214, 1995.

Whillans, I. M., and C. J. van der Veen, The role of lateral drag in the dynamics of ice stream B, Antarctica, *Journ. Glaciol., 43*, 231-237, 1997.

C. R. Bentley, Geophysical and Polar Research Center, University of Wisconsin, 1215 West Dayton Street, Madison, WI, 53706. (e-mail: bentley@geology.wisc.edu)

C. J. van der Veen, Byrd Polar Research Center and Department of Geography, The Ohio State University, 108 Scott Hall, 1090 Carmack Road, Columbus, OH 43210. (e-mail: vanderveen.1@osu.edu)

I. M. Whillans, Byrd Polar Research Center and Department of Geological Sciences, The Ohio State University, 108 Scott Hall, 1090 Carmack Road, Columbus, OH 43210. (e-mail: whillans+@osu.edu)

THE FLOW REGIME OF ICE STREAM C AND HYPOTHESES CONCERNING ITS RECENT STAGNATION

S. Anandakrishnan

University of Alabama, Tuscaloosa, AL, 35487

R.B. Alley

Pennsylvania State University, University Park, PA 16802

R.W. Jacobel

St. Olaf's College, Northfield, MN 55057

H. Conway

University of Washington, Seattle, WA 98195

The downglacier part of ice stream C, West Antarctica largely stagnated over the last few centuries, while upglacier regions continue to flow vigorously. We summarize hypotheses for explaining this behavior and suggest that the stagnation is due to a combination of water-diversion, thermal processes at the bed, and sticky spots. Stagnation likely occurred near Siple Dome before the entirety of the downglacier part slowed. Numerous data sets show that the slow-moving part of the ice stream is restrained largely by small, localized basal "sticky spots". The sticky spots are separated by extensive regions of soft till containing high-pressure liquid water. Near the transition from fast-moving well-lubricated ice to slow-moving ice, a hydrologic potential map indicates that basal water flowing in from the catchment is diverted away from the slow-moving ice to ice stream B. This diversion could have been caused by a flattening of the surface slope over time in response to the headward growth of ice stream C drawing down the inland ice. Previous mass-balance estimates indicate that the combined B-C drainage most likely is thinning slowly, similar to the rest of the Siple Coast.

1. INTRODUCTION

The dynamics of the West Antarctic Ice Sheet (WAIS) is dominated by fast-flowing ice streams that evacuate inland ice rapidly through three main ice-drainage systems to the ocean. The six ice streams that flow westward into one of these drainage systems, the Ross Sea Embayment (Plate 1), are separated by intervening ridges of slow moving ice. High shear strain rates at the lateral margins of the active ice streams result in long, linear, crevassed zones that are clearly visible from the air, in satellite images [e.g. *Bindschadler and Vornberger*, 1990], and in radio echo-sounding profiles [e.g. *Bentley*, 1987].

The first maps of the Ross ice streams came from airborne surveys and radar by *Robin et al.* [1970], who identified ice stream B and located several "pseudo ice shelves" either along ice stream C [interpre-

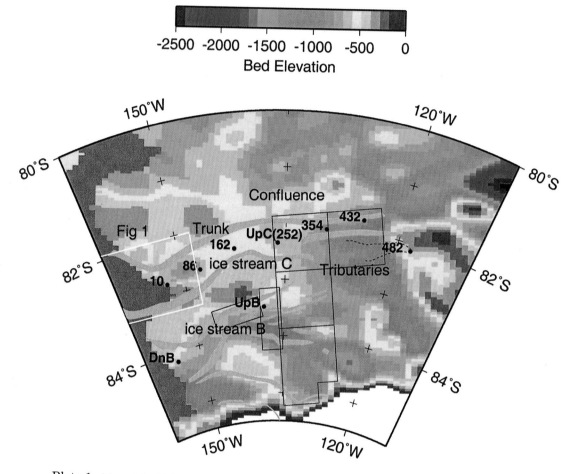

Plate 1. Map of the Siple coast region of West Antarctica showing the outlines of ice stream A, B, and C (from *Shabtaie et al.* [1987]) and the outline of the southern tributary of ice stream C (from *Hodge and Doppelhammer* [1996]; *Anandakrishnan et al.* [1998]). The bed elevation is based on radio-echo sounding data flown at a grid-spacing of 50 km (courtesy of J.L. Bamber). Flow is from the east to the Ross ice shelf on the west. The numbered sites on ice stream C are locations of microearthquake monitoring, where the site-number is the approximate distance in kilometers from the grounding line (which is from *Shabtaie et al.* [1987]). Upstream B camp (UpB) and Downstream B camp (DnB) on ice stream B are also shown. The boxes outlined in black are the region of radar coverage of *Retzlaff et al.* [1993], used by *Alley et al.* [1994] to calculate a hydrologic potential map. The box outlined in white is the region displayed in Fig. 1, and includes the southern flank of Siple Dome, the Duckfoot and the grounding line..

tation of *Hughes*, 1975] or along the edge of Siple Dome near ice stream C [interpretation of *Shabtaie et al.*, 1987]. After further radar surveys, *Rose* [1979] termed ice stream C "an enigma"— radar clutter indicated open crevasses, but visual observation showed an unbroken surface. He suggested that ice stream C was formerly active but had stopped; open crevasses that had formed during the active stage were now buried but still visible with radar. He also suggested that ice stream C is building up in its upper reaches possibly preparatory to resumption of fast flow, and that ice stream B is "now ...regaining ...lost territory" in the region between the heads of ice streams B and C owing to quiescence of C.

Extensive airborne radar with ground control [*Shabtaie and Bentley*, 1987; *Shabtaie et al.*, 1988; *Retzlaff et al.*, 1993] produced more-accurate maps of the ice thickness and bed elevations. Ice stream C extends more than 400 km from the grounding line near Siple Dome to the "onset region". The lower section of the ice stream has stagnated while velocities farther upstream (east of 125°W) are $\sim 100 \, \text{m} \cdot \text{a}^{-1}$ which, together with evidence of open crevasses and streaming flow from satellite images [*Hodge and Doppelhammer*, 1996; *Bell et al.*, 1998] suggests that the upper part of ice stream C may still be active. This is supported by recent interferometric RADARSAT data that indicate velocities up to $70 \, \text{m} \cdot \text{a}^{-1}$ in the source area [*Joughin et al.*, 1999]. Further, Joughin and others confirm and extend the identification by *Hodge and Doppelhammer* [1996] of two tributaries (C1 and C2) feeding the main stream. The northern tributary (C2), previously referred to as the limb [*Alley et al.*, 1994], has buried shear margins similar to those of the downglacier part of ice stream C, referred to as the trunk. The southern tributary, which was located using ground GPS measurements [*Anandakrishnan et al.*, 1998] and satellite interferometry, does not appear to have a continuous, buried shear margin (though individual buried crevasses do occur along the margins).

The tributaries come together just above the slow-flowing trunk near 132°W (about 50 km upstream from the old UpC camp—Plate 1) resulting in thickening there of $0.49 \, \text{m} \cdot \text{a}^{-1}$ [*Joughin et al.*, 1999]. This contrasts sharply with neighboring ice stream B, which is fast flowing along its entire length ($u \sim 400\text{--}800 \, \text{m} \cdot \text{a}^{-1}$) and thinning rapidly (estimated at $0.06\text{--}0.12 \, \text{m} \cdot \text{a}^{-1}$ [*Whillans and Bindschadler*, 1988; *Shabtaie and Bentley*, 1987, *Whillans et al.*, this volume]). This inverted flow pattern of fast-flowing tributaries converging on a slow-flowing trunk demonstrates that ice stream C is not in a steady-state configuration but it is still responding to the recent stagnation of the trunk.

The stagnation of the trunk of ice stream C has been variously interpreted as:

- evidence of stability in the ice sheet system: some negative feedback slowed ice motion that had become too vigorous, perhaps related to surging behavior in which fast and slow flow alternate because of internal dynamics [e.g. *Rose*, 1979; *Radok et al.*, 1987; *Retzlaff and Bentley*, 1993]. On the other hand, the rapid change might indicate that other such rapid changes, including ones that might destabilize the ice sheet, are possible [*Alley and Whillans*, 1991],

- a response to ongoing drawdown of the ice sheet [*Alley et al.*, 1994; *Anandakrishnan and Alley*, 1997b], possibly leading to its collapse [*Bindschadler*, 1997].

Understanding the change from fast to slow flow of ice stream C may be central to assessing the present stability of the WAIS.

2. TIME OF SLOW-DOWN

There is little question that ice stream C was once active and that the trunk region has now slowed greatly. Surface-based radar profiles show folding and deformation of the internal layers characteristic of fast ice flow [*Jacobel et al.*, 1993]. Airborne radar profiles reveal characteristic ice stream "clutter" (a prolongation of the surface echo due to buried crevasses and surface inhomogeneities) within the body of the ice stream [*Shabtaie and Bentley*, 1987]. Particularly strong clutter is characteristic of the boundaries of ice streams where high lateral shear strain rates produce a chaotic crevassed zone along the margins [*Raymond et al.*, this volume]. In contrast, the slower-flowing interstream ridges typically produce a short surface echo in the radargrams because surface inhomogeneities are primarily sastrugi rather than crevassing. The presence of clutter and buried paleo shear-margins are evidence that ice stream C was once similar to neighboring ice streams B and D, and flowed at speeds in excess of $120 \, \text{m} \cdot \text{a}^{-1}$ (necessary to maintain an active shear margin [*Scambos and Bindschadler*, 1993]).

Ground-based, high frequency radar has been used to measure the depth to the tops of buried crevasses on ice stream C [*Retzlaff and Bentley*, 1993; *Bentley et al.*, in press]. These measurements, when combined with age-depth profiles determined from snowpits and cores analysed for accumulation rate [*Whillans et al.*, 1987; *Whillans and Bindschadler*, 1988], yield an estimate of the time of slow-down.

Five radar profiles across the southern margin of the ice stream during the 1988-89 Antarctic field season showed depths to the tops of the shallowest buried crevasses ranging from ~ 7 m (near the upstream end at $120°W$) to ~ 20 m farther downstream [*Retzlaff and Bentley*, 1993]. Using estimates of the accumulation rate, they inferred the slow-down of the trunk region occurred 140 ± 30 a BP, and a more recent slow-down farther upstream. Recent work across the northern margin showed a similar pattern [BE Smith, pers. comm., 1998]. West of $125°W$, the crevasse tops are 20 to 30 m below the surface. Just east of $125°W$ depths are less than 10 m and open crevasses were observed farther to the east. These newer measurements (along with accumulation rate and density data) indicate that the trunk of the ice stream stagnated nearly synchronously at 150 ± 30 a BP.

We note that age-depth profiles are critical for determining the time of shut-down, and published measurements of the accumulation rate and density-depth profiles in the region are sparse [*Whillans and Bindschadler*, 1988]. Furthermore, analysis of the spacing between radar-detected internal layers indicates accumulation may vary by up to 30% within 10 km [BE Smith, pers. comm., 1998]. Nevertheless, the emerging picture is that the lower section of ice stream C stagnated nearly sychronously about 150 a BP, while the upper section (east of $125°W$) is still actively forming crevasses.

3. THERMAL CONDITIONS AT THE BED

It has been clear from the earliest work that the bed of ice stream C remains wet in most places. The pseudo ice shelves of *Robin et al.* [1970] were recognized in part based on the especially bright basal reflections in radar, indicating a wet bed. *Rose* [1979], *Shabtaie and Bentley* [1987] and *Shabtaie et al.* [1987] extended this work, demonstrating a wet bed with predominantly fresh rather than salt water. *Shabtaie et al.* [1987] suggested the possibility of centimeters-thick water or thicker in some places, with uncertainties caused by the numerous corrections required to estimate reflection coefficients from returned radar power. Recent reanalyses of these airborne radar data, which were collected during the 1987-88 field season and cover the downstream portions of ice streams B and C, show pronounced contrasts in the reflection-strength between the ice streams and ridge BC [*Bentley et al.*, 1998]. Reflection strengths are interpreted to confirm the idea that ice stream C is not frozen at the bed.

Seismic studies by *Atre and Bentley* [1993] similarly demand basal melting to explain the low acoustic impedance of the bed (any debris-rich frozen material would have given significantly higher acoustic impedances than observed in many places). The degree of basal lubrication demonstrated by *Anandakrishnan and Bentley* [1993] also was inconsistent with any model of widespread freezing to rigid basal material.

These observations suggest that although a thawed bed may be necessary for fast flow, it is not sufficient. The role of basal water has figured prominently in speculations on mechanisms for the slow-down of ice stream C. *Rose* [1979] calculated that either fast ice motion or high geothermal fluxes were needed to maintain the bed at the pressure melting point in steady state. *Shabtaie et al.* [1987] also noted that the surface of the trunk of ice stream C is terraced, and in places, the longitudinal slope reverses. They pointed out that using a one-dimensional model, as they did, instead of a more-realistic two-dimensional model makes it difficult to accurately model the basal hydraulic potential along the ice stream, but noted that the reversals in basal water flow may contribute to the observed spatial variability in the basal reflection coefficient.

4. HYPOTHESES FOR SLOW-DOWN

Several hypotheses for the shut-down of the lower trunk of ice stream C have been proposed:

Surging: The ice streams exhibit periodic dynamic oscillation [*Rose*, 1979; *Hughes*, 1975; *Lingle and Brown*, 1987] akin to that of some mountain glaciers [cf. *Kamb et al.*, 1985; *Clarke*, 1987a]. Questions are raised about this by the difficulty of modeling surges in the ice streams [*Radok et al.*, 1987]. Also, surging mountain glaciers spend most of their time in the slow-flow mode, whereas most of the Siple Coast ice streams are in fast-flow mode [*Clarke*, 1987b].

Surging via basal water feedbacks: In this model, ice flow speeds up until basal-water generation becomes sufficiently large that the water experiences the *Walder* [1982] instability, channelizes, lowers basal water pressures, and so slows or stops the ice [*Retzlaff and Bentley*, 1993, WB Kamb & HE Engelhardt, presented at Chapman Conference, Orono, ME, 1998]. Without rapid motion, the basal water channels could not be maintained, and basal water pressures eventually would rise and allow resumption of fast flow; interactions involving repeated capture of drainage basins by neighboring ice streams are suggested. The subsequent work of *Walder and Fowler* [1994] raises questions about the viability of this mechanism; on an unconsolidated sediment bed such as that indicated for ice stream C based on seismic results [*Atre and Bentley*, 1993] and direct coring [WB Kamb, WAIS Meeting, Sept. 1999], water channels are expected to show increasing water pressure with increasing flux. Emerging evidence that side drag is important or dominant in restraining active ice streams because basal lubrication is exceedingly efficient [*Echelmeyer et al.*, 1994; *Harrison et al.*, 1998; *Raymond*, 1996; *Whillans et al.*, 1993] also leads to questions of whether sufficient water could have been generated from the viscous dissipation of fast flow to allow the *Walder* [1982] instability. However, much of the evidence for dominant side drag and minimal bed drag is from ice stream B, which is known to have a smoother bed, hence potentially better basal lubrication, than the other Siple Coast ice streams [*Jankowski and Drewry*, 1981].

Loss of lubricating till: Several lines of evidence indicate that the soft tills known to exist beneath ice streams B and C are important in the rapid ice motion, through some combination of till deformation, burying of bedrock bumps, or allowing ploughing of controlling-obstacle-size bumps [e.g. *Alley et al.*, 1987; *Blankenship et al.*, 1987; *Brown et al.*, 1987; *Kamb and Engelhardt*, 1991]. Reduction or loss of that lubricating layer might slow or stop the ice motion [*Retzlaff and Bentley*, 1993]. The persistence of a widespread soft-sediment layer beneath the now-stagnant trunk of ice stream C [*Atre and Bentley*, 1993; *Anandakrishnan and Bentley*, 1993; *Anandakrishnan and Alley*, 1994, 1997b] argues against such a model, however. Similarly, evidence of a deep sedimentary reservoir upstream [*Anandakrishnan et al.*, 1998] suggests that the erosional source still exists. Finite-element model experiments by *Fastook* [1987] to validate an ice-piracy scenario versus a loss-of-till scenario suggest that the latter is more likely than the former.

Ice-shelf backstress: Assuming that ice stream subglacial sediments deform, and depending on the velocity-depth profile in the sediment [e.g. *Alley*, 1989], the rate of deposition could range from very small to quite large. It has been suggested that the lightly grounded "ice plain" at the mouth of ice stream B is the result of such deposition [*Alley*, 1989]. *Thomas et al.* [1988] suggested that deposition at the mouth of the ice stream may have increased grounding and backstress, stopping streaming flow. Ongoing grounding-line retreat (measured at about $30\,\text{m}\cdot\text{a}^{-1}$ between 1974 and 1984 [*Thomas et al.*, 1988]) in response to this stoppage might someday allow resumption of rapid flow. A potential difficulty with this hypothesis is that ice stream B maintains vigorous flow despite a large ice plain and despite the presence of Crary Ice Rise that provides significant restraint on flow [*MacAyeal et al.*, 1987, 1989]. Further, the results of *Anandakrishnan and Alley* [1997b] show that the grounding-line region of ice stream C provides little restraint on ice flow.

Ice piracy: "Piracy" is a concept borrowed from fluvial geomorphology defined as "the natural diversion of the headwaters of one stream into the channel of another stream having greater erosional activity and flowing at a lower level" [*Bates and Jackson*, 1980]. By analogy, if ice stream B in some fashion became better lubricated than ice stream C, hence faster flowing, the upglacier regions of ice stream B might thin, and ice would flow down the surface slope from the catchment of ice stream C into ice stream B. The surface of ice stream C then might flatten as ice flowing from its upper reaches was not replaced from the catchment, leading to stoppage. The biggest problem with this model is that the catchment of ice stream C does not appear to be feeding ice to ice stream B [*Shabtaie et al.*, 1988; *Retzlaff et al.*, 1993], and it is possible in fact that ice stream C has pirated the catchment of ice stream B [*Joughin et al.*, 1999].

Water piracy: Because the hydraulic potential of subglacial water is affected by bed elevation as well as ice pressure (and other factors such as degree of channelization of flow), water and ice flow need not be tightly coupled. The surface slope is about ten times more effective than the bed slope in controlling water flow direction, but steep transverse bedrock and flat ice surface slopes identified by *Rose* [1979], *Shabtaie and Bentley* [1987], and *Retzlaff et al.* [1993]

in the upglacier reaches of ice streams B and C suggest the possibility of water piracy, with lubricating water from the catchment of ice stream C diverted to ice stream B. A map of hydrological potential made from previously collected radar data [*Retzlaff et al.*, 1993] to test this idea [*Alley et al.*, 1994] shows that such water diversion probably is occurring, although the error bars include the small possibility that it is not. We favor this hypothesis as the cause of the shut-down of ice stream C [*Alley et al.*, 1994; *Anandakrishnan and Alley*, 1997a], though several difficulties remain [S Price and others, WAIS Meeting, Sept. 1999; see Sec. 6.1].

Thermal processes: Thermal processes have been suggested as controls on alternating fast and slow ice flow [*MacAyeal*, 1993b; *Payne*, 1995]. Thick ice traps geothermal heat and favors a thawed bed. However, rapid flow can bring cold ice near the bed through horizontal and vertical advection, and can thin ice so that the cold surface is closer to the bed. An oscillation has been modeled for the former ice sheet in Hudson Bay/Hudson Strait, linked to the Heinrich events of the North Atlantic [*MacAyeal*, 1993a, b]. A similar scenario is suggested for ice stream C [S Price, WAIS meeting, Sept. 1999; A Payne; WB Kamb & HE Engelhardt, Chapman Conference, Orono, ME, Sept. 1998]

Sticky spots: There is strong evidence for large spatial variation in basal drag of ice streams [see *Alley*, 1993; *MacAyeal et al.*, 1995]. If a localized region has a high basal shear stress, the water pressure in a distributed, connected basal water system will be reduced in that region. Thus, under an ice stream with sufficient water supply, sticky-spots would remain well lubricated and would not restrain flow. If the water supply were eliminated, as we hypothesize for ice stream C under the water-diversion scenario, these sticky-spots would play a significant role in restraining flow. The sticky spots would also be the first sites to freeze on in a thermal-shut-down scenario.

5. OBSERVATIONS AND DATA

5.1. Current flow pattern of ice stream C

Though the flow speeds of the trunk are low (comparable to ice-sheet or inter-ice-stream flow speeds), the ice stream is not frozen to its bed as discussed in Sec. 3. The trunk is flowing parallel to the main axis of the ice stream, towards the Ross Ice Shelf, though there appears to be locally divergent flow at UpC [HE Engelhardt, WAIS meeting, Sept. 1999]. The flow pattern of the tributaries feeding the ice stream is also complex. The southern tributary is located above a deep low-density sedimentary basin as determined by seismic, aeromagnetic, and gravity measurements [*Anandakrishnan et al.*, 1998; *Bell et al.*, 1998]. The velocities along a transverse line that crosses the southern margin of this tributary change from $12\,\text{m}\cdot\text{a}^{-1}$ to over $60\,\text{m}\cdot\text{a}^{-1}$ over a distance of 4.5 km, and distinct flow bands are visible in satellite imagery. This basin appears to control the position of the margin [*Anandakrishnan et al.*, 1998] and of the onset position of the ice stream [*Bell et al.*, 1998]. In addition, this basin could be a source of sediments for the ice stream subglacial till layer.

The northern tributary extends farther inland, following the Bentley Subglacial Trough. The onset location picked by *Hodge and Doppelhammer* [1996] from surface features does not correspond to a distinct change in flow speeds. Some component of the fast flow in the Bentley Trench is likely due to the thick ice there; it is unknown whether sediments are present there or not. The two tributaries coalesce upstream of the trunk (about 50 km upstream of station UpC) [*Joughin et al.*, 1999].

5.2. Shut-down of Siple ice stream and Duckfoot

At the eastern tip of Siple Dome (the ridge between ice streams C and D) ice stream C cuts across the older relict "Siple ice stream" [*Jacobel et al.*, 1996]. The age of stagnation for Siple ice stream is 420 to 470 a BP [*Conway and Gades*, in press, BE Smith, Chapman Conference, Orono, ME, Sept. 1998]. The cause of this shut-down is unknown, but we suggest that a similar water diversion occurred here as has been hypothesized for ice stream C. Because of the longer time since shut-down, the surface topography (and consequently the basal hydrologic potential) will be substantially different from those that existed four centuries ago.

The north margin of ice stream C runs along the flank of Siple Dome where satellite imagery (Figure 1) reveals a splayed pattern of margin scars and flowbands called the "Duckfoot" [RW Jacobel, TA Scambos, NA Nereson, and CF Raymond, Changes in the margin of ice stream C, *J. Glac.*, in review]. Though the underlying cause is not known, it appears that the north margin of C shifted inward with an accompanying change in flow direction. As it did so, ice from the area between the old and new mar-

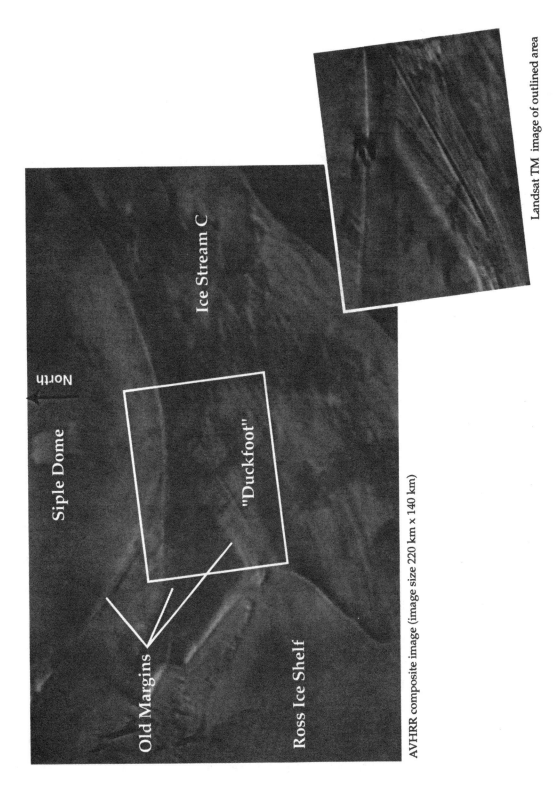

Fig. 1. An AVHRR (Advanced Very High Resolution Radiometer) satellite image of Siple Dome and inset Landsat image of the Duckfoot area. The edges of the main image are outlined in Plate 1. From Jacobel and others, in review.

gins did not stagnate immediately, and parts of it were sheared and folded by flow along the new direction of motion [Fig. 3 in *Jacobel et al.*, 1996] before stagnation. The lack of any substantial difference in the surface texture of the two regions suggests that the margin shift did not appreciably precede the shut-down of C, but it may be inferred that the outer margin is older because of the clear evidence of the inner one, and the presence of shearing between the two. Presumably all flow traces of the Duckfoot would have been transformed to normal flow striping if that area were part of an active ice stream for any significant time.

The longitudinally compressive strain rate observed in ice motion studies across the lower portion of the flank of Siple Dome [*Nereson*, in press] implies that this area must be thickening considerably, probably as a direct result of the shut-down and the thickening of ice stream C. The timing of shut-down of an ice stream may be determined by examining the rate of thickening of the adjacent ridge [*Nereson*, in press, Jacobel and others, in review]. According to this analysis, a wave of thickening which travels faster than the ice caused small topographic features associated with the former margin of an ice stream to be lifted onto the flank of the ridge. As the dome reaches a new steady state, ice flow carries the feature down slope. Models of this process suggest that the outermost Duckfoot scar is experiencing the beginnings of this rise. The pattern of thickening (inferred from the ice flow field for the south flank of Siple Dome) is consistent with a margin shut-down approximately 300 to 500 years before present.

This scenario proposes that the surface features on the north side of ice stream C are the result of a widening stagnation of a triangular-shaped area of ice just inside the older margin. The root causes of this stagnation are not yet clear, but some possibilities are [see Jacobel and others, in review]: (1) a general reduction in water pressure under ice stream C which caused the largest proportional increase in effective normal stress at the bed in this marginal zone because the ice was thinnest there; (2) reduced velocity in ice stream C, thus reducing the drag on this margin from the central parts of the ice stream; (3) onset of freezing associated with a general thinning of ice stream C which would have had the strongest and most rapid effect where the ice is thin; (4) water piracy that swept from the north to the south resulting in progressive shut-down.

5.3. Basal Environment of ice streams B & C

Ice stream B is thawed at the bed and fast-flowing along its entire length (see *Whillans et al.*, this volume). Much has been learned about the bed of ice stream B from seismic imaging, radar work, and direct drilling performed over the last decade. These experiments show that there is high water pressure at the bed and that there exists a low-strength subglacial till layer. The velocity is partitioned in some still-debated way between deformation in the till layer and sliding of ice over its base on the water layer [*Engelhardt and Kamb*, 1998, RB Alley, Continuity comes first: Recent progress in understanding subglacial deformation, Proc. of the Intl. Conf. on Deformation of Glacial Materials, London, Sept. 1999, in review].

Seismic imaging of ice stream B revealed a heterogenous bed. In particular, the till layer pinches out (or nearly so, within the resolution of the seismic experiment) along a longitudinal ridge. This region of higher strength material (and possibly other, similar regions where the till is absent) would present an obstacle to flow.

The bed of ice stream B appears to provide little resistance to flow, with lateral drag on the margins supporting most of the driving stress for ice flow [*Whillans and van der Veen*, 1997], though *Raymond et al.*, [this volume] suggest that ice streams D & E have significant support from the bed. Of the fraction of driving stress supported by the bed of ice stream B, most is supported on the till rather than on the rare sticky spots [*Alley*, 1993] despite the weakness of the till (yield strength of 2 kPa [*Kamb*, 1991]). The sticky spots are well-lubricated and at UpB camp, it is estimated that <13% of the basal shear force is supported by sticky spots.

This calculation is bolstered by the observation of rare basal microearthquake activity at UpB and no basal seismic activity at DnB [*Blankenship et al.*, 1987; *Anandakrishnan and Bentley*, 1993]. Seismic monitoring of ice stream B at UpB camp (1985–86) showed that the basal microearthquake activity was low (but significantly, non-zero). At UpB, six events were recorded that emanated from the bed of the ice (within the hypocentral depth determination error of ±15m) in 85 hours of monitoring [*Blankenship et al.*, 1987]. All the events were coincident in space (within the epicentral location ellipse of

±10 m) and occurred within a half-hour period. The events were low-angle thrust faults with slip in the direction of ice flow and of very small magnitude (seismic moment $\approx 10^6$ N·m). The interpretation of these events is of a transient increase in basal shear force on a local, more-competent portion of the bed, followed by fracture and slip [*Anandakrishnan and Bentley*, 1993].

We estimate that the stress drop for these events was approximately 10 kPa, which is estimated to equal or exceed the average basal shear stress of ice stream B. As stress drop is usually only a fraction of the total applied stress on the fault (between 1 and 10%), it is likely that the sticky spot was supporting the driving stress from some larger portion of the bed than simply that of the area of the sticky spot. That is, much or all the basal shear stress from that larger area is concentrated on the sticky spot and as a consequence the material fails. We caution that estimates of fault-plane area, stress drop, and the fraction of applied stress are strongly slip-model dependent and therefore inaccurate. Nonetheless, the presence of repeated fracture at a single spot at the bed that is induced by the shearing force of the ice, is evidence of at least one sticky spot beneath ice stream B. Other sticky spots (estimated to cover 2-3% of the bed at UpB [*Rooney et al.*, 1987; *Rooney*, 1988]) remained non-seismic and presumably well-lubricated throughout the seismic monitoring experiment.

5.4. Microearthquakes along ice stream C

To determine bed characteristics in the trunk and in the tributaries we measured the rate of basal seismicity along the ice stream. The hypothesis was that well-lubricated beds have low seismicity, but poorly lubricated beds that are not mostly-frozen would have high seismicity.

Surprisingly, the trunk of ice stream C was highly active seismically, with tens to hundreds of basal thrust-fault events recorded per day [*Anandakrishnan and Bentley*, 1993]. These events preferentially occured and recurred on localized sticky spots of order 10 m linear dimension, separated by order 100–1000 m. Quakes beneath ice stream C triggered other quakes on adjacent sticky spots, to distances as great as 1.5 km, and with time delays indicating propagation of the disturbance at approximately 1.9 m·s^{-1} [*Anandakrishnan and Alley*, 1994]. The microearthquakes were first observed in 1988 [*Anandakrishnan*, 1990; *Anandakrishnan and Bentley*, 1993] and remeasured in 1995 and 1996 [*Anandakrishnan and Alley*, 1997a].

Seismometers were deployed at approximately 90 km separations along the length of ice stream C, from just above the grounding line to above the onset of streaming flow. The sites are labeled by their distance from the grounding line (that is, site 10 is 10 km upstream of the grounding line, and so on; Plate 1). The rate of basal seismicity R (number of events per day) was low for the two sites on ice stream B (UpB and DnB), and for sites Km 482 and Km 432 (in the catchment of ice stream C, and in the uppermost part of ice stream C, respectively). There was a marked downglacier increase in seismicity between Km 432 and Km 354 and seismicity remained high from Km 354 down to the array closest to the grounding line at Km 10 (Fig. 2). Flow velocities were low on the ice sheet (Km 482); the ice flowed faster in the upper reaches of ice stream C (Km 432 to Km 354: $30 < u < 60$ m·a^{-1}) but nearly stagnated somewhere between Km 354 and Km 252 ($u < 10$ m·a^{-1} [*Anandakrishnan and Alley*, 1997a; *Whillans and van der Veen*, 1993]). Thus the pattern is clear: on the ice streams, low velocities are associated with high seismicity and vice-versa. The anomaly in this pattern is Km 354, which had a relatively high velocity but also had high seismicity. This site appears to be transitional between streaming and non-streaming ice and exhibits some of the qualities of each.

5.5. Thawed bed

The bed is thawed and exceptionally well lubricated by a soft till almost everywhere [cf. *Atre and Bentley*, 1993; *Bentley et al.*, 1998; *Anandakrishnan and Alley*, 1997b, WB Kamb & HE Engelhardt, Chapman Conference, Orono, ME, Sept. 1998], but with localized, poorly lubricated regions. The till is unfrozen and contains water at high pore pressures, but whether a distributed, connected water system exists under C (as exists under ice stream B [*Engelhardt and Kamb*, 1997]) is unknown. The observation of high seismicity from sticky spots at UpC over an extended period of time (1988 to 1996) suggests that there is not sufficient free water to flow down the local hydrologic potential gradient and lubricate the sticky spots. We measured different sticky spots in 1988 and in 1996 because the arrays were separated by about 10 km, but the presence of high-friction regions at the bed over an eight year period suggests that the basal water system is poorly developed.

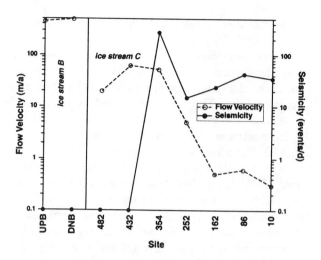

Fig. 2. Plot of ice-flow velocity u (m · a^{-1}) and seismicity R (day^{-1}) at the different array locations (from *Anandakrishnan and Alley* [1997a]). Note that the values of $R = 0$ are plotted at $R = 0.1$ because of the log scale.

The presence of a meters-thick layer of till under a different part of the ice stream was inferred by the transmission of tidal forcings upstream from the grounding line [*Anandakrishnan and Alley*, 1997a]. They discovered that the rate of basal seismicity was related to the tide beneath the Ross Ice Shelf. The seismicity at the grounding line is in phase with the tide and peaks at local low tide. The peak seismicity at a location 80 km inland from the grounding line (and on the ice stream) lagged the low tide by 13 hours. They successfully modeled the ice-stream system as an elastic beam resting on a viscous substrate and showed that a linear-viscous till was consistent with the measurements but a non-Newtonian rheology (high exponent in stress-strain relationship $\dot{\epsilon} \propto \tau^p$, $p \sim 100$) did not fit the data. We note that it is possible that till will behave as a high-exponent plastic material under high strain rates (and large total strain) as occurs under ice stream B, but will be linear-viscous under the lower strain rates at the base of nearly stagnant ice stream C [*Iverson et al.*, 1998, Alley, in review, see Sec. 5.3].

6. DISCUSSION

The near-stoppage of the trunk of ice stream C is clearly of considerable interest, and in Sec. 4 we identified 8 hypotheses for this shut-down. Of these, we favor water piracy. We believe that five are rendered highly unlikely by the data: there is as yet no evidence for classical surging behavior, soft till remains beneath the ice and is modeled to prevent the *Walder* [1982] water-system instability, ice plains are too well lubricated to be significant in the force budget, and ice piracy has not yet occurred. Thermal processes and sticky spots remain potential explanations, and we are intrigued by the possibility that water piracy, sticky spots, and thermal processes have all contributed to the changes in ice stream C, as discussed next.

6.1. Water diversion

As noted in Sec. 4 ("water piracy" hypothesis), *Alley et al.* [1994] hypothesized that ice stream C slowed because of loss of water that lubricates sticky spots where till is thin or absent. A hydrological potential map of ice streams B & C made to test this hypothesis indicated that water from the catchment area of ice stream C is being diverted to ice stream B. Despite the high data quality and tight flight-line spacing, the data errors might allow limited or zero water diversion, although strong water diversion is clearly the better interpretation.

The site of probable water diversion is near where the upglacier tributaries join the downglacier trunk (Plate 1). *Alley et al.* [1994] thus hypothesized that the tributaries remain active but aseismic because their sticky spots are water-lubricated, while the trunk nearly stopped and is seismically active on the sticky spots. The hydrologic potential is ten times as strongly affected by surface topography as by bed topography. Thus, water will tend to flow in the direction of the ice surface slope. However, with the low average surface slopes of the ice stream ($\alpha \sim 0.001$), the bed topography under the hypothesized diversion area becomes significant in affecting basal water flow. The large transverse bed slopes in the diversion zone would not control the basal water flow if the surface of the overlying ice were steep, as is common with ice-sheet ice. Under those conditions, which *Alley et al.* [1994] suggest existed in the past, the basal water would flow in the direction of the ice-sheet surface slope. Thus water from the catchment of ice stream C was directed towards the ice stream even in the presence of large transverse bed slopes. With Holocene warming and the subsequent drawdown of the ice sheet, ice stream C grew headward and the low ice-stream surface slopes impinged on this region of transverse bed slopes [*Alley and Whillans*, 1991]. This allowed the transverse bed slopes to dominate the hydrologic potential and the water from the catchment of ice stream C flowed towards ice stream B. The loss of this water layer was hypothesized to result in a loss of lubrication of sticky spots, an increased basal friction, and a stagnation of the ice stream below the water-diversion zone.

A bulge at the confluence of the tributaries is a result of the tributary ice "piling up" against a stagnant trunk as noted by *Alley et al.* [1994]. The profile of the confluence during the time of active trunk flow might have been concave up, preventing water diversion from the confluence area. Thus the suggestion of Price and others (WAIS meeting, Sept. 1999) is that the hydrologic potential that diverts basal water from ice stream C to B is a recent phenomenon and not the cause of the stagnation of the trunk [cf. *Alley et al.*, 1994]. We note, however that transverse ice stream surface elevation profiles are generally concave up in the downstream portions, but flat to convex up in the upper portions [see *Shabtaie et al.*, 1987, Fig. 3a]. Further, we note that even partial diversion of basal water from, e.g., one side of the ice stream to the other side would be nearly as effective at stagnating the ice stream as full diversion from ice stream C to ice stream B. Thus, a flattening of surface profile (a progressive change from longitudinally convex up to concave up during head migration of the ice stream) over a region of steep transverse bed slopes could funnel water to the southern half of ice stream C, possibly resulting in rapid stagnation. The present-day observation of water diversion to ice stream B would then be a later phenomenon. More work is needed on this hypothesis.

6.2. Thermal Processes

Price and others (WAIS meeting, Sept. 1999) argued in favor of a thermal model for the shut-down of ice stream C [cf. *MacAyeal*, 1993a; *Payne*, 1995]. Thermal surging is complicated by basal water transport in ice-contact systems or in subglacial till [*Alley and MacAyeal*, 1994]. Basal water can be considered to be stored thermal energy from beneath ice upglacier, and provides a heat source to any region where freezing is initiated through its latent heat. For deforming tills, the effectiveness of this heat source will depend in large part on the existence or absence of sticky spots of thin or absent till–without such sticky spots, the ice cannot freeze to bedrock until water in till pores is frozen, but sticky spots might allow freeze-on more quickly. Freeze-on in the presence of an active ice-contact water system likely requires that most or all of the water flux be frozen before the ice can freeze to its substrate, which could greatly suppress freezing-on for significant water fluxes. If water is not supplied from upglacier, freeze-on may contribute to consolidation of subglacial tills, strengthening the bed and resisting rapid ice motion [*Tulaczyk et al.*, in press]. The persistence of soft till beneath ice stream C (Section 5.5) demonstrates that this process has not proceeded far for much of the ice stream, but the process may have contributed to generation of the observed sticky spots, perhaps in regions of previously thin till over bedrock bumps.

Here, the water-piracy hypothesis and the thermal hypothesis may be complementary. Water piracy would have caused a significant loss of heat as well as lubrication to ice stream C. The stoppage of ice stream C may result from loss of lubrication of sticky spots, from incipient freeze-on to sticky spots (the intervening till remains soft), or from some combination of these end-members.

7. CONCLUSION

The rapid flow of Siple-coast-type ice streams depends on the bed of the ice presenting little or no resistance. Variability in bed properties such as bedrock knobs or non-uniform distribution of deformable till could present a significant resistance to flow unless these sticky spots are lubricated by water that decouples the ice. The Siple ice streams (with the exception of ice stream C) appear to receive sufficient water from their catchments and produce more water by melting of the bed due to fast sliding [*Engelhardt and Kamb*, 1997]. This basal water system efficiently nullifies the restraining forces of the sticky spots and allows the ice streams to maintain the high flow speeds observed. This thin water layer is hypothesized to lubricate sticky-spots under ice stream B and under the upstream portion of ice stream C (above the diversion zone). The sticky spots are regions of higher basal shear stress than their surroundings and are modeled to have lower water pressures. Thus, the well-connected basal water system can deliver lubricating water to the sticky spot, and reduce shear stress in a stable negative feedback mechanism.

We hypothesize that similar conditions currently exist under the upstream part of ice stream C, and existed under all of ice stream C as recently as 150 ± 30 a BP. We suggest that at that time, the headward growth of the ice stream brought low ice stream surface slopes over high transverse bed slopes resulting in a diversion of catchment water from ice stream C to ice stream B. As a consequence, the sticky spots were starved of lubricating water and could, and did, exert a restraining force on the ice stream. As a result, flow speeds of the lower part of the ice stream are less than $10 \text{ m} \cdot \text{a}^{-1}$, and the ice at UpC is thickening at a rapid rate of up to $0.49 \text{ m} \cdot \text{a}^{-1}$ [*Joughin et al.*, 1999]. An alternative suggestion is that the trunk freeze-on and stagnation occured first and the water diversion at the confluence is a consequence of that stagnation. A possible synthesis of these two primary hypotheses is that the water diversion occurred farther upstream (where transverse ice stream profiles might have been convex up even prior to stagnation). If so, we speculate that the regions downstream (including both the bulge and the trunk) were starved of water, but the trunk stagnated first and that the wave of stagnation will proceed upstream as more sticky-spots freeze on.

The mass balance of the combined ice stream B & C system (ice stream and catchment) appears to indicate slow thinning according to the best estimates available [*Shabtaie and Bentley*, 1987; *Shabtaie et al.*, 1988], though the few available point measurements show large spatial variability [e.g *Hamilton et al.*, 1998; *Whillans and Bindschadler*, 1988]. If the shutdown of ice stream C were a stabilizing influence, one might expect that C would be thickening and B remain in balance, resulting in a net thickening of the combined system. We suggest that the observed net thinning of ice stream B is possibly due to the extra basal lubrication provided by the water diversion from beneath C. Thus, the strong negative balance of ice stream B (as compared to approximately zero balance of ice streams D, E, and F [*Shabtaie and Bentley*, 1987]) is connected to general headward extension of the ice streams that resulted in the triggering of the shut-down of C.

The shut-down of ice stream C then is not an inherent feedback mechanism stabilizing the ice sheet, but a consequence of a particular combination of strong transverse bed slope in the onset region of the ice stream. To understand and predict the behavior of the other ice streams in the presence of ongoing headward migration, detailed knowledge of the basal environment (both topography and geology) is required.

Acknowledgments. A review paper necessarily draws upon the work of many people. We would like to thank T. Scambos for help with generating the satellite image (Fig. 1) and J. Bamber for providing the bed-elevations for Plate 1. We thank C. R. Bentley and R. Bindschadler for helpful reviews and we thank the NSF for financial support (OPP9725708 and OPP9996262 to RBA and SA, OPP9615347 to HC and OPP9316338 to RWJ). We used the GMT program of P Wessel and WHF Smith to generate Plate 1.

References

Alley, R. B., Water-pressure coupling of sliding and bed deformation: I. water system, *J. Glaciol.*, 35, 108–118, 1989.

Alley, R. B., In search of ice-stream sticky spots, *J. Glaciol.*, 39, 447–454, 1993.

Alley, R. B., and D. R. MacAyeal, Ice-rafted debris associated witht binge/purge osciallations of the Laurentide Ice Sheet, *Paleoceanography*, 9, 503–511, 1994.

Alley, R. B., and I. M. Whillans, Changes in the West Antarctic ice sheet, *Science*, 254, 959–963, 1991.

Alley, R. B., D. D. Blankenship, C. R. Bentley, and S. T. Rooney, Till beneath ice stream B, 3, Till deformation: Evidence and implications, *J. Geophys. Res.*, 92, 8921–8929, 1987.

Alley, R. B., S. Anandakrishnan, C. R. Bentley, and N. Lord, A water-piracy hypothesis for the stagnation of ice stream C, Antarctica, *Ann. Glaciol.*, 20, 187–194, 1994.

Anandakrishnan, S., Microearthquakes as indicators of ice stream basal conditions, Ph.D. thesis, University of Wisconsin—Madison, 1990.

Anandakrishnan, S., and R. B. Alley, Ice stream C, Antarctica, sticky-spots detected by microearthquake monitoring, *Ann. Glaciol.*, 20, 183–186, 1994.

Anandakrishnan, S., and R. B. Alley, Stagnation of ice stream C, West Antarctica by water piracy, *Geophys. Res. Lett.*, 24, 265–268, 1997a.

Anandakrishnan, S., and R. B. Alley, Tidal forcing of basal seismicity of ice stream C, West Antarctica, observed far inland, *J. Geophys. Res.*, 102, 15,183–15,196, 1997b.

Anandakrishnan, S., and C. R. Bentley, Microearthquakes beneath ice streams B & C, West Antarctica: Observations and implication, *J. Glaciol.*, 39, 455–462, 1993.

Anandakrishnan, S., D. D. Blankenship, R. B. Alley, and P. L. Stoffa., Influence of subglacial geology on the position of a West Antarctica ice stream from seismic measurements, *Nature*, 394, 62–65, 1998.

Atre, S. R., and C. R. Bentley, Laterally varying basal conditions under ice streams B and C, West Antarctica, *J. Glaciol.*, 39, 507–514, 1993.

Bates, R. L., and J. A. Jackson, *Glossary of Geology*, 2 ed., American Geological Institute, 1980.

Bell, R. E., D. D. Blankenship, C. A. Finn, D. L. Morse, T. A. Scambos, J. M. Brozena, and S. M. Hodge, Aerogeophysical evidence for geologic control on the onset of a West Antarctic ice stream, *Nature*, 394, 58–62, 1998.

Bentley, C. R., Antarctic ice streams: a review, *J. Geophys. Res.*, *92*, 8843–8858, 1987.

Bentley, C. R., N. E. Lord, and C. Liu, Radar reflections reveal a wet bed beneath stagnant ice stream c and a frozen bed beneath ridge bc, west antarctica., *J. Glaciol.*, *44*, 149–156, 1998.

Bentley, C. R., B. E. Smith, and N. E. Lord, Radar studies on Roosevelt Island and ice stream C, *Ant. J. U.S.*, in press.

Bindschadler, R., West Antarctic Ice Sheet collapse?, *Science*, *276*, 662–663, 1997.

Bindschadler, R. A., and P. L. Vornberger, Cover photograph, *Eos, Trans. Am. Geophys. Un.*, *71*, 1990.

Blankenship, D. D., S. Anandakrishnan, J. Kempf, and C. R. Bentley, Microearthquakes under and alongside ice stream B, detected by a new passive seismic array, *Ann. Glaciol.*, *9*, 30–34, 1987.

Brown, N., B. Hallet, and D. Booth, Rapid soft bed sliding of the Puget glacial lobe, *J. Geophys. Res.*, *92*, 8985–8998, 1987.

Clarke, G. C. K., Subglacial till: a physical framework for its properties and processes, *J. Geophys. Res.*, *92*, 9023–9036, 1987a.

Clarke, G. C. K., Fast glacier flow: Ice streams, surging and tidewater glaciers, *J. Geophys. Res.*, *92*, 8835–8841, 1987b.

Conway, H., and A. Gades, Glaciological investigations of Roosevelt Island and ice stream C, *Ant. J. U.S.*, in press.

Echelmeyer, K. A., W. D. Harrison, C. Larsen, and J. Mitchell, The role of the margins in the dynamics of an active ice stream, *J. Glaciol.*, *40*, 527–538, 1994.

Engelhardt, H., and W. B. Kamb, Basal hydraulic system of a West Antarctic ice stream: constraints from borehole observations, *J. Glaciol.*, *43*, 207–230, 1997.

Engelhardt, H., and W. B. Kamb, Basal sliding of ice stream B, West Antarctica, *J. Glaciol.*, *44*, 223–230, 1998.

Fastook, J., Use of a new finite element continuity model to study the transient behavior of ice stream C and causes of its present low velocity, *J. Geophys. Res.*, *92*, 8941–8950, 1987.

Hamilton, G. S., I. M. Whillans, and P. J. Morgan, First point measurements of ice-sheet thickness change in Antarctica, *Ann. Geophys.*, *27*, 125–129, 1998.

Harrison, W. D., K. A. Echelmeyer, and C. F. Larsen, Measurements of temperatures within a margin of ice stream B, Antarctica: implications for margin migration and lateral drag, *J. Glaciol.*, *44*, 615–625, 1998.

Hodge, S. M., and S. K. Doppelhammer, Satellite imagery of the onset of streaming flow of ice streams C and D, West Antarctica, *J. Geophys. Res.*, *101*, 6669–6677, 1996.

Hughes, T. J., The West Antarctic Ice Sheet: Instability, disintegration, and initiation of ice ages, *Rev. Geopys.*, *13*, 502–526, 1975.

Iverson, N. R., T. S. Hooyer, and R. W. Baker, Ring-shear studies of till deformation: Coulomb-plastic behavior and distributed strain in glacier beds, *J. Glaciol.*, *44*, 634–642, 1998.

Jacobel, R. W., A. M. Gades, D. L. Gottschling, S. M. Hodge, and D. L. Wright, Interpretation of radar-detected internal layer folding in West Antarctic ice streams, *J. Glaciol.*, *39*, 528–537, 1993.

Jacobel, R. W., T. A. Scambos, C. F. Raymond, and A. M. Gades, Changes in the configuration of ice stream flow from the West Antarctic Ice Sheet, *J. Geophys. Res.*, *101*, 5499–5504, 1996.

Jankowski, E., and D. Drewry, The structure of West Antarctica from geophysical studies, *Nature*, *291*, 17–21, 1981.

Joughin, I., L. Gray, R. Bindschadler, S. Price, D. Morse, C. Hulbe, K. Mattar, and C. Werner, Tributaries of West Antarctic ice streams revealed by RADARSAT interferometry, *Science*, *286*, 283–286, 1999.

Kamb, W. B., Rheological nonlinearity and flow instability in the deforming bed mechanism of ice stream motion, *J. Geophys. Res.*, *96*, 16,585–16,595, 1991.

Kamb, W. B., and H. Engelhardt, Antarctic ice stream B: Conditions controlling its motion and interactions with the climate syste, *IAHS Publ. No. 208*, pp. 145–154, 1991.

Kamb, W. B., C. Raymond, W. Harrison, H. Engelhardt, K. Echelmeyer, N. Humphrey, M. Brugman, and T. Pfeffer, Glacier surge mechanism: 1982-1983 surge of Variegated Glacier, Alaska, *Science*, *227*, 469–479, 1985.

Lingle, C. S., and T. J. Brown, A subglacial aquifer bed model and water pressure dependent basal sliding relationship for a west antarctic ice stream, in *Dynamics of the West Antarctic ice sheet*, edited by C. J. van der Veen and J. Oerlemans, D. Reidel Publ. Co., Dordrecht, Netherlands, 1987.

MacAyeal, D. R., Binge/purge oscillations of the Laurentide Ice Sheet as a cause of the North Atlantic Heinrich events, *Paleoceanography*, *8*, 775–784, 1993a.

MacAyeal, D. R., A low-order model of growth/purge oscillations of the Laurentide Ice Sheet, *Paleoceanography*, *8*, 767–773, 1993b.

MacAyeal, D. R., R. Bindschadler, S. Shabtaie, S. Stephenson, and C. Bentley, Force, mass and energy budgets of the Crary Ice Rise complex, Antarctica, *J. Glaciol.*, *33*, 218–230, 1987.

MacAyeal, D. R., R. Bindschadler, S. Shabtaie, S. Stephenson, and C. Bentley, Correction to: Force, mass and energy budgets of the Crary Ice Rise complex, Antarctica, *J. Glaciol.*, *35*, 151–152, 1989.

MacAyeal, D. R., R. A. Bindschadler, and T. A. Scambos, Basal friction of Ice Stream E, West Antarctica, *J. Glaciol.*, *41*, 247–262, 1995.

Nereson, N. A., The evolution of ice domes and relict ice streams, *J. Glaciol.*, in press.

Payne, A. J., Limit cycles in the basal thermal regime of ice sheets, *J. Geophys. Res.*, *B100*, 4249–4263, 1995.

Radok, U., D. Jenssen, and B. McInnes, On the surging potential of polar ice streams. Antarctic surges – a clear and present danger?, *Tech. rep.*, Washinton, DC, U.S. Department of Energy. (DOE/ER/60197-H1), 1987.

Raymond, C. F., Shear margins in glaciers and ice sheets, *J. Glaciol.*, *42*, 90–102, 1996.

Retzlaff, R., and C. R. Bentley, Timing of stagnation of ice stream C, West Antarctica from short-pulse-radar studies of buried surface crevasses, *J. Glaciol.*, *39*, 553–561, 1993.

Retzlaff, R., N. Lord, and C. R. Bentley, Airborne radar studies: ice streams A, B, and C, West Antarctica, *J. Glaciol.*, *39*, 495–506, 1993.

Robin, G. d. Q., C. Swithinbank, and B. Smith, Radio echo exploration of the Antarctic ice sheet, *IASH*, *86*, 523–532, 1970.

Rooney, S. T., Subglacial geology of ice stream B, West Antarctica, Ph.D. thesis, University of Wisconsin-Madison, Madison, Wisconsin, 1988.

Rooney, S. T., D. D. Blankenship, R. B. Alley, and C. R. Bentley, Till beneath ice stream B, 2, structure and continuity, *J. Geophys. Res.*, *92*, 8913–8920, 1987.

Rose, K., Characteristics of ice flow in Marie Byrd Land, Antarctica, *J. Glaciol.*, *24*, 63–74, 1979.

Scambos, T. A., and R. A. Bindschadler, Complex ice stream flow revealed by sequential satellite imagery, *Ann. Glaciol.*, *17*, 177–182, 1993.

Shabtaie, S., and C. R. Bentley, West Antarctic ice streams draining into the Ross Ice Shelf: Configuration and mass balance, *J. Geophys. Res.*, *92*, 1311–1336, 1987.

Shabtaie, S., I. M. Whillans, and C. R. Bentley, The morphology of ice streams A, B, and C, West Antarctica, and their environs, *J. Geophys. Res.*, *92*, 8865–8883, 1987.

Shabtaie, S., C. R. Bentley, R. A. Bindschadler, and D. R. MacAyeal, Mass balance studies of ice streams A, B, and C, West Antarctica, and possible surging behavior of ice stream B, *Ann. Glaciol.*, *11*, 137–149, 1988.

Thomas, R. H., S. N. Stephenson, R. A. Bindschadler, S. Shabtaie, and C. R. Bentley, Thinning and grounding-line retreat on Ross Ice Shelf, Antarctica, *Ann. Glaciol.*, *11*, 165–172, 1988.

Tulaczyk, S., W. B. Kamb, and H. F. Engelhardt, Basal mechanics of ice stream B, West Antarctica: 2. Undrained plastic bed model, *J. Geophys. Res.*, in press.

Walder, J., Stability of sheet flow of water beneath temperate glaciers and implications for glacier surging, *J. Glaciol.*, *28*, 273–293, 1982.

Walder, J., and A. Fowler, Channelised subglacial drainage over a deformable bed, *J. Glaciol.*, *40*, 3–15, 1994.

Whillans, I. M., and R. A. Bindschadler, Mass balance of ice stream B, West Antarctica, *Ann. Glaciol.*, *11*, 187–193, 1988.

Whillans, I. M., and C. J. van der Veen, New and improved determinations of velocity of ice streams B and C, West Antarctica, *J. Glaciol.*, *39*, 483–490, 1993.

Whillans, I. M., and C. J. van der Veen, The role of lateral drag in the dynamics of ice stream B, Antarctica, *J. Glaciol.*, *43*, 231–237, 1997.

Whillans, I. M., J. Bolzan, and S. Shabtaie, Velocity of ice stream B and C, Antarctica, *J. Geophys. Res.*, *92*, 8895–8902, 1987.

Whillans, I. M., M. Jackson, and Y.-H. Tseng, Velocity pattern in a transect across ice stream B, Antarctica, *J. Glaciol.*, *39*, 562–572, 1993.